Venomous Reptiles of the United States, Canada, and Northern Mexico

Venomous Reptiles of the United States, Canada, and Northern Mexico

VOLUME 1: *Heloderma, Micruroides, Micrurus, Pelamis, Agkistrodon, Sistrurus*

Carl H. Ernst and Evelyn M. Ernst

THE JOHNS HOPKINS UNIVERSITY PRESS
Baltimore

Printed in the United States of America on acid-free paper
9 8 7 6 5 4 3 2 1

The Johns Hopkins University Press
2715 North Charles Street
Baltimore, Maryland 21218-4363
www.press.jhu.edu

Library of Congress Cataloging-in-Publication Data

Ernst, Carl H.
Venomous reptiles of the United States, Canada, and northern Mexico / Carl H. Ernst and Evelyn M. Ernst.
p. cm.
Includes bibliographical references and index.
ISBN-13: 978-0-8018-9875-4 (v. 1 : alk. paper)
ISBN-10: 0-8018-9875-7 (v. 1 : alk. paper)
1. Poisonous snakes—North America. 2. Heloderma—North America. I. Ernst, Evelyn M. II. Title.
QL666.O6E773 2011
597.9′165097—dc22 2010036966

A catalog record for this book is available from the British Library.

Special discounts are available for bulk purchases of this book. For more information, please contact Special Sales at 410-516-6936 or specialsales@press.jhu.edu.

The Johns Hopkins University Press uses environmentally friendly book materials, including recycled text paper that is composed of at least 30 percent post-consumer waste, whenever possible. All of our book papers are acid-free, and our jackets and covers are printed on paper with recycled content.

With love to our grandchildren,

Emma, Luke, Wells, and Gus

Contents

Preface

The biology of many of the venomous reptiles of North America is still poorly known. It is the primary purpose of this book to provide a synthesis of what is known of the natural history of each venomous lizard, elapid snake, and pitviper within the United States, Canada, and northern Mexico.

The literature on venomous reptiles is enormous. It includes thousands of specialized papers on such topics as their behavior, biochemistry, ecology, genetics, morphology, physiology, and venoms and the symptoms and treatment of envenomations. Our interests, however, lie principally in natural history. Many more scientific papers and reports are now available concerning the biology of the venomous reptiles within the area covered than were available prior to the publication of Ernst (1992) or Ernst and Ernst (2003). These include edited volumes on pitvipers (Campbell and Brodie 1992), vipers (Schuett et al. 2002), and rattlesnakes (Hayes et al. 2008), and individually written books on coralsnakes (Roze 1996) and the venomous reptiles of the Western Hemisphere (Campbell and Lamar 2004). In the preparation of this book, we reviewed these volumes and more than 3,000 original papers on venomous reptiles, and have added our own personal observations to them in the discussions in the text. The cutoff date for listing references in the bibliography of volume 1 was 30 June 2009, that of volume 2 was 30 June 2010; however, a few colleagues sent us some of their papers that were scheduled to be published later, and these are also included.

A second major purpose of the book is to present a detailed overview of what constitutes reptile venoms, and the effects, treatment, and prevention of bites by these animals. The amount and scope of the literature concerning these topics, especially on venomous snakes, have increased tremendously since 1992. The discussion of reptilian venoms is rather technical, so some of the data on venoms can be confusing to one lacking the necessary biochemical background. We have tried to simplify as much as possible, and have included citations for the most important papers on venoms for both general and advanced readers to examine for further explanations and information. It is hoped that the nonscientific reader will bear with us on this. Readers deeply interested in reptilian venoms are encouraged to examine the journal *Toxicon*.

Third, we emphasize conservation of these animals, as their survival is of utmost importance to us. The sizes of venomous reptile populations are decreasing at an alarming rate in Canada, Mexico, and the United States. While some persons may consider this a blessing, herpetologists and other naturalists see the decrease as another evidence of human failure to maintain a balance between nature and the use of natural resources for human living and development. If this trend continues, some species will probably be extirpated—a great loss to the North American system of biological communities. In the near future, wise decisions regarding the welfare of the remaining populations of venomous reptiles will have to be made if they are to survive. To conserve these beasts, we must have adequate knowledge of their life histories so that critical facets of their ecology and behavior can be identified. It is a major purpose of this book to present the current knowledge of the biology of each species living in North America (as defined in the introductory chapter), and to point out critical gaps in this knowledge worthy of further study.

Since the data included are extensive, this synthesis is to be published in two volumes. The present volume includes chapters on venom, envenomation and its treatment, conservation of our venomous reptiles, the family and species accounts of the venomous lizards (Helodermatidae), coralsnakes and seasnakes (Elapidae), the account of the family Viperidae, and the species accounts of the genera *Agkistrodon* (American moccasins) and *Sistrurus* (pygmy rattlesnakes). The second volume is to cover the biology of the rattlesnakes of the genus *Crotalus*. Each volume has its own glossary of scientific names, bibliography, and index to common and scientific names.

The family chapters include discussion of fossil history and possible evolution, characteristics, the morphological apparatus (fangs and venom glands) and the mechanism involved in envenomation, other pertinent morphology and physiology, and a key to the genera of that family living within southern Canada, the United States, mainland Mexico north of 25° latitude, Baja California, and the islands of the Gulf of California and off the Pacific Coast that include the snakes.

Each species has its own account, which includes a detailed description, geographic variation and confusing species, its karyotype and fossil record (if known), distribution, and habitat preferences. These are followed by a detailed life history, a description of its venom delivery system, and discussions of its venom and envenomation symptoms and severity.

Many people have contributed to publication of this book by giving advice, data, encouragement, materials, or specimens through the years: Daniel Beck, William Brown, Vincent J. Burke, Roger Conant, Terry Creque, Bela Demeter, Henry Fitch, Oscar Flores-Villela, J. Whitfield Gibbons, Howard Gloyd, Steve Gotte, Carl Kauffeld, Jeffery Lovich, Roy McDiarmid, John Orr, Robert Reynolds, Howard Reinert, Richard Seigel, Addison Wynn, and George Zug. William Brown graciously introduced CHE to his timber rattlesnake research dens. Over the years these students of CHE helped with data, specimen collections, and field studies: Thomas Akre, Timothy Boucher, Timothy Brophy, Christopher Brown, Terry Creque, Sandra D'Alessandro, Lee French, Dale Fuller, Steve Gotte, Kerry Hansknecht, Traci Hartsell, Blair Hedges, Arndt Laemmerzahl, Jeffery Lovich, Peter May, John McBreen, John Orr, Carol Robertson, Steven Sekscienski, James Snyder, Sheila Tuttle, John Wilder, James Wilgenbusch, Gordon Wilson, and Thomas Wilson. The following persons supplied specimens or photographs: the family of the late Roger W. Barbour, Richard Bartlett, Ted Borg, Jeff Boyd, Eric Dugan, Dale Fuller, J. Whitfield Gibbons, Steve Gotte, Paul Hampton, James Harding, Blair Hedges, Alex Henderson, Jr., Jeffrey Lovich, Robert Lovich, Barry Mansell, Peter May, Liam McGranaghan, Bradley Moon, Ali Rabatsky, and John Tashjian. Special thanks go to Richard Greene, Polly Lasker, Martha Rosen, Courtney Shaw, and David Steere, of the Natural History Library of the Smithsonian Institution, and to the library staffs at Franklin and Marshall College and Millersville University of Pennsylvania for their help in acquiring the necessary literature.

Introduction

Although organized in a taxonomic manner for ease of presentation, this book is neither a taxonomic nor a distributional treatise. Its major purposes are threefold. The first is to provide a better understanding of the life histories of our venomous reptiles, and to point out, through omissions, the important aspects of their biology that need to be studied if we are to conserve them. The second major purpose is to present a detailed explanation of what constitutes a reptilian venom and to present its various characteristics (chemical, medical, pharmacological, and physiological), the features of each species' venom as far as is currently known, and the modern treatment of reptile envenomations.

Finally, we hope to present a balanced picture of the lives of venomous reptiles found in the United States, Canada, and northern Mexico to dispel human fears of these magnificent creatures. We believe that if more people learn the many fascinating aspects of reptilian biology they will become more interested and appreciative of them, and, particularly, in the protection of these shy creatures. The creation of such an attitude, not only toward reptiles but also toward all of our dwindling wildlife resources, is a major goal of this book.

More than 2,800 species of snakes and about 7,200 species of lizards exist in the world today. Only two species of lizards (*Heloderma horridum* and *H. suspectum* that live in North America), and about 600 species of snakes (about 21% of the total number; occurring in the four families: Atractaspididae, Colubridae, Elapidae, and Viperidae) are what humans usually consider dangerously venomous. In addition, the saliva of the Komodo Dragon (*Varanus komodoensis*)—and possibly other monitor lizards—has shown some venom characteristics (Fry 2009; Sanders 2009).

Venomous reptiles, especially snakes, are among the least understood of all animals on earth. The public, as a whole, totally misunderstands them (see Ernst and Zug 1996 or Zug and Ernst 2004 for discussions of the popular misbeliefs regarding snakes in general, and venomous ones in particular). Everyone recognizes a snake, and many believe that all snakes are venomous. Unfortunately, this misconception has led to the death of numerous harmless, useful snakes at the hands of humans.

This is in large part because most persons fear them. Such a fear (ophiophobia) is probably because some lizards and snakes are dangerous, and their bites can cause severe systemic damage or even death. Fortunately, in North America, such reptiles are in the minority, and are seldom encountered unless sought out by the human; so our snake fauna should be respected, not feared (if in doubt, leave the animal alone). Second, mythology has portrayed the snake as a symbol of evil or as Satan's helper (i.e., the biblical Adam and Eve story). This is also totally wrong. The "sins" committed by humans are their own!

Venomous reptiles are an integral part of the natural history and folklore of North America. However, if current trends continue through the twenty-first century, they may totally disappear from our wilds. Populations of the Gila monster (*Heloderma suspectum*) and the rattlesnakes *Crotalus horridus* (now extirpated in Canada), *C. willardi*, and *Sistrurus catenatus* have already been so severely reduced as to be officially listed as "threatened" or "endangered" on some international, national, and state or province lists. Although natural events, like brush fires, floods, or landslides, may on occasion kill large numbers of these animals, human interference in some form is the usual cause of their decline (see the chapter "Conservation of North American Venomous Reptiles" and the various species accounts for details).

If venomous reptiles are to remain a significant part of our fauna, we must initiate conservation measures. Although we do not know all that is needed about the biology of most species to formulate effective conservation plans, certain needs are obvious. The waterways and lands harboring important populations should be protected from undue human disturbance and pollution. We must continue to move away from the use of dangerous residual pesticides that may be fatal to wildlife. States and provinces must pass, and, particularly, enforce legislation controlling the capture of these animals in the wild and their sale to the public, and definitely ban bounties on them and "snake roundups."

To achieve our purposes, the book is basically divided into several sections. The first, a suite of three chapters, addresses venom, envenomation by North American reptiles, and the treatment of their bites. The second section is a chapter discussing the current problems facing our venomous species and the possible methods for their conservation. If we are to conserve North America's venomous reptiles, a more thorough knowledge of their life requirements and behaviors is a necessity. The third, largest section of the book, is one on their identification (see Ernst 1992 and Ernst and Ernst 2003 for detailed discussions of the morphology of helodermatid lizards and snakes), and the 34 individual species accounts (2 lizards, 32 snakes). A detailed bibliography through June 2009 is also included in volume 1. The cutoff date for literature in the bibliography of volume 2 is June 2010.

The geographic area covered in this book includes the ranges of venomous reptiles from their northernmost distribution in southern Canada, through the United States, to 25°N latitude in continental Mexico, the Baja California Peninsula, and those islands in the Gulf of California and off the Pacific Coast of Baja California where they occur. To expand the coverage to include in the same detail the many venomous snake species occurring farther south in Mexico and Central America is beyond the scope of this publication. For details on such species, we refer the reader to their excellent coverage in Campbell and Lamar (2004).

Before continuing, it must be stated that this book presents only North American reptiles belonging to the lizard family Helodermatidae and the snake families Elapidae and Viperidae, which are considered dangerously venomous. Rear-fanged (*opisthoglyphous*, fig. 1) snakes of the family Colubridae also use oral secretions to secure their prey. Of these, the African bird or twig snake, *Thelotornis kirtlandii* (Broadley 1957; Fitz Simons and Smith 1958), and boomslang, *Dispholidus typus* (Broadley 1957; Pope 1958), and the oriental natricine tiger snake, *Rhabdophis tigrinus* (Mittleman and Goris 1978; Ogawa and Sawai 1986), have delivered fatal bites to humans; and the North American garter snakes *Thamnophis elegans terrestris*, *T. e. vagrans*, and *T. sirtalis* (Jansen 1987; McKinstry 1978; Nichols 1986; Vest 1981), ring-necked snake, *Diadophis punctatus* (Anton 1994; Myers 1965; Shaw and Campbell 1974), and plains hog-nosed snake,

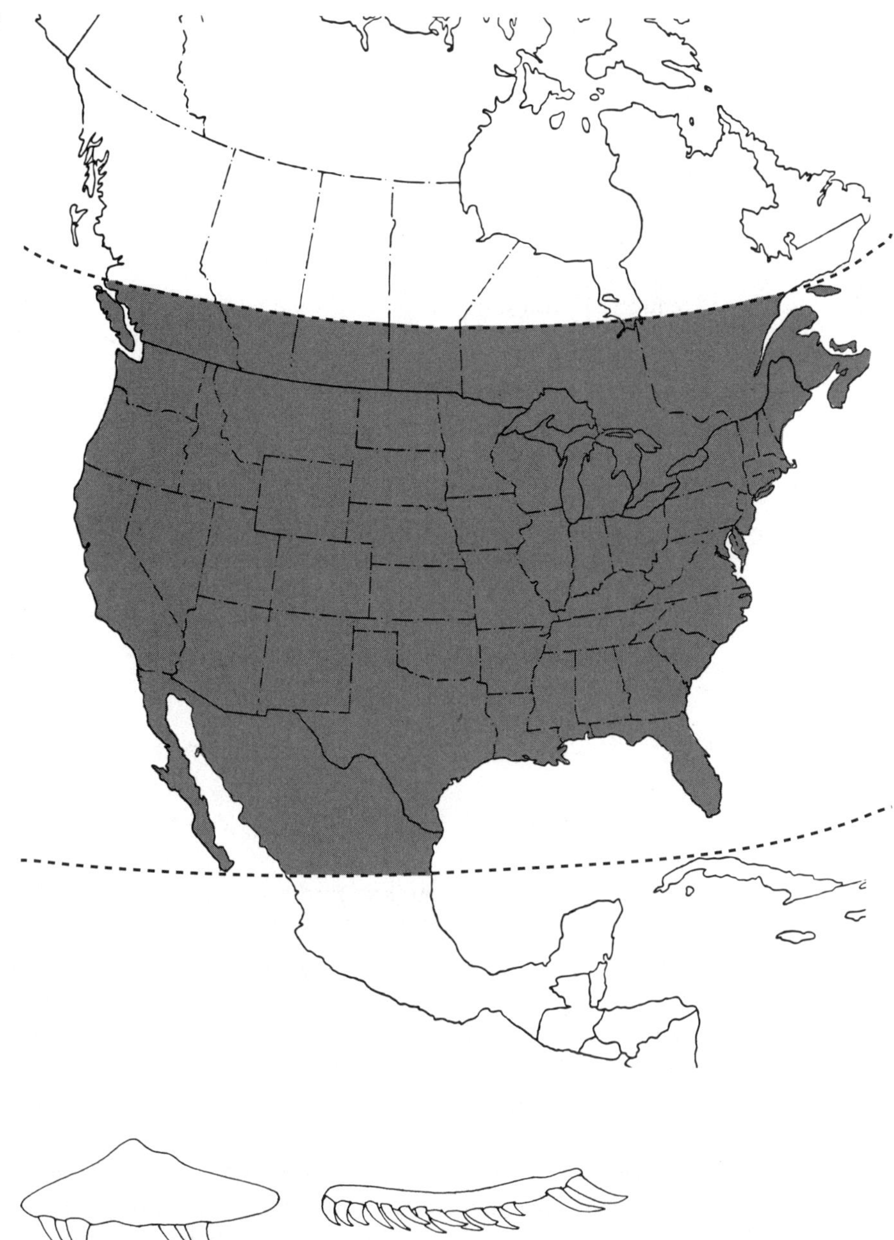
Geographic area of North America covered in this book.

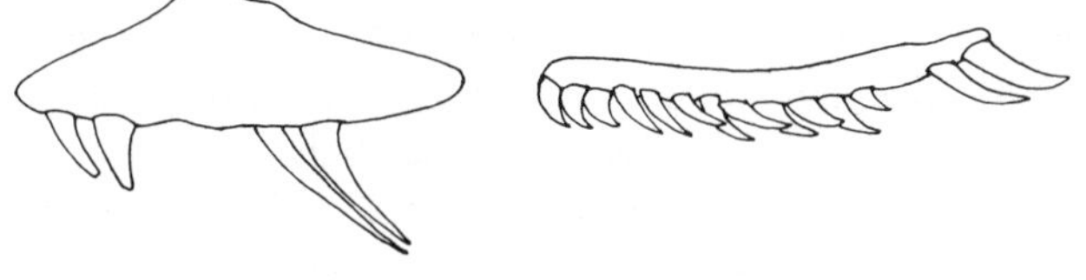
Figure 1. Two examples of the opisthoglyphous maxillary dentition of colubrid snakes; note the enlarged posterior teeth.

Heterodon nasicus (Morris 1985), among others (see Ernst and Ernst 2003), have also caused human envenomations. Additional data on venomous colubrids are included in Fry et al. (2003c), Harding (1984), Hill and Mackessy (1997), Mackessy (2002), McKinstry (1978), Minton (1976, 1979, 1986, 1990a, 1996), and Minton and Weinstein (1987).

Several aspects of the biology of our venomous reptiles need to be explained so that the reader will have a better understanding of these when they appear in the individual species accounts. The common names used for the amphibians and reptiles mentioned in the text are adapted from Crother (2000) and Liner (1994).

Karyotype: The karyotype consists of the number and physical appearance of its diploid chromosome pairs, and its terminology needs some explanation. The karyotypical morphology of a species gives evidence of its evolution and systematic position, as well as relationships with other species within its genus. Each chromosome has a small, unstaining, constricted region called a centromere, which may be located at different positions in different pairs of chromosomes. Its position is used to designate the chromosome as either metacentric, having the centromere located near the middle so the two arms are of equal or near equal length; acrocentric, having the centromere near one end of the chromosome so that the two arms are of unequal length (one long, one short); or telocentric, where the centromere is very near one end of the chromosome. Snakes possess sex chromosomes: in females the sex chromosomes are of unequal length (heteromorphic) and are termed ZW; in males the sex chromosomes are of equal or near equal length (homomorphic) and are termed ZZ. A nucleolus organizer region (NOR) is present on a pair of chromosomes.

Thermal ecology: The behavior and physiology of our venomous reptiles are basically driven by the ambient or environmental temperature. Most are active when the temperatures of their surroundings are between 15 and 30°C. Because our reptiles are conformers, with a body temperature generally matching that of their external surroundings, they are referred to as *ectotherms*; their body temperatures vary, so they may also be termed *poikilotherms*. In the text, where possible, a discussion of the thermal ecology of each species is presented. In these discussions, the following abbreviations are used for the sake of brevity to denote various temperatures: environmental temperature (ET), air temperature (AT), surface or soil temperature (ST), body temperature (BT), cloacal temperature (CT), critical thermal maximum (CT_{max}), critical thermal minimum (CT_{min}), and incubation temperature (IT).

Movement and space relationships: Venomous reptiles, like other vertebrates, are normally active in only a relatively small portion of the available habitat. This area is termed their *home range* or *activity range*, and all life behaviors are normally carried out within it. However, some do not have suitable overwintering habitat within their home range, and so must make a fall movement beyond the limit of their summer home range to a distant hibernaculum, and return from it the following spring. Such one-way movements may be relatively short, 100–300 m, but sometimes they are of several kilometers.

Reproduction: All reptiles have internal fertilization at the upper end of the oviduct and direct development (no metamorphic stage occurs). The young either hatch out of shelled eggs laid by *oviparous* species (Elapidae, Helodermatidae) after a suitable *incubation period* (IP), or are born alive in *ovoviparous* species (Viperidae) after a *gestation period* (GP). In ovoviviparous snakes, the embryos (covered with a membranous amniotic sac) develop within the female's reproductive tract, then pass out of her cloacal vent (*parturition*). No placental connection is formed between the embryo and the female, so no nourishment, other than the yolk within the original yolk sac, is provided.

In the text, when possible, the following reproductive parameters are provided for each species: size and age at maturity; gametic and hormonal cycles of both sexes; the mating season(s); courtship and mating behaviors; the season of egg laying (*oviposition*) of oviparous species; the season of birth (parturition) of ovoviparous species; and descriptions of eggs, hatchlings, or neonates. The *relative clutch mass* (RCM), the total clutch mass divided by the female postparturient mass (Barron 1997; Seigel and Fitch 1984), is presented where known.

Diet and feeding behavior: Our venomous reptiles are carnivores that consume other animals. With so many different species adapted to various habitats, these North American species take a wide variety of prey, and a prey list is presented for each species. Venomous reptiles inject venomous saliva into the prey. Once envenomated, snakes usually release the prey and later follow their odor trail; they finally swallow their prey when they are dead or incapacitated. The helodermatid lizards normally continue to hold their prey and chew the venom into the wounds. A description of the venom delivery system is included in each species account, and notes are presented on the feeding behavior of each reptile.

Predation: Venomous reptiles, like all other vertebrates, have their own predators. Those animals that prey on a venomous species are listed, and its means of defense are described.

Venoms: A venom is a highly toxic poison that one animal injects into another. The venom of snakes is produced in modified salivary glands (Duvernoy's glands, or venom glands) located toward the rear of the upper jaw, or, in the two helodermatid lizards, in modified salivary glands in the gums of the lower jaw. These glands are ducted to either the base of a grooved tooth or into the hollow center of a special injecting fang (see the family accounts of Elapidae, Helodermatidae, and Viperidae for detailed descriptions). Each venomous species has a unique venom composition with different amounts of toxic and nontoxic compounds. The chemistry of such venoms is complicated, but, overall, venoms are about 90% protein (by dry weight), and most of the proteins are enzymes. About 25 different enzymes have been isolated from snake venoms. Ten of these occur in the venom of most snakes. Proteolytic enzymes (involved in the breakdown of tissue proteins), phospholipases (either mildly or highly toxic to muscles and nerves), and hyaluronidases (dissolve intercellular materials and speed the spread of venom through the prey's tissue) are the most common types. Other enzymes are collagenases (which break down connective tissues), ribonucleases, deoxyribonucleases, nucleotides, amino acid oxidases, lactate dehydrogenases, and acidic or basic phosphatases, all of which disrupt normal cellular function. (See the chapter "Venom" and Ernst and Zug 1996; Russell 1983; Tu 1977; and Zug and Ernst 2004 for more complete discussions of venom chemistry.) Venoms evolved first as a prey-capture device, and second as a defensive tool.

Because North American snakes of the families Elapidae and Viperidae and the lizards of the family Helodermatidae possess dangerous venoms, where information is available, the toxicity and symptoms in human envenomations are presented for each species in these families.

Populations: The population dynamics, including sizes, sex (male to female) ratio, juvenile to adult ratio, size and age classes, conservation issues, and survival status are given, as far as are known, for each species.

Abbreviations

AT	air temperature
BM	body mass
BT	body temperature
CITES	Convention on International Trade in Endangered Species
CT	cloacal temperature
CT_{max}	critical thermal maximum
CT_{min}	critical thermal minimum
D	Dalton
DNA	deoxyribonucleic acid
DOR	dead on road (road-killed)
ESA	Endangered Species Act (United States)
ET	environmental (ambient) temperature
FL	fang length
FL_{max}	maximum recorded fang length
FMR	field metabolic rate
GP	gestation period
HD	head depth
HL	head length
HW	head width
IP	incubation period
IT	incubation temperature
IUCN	International Union for Conservation of Nature
LD_{25}	venom dosage at which 25% of the envenomated animals die
LD_{50}	venom dosage at which 50% of the envenomated animals die
mtDNA	mitochondrial DNA
MYP	million years before present
N	Newton
nDNA	nuclear DNA
NOR	nucleolus organizer region
Q_{10}	temperature coefficient
RCM	relative clutch/litter mass

RNA	ribonucleic acid
rRNA	ribosomal RNA
SMR	standard metabolic rate
ST	surface temperature (soil temperature)
SVL	snout-vent length (measured from tip of snout to anal vent)
TBL	total body length (measured from tip of snout to tip of tail; from tip of snout to base of rattle of rattlesnakes)
TBL_{max}	maximum recorded total body length
TL	tail length
USFWS	U.S. Fish and Wildlife Service
WT	water temperature
YBP	years before present

Venomous Reptiles of the United States, Canada, and Northern Mexico

Venom

This chapter deals exclusively with snake venoms. Those produced by helodermatid lizards are discussed in the chapters on *Heloderma horridum* and *H. suspectum*. The venom chemistry of an individual snake species is discussed in its account.

Venom is a highly toxic poison, but the term *poison* is a broad, misleading one that includes any harmful or fatal substance that enters the body. It is particularly used for those substances that enter the body by absorption through the skin, inhalation while breathing, or swallowing. Venom is a particular type of poison produced by a phanerotoxic animal (like a snake, helodermatid lizard, scorpion, hymenopteran insect [ant, bee, wasp], spider, stingray or some other fish, and the duckbill platypus) and injected into another animal through a bite or sting.

Snake venom is a modified, highly toxic chemical compound secreted by special oral glands related to the salivary glands of other vertebrate animals, in particular, modified Duvernoy's glands in snakes. It first evolved as a means of immobilizing prey and making handling of them easier, and, although a very good one, is secondarily used for defense. In most species, it initiates the digestive process by breaking down the other animal's tissues. It is capable of producing one or more harmful changes in several organ systems or tissues, and is able to promote these changes simultaneously (Russell and Brodie 1974).

The venom of each species is unique and genetically determined, and contains a mixture of different components and amounts of toxic and nontoxic biochemical compounds. Closely related snakes, like those belonging to the same genus or family, have more similar venoms than those in other genera or families. Probably, venoms evolved several times in those families of snakes containing venomous species, as is witnessed to by their diverse chemistry, functions, genetics, and delivery systems (Graham et al. 2008; Juárez et al. 2008; Lynch 2007; McCue 2005; Pook and McEwing 2005; Powell and Lieb 2008). Over time snake venom has reached a high degree of development, and such evolution may have been triggered by competition among species, changes in prey availability or preference, or even the development of immune mechanisms by prey.

The venom produced by a particular species of snake is usually classified as neurotoxic (affecting primarily muscles or the transmission of nerve impulses to them) or hemotoxic (affecting the blood components, circulatory vessels, or heart). Russell (1983, 1984) has objected to this broad classification, noting that the venom of any particular snake species is a mixture that contains more than one toxin and may produce clinical symptoms in both categories. He correctly suggested (1984) that a better system would be to classify snake venoms as either cardiotoxins (cytotoxins in certain cobra venoms affecting the heart), hemotoxins, myotoxins (affecting mainly nerve impulse conduction to skeletal muscles, leading to muscular degeneration), or neurotoxins (which block nerve impulses to muscles and affect the production of the neurotransmitters acetylcholine and norepinephrine).

For the sake of simplicity, because the combination of toxins in any snake's venom has a more potent effect than the sum of their individual actions, and to avoid confusing untrained medical personnel, the terms *hemotoxic* (most North American pitvipers) and *neurotoxic* (coralsnakes, *Pelamis platura*, and some *Crotalus* rattlesnakes) are still the commonly used terms for the most notable effects of snake venoms in the Americas.

CATEGORIES OF TOXIC VENOM COMPONENTS

Toxic venom components have been broadly categorized by their actions and how they work to disrupt normal function (Ernst and Zug 1996; Zug and Ernst 2004). Following is an annotated list of these categories.

Cardiotoxins, associated mainly with cobras and vipers, have variable effects. They are direct membrane-active, lytic factors that are highly hydrophobic, basic, single, short-chained polypeptides closely related to the α-neurotoxin that binds to nicotinic acetylcholine receptors (nAChRs). Some depolarize contracture of cardiac, smooth, and skeletal muscles and alter heart contraction, causing heart failure. They have actions similar to those of phospholipase A_2 myotoxins, which rapidly lyse sarcolemma, clump myofibrils, and hypercontract muscle cells. Their actions differ, however, from the skeletal muscle necrosis induced by crotamine and myotoxin α, which is much slower and consists of mitochondrial and sarcoplasmic reticulum swelling, myofibril degeneration, and lack of sarcolemma or transverse tubule damage. The end result is a disturbance of the cell membrane of some heart cells, causing tachycardia, hypotension, and variances in heart rate and rhythm, all leading to cardiac arrest (Backshall 2008; Fletcher et al. 1996; Koh et al. 2006).

Coagulation-retarding compounds, produced by some elapids and pitvipers, prevent blood-clotting.

Cytolysins destroy white blood cells, and are found in pitviper and viper venoms.

Hemorrhagic toxins (hemorrhagins) are proteases found in the venoms of pitvipers, vipers, and the king cobra that cause hemorrhages and damage blood vessels. Almost all are metalloproteinases (discussed later) with molecular weights in the range 62,000–68,000 Daltons (D). They attack and destroy capillary walls, causing hemorrhaging near and distant from the bite site. Hemolysins, in the venom of elapids, pitvipers, and vipers, destroy red blood cells (Johnson and Ownby 1993b; Mashiko and Takahashi 1998; Shu et al. 1988; Tan and Ponnudurai 1992).

All require metal cations as cofactors. They can be divided into four groups according to their molecular weights (D) and domain compositions: (1) small, 15,000–30,000; (2) medium, 30,000–50,000; (3) large, 50,000–80,000; and (4) extra large, >80,000 (Takahashi and Mashiko 1998).

Myotoxins are muscle impairers that are principally found in pitviper and viper venoms (Bober et al. 1988), but are also in the venoms of some elapids, especially those of seasnakes (which contain high amounts). They are basic and can be categorized into three types: (1) small, basic polypeptides, like myotoxin α and crotamine, with molecular weights around 5,000 D; (2) large, 12,000–16,000 D, cardiotoxins from cobra venom; and (3) phospholipase A_2 toxins, like crotoxin.

All three types of myotoxins cause depolarization and contraction of skeletal muscle cells, and can cause permanent tissue damage (myonecrosis), even to the point of loss of use of limb parts in severe cases (the muscle usually regenerates in mild envenomations) or death (Kitchens et al. 1987). In contrast, the damage induced by the small, basic polypeptide myotoxins is different from that caused by the other two types of myotoxins; small, basic myotoxins do not appear to lyse the sarcolemma, whereas the other two myotoxins cause rapid lysis of sarcolemma. The three myotoxins are similar in being highly basic proteins with a large portion of their surface charge positive, and having a considerable ß-sheet structure that may be involved in interaction with the membrane (Ownby 1998).

All known low molecular weight, basic myotoxins have been isolated from the venom of rattlesnakes of the genus *Crotalus*. Included in this group are: (1) CAM (cell adhesion molecules; 45 amino acids, 5,202 D) from *C. adamanteus* (Samejima et al. 1991); (2) crotamine (42 amino acids, 4,890 D) from *C. durissus terrificus* (Laure 1975); (3) myotoxin I (43 amino acids, 5,061 D) from *C. oreganus concolor* (Bieber et al. 1987); Griffin and Aird (1990) reported a 45 amino acid small myotoxin from *C. o. concolor* that is similar to myotoxin I, but has an additional C-terminal aparaginyl-alanine; (4) myotoxin II (43 amino acids, 5,034 D), also from *C. o. concolor* (Bieber et al. 1987); (5) myotoxin α (42 amino acids, 4,828 D) from *C. viridis viridis* (Fox et al. 1979); (6) myotoxins 2 and 3 (42 amino acids, 5,246 D) from *C. v. viridis* (Griffin and Aird 1990); and (7) peptide C (43 amino acids, 4,989 D) from *C. helleri* (Maeda et al. 1978). Exactly how these low molecular weight compounds work is not totally understood, but it is known that crotamine and myotoxin α cause an increase in the sodium channel or its modulator for sodium ions (Chang and Tseng 1978; Meier and Stocker 1984).

Neurotoxins are found in the venoms of elapids, vipers, tropical rattlesnakes, and populations of the North American rattlesnakes *Crotalus helleri, C. lepidus, C. oreganus, C. scutulatus, C. tigris,* and *C. totonacus*.

Venoms of terrestrial elapids, especially those from Australia, and seasnakes are rich in neurotoxins, like presynaptic ß-bungarotoxins from kraits and taipoxin from the Australian taipans (*Oxyuranus*), and muscarine toxins that bind specifically to muscarine acetylcholine receptors (mAChRs) from elapids (Harvey et al. 2002). Elapid venoms include both short-chain toxins and long-chain toxins. The actions of both are postsynaptic (Australian elapids, cobras, kraits, seasnakes). Not all neurotoxins are strong; weakly postsynaptic ones are found in some Australian elapids, cobras, and kraits.

Potency of neurotoxins lies in their affinities toward the biomolecules involved in neuromuscular transmission. Their neuromuscular and physiological pathological effects result from interactions with various microcompartments that have similar masses, amino acid structures, and disulfide bridges in normal bonds. They block the transmission of nerve impulses to muscles, particularly those associated with the diaphragm and breathing.

Neurotoxins are categorized as either α-neurotoxins (postsynaptic) that bind to nAChRs, or ß-neurotoxins (presynaptic) that cause the release of acetylcholine. Neurotoxins can also be separated as to whether their active site is at the skeletal-muscular junction, or at sites other than the neuromuscular synapse (both presynaptic and postsynaptic) (Koh et al. 2006). Gawade (2004) has proposed a complicated pharmacological classification of neurotoxins, and we refer the reader to his paper for details.

The term *three-fingered toxins* (3FTx) is often used in reference to some low-potency venom α-neurotoxins found in the family Elapidae. It describes the configuration of the neurotoxin with regard to the positioning (extensions) of the structural amino acid residues located on the central loop of the flat molecule directed toward the observer when viewed. The residues appear as three adjacent loop-like projections that emerge from the small, globular, hydrophobic core of the molecule that is cross-linked by four conserved disulfide bridges (see diagrams in Nirthanan et al. 2003 and Tamiya 1980). More than 60 amino acid residues with disulfide bridges are involved (Endo and Tamiya 1991; Kini 2002; Tsetlin 1999). A particular group of these enzymes, termed *nonconventional toxins* by Nirthanan et al. (2003), contains 62–68 amino acids and 5 disulfide bonds, targets low-affinity muscle receptors, and antagonizes neuronal α-7 nicotinic acetylcholine receptors (Servent et al. 1997; Tsetlin and Hucho 2004). Overall, the 3FTx aids communication among nerve cells, but sometimes it binds too closely to nerves and causes muscle paralysis.

Fry et al. (2003b) reviewed the possible evolution of three-fingered toxins within the Elapidae, and found a basal split between an Asian-African-American group and an Australian seasnake group.

Proteolysins, found mostly in viper and pitviper venoms, dissolve cells and tissues at the bite site, causing local pain and swelling.

Sarafotoxins are three highly toxic compounds contained in the venom of the African burrowing asp, *Atractaspis engaddensis* (LD_{50}, 0.06–0.075 mg/g in mice). Each is composed of a single-chain isopeptide with 21 amino acid residues cross-linked by 2 disulfide bridges. They apparently directly attack the heart, causing a rapid constriction of the coronary artery, severely blocking atrioventricular transmission and having a slower but very positive inotropic effect that cannot be blocked by either α- or ß-adrenergic receptor blockers. These toxins have a high affinity to the membranes of the heart atria and brain, where they act directly on phosphoinositide phosphodiesterase or a novel receptor to hydrolyze phosphoinositides (Kloog et al. 1988; Kochva et al. 1982; Weiser et al. 1984; Wollberg et al. 1988).

Thromboses, produced by some pitvipers and vipers, coagulate blood and foster clot formation throughout the circulatory system.

COMPOSITION OF SNAKE VENOMS

The pathology following a snake envenomation is the result of mechanisms of the multiple components in the venom, but chiefly destructive peptides (Straight et al. 1991). The clinical conditions are not only dependent on the qualitative composition of the venom compounds involved, but also on their quantitative distribution (Meier and Stocker 1995).

The chemistry of snake venoms is very complex. They are complicated mixtures, mainly composed of proteins (normally about 75–77%, but up to 93% by dry weight); rattlesnake venoms contain 225–250 mg/mL (Mackessy 2008), most of which are toxic enzymes. The venom of most species contains at least 26 enzymes, of which no less than 10 are found in all snake venoms (Russell 1984). The percentage of solids in the whole venom of adult snakes ranges from 3–55% in the Elapidae (including seasnakes), 10–37% in Crotalinae, and 9–32% in Viperinae, and the percentages of proteins in these solids are 75–91%, 62–87%, and 64–74%, respectively (Brown 1973).

The most common types are proteolytic enzymes, phospholipases, and hyaluronidases (all discussed individually later). Hyaluronidases dissolve intercellular materials and speed the spread of venom through the prey's tissue. Other enzymes include collagenases (in pitvipers), which promote the breakdown of the protein collagen, an important structural component of connective tissues. Deoxyribonucleases and ribonucleases, nucleotidases, amino acid oxidases, lactate dehydrogenases, and acidic and basic phosphatases all disrupt normal cellular function and cause the collapse of cell metabolism, shock, and death.

Even so, not all of the toxic chemical compounds in snake venoms are enzymes. Polypeptide toxins (some of which are among the most lethal compounds in snake venom), glycoproteins, and low molecular weight compounds are also present in the venoms of some colubrids and elapids. Those of elapids contain higher quantities of polypeptides, but fewer enzymes than either pitviper or viper venoms. Lipids compose <0.5% of venoms. The balance of dry venom weight is mostly by inorganic compounds (Elliott 1978).

Venom enzymes by themselves have a low toxicity or are nontoxic, and the lethality of venoms containing only enzymes appears to be due to the combined effects of these enzymes and substances released by the envenomated animal. The only venom constituents that are truly lethal by themselves are nonenzymatic neurotoxins, cardiotoxins, and hemorrhagins (Van Mierop 1976a).

Highly toxic venoms lack high amounts of metalloproteases, which cause much structural damage, while those with potent metalloprotease activity generally lack presynaptic neurotoxins found in the most deadly venoms. Four classes of metalloproteases are found in rattlesnake venoms (Fox and Serrano 2005; Mackessy 2008). Mackessy (2008) has termed those venoms with higher levels of metalloproteases (see below) and moderate toxicity *Type I venoms,* while those with low metalloprotease activity and higher toxicity he termed *Type II venoms*. Different individuals or populations of a species may contain different amounts of these types of venom.

The viscosity range of fresh venom is normally 1.5–2.5 (Devi 1968). Distilled water in low concentration will only dissolve a portion of dried venom, but more can be dissolved in solutions containing >0.9% NaCl.

The roles and functions of other venom components are largely unknown. Several things complicate the overall action of a venom. Composition of the venom can vary seasonally between species over a yearly period (Gregory et al. 1984; Gregory-Dwyer et al. 1986). The venom ejected must be replaced after each envenomation, and full replacement takes time. An envenomation from a snake that has not recently fed, like one just emerging from hibernation, is more dangerous than a bite from one that has recently fed, because it has more venom to inject.

The chemistry and potency of venom can, but not always, also vary during ontogenetic development of an individual species (Fiero et al. 1972); the venom of neonates and small juveniles appears to be more potent than that of adults of their own species. Why this is so is unclear, but it may reflect a change in prey size or type. Both qualitative and quantitative differences are also found within populations of the same species (see the accounts of *Agkistrodon contortrix, Crotalus horridus, C. lepidus, C. oreganus,* and *C. scutulatus*). The venom from littermates can also differ. A particular snake's venom may be more specific to one prey group than to others; a thrombin-like serine proteinase (batroxobin, Defibrase) in the venom of the Brazilian lancehead *Bothrops moojeni* is less toxic to mammals but highly lethal to birds (Stocker 1986). In addition, the physiochemical properties of venom or its fraction, pH, vehicle, particle size, and concentration all affect prey in different ways.

Venom is rather stable, and can be stored without losing its potency for a relatively long period under the right conditions (Russell 1983; Russell et al. 1960).

The following list of venom components and data concerning them were compiled from Elliott (1978), Mackessy (2008), Mebs (1998), Meier and Stocker (1995), Russell (1983, 1984), Russell and Brodie (1974), Tu (1977), Van Mierop (1976a), Werman (2008), and other recent papers. However, these authors approach the discussion of snake venom in different ways; some emphasize the venom's chemical structure, while others concentrate on its physiological actions. This causes confusion, and makes it difficult to organize a discussion of venoms in a structured manner, so please note that some necessary overlap of venom components occurs between the various categories presented.

Enzymes: These are proteins that do local damage, and all snake venoms have them. They differ in mixture and proportion, initiate physiologically disruptive or destructive processes, and have been classified either by their chemical composition or by their physiological action.

Acetylcholinesterase (AChE, Acetylcholine acetyl) has a molecular weight of either 126,000 or 130,000 D and occurs in small, highly concentrated amounts almost exclusively in elapid venoms; but, surprisingly, it is absent from the venom of *Micrurus fulvius* (McCollough and Gennaro 1970). It catalyzes the hydrolytic process converting acetylcholine to choline and acetic acid, but does not appear to cause the nerve-muscle blocking effect of elapid venoms.

Adenosine triphosphatase (ATPase, ATP pyrophosphohydrolase) is found in elapid and viper venoms. It releases a mole of orthophosphate from each mole of ATP through dialysis, which may result from the sequential action of omega-exonuclease and 5′-nucleotidase (Elliott 1978; Zeller 1950a, 1950b).

Alkaline phosphatases are basic enzymes that hydrolyze monophosphate esters with an optimum pH of 8.0–10.5, depending on the substrate. They are composed of 5′-nucleotidase, glycerophosphatase, phosphodiesterase, and phosphomonoesterase (Elliott 1978; Suzuki and Iwanaga 1958a, 1958b).

Arginine ester hydrolase is a noncholinesterase commonly occurring in venoms of pitvipers (at least three are present) and vipers, but not in those of elapids, except possibly that of the king cobra (Deutsch and Diniz 1955). This enzyme's activity differs from those caused by proteolytic enzymes, but can be associated with proteases having very high specificity. It hydrolyzes the ester or peptide bonding to which argine is attached to the carboxyl group (Tu 1977). Possibly it plays a role in preventing blood-clotting and promoting hemorrhage.

Bradykinin-releasing factor (kinogenase), an important component of snake venoms, is a specific protease that hydrolyzes two particular peptide bonds in bradykinogen to release bradykinin.

Deoxyribonuclease (DNase) and Ribonuclease (RNase). DNase is an endonuclease that attacks DNA and predominantly produces tri- or higher oligonucleotides terminating in 3-monoesterified phosphate. The exact mechanism of this action is not fully understood, and it is possibly catalyzed by phosphodiesterase. Some crotalid venoms contain more than one DNase with optimum pHs of 5 or 9 (Iwanaga and Suzuki 1979; Mackessy 1998b).

RNase is a hydrolyzing endopolynucleotidase that works specifically on the pyrimidines of RNA containing pyrimidyladenyl bonds. It is found in elapid venoms, has a relatively low molecular weight of around 16,000 D, and has optimum pHs of 7 and 9. According to Mackessy (1998b), the "nonspecific" RNase activity present in many venoms is likely due to phosphodiesterase, or possibly nonspecific endonuclease activity.

Glutamic-pyruvic transamine (L-Alanine: 2-Oxoglutarate aminotransferase, Transaminase), a nonhydrolytic enzyme, is found specifically in the venoms of some elapids and vipers (Tsai 1961).

Glycerophosphatase is present in the venoms of some elapids and Asian pitvipers. It suppresses the activity of 5′-nucleotidase (Suzuki and Iwanaga 1958a).

Hyaluronidase (Hyaluronate glycanohydrolase, Mucinase) is a high molecular weight, proteolytic enzyme that is often referred to as the "spreading factor" (Duran-Reynals 1939); it decreases the viscosity of connective tissues by catalyzing cleavage of the internal glycosidic bonds, in particular, mucopolysaccharides, allowing other venom components to penetrate the affected tissues. It plays some role in edema production, but none in hemorrhage activity (Tu 1977). All snake venoms contain this enzyme in varying amounts, but less occurs in elapid venoms. Antivenom acts to inactivate the spreading action of venom.

Lactate dehydrogenase (L-Lactate: NAD + oxidoreductase) has been found in some elapid venoms; it reverses the conversion of lactate to pyruvic acid.

L-Amino acid oxidase (L-amino acid: O_2) is another high molecular weight enzyme found in all snake venoms (although apparently absent from those of young and newborn pitvipers and vipers; Van Mierop 1976a). It is nonhydrolytic, has a high acidic amino acid content, and is the most active of the amino acid oxidases in snake venom. It catalyzes the oxidation of L-α-amino and α-hydroxy acids to produce a group of homologous enzymes (including α-ketoacids) with molecular weights of 85,000–150,000 D (most commonly 130,000). Its toxicity is very low, and it does not affect muscle, neuromuscular, or nerve transmission (Russell et al. 1963b).

The protein riboflavin, occurring in the form of flavine mononucleotide, is an integral part of L-amino acid oxidase (Iwanaga and Suzuki 1979; Meier and Stocker 1995). The enzyme is responsible for the yellow color of venom (note, however, that some snakes' venoms are clear to whitish, and that both white and yellow venoms may occur within the same species; Johnson et al. 1987).

NAD-Nucleotidase (NAD glycohydrolase, NADase), present in the venoms of some elapids, pitvipers, and vipers, breaks down nicotinamide adenine dinucleotide (NAD) by catalyzing the hydrolysis of the nicotinamide N-ribosidic linkage to produce nicotinamide and adenosine diphosphate riboside. It loses its activity at 60°C, and has an optimum pH range of 6.5–8.5. Although present in the venoms of *Agkistrodon contortrix* and *A. piscivorus*, it is not in those of *Crotalus* rattlesnakes.

NAD-nucleotidase is not to be confused with nucleotide pyrophosphate, which cleaves NAD at a different active site on the molecule (Tu 1977).

Nonspecific phosphatases, which probably occur in all snake venoms, result from the cleavage of nucleic acid into 3′, 5′-mononucleoside fragments. They produce little exonuclease activity, and show no 5′-nucleotidase activity. More than one of these enzymes, all working at different active sites, may occur in a particular species' venom.

Phosphodiesterase (Orthophosphoric diester phosphohydrolase), sometimes termed *venom exonuclease,* is ubiquitous in snake venoms, and some contain more than one form. Stable at pHs of 6–9, the molecular weight is >75,000 D, probably mostly around 130,000 D; those of *Crotalus* rattlesnakes have a molecular weight of 115,000 D (Gulland and Jackson 1938a; Mackessy 1998b; Russell et al. 1963a; Sinsheimer and Koerner 1952; Tu 1977; Wechter et al. 1968; Williams et al. 1961).

Phosphodiesterases act as orthophosphoric diester phosphohydrolases that cleave phosphomonester and phosphodiester bonds to release 5′-mononucleotide from DNA and RNA. The enzymes separate a molecule of phenyl from diphenylphosphate to release 5′-nucleotides from DNA and RNA. They also hydrolyze nucleoside *p*-nitrophenylphosphate, nucleoside 2,4-dinitrophenol phosphate, nucleic acid, nucleotides with specific cyclic phosphate termini, oligonucleotides containing α-phosphate, and the phosphate bonds of poly(ADP-ribose). These actions result in cardiovascular changes similar to those produced by raw (crude) venom, but no neuromuscular action in mammals. The LD_{50} is 3.08–3.48 mg/kg, about twice that of raw venom. Mg^{2+} is required for activation, but Cu^{2+} can inactivate them, and thymine diemers block their actions. Phosphodiesterases are not affected by Ba^{2+}, Ca^{2+}, Co^{2+}, Fe^{2+}, Mn^{2+}, NH_4^{2+}, Ni^{2+}, Pb^{2+}, Sn^{2+}, SO_4^{2+}, and Zn^{2+} (Beasley et al. 1993; Elliott 1978).

Using disc electrophoresis, Russell et al. (1962) estimated that the enzyme composed <5% of crotalid venom and contributed about 2% of the lethal effect of raw venom. However, Russell (1983) admits that in 1962 they were using an unpure product, and that even in later studies they were unable to isolate the purified enzyme.

Diphosphoesterases are used extensively to cleave bonds during molecular DNA and RNA research.

Phospholipase A_2 (PhA_2, PLA_2, Phosphate acyl hydrolase, Lecithinase) is perhaps the most intensively studied group of all β neurotoxic components in snake venom. They constitute a large family of enzymes that catalyze the hydrolysis of the *sn*-2 ester bond of phospholipids to produce free fatty acids and lysophospholipids (1–4) (Ambrosio et al. 2005). Functioning as myotoxins, they cause presynaptic neurological damage, and are one of the most potent of all compounds in snake venom. The higher the concentration of PhA_2, the more potent the venom (Breithaupt and Habermann 1973). The North American snakes with the most toxic PhA_2 venoms are the coralsnakes and seasnakes (*Micruroides, Micrurus, Pelamis*) and several rattlesnakes of the genus *Crotalus*.

Perhaps the most famous of these PhA_2 presynaptic myotoxins is Mojave toxin, found in the venom of *Crotalus scutulatus* and to a lesser extent in the venoms of some other southwestern *Crotalus* that contain subunits of the toxin (Henderson and Bieber 1985, 1986; Powell et al. 2008). PhA_2 is universally found in variable amounts in the venoms of elapids (including seasnakes), but also occurs in varying concentrations in venoms of some colubrids, pitvipers, and vipers. The activities of PhA_2 do not differ between the major groups of venomous snakes (McCue 2005). Other members of the neurotoxic complex include canebrake toxin, concolor toxin, crotoxin and vergrandis toxin in *Crotalus,* and sistruxin in the *Sistrurus* rattlesnakes (Werman 2008).

In addition to neurotoxic and myotoxic biological effects, research has also shown that PhA_2 plays a role in anticoagulation (see diagrams in Zingali 2007), artificial membrane disruption, cytotoxic effects, hypotension, induction of edema, and platelet aggregation inhibition; it also has some antibacterial, antimalarial, antiparasitic, and antitumor properties (Soares and Giglio 2003).

The principal activator of PhA_2 is Ca^{2+}, but albumin also stimulates it. It may be inhibited by Al^{3+}, Ba^{2+}, Ca^{2+} (in some cases), Cd^{2+}, Co^{2+}, Cu^{2+}, Fe^{3+}, Hg^{2+}, Pb^{2+}, $(PO_4)^{3-}$, Zn^{2+}, some peptides, and other chemicals (Elliott 1978; Meier and Stocker 1995; Rosenberg 1987, 1988, 1990).

The enzyme catalyzes the hydrolysis of one of the fatty acid ester bonds in diacyl phosphatides, specifically lecithin (lecithin + $H_2O \rightarrow$ unsaturated fatty acid + lysolecithin), and causes a slow decline in the contraction of diaphragmatic muscles (see Dijkstra et al. 1981; Elliott 1978; and Tu 1977 for thorough discussions of biochemical reactions involving PhA_2).

Structural damage in the central nervous system possibly results from the demyelinating actions of PhA_2 on the spinal cord and medulla, and because of its increasing the permeability of the blood-brain barrier. Release of physiologically active compounds by this enzyme probably depends on its disruption of membranes to release cellular acetylcholine, histamine, kinins, and serotonin. In addition, this enzyme uncouples oxidative phosphorylation, thus inhibiting cellular oxidative respiration and metabolism by disrupting the mitochondrial electron transport chain. There is also evidence that PhA_2 causes vascular permeability to protein and red blood cells. Peptide I in the venom of *C. helleri* concentrates hemoglobin, significantly lowering blood pressure accompanied with lowered concentrations of the blood serum level and of both lactate and protein in the blood (Schaeffer et al. 1979).

Several forms of PhA_2 exist; these have similar amino acid sequences, but different pharmacological actions (Arni and Ward 1996; Elliott 1978; Faure 1999; Kini 2005; Omori-Satoh et al. 1975; Powell et al. 2008; Rosenberg 1987, 1988, 1990; Russell 1983; Six and Dennis 2000; Tu 1977). One such difference occurs at position 49 on the amino acid chain, where either aspartic acid (D49 enzymes) or lysine (K49 enzymes) may be present (Fletcher and Jiang 1998; Lomonte et al. 2003a; Maraganore et al. 1984; Ownby et al. 1999; Tsai et al. 2001). The forms of PhA_2 also differ in stability, particularly related to substrate preference and to enzyme inhibition or stimulation in the presence of various ions. The molecular weights vary from 29,000 to 30,000 D in North American *Crotalus* (Hachimori et al. 1971; Nair et al. 1979; Wells and Hanahan 1969). Two NH-amino acid terminals are present, indicating nonidentical dissociable monomers (Nair et al. 1979).

There are also differences in pH. It is composed of two subunits, an acidic PhA_2 subunit and a basic PhA_2 subunit. Both subunits must be present to produce the toxin (Powell et al. 2008). The amount of these subunits varies geographically in the venoms of several species of *Crotalus* (Powell et al. 2008; also see individual species accounts). Basic PhA_2 enzymes are more lethal than acidic ones (Russell 1983). So it is difficult to generally describe the enzyme's biological properties (Elliott 1978; Nair et al. 1979; Rosenberg 1979, 1987, 1988, 1990; Russell 1983; Tu 1977).

Lynch (2007) has studied Darwinian adaptive evolution and neofunctionalization of the PhA_2 genes in snake venoms, and found common occurrences in the genes. He proposed that genomic complexity through gene duplication can lead to the adaptation of a phospholipase arsenal in response to novel prey species after niche shifts and to phenotypic complexity in venom composition. The regions on the surface of PhA_2 enzymes identified under such selection are potential sites for structural-based antivenom development.

Werman (2008) discussed the general evaluation of β-neurotoxic PhA_2 in the family Viperidae.

Phospholipase B (Lysolecithin acyl hydrolase) works in tandem with PhA_2 in the venoms of some Australian elapids (Doery and Pearson 1961; Iwanaga and Suzuki 1979). The higher the activity of this enzyme, the greater the ability of snake venoms to hydrolyze lysophosphatides.

Phospholipase C (Phosphatidylcholine cholinephosphohydrolase, α-Phospholipase C, Lecithinase) catalyzes the hydrolysis of the linkage between glycerol and phosphate in lecithin and other phosphates, causing the disruption of cell membranes, particularly those of blood cells. It degrades phosphatidyl ethanolamine and phosphatidyl serine.

Phosphomonoesterase (Phosphatase) is present in venoms in all snake families except the Colubridae. Some venoms contain nonspecific as well as specific types. Nonspecifically, it exists in two forms, separated by their optimal pHs of 5.0 and 8.5, respectively, which exhibit the characteristics of orthophosphoric monoester phosphohydrolase. Many venoms contain both of these, but others have only one. Their pharmacologic characteristics are essentially unknown.

5′-nucleotidase (5′-ribonucleotide phosphohydrolase) is a small (around 10,000 D), very specific, very active phosphomonesterase, found in all snake venoms, but in greater amounts in crotalid and viperid venoms than in those of elapids. It attacks DNA and RNA, specifically hydrolyzing the phosphate monoesterase link at the 5′ position. It is active over a broad pH range of 5.8–9.0 in various venoms, and is of low lethality (Gulland and Jackson 1938b; Mebs 1970b; Tu 1977). Its activity is suppressed by Zn^{2+}, glycerolphosphatase, and nonspecific monophosphatase, but activated by nucleotides (mono-, di-, or tri-) bearing a 2′ or 3′ phosphate group.

Proteolytic enzymes (proteases, proteinases) or metalloproteinases are toxic enzymes found in the venoms of pitvipers and vipers, but only at very low concentrations at best in elapids, and not at all in the seasnake *Pelamis platura*. Snake venom proteinases that break the bonds in fibrinogen can be divided into thrombin-like enzymes (thrombin proteases), fibrinogenolytic enzymes, and plasminogen-activating enzymes (Siigur and Siigur 1992; Swenson and Markland 2005).

Proteinase, N-benzoyl-L-arginine ethyl esterase (BAEEase), and p-tosyl-L-argenine methyl esterase (TAMEase) hemorrhagic activities are all present in the venoms of *Crotalus basiliscus, C. molossus*, and *C. scutulatus*. The two serine protease arginine esterases are not affected by ethylenediaminetetraacetic acid (EDTA), while most proteinases are inhibited by it (Stegall et al. 1994).

Meier and Stocker (1995) have suggested the following classifications for this group. All enzymes that participate in protein degradation are termed *proteases*. Endopeptidases acting on the interior of peptide chains are termed *proteinases*; the activity of endopeptidase is considerably higher in pitviper venoms than in those of elapids (Oshima et al. 1969). Examples are aspartic proteinases, cysteine proteinases, and metalloproteinases (see Mashiko and Takahashi 1998; Muniz et al. 2008; Ramos and Selistre-de-Araujo 2006; Rodríguez-Acosta et al. 2007; and Takahashi and Mashiko 1998 for reviews of their structure, function, and classification), and serine proteinases.

Serine proteinases are hydrolytic, and use the hydroxyl group of serine 195 to attack and cleave peptide bonds (Castro and Rodrigues 2006; Krem and Di Cera 2001; Yousef et al. 1995, 2004). Exopeptidases (dipetide and tripeptide hydrolases), in contrast, attack peptide chains at their terminal ends. All of the preceding are heat-labile proteins that are responsible for anticoagulant, fibrinolytic, and antithromboplastic activities, and enhancement of plasminogen activation of snake venoms. However, they do not release bradykinin (Van Mierop 1976a).

Included are hemorrhagins and serine kallikreins (kinin-releasing enzymes), which are very active in pitviper venoms, less active in viper venoms, but of very low or no proteolytic activity in elapid venoms. These are very proteolytic active, and produce much tissue destruction. The kallikrein-like enzymes act on plasma kinogen to release the potent vasodepressors bradykinin and/or kallidin. They also hydrolyze the vasodepressor angiotensin II and its precursor angiotensin I. Although with no ability to stimulate fibrinogen clotting, these enzymes do degrade the peptide chains of fibrinogen. The sequence of their actions is very similar to those of thrombin-like enzymes, but the two enzyme groups differ in substrate specificities (Komori and Nikai 1998; Mebs 1998; Minton 1956). Arginine esterase, kininase (which degrades kinin), kininogenase (which releases kinin), and tyrosine esterase are present in hemotoxic rattlesnake venoms (Al-Joufi and Bailey 1994).

The roles of proteases and hemorrhagins are separable (Toom et al. 1969); these enzymes do not induce clotting and have no hemorrhagic properties. Generally, venoms with high proteolytic activity (like that of *Agkistrodon piscivorus*) are poor coagulants, while those with low proteolytic activity (like that of *Crotalus adamanteus*) are very coagulant. Venom from *Agkistrodon contortrix* is highly proteolytic but nonhemorrhagic. Proteases generally do not contribute much to the toxicity of even highly proteolytic venoms (Van Mierop 1976a).

Some crotalid and viperid proteolytic enzymes inactivate the complement activity of serum and degrade complement factors. These degradations seem nonspecific (Meier and Stocker 1995; Vogel 1991; Vogt 1990).

The venoms of 10 species of rattlesnakes, including species in both *Crotalus* and *Sistrurus*, contain the low-affinity peptide inhibitors pEQW and pENW that affect endogenous metalloproteases and prevent autolysis of the venom. The former inhibits the major venom metalloproteases Cvo, PrV, and cromipyrrhin, and also stabilizes cromipyrrhin against autoproteolysis under extreme heat conditions. Such peptides may prove very useful in designing similar low-affinity peptide inhibitors in protein drugs that will increase their stability and/or allow their storage under less stringently controlled conditions (Munekiyo and Mackessy 2005).

In addition, the tissues of some venom-resistant animals contain natural inhibitors (TIMP) of metalloproteinases in snake venoms (Pérez and Sánchez 1999). They constitute another possible field of research in medicine.

Proteolytic enzymes have molecular weights of 20,000–95,000 D, are composed of as many as 18 amino acids, and contain trypsin-like enzymes that digest peptides and proteins (Tu 1977). Unfortunately, many are yet to be biochemically or pharmacologically characterized. At least five such enzymes may be present in a species' venom. Metals are intrinsically involved in the activity of some venom proteases, which may be inactivated with EDTA and other reducing agents.

Zinc ions are usually present, and appear to be the necessary catalytic metal. Removal of zinc ions inhibits the peptidase A activity of *Agkistrodon piscivorus* venom

(Wagner and Prescott 1966; Wagner et al. 1968). Removal of calcium ions also causes structural changes in the enzyme (Russell 1983). Addition of magnesium and zinc ions restores proteolytic activity in certain snake venoms after treatment with EDTA (Friederich and Tu 1971).

Collagenase is a specific proteinase found in many pitviper (Hadidian 1956; Simpson and Rider 1971; Simpson et al. 1973) and viper venoms, and is more potent in the Crotalinae than in the Viperinae. It digests mesenteric collagen fibers but not protein. Its activity appears identical to that of Zn-metalloproteinases (hemorrhagins and nonspecific proteinases) and may not represent an entirely distinct enzyme activity (Takeya and Iwanaga 1998; Tu 1977).

Elastase, often classified as proteolytic, may also not represent a distinct enzyme activity. Its action seems identical to that of venom Zn-metalloproteinase (hemorrhagins and nonspecific proteases). It is present in the venoms of both pitvipers and vipers, but not in elapid venoms (Mebs 1998; Tu 1977).

Thrombin-like enzymes. Venoms of crotalids (Bonilla 1975; Castro and Rodrigues 2006; Herzig et al. 1970; Markland and Damus 1971; Sant' Ana et al. 2008) and viperids contain significant amounts of these serine protease enzymes, which paradoxically inhibit blood-clotting in vivo, but promote it in vitro. The venoms of elapids contain little or no thrombin-like enzymes. Almost all that have been identified are glycoproteins, which predominantly release fibrinopeptide A (Markland 1998a; Markland and Pirkle 1977), and work differently than thrombin, which releases both fibrinopeptides A (FPA) and B (FPB). It activates the blood-clotting factors V, VIII, and XIII; cleaves the prothrombin fragment; and aggregates and releases platelets (Brinkhous and Smith 1988; Meier and Stocker 1995; Smith and Brinkhous 1991).

That the venoms of certain snakes coagulate blood has been known for more than two centuries (Fontana 1795; Mitchell 1860; Mitchell and Reichert 1886). Since the late 1960s, several venom procoagulants have been isolated and identified (Esnouf and Tunnah 1967; Holleman and Weiss 1976; Markland and Damus 1971; Meaume et al. 1966; Pirkle and Stocker 1991; Schieck et al. 1972a, 1972b; Siigur and Siigur 2006), and the mechanisms of their actions on blood-clotting have been outlined (Kornalik and Hladovec 1975; Markland and Pirkle 1977).

A venom may act as a coagulant if it contains factor X activator, can activate factor V or prothrombin, and either has a thrombin-like activity or causes thromboplastin-like activity (Tu 1977).

The normal mechanisms associated with blood-clotting, and the roles of venom procoagulants, as outlined by Elliott (1978), Russell (1983), Siigur and Siigur (2006), Tu (1977), and White (2005), are as follows. Clotting proteins occur in the blood as unactivated (zymogen) precursors that are activated either by surface contact, as is factor XII, or by proteolysis. The proteolyzed, activated clotting proteins then appear to activate other zymogen precursors in a domino effect (Joseph et al. 2002). Eventually this process leads to the generation of large amounts of thrombin, either by an intrinsic process beginning with the activation of small quantities of factor XII or by the more rapid extrinsic pathway, which is initiated by the liberation of tissue factor. The final clotting enzyme, thrombin, releases FPB from the clotting protein fibrinogen, converting it to a fibrin clot (see also Ahmed et al. 1990a, 1990b; Dyr et al. 1989a, 1989b, 1990; Guan et al. 1991; Mackessy 1993; Manning 1995; Markland 1991; and Swenson and Markland 2005), and activates factor XIII, which converts the noncovalently, cross-linked fibrin clot into a stabilized, covalently cross-linked clot. Thrombin is also responsible for the activation and inactivation of factors V and VIII, and probably plays a controlling role for factor VII, thus affecting the extrinsic pathway.

According to Russell (1983; see also Tu 1977 and White 2005), six sites occur in the blood coagulation process where venom procoagulants can interact; included are activation of factors V, IX, and X, inactivation of factor IX, direct activation of prothrombin, and prothrombin activation requiring factor V. These actions are found throughout the major families of venomous snakes, and some snake venoms contain more than one broad spectrum procoagulant or anticoagulant protease to account for coagulant, anticoagulant, and fibrolytic activities.

A single venom may activate both blood-clotting factors IX and X, and possibly a different protease could activate factor V, convert plasminogen to plasmin, activate factor VIII, and then degrade it during a long-lasting proteolysis. Russell (1983) thought that this could help explain why some snake venoms are coagulant at high concentrations, presumably by activation of specific blood-clotting proteins by components present in low concentrations at which there may be extensive proteolyses and degrading of blood-clotting factors by nonspecific proteolytic enzymes. The same protein may have both coagulant and anticoagulant properties, as with venom thrombin-like, procoagulant enzymes that prevent blood-clotting by removing fibrinogen (thrombin releases four fibrinopeptides from fibrinogen; Elliott 1978).

A venom should not be classified as a factor X-activating toxin or as a thrombin-like one because its component that activates factor X, as well as that which converts fibrinogen to fibrin, may have other procoagulant characteristics.

It is wrong to classify a snake's venom as either a coagulant or an anticoagulant, because it possibly produces both activities (Rosenfeld et al. 1968; Russell 1983). Some coagulate blood at low concentrations, but are anticoagulant at high concentrations, particularly those with thrombin-like enzymes (procoagulant in vitro, anticoagulant in vivo).

A snake's venom may be anticoagulant if it either produces fibrogenolytic or fibrinolytic activities, inhibits or destroys any of the blood coagulation factors (II, III, V, or thrombin), or activates plasminogen to release plasmin (Tu 1977).

Venom anticoagulant activities include those that inhibit one or more of the blood-clotting proteins or prevent activation of one of the clotting proteins (all families of venomous snakes); those with fibrinolytic or fibrinogenolytic activity that works by direct action on fibrin or fibrinogen (pitvipers, vipers); those that cause activation of fibrinolytic mechanism by direct action on plasminogen or by activation of a plasma proactivator of plasminogen (some elapids, vipers); or those that inhibit clot formation by direct action of the venom anticoagulant with phospholipids (which can be catalyzed by phospholipase or some other venom component).

Certain venom proteins, with or without enzymatic activity, bind blood coagulation factors and produce anticoagulation: anticoagulant phospholipase A_2s, which bind factor X and impair the formation of the prothrombinase independently from their enzymatic activity; factor IX and X binding proteins, which belong to the C-type lectin-like proteins and interact with the gamma-carboxyglutamic acid (Gla) rich domain of factor X/Xα and/or IX/IXα (note, Gla is also found in factor VII); prothrombin-

and thrombin-binding proteins, which are C-type lectin-like molecules that bind either to prothrombin (impairing thrombin formation) or to thrombin (inhibiting its activities), like blood platelet activation, clotting of fibrinogen, and the like; and L-amino oxidases with anticoagulant activity (Mashiko and Takahashi 1998; Willis et al. 1989; Zingali 2007). See also the later discussion of venom cations.

Nerve growth factors (NGFs): NGFs are found in the venoms of some snakes in the Elapidae, Crotalinae, and Viperinae. They induce and enhance the growth of nerve tissues and support their well-being through trophic effects. These multifunctional proteins function as strong mast cell degraders that also affect the immune system, and probably contribute to the overall toxicity of snake venoms (Hogue-Angeletti and Bradshaw 1979; Meier and Stocker 1995).

Enzyme inhibitors: Inhibitory compounds that either stop or limit the actions of specific venom enzymes are also present in snake venoms (Meier and Stocker 1995).

Acetylcholinesterase inhibitor activity has been found in some elapids' venoms.

Angiotensinase inhibitor prevents the conversion of angiotensin II from angiotensin I. Apparently, it is identical to the bradykinin-potentiating factor, which increases the toxicity of bradykinin (Tu 1977).

Bradykinin inhibitor is a novel basic peptide that has been isolated from venoms of *Agkistrodon bilineatus, Crotalus viridis*, and *Lachesis muta* (Graham et al. 2005). A chain of 11 amino acids attached to OH^- compose its primary structure, and its nonprotonated molecular mass is 1,063.18 D.

Citrate is another compound that inhibits several enzymes and stabilizes venom (Francis et al. 1992; Freitas et al. 1992).

Phospholipase A_2 inhibitors (PhA_2 inhibitors) are polypeptides with low molecular weights (1,500–5,000 D) found in the venoms of a number of elapids and tropical pitvipers (Meier and Stocker 1995). Also, the presynaptic viper neurotoxin vipoxin contains a toxic basic protein with weak PhA_2 inhibitory activity and another nontoxic proteinaceous acidic inhibitor (Bragança et al. 1970a; Vidal and Stoppani 1971).

Platelet aggregation inhibitors, present in the venoms of some elapids, pitvipers, and vipers, are serine proteinases that prohibit the binding of adhesive proteins to the glycoprotein complex on the surface of blood platelets, preventing fibrinogen binding and the formation of blood clots (Oyama and Takahashi 2007). A potent platelet inhibitor, crotovirin, occurs in the venom of *Crotalus viridis* (Liu et al. 1995).

Proteinase inhibitors are polypeptides composed of 52 to 62 amino acids cross-linked by 2–3 disulphide bridges found in some snakes in all 3 major venomous families (Meier and Stocker 1995; Tu 1977).

Tripeptide inhibitors decrease the potency of metalloproteases and some other enzymes to stabilize venom components (Mackessy 2008; Munekiyo and Mackessy 2005).

Other venom components: *Amino acids*—Free amino acids, in the form of either purines or pyrimidines (alanine, aspartic acid, cysteine, glutamic acid, glycine, histidine, hypoxanthine, inosine, isoleucine, leucine, lysine, phenylalanine, proline, serine, spermine, threonine, tyrosine, and valine), have been found in pitviper and viper venoms in various, usually small, amounts. By themselves, they apparently do not contribute to the venom's toxicity (Bieber 1979; Tu 1977).

Aird (2002, 2005) suggested that the purines adenosine, quanisine, and isosine play a central part in the envenomation strategies of most venomous snakes by apparently binding to other toxins, which then serve as molecular chaperones to deposit the bound purines at specific subsets of purine receptors. These amino acids constitute the perfect multifunctional toxins, participating simultaneously in all envenomation strategies. Because they are endogenous regulatory products in all vertebrates, it is impossible for any prey species to develop resistance to them. Purine amino acid generation from endogenous precursors in prey explains the presence of many formerly unexplained enzyme activities in snake venoms, like ATPase, endonucleases (including ribonuclease), 5′-nucleotidase, NADase, phophodiesterase, and phosphomonoesterase. Cytotoxins, hesperinase, myotoxins, and PhA_2 also help liberate purines.

Adenosine contributes to prey immobilization by activation of neuronal adenosin A_1 receptors, suppressing acetylcholine release from motor neurons and excitatory neurotransmitters from central sites. It also enhances venom-induced hypotension by activating A_2 receptors in the circulatory system. Adenosine and inosine activate mast cell A_3 receptors, liberating vasoactive substances and increasing vascular permeability. Guanosine most likely also contributes to hypotension by adding to vascular endothelial cGMP levels by an unknown mechanism.

The assignment of pharmacological activities (like transient neurotransmitter suppression, histamine release, and actinociception) to many proteinaceous toxins is most likely erroneous. Instead, these effects are probably due to purines bound to the toxins and/or free purines in the venom. Some purines, functioning as *nucleosides* or *nucleotides*, are normally absent from most pitviper venoms, but present in those of elapids and vipers. Free purines, principally adenosine, guanosine, and inosine, may make up as much as 8.7% of their solid components and are more abundant than many proteinaceous toxins in the venoms of several elapids and vipers, but, of the pitvipers, only in the venom of the rattlesnake, *Crotalus adamanteus*. Hypoxanthine is a minor constituent (<60 αg/g) in about half of elapid, pitviper, and viper venoms, and adenosine monophosphate has been tentatively identified in only three elapid and two viperid venoms. Other possible actions of purines are discussed by Aird (2002).

The pyrimidines cystidine and the always more abundant uridine are found in most elapid and viper venoms, but not in pitvipers. Thymidine is also absent from pitviper venoms (Aird 2005; Bieber 1979; Elliott 1978).

Biogenic amines—Several nonpeptide amines have been isolated from snake venoms. Acetylcholine is found mostly in elapid venoms, but also in trace amounts in those of *Agkistrodon contortrix, A. piscivorus, Crotalus horridus*, and *Sistrurus miliarius* (Jaques 1955; Schwick and Dickgiesser 1963). Bradykinin probably contributes to pain developing at the bite site during pitviper and viper envenomations. Venoms of *Crotalus adamanteus* and *C. atrox* may contain small amounts of serotonin (5-hydroxytryptamine), although this is questionable; positive indication of serotonin activity has, however, been recorded from the venoms of *Agkistrodon contortrix, A. piscivorus*, and *Sistrurus miliarius* (Elliott 1978; Tu 1977). Zarafonetis and Kalas (1960b) isolated catecholamines from the venoms of *Agkistrodon piscivorus, Crotalus adamanteus*, and *C. atrox*. These probably have adrenaline and noradrenaline as major components (Anton and Gennaro 1965; Elliott 1978).

Carbohydrates in snake venoms occur as glycoproteins. Most sugars are either in the form of amino sugars (glucosamine), neutral sugars (fructose, galactose, glucose, mannose), or sialic acid. L-fructose, D-galactose, and D-mannose are found in the venoms of North American *Agkistrodon piscivorus* and *Crotalus adamanteus* (Bieber 1979; Ogilvie and Gartner 1984; Oshima and Iwanaga 1969; Tu 1977). Mucopolysaccharides have also been found in the venom of *Agkistrodon piscivorus*. Protein lectins in the venom serve to attach sugars as a covering that protects the underlying tissue in the lumen of the venom gland of pitvipers and vipers.

Cysteine-rich secretory proteins (CRISPs) are widely distributed in the venoms of snakes in the Colubridae, Crotalinae (*Agkistrodon piscivorus, Crotalus atrox*), Elapidae, and Viperinae. CRISPs are single-chain polypeptides with molecular weights of 20,000–30,000 D. Their sequences are present throughout the protein, and all 16 cysteines are strictly conserved, 10 of which are clustered in the C-terminal third of the molecules. Functionally, CRISPs inhibit smooth muscle contraction and cyclic nucleotide-gated ion channels (Yamazaki and Morita 2004).

Disintegrins (Integrin antagonists) are statin protein compounds that act as antagonists of cell adhesion and migration by binding integrins (substances that promote the binding of molecules), and thus blocking their binding to receptors. Chiefly, they prevent the binding of molecules to cell membrane receptors by attacking the integrin proteins on the surface of blood platelets (Calvete et al. 2005).

Several have been identified in snake venoms, like lebestatin, a short disintegrin in the venom of *Macrovipera lebetina*, and echistatin in the venom of *Echis* vipers. They are composed of a loop of amino acids that adhere to the cell membrane of platelets and inhibit fibrinogen from binding with them. This action prevents the blood from clotting, and causes rapid bleeding. After the antagonist is removed, some receptors remain activated and the clotting process can resume. Full disintegrin activity requires an axillary site that includes carboxyl-terminal nine amino acid residues (Hantgan et al. 2004; Kallech-Ziri et al. 2007).

In the future, disintegrins may play an important role in the development of new drugs for treating blood-clotting disorders and strokes, preventing heart attacks, and possibly also treating some malignancies.

Ions are inorganic in nature, are charged electrically, and function as necessary cofactors enabling enzyme reactions to occur. Particularly chloride (Cl^-), phosphate $(PO_4)^{3-}$, and sulfate $(SO_4)^{2-}$ ions are present in many venoms.

Negative ions, or anions, include phosphate as a significant component; total phosphorus values of 9–14% and values of 5–6% of inorganic phosphorus may be present. The pyrophosphate found in the ash of snake venom results from inorganic and organic, hydrolyzable phosphates; most of the sulfate in venom ash originates from cysteine and cystine sulfur. The chloride content is $<0.3\%$ (Elliott 1978; Russell 1983).

Ions with a positive charge are termed *cations,* and occur in snake venoms in the form of a wide range of metallic ions, depending on the method of analysis and venom condition. They serve as cofactors to certain proteases to form metalloproteinases. The action of most of these is directed toward extracellular matrix components, and this effect is thought to result in bleeding because of disruption of the basement cell membrane in capillary walls (García et al. 2004; Mashiko and Takahashi 1998).

Zinc (Zn^{2+}) is the most important cation involved in snake venoms. Analysis of venoms by several methods indicates that zinc in its various forms makes up 0.31–0.56% of venom dry weight in venomous colubrids, 0.001–0.57% in elapids (0.01% by atomic absorption in *Micrurus fulvius*; Kumar et al. 1973), and 0.04–0.19% in vipers

(Elliott 1978). Pitviper venoms probably contain percentages comparable to those found in vipers. Zinc may switch on or off the activity of some toxic enzymes, and is probably present in high amounts in all snake venoms. It may also serve as a reversible enzyme inhibitor, protecting the venom gland from being attacked by its own product. Reactivation and dilution of zinc to an inactive level probably occur when the venom is injected into prey (Delezenne 1919; Fleckenstein and Gerkhardt 1952; Fleckenstein and Jaeger 1952).

Calcium (Ca^{2+}) occurs at concentrations of <0.02–0.6% of venom dry weight; those of most pitvipers and vipers examined have contained from 0.2 to 0.4% calcium (Elliott 1978).

Cobalt (Co^{2+}) catalyzes acetylcholine activity in elapid venoms (Kumar et al. 1973).

Copper (Cu^{2+}) has been found in very low amounts, 0.001–0.02% of venom dry weight (*Sistrurus miliarius*), in only a few snake venoms; most species contain none (Elliott 1978).

Iron (Fe^{3+}?) is found in the venoms of only a few elapids and vipers (Friederich and Tu 1971; Vidal Breard 1950).

Magnesium (Mg^{2+}) is present in amounts ranging from 0.01% in *Crotalus adamanteus* to 0.14% in *C. durissus* (Elliott 1978).

Manganese (Mn^{2+}) is only known from the venom of the puffadder, *Bitis arietans*, and the cobra, *Naja naja* (Elliott 1978).

Nickel (Ni^{2+}) is present in trace amounts in some Asian pitviper (*Trimeresurus*) venoms (Tu 1977).

Potassium (K^{+}) is found at dry weight concentrations of <0.01–2.9%, and is normally higher in pitviper venoms than those of either elapids or vipers, but much overlap occurs (Elliott 1978).

Sodium (Na^{+}) has a dry weight range of 0.64–6.0% in snake venoms (Russell 1983). Regional differences in the content percentage occur in *Crotalus horridus* (Friederich and Tu 1971).

Cations of bismuth (Bi), gold (Au), molybdenum (Mo), platinum (Pt), selenium (Se), and silver (Ag) do not occur in amounts detectable by atomic absorption (Friederich and Tu 1971).

Lipids are composed of fatty acids and glycerol, and most in snake venoms probably are present as organic phospholipids or triglycerides. Citrate is the major organic acid in the venoms of *Agkistrodon contortrix, A. piscivorus, Crotalus adamanteus, C. atrox* (>5% of dry residue), *C. horridus, C. viridis*, and *Sistrurus miliarius* (Freitas et al. 1992). Analysis of venoms from the cobra *Naja naja* and those of the North American pitvipers *Agkistrodon contortrix, A. piscivorus, Crotalus adamanteus, C. atrox, C. horridus, C. viridis*, and *Sistrurus miliarius* has revealed small quantities of lipids containing the fatty acids arachidonic, capric, caprylic, lauric, linoleic, myristic, oleic, palmitic, palmitoleic, and stearic. These occur as cholesterol, cholesterol esters, diglycerides, free fatty acids, hydrocarbons, monoglycerides, phospholipids, or triglycerides. Venom lipids may play a role in enhancing neurotoxicity in some viper venoms (Cooke et al. 1986; Meier and Stocker 1995; Tu 1977).

Peptides—Some neotropical pitviper venoms contain several proline-rich peptides that potentiate bradykinin, and pyroglutamylpeptides occur in the venoms of some Asian pitvipers (Tu 1977).

Nonenzymatic polypeptides in some venoms may have a lethal index from 2 to 10 times greater than the raw venom, while in other venoms their index is less or unknown (Russell 1984).

Slotta and Fraenkel-Conrat (1938a, 1938b) discovered the protein crotoxin, a highly virulent, 18 amino acid polypeptide with a molecular weight of 30,000 D, in the venom of the neotropical rattlesnake *Crotalus durissus terrificus*. In addition to crotoxin, the polypeptide also contained hyaluronidase, phospholipase, and possibly several other toxic enzymes. Crotoxin is about 15 times as toxic as the snake's crude venom, and has neurotoxic and indirect hemolytic activities, but also stimulates smooth muscle. It lacks 5′-nucleotidase and proteolytic activities, and does not coagulate blood.

The molecule was separated into two fractions after PhA_2 was removed. One, a general toxin known as crotactin, was found to be more lethal than crotoxin. The other fraction was possibly crotamine, which was further separated by Slotta and Fraenkel-Conrat into several fractions.

Further study by Gonçalves (1956) on the venom of *Crotalus d. terrificus* produced three biologically active fractions. One of these was crotamine (10,000–15,000 D); the other two were a proteolytic enzyme and a neurotoxin that corresponded to crotoxin. Crotamine caused respiratory difficulties, paralysis of the hindquarters, and death when injected into mice. The proteolytic enzyme hydrolyzed denatured hemoglobulin, clotted fibrinogen, and freed bradykinin. Crotoxin lysed blood cells, and caused more rapid breathing, paralysis, and death in mice.

According to Russell (1983), the term *crotoxin* has been retained, in one form or another, in the literature for [at least] 17 different components of the venom of *C. d. terrificus* (obviously, this has caused much unnecessary confusion). He presented a table of the various derivatives of the name used since 1974, and suggested that the term *crotoxin* be limited to the historically significant polypeptide first isolated by Slotta and Fraenkel-Conrat in 1938.

Subsequent studies of venoms of rattlesnakes by Bonilla and Fiero (*Crotalus horridus* and *C. viridis*, 1971), Dubnoff and Russell (*C. helleri*, 1970), Glenn and Straight (1982) and Glenn et al. (1972) (*C. atrox*), Jiménez-Porras (*C. atrox*, 1961), Pattabhiraman and Russell (*C. helleri*, 1973), Pattabhiraman et al. (*C. helleri*, 1974), and Russell et al. (*C. helleri* and *C. scutulatus*, 1976) identified other highly toxic peptides. The properties and activities of these are discussed in detail by Russell (1983) and Schaeffer et al. (1979).

Medical uses of venom components: Snake venoms are potentially very useful for therapeutic medical use because of their great variety of compounds and the biological activities of each compound, as first proposed by Calmette et al. (1933) and Macht (1940). The specific antiarthritic, anticoagulant, antineoplastic, antiparalytic, cardiotonic and antiarrythmic, and fibrinogenolytic and fibrinolytic actions of these components have led to research into their possible use in treating various medical conditions.

Nowhere is the whole venom used as a medicine; instead, specific compounds are extracted and used. Although, to date, relatively few drugs have been manufactured from snake venoms, the potential is great.

Stocker (1990a) edited an excellent review of the medical aspects of snake venoms known to that time. It contained detailed information concerning the physicochemical, biochemical, and pharmacological characteristics of biomedically relevant snake venom proteins, and their practical use and potential in medical treatments. Its chapters included sections concerning neurotoxins and their application in neurophysiological and muscular research (Mebs 1990); venom proteins affecting the complement system (a system of lytic substances in normal blood serum that combines with the antigen-antibody complex to cause lysis when the antigen is in an acellular form; Vogt 1990), in addition to the large field of venom components affecting the hemostatic and fibrinolytic system (Furukawa and Ishimaru 1990; Markland 1998a, 1998b; Stocker 1990b); and the effect of venom components on tumor growth (Markland 1990). For more updated information on the uses of snake venom components in medicine, see Koh et al. (2006), Ohta et al. (1991), and Pal et al. (2002), and the Public Broadcasting Service (PBS) film *The Venom Cure* in the *Nature* television series.

Space is too limited in this book to duplicate in detail the data presented in the above publications, but we have tried to provide an overview of some of the current and potential uses of various snake venom components in medicine. The role of snake venom in the production of antivenom is described in the chapter "Treatment of Envenomation by Reptiles," and is not included here.

Following is a list of medical disorders for which venom compounds are being used as treatments, or for which research is currently being conducted on possible treatment by venom components. Most of these components have already been discussed in this chapter.

Arthritis: Proteinase inhibitors from snake venoms bind to protease enzymes to prevent them from reacting with and cleaving bonds of hormones and other proteins when they are no longer needed, preventing inflammation, and have been used to prevent immune complex arthritis in experimental animals (Henson and Cochrane 1975).

Fibrin is usually found in large amounts in arthritic joints, and sometimes contributes to acute and chronic arthritis (Pal et al. 2002). So those previously discussed enzymes that reduce the amount of fibrin are potential treatments for rheumatoid arthritis.

Cancer: The use of venom enzymes in the treatment of malignancies is a prospectively new and positive approach. Their possible such use has been known since the 1930s (Calmette et al. 1933). Most cancers do not become mortal until they spread (metastasize) via the circulatory system to new parts of the body. This distribution depends on the malignant cell's ability to bind onto neighboring normal cells and promote the formation of new blood or lymph vessels that supply them with the nutrients and oxygen needed for growth and also provide a highway for the cancer cells to spread throughout the body.

The metalloproteinase disintegrin, contortrostatin (CN), isolated from the venom of *Agkistrodon contortrix,* stops tumor cells from sticking to other cells, thus preventing metastasis (Markland et al. 2005). It also has antiangiogenic properties that starve the tumor by preventing new feeding blood vessels from forming. It has been successful in treating breast cancer in laboratory mice (*Mus musculus*, 60–70% reduction rate and a 90% reduction in tumor metastases to the lung; Finn 2001; Minea et al. 2005; Swenson et al. 2005). Although more research and clinical tests must be performed, and a synthetic substitute produced, it could possibly be used to treat these devastating cancers in humans in the future. More recently, venom components from the African desert viper, *Cerastes cerastes*, killed approximately 55% of mouse mammary tumor cells within 48 hours (El-Refael and Sarkar 2009).

In addition, VRCTC-310, a combined mixture of cardiotoxin and crotoxin, is currently being used in early clinical trials for the treatment of some malignant tumors (Markland 1986). Also, a venom fraction from *Crotalus durissus* containing crotamine and crotoxin has caused tumor regression in laboratory animals (Hernandez-Plata et al. 1993), as has also the disintegrins crotatroxin 1 and 2 from *C. atrox* venom (Sánchez et al. 2006).

Experiments using the disintegrin, eristostatin, have shown that it prevented metastasis of melanoma cells injected into mice, and experiments on the use of fibrolase to treat glioma (a tumor composed of neuroglia cells) and prostate cancer are currently underway.

Research in China has revealed that other enzymes may also prove useful in treating cancer. Cobra venom factor (CVF), a cytotoxin with no phospholipase activity from the Chinese cobra, *Naja naja atra*, is toxic to human cancer cell lines; the peptide rhodopsin (rhodostomin) from the venom of *Calloselasma rhodostoma* inhibits thrombin from enhancing metastasis; and saxatilin from *Gloydius saxatilis* helps control lung cancer cells (Chiang et al. 1996; Jang et al. 2007; Yeh et al. 2001; Zhang and Wu 2003; Zhong et al. 1993). Other studies on the anticancer components of snake venoms are discussed in Pal et al. (2002).

Cardiovascular disorders: The major cardiovascular symptoms of snake envenomation are reduced coagulability of the blood, leading to bleeding or even severe hemorrhage; bleeding due to necrosis of blood vessel walls; secondary effects of blood loss, like shock, or secondary organ damage (hemorrhage of the anterior pituitary or intracerebrum, or renal damage); and pathological clotting (thrombosis), particularly pulmonary embolism (Koh et al. 2006). So far, venom components have mostly been used in the treatment of blood malconditions (hemorrhage, clotting), drastic changes in blood pressure, or heart failure.

One of the first such uses was to treat a case of malarial [?] hemorrhaging in tropical America with *Bothrops* antivenom, which stopped the bleeding from the gums and nose in two hours. Although the patient passed blood in his urine for 2 more days, he was released bleeding-free after 13 days (Taylor 1929).

Coagulant components in snake venoms may prove useful in treating excessive bleeding. The drug hemocoagulase produced from the venom of *Bothrops atrox* has a thrombin-like effect and is used to treat hemorrhaging. Coagulant enzymes from the Asian viper, *Daboia russelii*, have been used to control some forms of hemophilia, and are currently used as a diagnostic tool to determine deficiencies in clotting factors in hemophiliacs.

More research has been conducted on the possible uses of snake venom anticoagulants to control plaque buildup and blood-clotting than on their ability to control bleeding. Snake venoms are rich in prothrombin activators. These proteins are very useful in determining blood-clotting mechanisms (previously discussed), those responsible for platelet activation, and the structure-function relationships of blood-clotting factors and platelet glycoproteins. Some, like dysprothrombinaemias, are even used in prothrombin and lupus anticoagulant (LAs) assays, and also to assay protein C and activated protein C resistance. A C-type lectin-like protein isolated from the venom of

Bothrops jararaca has been used to study the von Willebrand congenital blood-clotting disease (Marsh and Williams 2005).

Some venom thrombin-like enzymes (SVTLEs) are intermediate between true coagulants and the anticoagulants that produce effects similar to those of thrombin. These usually cleave FPA alone, but only a few cleave FPB; so, without cleavage of both FPA and FPB, they do not activate clotting factor XIII and the clots easily disintegrate. The SVTLEs batroxoben and reptilase, from the venom of some South American *Bothrops*, and ancrod (Viprinex) from *Callosellasma rhodostoma* are powerful anticoagulants that have been used to dissolve vascular blood clots and to treat heart attacks, strokes, sickle cell anemia, and priapism (an abnormal, continuous, and painful erection of the penis due to disease). Ancrod is particularly effective if administered within three hours of a stroke, and has been successfully used to prevent clotting during cardiac bypass surgery and catherizations (Lathan and Staggers 1996). Aggrastat (tirofiban) is a C-type lecithin-like protein inhibitor made from hemorrhagins in the venom of *Echis carinatus* that has been used to treat myocardial infarction by dissolving the clot in the coronary artery and to relieve angina (Koh et al. 2006). Recently, a metalloproteinase, kistomin, has been found to retard platelet aggregation by inhibiting the interaction of collagen and platelet glycoprotein VI (GPVI) and glycoprotein Ib-von Willebrand factor interaction, providing an alternative strategy for developing new antithrombotic agents (Hsu et al. 2007, 2008).

The fibrino(ogen)olytic metalloproteinases and serine proteases of the fibrinogenase group of snake venom enzymes are capable of cleaving more than one specific fibrinogen chain and break down fibrin-rich blood clots. Venoms from *Agkistrodon contortrix* and *A. piscivorus* cleave the α- and ß-chains of fibrin (Egen et al. 1987). Much research regarding their mechanism and possible use in dissolving clots has been conducted on these anticoagulants in recent years. Other possible snake venom clot-dissolvers being researched include atroxase from *Crotalus atrox* (Willis and Tu 1988) and fibrinogenase (lebetase) from *Macrovipera lebetina* (Gasmi et al. 1991, 1997).

Another approach in treating blood clots is to prevent platelet adhesion and clumping. CRISPs, PhA_2, some disintegrins and serine proteinases, and zinc metalloproteinases inhibit platelet adhesion and aggregation, and have interesting potential for preventing and treating blood clots. Crotovirin and Integrilin (eptifibatide) from the venoms of *Crotalus viridis* and *Sistrurus miliarius*, respectively, are used to dissolve blood clots and treat acute coronary syndrome.

C-type lectin-like venom proteins bind to a wide range of anticoagulants and to platelet receptors and display both anticoagulant and platelet-modulating activities. They activate platelets by either binding von Willebrand factor or specific molecules that inhibit platelet functions (Koh et al. 2006). Included are aggretin from *Callosełasma rhodostoma*, convulxin from *Crotalus durissus*, echicetin from *Echis carinatus*, and EMS16 in the venom of *Echis multisquamatus*. Exanta (ximelagatran), from cobra venom, is a thrombin inhibitor anticoagulant that has been used as a blood thinner.

Envenomation by *Bothrops moojeni* (possibly also *B. jararaca* and *B. jararacussu*) causes a sudden, massive drop in blood pressure. Research on the venom revealed it contained a protein that blocks the action of angiotensin-converting enzyme (ACE). This led to the first venom-based drug, captopril (commercially sold as enalapril), to lower high blood pressure and later to the first oral angiotensin-converting enzyme inhibitor (ACE inhibitor), now widely used to treat high blood pressure and angina. Research also continues on the possible use of cardiotoxin from cobra venoms in the treatment of cardiovascular disorders (Chang et al. 2000).

Ecarin, a prothrombin activator in the venom of *Echis carinatus*, and protac, a protein C activator from *Agkistrodon contortrix*, have been used in the clinical diagnosis of blood disorders.

Microbial infections: Venoms are currently being investigated for their potential as antibacterial and antiviral agents. If the specific venom components with antibacterial actions can be identified, they may possibly lead to new drugs to treat such diseases as botulism, hepatitis, malaria, scarlet fever, tetanus, and trachoma. L-amino acid oxidase apparently has some antibacterial properties (Svoboda et al. 1992).

Despite having rich oral and fang microbial contamination, pitviper envenomations seldom are accompanied by bacterial infection. Venoms from the rattlesnakes *Crotalus atrox, C. helleri*, and *C. horridus* have antimicrobial properties against the aerobic bacteria *Citrobacter* sp., *Enterobacter* sp., *Morganella morganii*, *Proteus* sp., *Pseudomonas aeruginosa*, and *Staphylococcus* sp., although eight species of anaerobic bacteria were resistant. The venoms were particularly active against gram negative aerobic bacteria (Talan et al. 1991). It is already known that cobra venom factor (CVF) helps activate the attacks of other drugs on microbes.

Nervous disorders: Treatment of some brain tumors, Alzheimer disease, neural trauma, and Parkinson's disease has met with only limited success when nonspecific inhibitors have been used. PLA_2 enzymes are thought to be crucial in the inflammatory process, like that which occurs in the above chronic neurological disorders, so the use of specific inhibitors (PLI) may prove more useful in the future than current pharmacological drugs.

Components from the very potent neurotoxic venom of African mambas (*Dendroaspis*) are being researched to determine their specific actions on nerve receptors in hopes of finding cures for the above mentioned neurological conditions. Unfortunately, so far the efforts have been slowed by an inability to find the right compound(s) that act solely on only one type of neural receptor.

Muscular paralysis can be relieved by some snake venom neurotoxins specifically targeting nicotinic acetylcholine receptors (nAChRs) in neuron and neuromuscular junctions. The venom phospholipase, notexin, is a possible future treatment for muscle cell-caused ptosis (Senior 1999).

Other nervous conditions in which snake venoms have been used in research for treatments include epilepsy, multiple sclerosis, myasthenia gravis (Lou Gehrig's disease), poliomyelitis, and visual disorders like cataracts, conjunctivitis, and neuritis.

Pain: Pain-killing components from the venoms of several snakes, particularly those in cobra (cobratoxin, commercially sold as Cobroxin) and *Crotalus durissus* venoms, act as analgesics and may potentially be used to give relief to humans experiencing discomfort from a number of medical conditions (Chen 1996; Chen et al. 2006; Giorgi et al. 1993; Picolo et al. 1998; Pu et al. 1995; Xiong et al. 1992). Nyloxin, manufactured from cobra venom, is used to reduce severe arthritis pain.

Other uses: Some venom components have been found to be useful as a three-dimensional template in the construction of small synthetic molecules that mimic the actions of other toxic venom components (Harvey et al. 1998).

As can be seen, snake venoms have great potential for medical use because of their many varieties of structure and activity. Research will continue on the possible uses of venom components in developing therapeutic drugs to treat such conditions as AIDS, asthma, cancer, and other serious diseases. The papers mentioned in this brief discussion of medical uses of snake venoms are only a small fraction of the many concerned with their possible uses in the treatment of many types of human disorders.

Envenomation by North American Reptiles

Although as many as 45,000 human envenomations may occur in the United States each year (Russell et al. 1975), only about 7,000 to 8,000 of these are by what are normally considered venomous reptiles, with usually less than 10 of these fatal. Hutchison (1929) reported 30 deaths occurred in 607 bites in the United States in 1928, but Bellman et al. (2007) reported a reduction in snakebite deaths to only 142 during the period 1979–2005.

More than 27,000 such envenomations occur in Mexico annually, with 120 being fatal (see below). This figure includes envenomations by all species of venomous Mexican snakes, not only the several species of rattlesnakes that live there or only the venomous snake living in the geographic area covered in this book.

Canada has only three venomous snakes, the rattlesnakes *Crotalus oreganus, C. viridis,* and *Sistrurus catenatus* (the timber rattlesnake, *C. horridus*, has been extirpated there). Between 2,000 and 3,000 rattlesnake bites are reported annually in Canada, an incidence of approximately 0.79 bites per 100,000 population (Kasturiratne et al. 2008). Normally less than one of these envenomations is fatal.

More bites than these occur in the three countries; however, some go unreported or untreated, particularly in rural Mexico, and the incidence of bites and fatalities in the three countries varies from year to year. Unfortunately, most of these estimates are based on old data; but, fortunately, with new treatment techniques the mortality rate is much lower today than in the past (see "Treatment of Envenomation by Reptiles"). In Mexico and the United States, most human envenomations are by snakes, but some involve the helodermatid lizards *Heloderma horridum* and *H. suspectum*.

Envenomation Classification: Bites by venomous reptiles fall under two categories. *Legitimate* envenomations (termed *accidental* by Dart et al. 1992b) are those that occur when and where the victim was unaware of the presence of the snake. These total about 3,000 each year, and are those defensive bites made in the wild when the reptile first perceives it is in danger because of the close proximity of a human. *Illegitimate* bites (termed *nonaccidental* by Dart et al. 1992b), numbering about 4,000 annually, are inflicted by the reptile when it is purposely handled or disturbed by a human. These also include reflex envenomations by supposedly dead or decapitated snakes (Carroll et al. 1997; Ditmars 1936; Suchard and LoVecchio 1999).

Bites by copperheads (*Agkistrodon contortrix*) and rattlesnakes (usually *Crotalus horridus*) that occur during services involving snake-handling cults at some Appalachian churches and those elsewhere fall under the category of illegitimate bites. The handling of venomous snakes by members of these cults stems from the biblical statement in Mark 16:17–18 that true believers may pick up snakes and drink deadly things but not be hurt. This assumes that no one is ever bitten, or certainly dies from snake bite, who practices this behavior in church. Such bites are rarely treated. This is pure foolishness—one does not tempt fate!

According to Russell (1983), there were approximately 40 deaths and the bites numbered in the hundreds between 1910 and 1977. Klauber (1972) reported 100 persons were bitten, 7 fatally, between 1934 and 1948, and thought the number of bites to be a gross underestimation as some preachers claimed to have been bitten 250 to 400 times. He reported two envenomation deaths during church services occurred in the southeastern states in 1954, one in Georgia and another in Alabama; two deaths were recorded in 1955 (one each in Alabama and Florida); and a case of envenomation was reported in South Carolina in 1953. Two bites, one fatal, were recorded in California in 1954. A death occurred in Kentucky as late as 2006. If the reader is interested in more information on snake-handling cults in the United States, we recommend the discussions in Baird (1946), Callahan (1952), Kerman (1942), Klauber (1972), La Barre (1969), Pelton and Carden (1974), Russell (1983), Sargant (1949), and Whitehead (1940).

Today, illegitimate envenomations probably make up the majority of envenomations in the United States; however, some of these may go unreported. In contrast, between January 1986 and December 1988, 86 legitimate bites, 34 illegitimate, and 11 of unknown circumstances were recorded in the Western Envenomation Database (Dart et al. 1992b).

Distribution of Bites: Most reports of reptile envenomations in the United States are from the southern and southwestern states, and by pitvipers. Those by the three coralsnakes and the Gila monster occur less frequently; Seifert (2007a) reported 382 bites during 2001–2005 by all species of coralsnakes combined within their total range in the Americas. Twelve (2.4/year) were by *Micruroides euryxanthus*, 82 (16.4/year) each by *Micrurus fulvius* and *M. tener*, and only 1 (0.2/year) by *M. distans*; no fatalities were reported, compared to a 0.06% rate for pitvipers. Seifert's numbers may be low, however, as Morgan et al. (2007) reported that 100 coralsnake bites occurred in Florida and Texas in 2003–2005.

The estimated venomous snakebites per 100,000 population and annual total bites for the region in the United States based on data from Gomez and Dart (1995), Parrish (1966), and Wingert and Wainschel (1975) are New England Region—0.07 per 100,000 (Connecticut, 0.16; New Hampshire, 0.16), 7.6 total annual bites; Middle Atlantic Region—0.4 (Pennsylvania, 0.65; New Jersey, 0.41; New York, 0.22), 136 bites; East North Central Region—0.55 (Indiana, 0.97; Michigan, 0.74), 198 bites; West North Central Region—2.88 (Missouri, 5.42; Kansas, 5.28; South Dakota, 4.56; Nebraska, 3.26), 443 bites; South Atlantic Region—9.24 (North Carolina, 18.79; Georgia, 13.44; South Carolina, 7.72; Florida, 7.07), 2,402 bites; East South Central Region—5.53 (Mississippi, 10.83; Alabama, 6.37; Kentucky, 4.71), 666 bites; West South Central Region—13.3 (Arkansas, 17.19; Texas, 14.70; Louisiana, 10.25; Oklahoma, 8.85), 2,255 bites; Mountain Region—4.52 (Arizona, 7.83; New Mexico, 7.47; Wyoming, 6.36; Montana, 5.63), 310 bites; and Pacific Region—1.28 (California, 1.41; Washington, 0.88; Oregon, 0.79), 260 annual bites.

In what was one of the first comprehensive surveys of venomous snakebites in the continental United States, Hutchison (1929) examined 607 case reports from 1928, including 31 fatalities (Texas, 9; Florida, 3; New Mexico, 3; California, 2; Colorado, 2; Georgia, 2; South Carolina, 2; and Arizona, Iowa, Louisiana, Mississippi, Montana,

North Carolina, Pennsylvania, and Wyoming, 1 each). With improvements in treatment, only 97 persons died from venomous snakebites in the combined years 1979–1998: Texas, 17; Florida, 14; and Georgia, 12. No deaths were reported from 24 states and the District of Columbia.

In 1986–1988, 132 venomous snakebites were recorded in the Western Envenomation Database: Arizona, 67; Oklahoma, 15; California, 8; Tennessee and Texas, 4 each; 19 other states, 29; and unknown distribution, 4 (Dart et al. 1992b).

Christopher and Rodning (1986) treated 18 nonfatal pitviper bites in Alabama between January 1980 and September 1984. Between March 1979 and December 1986, 93 rattlesnake bites were treated in Arizona, with no reported deaths (Curry et al. 1989). McCollough and Gennaro (1968) estimated there were 200 venomous snakebites in Florida each year with approximately 3 fatalities (*Sistrurus miliarius* is responsible for about 44% of the bites there each year; Tu 1977).

In the years 1982–2002, 31 venomous snakebites, including only 5 accidental bites by *Crotalus horridus*, occurred in Minnesota, with 16 (52%) involving nonnative species including 2 by exotic species, but no bite proved fatal (Keyler 2005). Wyoming experiences about 6 bites by *C. o. concolor* or *C. v. viridis* each year (Christie 1994).

Downey et al. (1991) reported 36 envenomations by rattlesnakes in New Mexico during 1979–1989, but mentioned no fatalities. Fifty-five snake envenomations (including nonnative captive species, an incidence of 0.7 bites per 100,000 population annually) occurred in Utah during 69 months between January 1985 and September 1989, resulting in only one death (Plowman et al. 1995). Straight and Glenn (1993) had already reported five deaths from bites by *Crotalus o. lutosus* in Utah during the period 1931–1987, but none from bites caused by the Gila monster. Three of the fatalities involved extenuating and/or unusual circumstances, and one was an apparent homicide.

During 1979–2005, the 5 states with the highest mortality rates were Texas (15% of national total), Georgia (13%), Florida (12%), Arizona (8%), and California (7%) (Bellman et al. 2007).

These statistics indicate that normally fewer than five deaths occur annually nationwide from bites by venomous reptiles.

The annual mortality rate from venomous snakebites (all species combined) in the 10 Mexican states reporting the most deaths during 1960–1974 were Oaxaca, 32.7 (22.0% of total fatalities); Veracruz, 22.7 (15.7%); Puebla, 12.7 (8.5%); Chiapas, 11.7 (7.8%); Hildalgo, 10.5 (7.1%); San Luis Potosí, 8.3 (5.6%); Tabasco, 7.7 (5.2%); Michocan, 5.0 (3.4%); Yucatán, 5.0 (3.4%); and Jalisco, 4.0 (2.7%) (Gomez and Dart 1995; Juliá-Zertuche 1981). The total fatalities per year in these 10 states is about 120; approximately 149 reported fatalities from snakebite occur annually in all of Mexico (Juliá-Zertuche 1981; Tay et al. 1980a, 1980b), although probably more bites and fatalities are unreported in rural areas. *C. atrox* is the snake responsible for most serious envenomations and deaths in northern Mexico (Minton and Weinstein 1986).

Reptile Species Involved in Envenomations: Of the 4,408 venomous snakebites in the United States in 1991, rattlesnakes (*Crotalus, Sistrurus*) accounted for 13.7%; copperheads (*Agkistrodon contortrix*), 8.8%; cottonmouths (*A. piscivorus*), 1.5%; unknown crotalids, 0.2%; and coralsnakes (*Micruroides, Micrurus*), 0.6% (American Association of Poison Control Centers).

In 2001, 6,440 snakebites were reported in the United States: native venomous species, 2,217; exotic venomous species, 96; nonvenomous native species, 2,129; nonvenomous exotic species, 166; and unknown species, 1,922 (Litovitz et al. 2002).

Between January 1986 and December 1988, 132 snake envenomations were recorded in the Western Envenomation Database; 106 (80.3%) were by species of *Crotalus*, 1 (0.07%) by *Sistrurus*, and 26 (18.9%) by other crotalids (Dart et al. 1992b).

No fatalities have been reported from Gila monster bites in the United States since the early 1950s (Norris and Bush 2007).

The reptiles of greatest medical importance in the United States are the rattlesnakes *C. adamanteus, C. atrox, C. helleri, C. horridus, C. oreganus. C. scutulatus, C. viridis,* and *S. miliarius*. Although not as deadly as the aforementioned rattlesnakes, the copperhead, *Agkistrodon contortrix,* is also of importance because of its number of annual envenomations. In 1928 and 1929, the species responsible for human envenomations were, in order of frequency, *Agkistrodon contortrix* (1928, 171; 1929, 137 cases), *Crotalus atrox* (100, 94), *A. piscivorus* (43, 39), *C. horridus* (43, 31), *C. viridis* (37, 64; including *C. helleri* and *C. oreganus*), *C. oreganus* (including *C. helleri*, 27 in 1928), *Sistrurus miliarius* (18, 22), *C. adamanteus* (12, 14), *C. cerastes* (5, 2), and *S. catenatus* (2, 2). The species identified in fatal cases during those combined years were *C. atrox* (14), *C. horridus* (9), *C. viridis/helleri/oreganus* (8), *A. piscivorus* (5), *C. adamanteus* (3), and *C. cerastes* (1) (Hutchison 1929, 1930).

Stickel (1952) gave the following bite and mortality percentages for snakebites in the United States for the 8-year period 1927–1934: *A. contortrix*, 691 bites (0.9% fatal); *C. atrox*, 411 (11%); unidentified snake, 339 (5%); *C. horridus*, 215 (6.7%); *A. piscivorus*, 205 (4%); *C. viridis*, 154 (6.5%); *C. helleri* and *C. oreganus* combined, 134 (6.7%); *S. miliarius*, 73 (0%); *C. adamanteus*, 70 (27%); *S. catenatus*, 22 (0%); *C. cerastes*, 17 (6%); *M. fulvius*, 6 (unknown); *C. lepidus*, 2 (0%); and *C. scutulatus*, 1 (0%). Bellman et al. (2007) noted that in 27 fatal cases in 1979–2007, 19 (70%) involved rattlesnakes and 4 (15%) were by exotic nonnative snakes.

The most severe Georgia envenomations in 1927–1984 were by *C. adamanteus* (68 bites, 36 severe cases), *A. piscivorus* (65, 2), *M. fulvius* (4, 1), other rattlesnakes (31, 6), and unidentified snakes (27, 1); the only fatalities were by *C. adamanteus* (8) (Watt 1985). Since 1966, Glass (1976) treated 200 pitviper bites in Texas (1 death by a "small rattlesnake") and performed surgery on 140 persons. Of those bites requiring surgery, *C. atrox* accounted for 112; *A. contortrix*, 24; and *A. piscivorus,* only 4.

Although *A. contortrix* may cause the most envenomations yearly, its venom is relatively weak and normally only causes moderately severe bites at best.

The Victim: Most human envenomations occur in the 10–39 year age classes; Curry et al. (1989) reported that 74.4% of snakebite victims are 18 to 50 years old. Hutchison (1929) reported the following number of bites from 571 cases in 1928 by 5-year age group: <2 years, 9 envenomations; 2–5, 53; 5–10, 89; 10–15, 93; 15–20, 54; 20–30, 85; >30, 152; and age not stated, 36. Glass (1976) treated 33 victims 1–9 years old, 34 10–19 years old, 70 20–70 years old, and 3 over 70 years old.

The number of envenomations by age group in 131 snakebites recorded in the Western Envenomation Database during 1986–1988 were 0–9 years, 29 (22.1%); 10–19, 21 (16%); 20–29, 21 (16%); 30–39, 25 (19.1%); 40–49, 13 (9.9%); 50–59, 9 (6.9%); 60–69, 9 (6.9%); and 70–79, 4 (3.1%) (Dart et al. 1992).

Downey et al. (1991) reported that in 36 New Mexico envenomations by snakes during 1978–1989 12 (33.3%) were to children <11 years old, 6 (16.7%) to 11- to 20-year-olds, 6 (16.7%) 21- to 30-year olds, 5 (13.9%) to 31- to 41-year-olds, 5 (13.9%) to 41- to 50-year-olds, and 1 each for 51- 60-year-olds and 71- to 80-year-olds (combined, 5.5%).

Males are bitten in a higher percentage than females in all age groups, except possibly that of very young children (Minton 1980a). In the 571 1928 cases examined by Hutchison, 395 (69.2%) involved males and 176 (30.2%) females; of those victims under 10 years of age, the ratio was closer, 91 (60.3%) male and, 60 (39.7%) female. Seifert (2007a) reported that males were involved in 81% of the elapid cases in 2001–2005. Fatalities in 1979–2005 were 94% (133 / 142) male; 52% (73 / 141) were individuals between 25 and 54 years of age (Bellman et al. 2007). Weber and White (1993) reported on 18 Texas snakebite patients who were 12 years old or younger, and 1 child each of ages 13, 14, and 16 years; 13 rattlesnakes, 4 copperheads, and an unidentified snake were involved. Boys constituted 67% of the patients and girls 33%. Males made up 50% of the snakebite victims reported by Offerman et al. (2002), 69% by Glass (1976), 73% by Plowman et al. (1995), 79% by Spiller and Bosse (2003), and 87.2% by Curry et al. (1989). Parrish (1966) estimated the percentage of bites by sex per 100,000 population in the United States to be 10.86 for males and 4.92 for females. Males appear to be more adventurous with snakes than females.

Children may suffer as much as 20% of annual rattlesnake envenomations within the geographic range covered by this book. Case information in the database of the American Association of Poison Control Centers (AAPCC) suggests that more than 50% of all reported cases involve pediatric patients (defined as 13 years or younger). The fatality rate from snakebites is highest in children and adults over 50 years of age. In 12 Southern California pediatric envenomations by rattlesnakes (*Crotalus cerastes, C. helleri, C. ruber, C. scutulatus*) in 2001, the victims' ages ranged from 14 months to 13 years, with an average of 6.9 years (Offerman et al. 2002). Symptoms become more severe in children, and they usually require a hospital stay after envenomation; 13 (72%) of the cases reported by Weber and White (1993) developed systemic manifestations.

Envenomation of a pregnant woman can be particularly serious and result in the death of the fetus (Zugaib et al. 1985). However, if treated with antivenom as soon as possible, both the mother and baby should survive (Malz 1967; Pantanowitz and Guidozzi 1996). However, fatalities of newborns do occur (Entman and Moise 1984).

When Do Most Legitimate Envenomations Occur? Illegitimate envenomations can occur at any time of the year, but legitimate ones are most frequent in the late spring and from July into autumn when humans are outdoors more often; few envenomations occur in the winter (Curry et al. 1989; Downey et al. 1991; Hutchison 1929, 1930; Parrish 1966). The high incidence periods encompass the seasons when the male reptiles have high testosterone levels and are moving more in search of females to mate. At such times they come into contact with humans more often (Cardwell et al. 2005).

Where Bitten: Approximately 85% of legitimate snakebites occur below the knee; most illegitimate bites are to the upper extremities.

Hutchison (1929) reported the 514 bites in 1928 as occurring on the foot, 142 (27.6%

of incidence) times; ankle, 66 (12.8%); shin, 88 (17.1%); thigh, 1 (0.19%); finger, 123 (23.9%); hand, 66 (12.8%); wrist, 3 (0.58%); forearm, 19 (3.7%); upper arm, 1 (0.19%); torso, 2 (0.4%); and head, 3 (0.58%). In 1930, he reported the details of an additional 431 bites that occurred in 1929: lower extremities, 240 (55.7%); upper extremities, 186 (43.2%); head, 5 (1.16%); and none to the torso.

In 140 snakebites, Glass (1976) treated 90 (64.3%) lower-extremity wounds, 48 (34.3%) on the upper extremities, and 2 (1.4%) to the torso. In 80 snakebite cases noted by Spiller and Bosse (2003), the bite was to the toe 3 times (3.8%), foot, 21 (26.3%), leg, 17 (21.3%), finger, 14 (17.5%), hand, 20 (25%), arm, 4 (5%), and face, 1 (1.3%). Location data presented by Curry et al. (1989), Grace and Omer (1980), Huang et al. (1981), and Parrish (1966) also indicate that most bites occur on the lower extremities.

However, the incidence to body parts in 131 snakebite cases reported by Dart et al. (1992b) was somewhat different, with a shift away from the lower extremities: toe, 8.4%; foot, 18.3%; ankle/calf, 26.7%; knee/thigh, 6.9%; fingers, 30.5%; hand, 6.1%; arm/shoulder, 1.5%; torso, 0.8%; and neck/head, 0.8%. Downey et al. (1991) also reported more bites to the upper extremities than to the lower ones, but both of these papers included data from illegitimate bites.

Bites to fingers or toes usually produce less severe clinical manifestations, probably because the distal region of a digit has an area of smaller volume that limits the quantity of venom injected; digits also have reduced circulation compared to other parts of the body so the venom is carried toward the body at a slower rate. Rattlesnake envenomations above the digits are almost twice as severe as those to the digits (Moss et al. 1997).

In a series of rattlesnake bites examined by Grace and Omer (1980), there were 31 involving the upper extremity and hand with a clinical complication rate of 32%, ranging from coagulopathies to Volkmann's contracture (degeneration, contracture, fibrosis, and atrophy of a muscle injuring the local blood supply). The most frequent injury was tissue necrosis with functional joint stiffness and loss of sensibility.

Symptoms of Reptile Envenomation: Venomous snakes can meter the amount of venom they release in a single bite. This ranges from no venom (a dry bite) to almost all of the venom contained within the two venom glands (see below). Dry bites, those resulting in no injection of venom and not producing clinical symptoms, make up about 24 to 50% of reported bites. However, when symptoms of envenomation occur they can be very serious and require medical treatment, particularly if the bite is by one of those most dangerous species previously listed.

Envenomation may be either hemotoxic, principally attacking the circulatory system, or neurotoxic, principally affecting the nervous system. Symptoms characteristic of crotalid hemotoxic envenomation include intense pain, edema, weakness, swelling, numbness or tingling, rapid pulse, blood blisters on the skin (ecchymoses), muscle contraction (fasciculation), heightened sensitivity, unusual metallic taste, vomiting, confusion, and either hemorrhaging or blood coagulation (Burgess and Dart 1991; Bush and Jansen 1995; Juckett and Hancox 2002; Keyler 2008; Parrish and Hayes 1970).

A grading scale has been proposed for severity of crotalid hemotoxic envenomation based on the amount of venom injected: Grade 0 (no envenomation, a dry bite)—no severity; minimal pain or tenderness; fang punctures or abrasions; edema/erythema 2.5 cm/12 hours; no systemic manifestations. Grade I (mild envenomation)—minimal severity; intense pain; fang punctures and tenderness; edema/erythema <2.5–12 cm/12 hours; numbness or tingling about the mouth; no systemic manifestations; normal laboratory parameters. Grade II (moderate envenomation)—moderate severity; severe pain; fang punctures and tenderness; edema/erythema 15–30 cm/12 hours; some blood ooze at the bite site; mild systemic manifestations; mild coagulopathy; mildly abnormal laboratory parameters; often rapid progression from Grade I. Grade III (severe envenomation)—above moderate severity; intense pain; edema/erythema >30 cm/12 hours (often of entire extremity); coagulopathy; severe systemic manifestations; very abnormal laboratory parameters; signs of Grades I and II appear in rapid progression, with immediate systemic signs and symptoms. Grade IV (life-threatening)—very severe; multiple envenomation; very intense pain; edema/erythema beyond involved extremity; severe coagulopathy; very severe systemic manifestations; profoundly abnormal laboratory parameters; local reaction develops rapidly with blood blisters, necrosis; reduced blood pressure and muscle contraction may obstruct venous and even arterial flow (Christopher and Rodning 1986; Dart et al. 1996; Juckett and Hancox 2002).

Dart et al. (1996) proposed a more detailed point list in relation to the above snake bite severity score to further aid physicians in determining the victim's proper envenomation grade. We encourage the reader to review their proposal.

Bites by the coralsnakes (*Micruroides, Micrurus*), seasnake (*Pelamis platura*), and some rattlesnakes (*Crotalus helleri, C. oreganus concolor, C. scutulatus, C. totonacus*, and possibly *C. tigris*) produce neurotoxic symptoms: minimal pain, drooping eyelids, contraction of pupils, double vision, weakness, numbness at the bite site, confusion, double vision, difficulty swallowing, sweating (sometimes profuse), salivation, diminished reflexes, harsh-sounding inhalation, breathing difficulty (possibly leading to respiratory paralysis), seizures (in children, probably because of cerebral hypoxia), muscle paralysis, hypotension, rapid heartbeat, and cardiac arrest (Banner 1988; Bush et al. 1997; Dart and McNally 2001; Juckett and Hancox 2002; Seifert 2007a).

The incidence of secondary bacterial infection at the fang puncture site is usually less than 10%. The bacteria isolated from infected wounds include *Acinetobacter, Citrobacter, Pseudomonas, Staphylococcus aureus, S. epidermidis*, and *Streptococcus* (group A); most of these can be found in the snake's mouth or on the victim's skin (Downey et al. 1991).

Legitimate bites by *Heloderma horridum* and *H. suspectum* are very rare. Most envenomations involve handling captives, and probably many of these bites are not reported. Their venom delivery system is not as efficient as that of either vipers or elapids. This is good, because their venom is highly toxic and painful symptoms may occur within seconds. The severity of the envenomation depends on several factors: the size and health of the lizard, the length of time the lizard is allowed to bite and chew, which teeth make contact with the skin, if the bite is through clothing, and how much the lizard has been agitated before the bite (Hooker and Caravati 1994). Envenomation is significant in about 70% of bites (Bou-Abboud and Kardassakis 1988), but is rarely severe (Ernst 1992). Refer to the family account of Helodermatidae and the two species accounts for specific details on the venom delivery system and venom.

The clinical manifestations reported for helodermatid lizard bites include bleeding, particularly if the animal is torn away from the bite site (some teeth may be left in the wound after the lizard is detached); severe local pain (possibly more intense than that produced by a pitviper bite), sometimes extending up the injured limb; significant swelling and edema at and near the bite site that may extend up the limb involved; numbness (possibly caused by tissue swelling); cyanosis or ecchymosis around the bite site; blood vessel spasms at the bite site; swelling and tenderness of local lymph glands; weakness and dizziness; chills; vomiting; profuse sweating; muscle fasciculations (rarely); breathing difficulties; hypotension; a rapid heartbeat; and shock (Albrittton et al. 1970; Bou-Abboud and Kardassakis 1988; Hooker and Caravati 1994; Mebs 1995; Norris 2004; Norris and Bush 2001, 2007; Russell 1983; Russell and Bogert 1981; Strimple et al. 1997). Russell and Bogert (1981), however, reported that no respiratory problems occur in human envenomations. Symptoms of bites by the individual venomous reptiles are presented in those species accounts.

Avoidance of Venomous Snakebite: Common sense prevails here. The three cardinal rules for avoiding a venomous reptile bite are as follows: (1) Gain as complete a knowledge as possible of the species of venomous reptiles that may occur in the target area. This includes their descriptions, habits (including the time of day they are most active), and habitat preferences. (2) Do not handle venomous reptiles, even "dead" ones. Most persons are bitten because they want to get a closer look at the animal or try to kill it. It is very important to teach children not to harass venomous reptiles. (3) Watch where you place your hands and feet when in a venomous reptile's habitat (look before you reach or step!).

In addition, stay out of places where the vegetation or other habitat features prevent good visual surveillance. Wear protective clothing where venomous reptiles are known to occur. Because most snakebites occur below the knee, leather boots that are at least higher than the ankles should be worn (flip-flops, sandals, and sneakers don't cut it; nor do thin leather shoes or boots). This is extremely important when walking through high grass or brush where the ground ahead cannot be seen.

If you sight a dangerous reptile, give it a wide berth; North American snakes can strike 33–50% the length of their bodies.

Other important safety measures include not camping in areas of known snake activity; sleeping on a cot rather than on the ground in a tent when there; and thoroughly examining bedding and haversacks that have been left on the open ground before further use. A little thought goes a long way in protecting oneself from venomous reptile bites!

The references mentioned in this chapter are but a small percentage of the many papers that have been published regarding reptile envenomations in North America. If further interested, we encourage readers to make their own electronic searches on this topic.

Treatment of Envenomation by Reptiles

The factors that most reduce mortality from venomous reptile bites are rapid transport, intensive care, and the use of antivenom. Those that increase the likelihood of a more severe envenomation are dangerous snake species, large snake size, small patient size, prolonged fang contact (or multiple sets of fang punctures), previous envenomations (treated or not) or exposure to venomous reptiles, and delays in transport for medical help (Bush 2005).

Some reptiles (like the lizards *Heloderma horridum* and *H. suspectum* and the coralsnakes *Micruroides* and *Micrurus*) do not immediately release their hold after biting a person, but instead continue to chew. The longer this occurs, the more venom is deposited into the wound; the reptile must be removed at once (see Beck 2005 for suggested methods to remove *Heloderma*). Also, remove any teeth or fangs that remain in the wound after a reptile releases its hold.

Although the chance of being bitten by a venomous reptile in North America is low if one is not provoking or handling it, all bites should be examined by a qualified physician, preferably at a medical facility emergency room. This is important because some symptoms and serious conditions resulting from a bite may not be immediately evident, thus luring the one bitten into a false sense of security. Persons who have experienced previous envenomation(s) or exposure to venom through ingestion or inhalation should consider carrying an injectable epinephrine kit or pen in case of an immediate severe allergic reaction (Zozaya and Stadelman 1930). Also, antivenom could be carried into the field, if affordable (at present antivenom is extremely costly). Injection of antivenom in the field may lead to dangerous side effects, so inoculations should be left to a qualified physician when the victim reaches a medical facility.

In addition, it is very important that the reptile in question be accurately identified. If possible, a photograph (preferably digital or Polaroid) should be taken of the offending reptile. If a photograph is not possible without the chance of being bitten again, the reptile or its remains should be taken to the medical facility along with the patient. Just to say the snake was a "rattlesnake" in western Texas, Arizona, Southern California, or some parts of Mexico is inadequate as the different species in these and other areas may have either hemotoxic or neurotoxic venoms that must be treated differently. Also, as related above, some symptoms of envenomation may not occur until somewhat later, especially in mild neurotoxic bites. It is better, of course, to avoid a second bite if the reptile is difficult to capture. Neither capture of the reptile or first aid attempts should delay transport of the victim to a hospital.

It is also very important to record the time of the bite so that the progressive rate of symptoms may be better monitored.

Immediate First Aid: As stated above, the first priority is to transport the bitten person immediately to a medical facility! If the victim collapses or experiences hypotension before reaching there, he or she should immediately be put in a reclining position. Other important first steps are as follows.

The patient must be reassured and kept as calm as possible during the journey to the medical facility. Avoid all excessive activity, which may stimulate the heart to a greater beat or stroke rate, or cause muscle contraction around the wound site. These promote more rapid and greater blood flow, which moves the venom from the wound site toward the heart at a quicker rate. All tightly constricting objects, like clothing, rings, or other jewelry (including pierce rings), should be removed to prevent complications from swelling.

The wound site should be washed with soap and water and wiped clean, and any apparent helodermatid teeth or snake fangs remaining in the wound should be removed at once. The bitten limb should be immobilized, using a splint if possible, and positioned below the level of the heart (Leopold et al. 1957; McKinney 2001). The border of advanced swelling should be marked and timed every 15 minutes (McKinney 2001; Wingert and Chan 1988).

If the bite is by a suspected neurotoxic reptile, wrap the limb in a pressure bandage to help localize the venom. This measure has proven effective for bites by very neurotoxic Asian and Australian elapid snakes, and can be used effectively for North American coralsnake bites and neurotoxic rattlesnakes (Bush et al. 1997; Norris 2008). It may also help in somewhat slowing the distribution of venom in hemotoxic pitviper bites (Burgess et al. 1992; Bush 2005; Bush et al. 2005; Campbell and Lamar 2004), although this has been questioned (Sutherland and Coulter 1981). Once a pressure wrap has been applied, it should not be removed until antivenom is administered, as this may cause a "bolus effect" (Bush 2005).

If possible, an intravenous fluid, like a physiological saline solution or Ringer's lactate, should be administered. If not possible in the field, this should be done at the medical facility, and may also involve the intravenous administration of electrolyte or colloid solutions (shock should be treated with a plasma expander rather than a saline solution; Wingert and Chan 1988) or possibly steroids (Whitley 1996).

If needed, airway clearance and support should be administered.

Should venom be sprayed into the eyes, they should be irrigated as soon as possible. A physiological salt solution is best for this, but water (either pure or tap) can also be used to wash away the venom. Some irritation and redness may result, but there has been no permanent ophthalmic nerve damage reported in such cases involving North American pitvipers. However, the neurotoxic venom of some elapids is known to do severe damage to the optic nerve.

First Aid Procedures Not to Be Used: Several first aid measures used in the past have proven either useless or to cause more harm than good.

Although formerly promoted (Oldham et al. 1983; Russell 1983; Seelex 1963; Snyder and Knowles 1988), the traditional cut-and-suck first aid methods are now considered of doubtful help (Bush 2005). These include the use of the Cutter's Snakebite Kit, Venom Ex®, the Sawyer Extractor Vacuum Pump (or "EXTRACTOR"), and other popular sucking devices. Although use of the "EXTRACTOR" has shown mixed results (see Bush et al. 2000), tissue damage can result from the suction applied. Such procedures involve cutting and the use of tourniquets that constrict blood circulation. Cutting increases bleeding and causes blood loss, and is particularly detrimental if the venom contains nonclotting enzymes. Cutting can also damage local nerves and tendons, and open the wound to possible bacterial infection (Reid and Theakston 1983). Tight constriction bands may cause a pooling of the venom in the vicinity of the wound, which can cause more severe tissue damage.

Oral sucking of the wound should be avoided at all times, as venom can be transferred to cuts or other lesions in the mouth, resulting in a second envenomation of the person giving the first aid.

Do not give the victim alcohol or any type of stimulant!

Do not administer NSAIDS (nonsteroidal anti-inflammatory drugs), like aspirin or other painkillers that contain ibuprofen. These drugs enhance bleeding by slowing clotting and increase blood loss in hemotoxic envenomations.

Do not treat the wound with ammonia or other household chemicals, gasoline or kerosene, gun powder, potassium permanganite, salt, strychnine, tar products, or turpentine. These are old treatments that have no positive value, but may cause lasting tissue damage around the wound.

Do not cauterize the wound. This will not prevent the spread of the venom, especially in neurotoxic bites, and will cause severe tissue damage and facilitate possible infection. Also, do not heat the wound or bitten limb, as this may dilate blood vessels and promote the distribution of the venom.

Cryotherapy adds to the morbidity of the bite and should not be used (Wagner and Golladay 1989). Do not cool the wounded area with ice, which may lead to tissue loss through necrosis. This is caused by contraction of blood vessels that restricts the outward distribution of the venom, resulting in a localized bolus pooling of venom at the wound site.

The data from several studies and reports concerning the use of Stungun and other forms of electroshock in the treatment of snakebite have been inconclusive; they seem to have little value in treating bites by North American pitvipers. Until the results of comprehensive controlled experiments are available, such treatment should be avoided (see review in Hardy 1992).

See Babcock (1928) for a report on snakebite treatment during the colonial period in Massachusetts and other now shunned remedies used in the past.

Treatment at a Medical Facility: When the bite victim arrives at the medical facility, regardless if the envenomation is by a snake or a helodermatid lizard, his or her age, sex, and alcohol and possible drug levels should be recorded (many illegitimate envenomations involve alcohol or drug use, which can complicate treatment).

The patient's physiological baseline, involving circulatory, hematological, renal, and respiratory levels, should be determined (see Wingert and Chan 1988 for required tests). The medical staff should monitor and periodically record all vital signs, and closely observe the person for the major signs of envenomation listed in the chapter "Envenomation by North American Reptiles" to determine the severity of the envenomation. All bites by North American pitvipers begin with few clinical effects, which may worsen over time and ultimately become severe. If the bite victim is unconscious and no pulse can be detected, cardiopulmonary resuscitation should be started immediately.

Some snakebites are dry ones in which only fang punctures occur and no venom has been released; but if signs of envenomation are evident, the staff should immediately contact the nearest poison control center for expert advice on the proper treatment. Unfortunately, because reptile envenomations are relatively rare, few local medical personnel are experienced in treating them.

The adult patient should typically be observed for at least 8 hours and children for up to 24 hours, because the onset of some symptoms, particularly for those resulting from neurotoxic envenomation, may not occur until hours after the bite. Unfortunately, if the reptile in question has not been identified, major treatment must await the appearance of symptoms, and by that time major damage possibly has occurred.

All bandages or coverings should be removed immediately, and swabs made of them and the skin at the bite site to determine if any venom has flowed from the fang punctures onto them. Like all puncture wounds, bites must be thoroughly cleaned and sterilized. A tetanus antitoxin should be administered. Also, a broad-spectrum antibiotic should be injected to counteract any possible infectious bacteria (including that of gas gangrene) that may have been introduced into the wound (Bush et al. 1997).

Persistent vomiting can be treated with an injection of chlorpromazine, 25–50 mg for adults, 1 mg/kg for children. Shock, fainting, angioedema, and other autonomic symptoms can be treated by an intramuscular injection of 0.1% epinephrine (0.5 mg for adults, 0.2 mg/kg for children), and an intravenous injection of an antihistamine (like chlorpheniramine maleate; 10 mg for adults, 0.2 mg/kg for children) (Campbell and Lamar 2004; Norris 2008).

The intracompartmental pressure within the advancing swollen area of a bitten limb can be measured with a Stryker Monitor. Increased intracompartmental pressure is determined at >30 mm Hg, and, if this high, will require an injection of antivenom (Shum et al. 2006).

Neurotoxic envenomations by the *Micrurus* coralsnakes and some *Crotalus* rattlesnakes produce particularly dangerous symptoms. If signs of respiratory failure are present, these must be aggressively treated with endotracheal intubation to prevent aspiration. Breathing distress and cyanosis should be treated by clearing the airways and administering oxygen if necessary. Both cardiac and pulse rates should be constantly monitored, as well as the blood oxygen level. If these rates and levels fall, the appropriate treatment should be started at once. In such cases antivenom is required (see the following discussion on antivenom).

If hematologic abnormalities occur, treatment with transfusions of fresh-frozen plasma or platelets is recommended. If hypotension persists, dopamine can be administered (Burch et al. 1988).

Fasciotomy (surgical incision and division of the fascia) or amputation of an injured digit or limb should be avoided unless absolutely necessary! Such a procedure normally results in more damage to the limb than would naturally occur from venom-caused necrosis or swelling (see the discussions in Hall 2001 and Shum et al. 2006).

Envenomation by *Heloderma horridum* and *H. suspectum* requires somewhat different treatment, although similar to that for snake envenomation. The wound area should be soaked for at least 15 minutes, 3 times within 24 hours for several days postenvenomation in Burow's solution (a 1:20 aluminum acetate solution). After the first 24 hours, the digits of the wounded limb should be exercised while the limb is undergoing the soaking procedure. After each treatment with Burow's solution, the wound area should be covered with a sterile dressing. Antitetanus serum and a broad-spectrum antibiotic should be administered. Bites by these lizards may be very painful;

patients should take aspirin or codeine, or morphine if the pain is extremely severe. Five milligrams of diazepam reduces the need for any analgesic, and may also help calm the victim. Nerve blocks are of questionable use, and antihistamines and corticosteroids are usually of no use.

It is generally recognized that treatment with antivenom (serotherapy) is the only specific treatment to counteract snake venom, and it should be given by a knowledgeable medical person only to bite victims with symptoms or signs of envenomation. If the patient experiences acute hypotension, an appropriate electrolyte solution, as determined by blood chemistry and urine output, should be administered. Severe cases may require albumin, plasma, or whole blood transfusion, and antishock drugs and oxygen should be readily available (Russell 1983).

The administration of polyvalent immunoglobulin antivenom is the most recommended treatment for severe envenomations by venomous snakes, and its intravenous injection is the most effective route of administration (see the following discussion on how antivenom works).

What Is Antivenom? Antivenom, also known by the French name *antivenin*, is an artificially produced, commercial, mammalian, immunoglobulin serum that neutralizes the effects of antigens (toxic proteins) injected during the bites of venomous reptiles.

How Is Antivenom Produced? Antivenom is produced from fresh venom extracted from the living reptile either by electrical stimulation or by manual "milking," and is usually obtained from captive animals every 20 to 30 days.

During electrical stimulation, the reptile's body is restrained and it is held behind its head and induced to bite through a rubber diaphragm covering a sterile collection vessel. Electrodes are then touched to the opposite sides of the reptile's head, causing the muscles involved with venom discharge to contract and squeeze the venom from the gland and into the vessel.

During manual milking, the reptile is similarly restrained, held, and induced to bite the thin diaphragm covering a sterile collecting vessel (do Amaral 1928a). The handler then applies pressure to the reptile's venom glands with his or her fingers. The pressure is continued until the last of the reptile's venom is discharged into the container.

The venom is freeze-dried (the preferred method, although drying with the help of reagents or a vacuum is sometimes used). Dried venom, usually from only one species, is mixed with a saline solution and injected in a nonlethal dose into either a healthy domestic horse (*Equus caballus*; normally 7 to 8 years old) or a domestic sheep (*Ovis aries*) at regular intervals until the injected mammal builds up the necessary antibodies (specialized proteins that produce immunity; see the chapter "Venom") and becomes immune to the venom mixture. The venom dosage injected is slowly increased over time to produce a greater degree of immunity. When injected into humans, these antibodies in turn neutralize the same species' venom injected into the immune animal, and possibly those of others.

The antibodies are obtained by regularly extracting a small amount of blood from the jugular vein of the immunized mammal (normally about 6 to 8 liters from horses, less from sheep). The horse blood is combined with a sodium citrate solution to buffer it and prevent coagulation and degradation. The *Y*-shaped globulin (IgG), to which two arms of antibodies and a separate long tail are hinged to the main molecule, is separated out and purified IgGs produced from the inoculations of venom from several species are then mixed together and freeze-dried (Bon 1991; Cope 1979; Dart and McNally 2001; Githens 1931; Klauber 1972). This freeze-dried compound, when mixed in solution, composes the antivenom to be administered to the bite victim.

Early production of antivenom using sheep serum basically follows the process related above with some exceptions. The IgG precipitate is separated out with the aid of sodium sulphate (Russell et al. 1985), and isolated through digestion with papain to produce antibody fragments (Fab, Fc). The IgG is then purified by ion exchange and affinity chromatography to eliminate most of the more immunogenic Fc antibody portion and the long tail that promotes allergic reactions, then mixed and freeze-dried. When tested in an animal lethality model, the polyvalent ovine antivenom was 3.0 to 11.7 (mean, 5.2) times more potent than the equine antivenom against 14 different pitviper venoms (Consroe et al. 1995).

Development of ovine antivenom in this way has several serious drawbacks. It takes a long time to produce a single batch (eight weeks); production utilizes the entire capacity of the factory; and the process is complex, difficult, and labor intensive. O'Donovan (2006) has suggested several improvements to increase the volume of production and to speed it up: substituting affinity chromatography capture techniques; optimizing the digestion step by outer limit experimentation that could reduce the enzymatic input by >50%; reengineering the chromatography processes to create a shorter cycle period involving faster recovery, cleaning, and reuse times; changing the formula of the buffer solution used to create more rapid reconstitution; shortening the lyophilization cycle; and shortening the entire single-batch cycle to <20 days by reducing the number of hands-on steps.

Some mammals, like the California ground squirrel (*Spermophilus beecheyi*) and North American opossum (*Didelphis virginiana*), are naturally resistant to the venom of local rattlesnakes (see the accounts of *C. oreganus* and *C. helleri*; the papers by Biardi [2000, 2006, 2008] and Biardi et al. [2005]; and McKeller and Pérez 2002). The concept of developing a human antivenom from the serum of these mammals is promising, but is only in the experimental stages, as is also the possibility of antivenoms produced through genetic engineering (Nkinin et al. 1997).

The Worldwide Inventory of Snakebite Antivenom: A global shortage exists in antivenom supplies, so development of any way to speed up production would be welcome. Current antivenom production provides only about 1% needed, and the price of a single vial ($50–$150) is cost prohibitive in most third world countries. The average dose of antivenom for complete treatment is usually at least three vials, but may be much higher in severe cases (Warrell 2007). The cost of a course of the currently used CroFab antivenom (see below) in the United States may run more than $10,000, and so is stocked by relatively few pharmacies.

Less than 30 laboratories around the world produce antivenoms, principally for the venomous snakes but very rarely for the helodermatid lizards in their region, and only 34 facilities in 21 countries [6 in the United States, none reported for either Canada or Mexico] extract snake venom for medical use (Powell et al. 2006). Only one pharmaceutical company currently produces antivenom in the United States, and, although three additional laboratories produce antivenoms in Mexico (Campbell and Lamar 2004), no antivenom is produced in Canada.

To help counteract the global shortage of snakebite antivenom, the World Health Organization (WHO) initiated programs in 2007 to ease access to antisnakebite serums in rural areas, but these will take some time to complete.

Venomous snakes are often genetically distinct in different countries or even among different regions of the same country (see the discussions in the accounts of *Crotalus helleri, C. lepidus, C. oreganus*, and *C. scutulatus*; and Ernst and Ernst 2003). So the antivenoms produced in one country, even an adjacent one, may be ineffective in neutralizing the venom of the same or related species elsewhere (Fry et al. 2003a). This is particularly true in Asia, where antivenoms produced from local populations of the Indian cobra, *Naja naja* (Warrell 1997), Malayan pitviper, *Calloselasma rhodostoma* (Daltry et al. 1997), and Russel's viper, *Daboia russelii* (Belt et al. 1997), are only effective against the venom from specific populations.

Types of Antivenom: As discussed above, reptile antivenoms are produced from either horse (equine) serum or sheep (ovine) serum. Antivenoms may also be classified as either monovalent, which are made from the venom of a particular species to neutralize its venom, or polyvalent, which are made by mixing the venoms of several species and are effective against several snake venoms.

Equine antivenom—This type of antivenom, termed Nearctic Crotalidae Polyvalent or Antivenom Crotalidae Polyvalent (ACP), was first manufactured in the United States in 1954 by Wyeth-Ayerst Laboratories, the pharmaceutical division of American Home Products Corporation, and made available through Wyeth Inc., Philadelphia, Pennsylvania, and, in Canada, by its subsidiary John Wyeth & Brother, LTD, Walkerville, Ontario. Wyeth was formally acquired by Pfizer, Inc. in October 2009.

Until the end of the twentieth century it was the only antivenom available for venomous North American pitvipers, and was normally quite successful in counteracting the effects of their bites, reducing the human mortality rate from about 5 to 25% prior to its use to less than 1% when used (Dart and McNally 2001). In addition, envenomated laboratory mice (*Mus musculus*) and rats (*Rattus norvegicus*) receiving ACP have vastly increased survival rates (Cope 1979; Russell et al. 1973).

ACP was a partially purified polyvalent mixture of both the Fab and Fc antibody fragments along with the long tail of IgG produced in horses immunized against the venoms of *Crotalus adamanteus, C. atrox, C. durissus terrificus*, and *Bothrops atrox*. It was effective for weeks against a wide range of hemotoxic pitviper venoms, but not against the venoms of coralsnakes and only partially so against the venom of neurotoxic rattlesnakes. A vial of the yellow antivenom consists of IgG with a molecular weight of 150,000 D, a total antibody content of 18.9%, 2.1 g of total protein, and 120 mg (6%) of albumin (Dart and McNally 2001). Envenomations have required as many as 118 vials of ACP, but usually about 18 (12–25) in severe cases (Jurkovich et al. 1988; Offerman et al. 2001).

Unfortunately, the allergy-producing elements left in ACP during the incomplete purification stage produced acute hypersensitivity reactions in up to 75% of persons treated with the antivenom, mostly because of allergies to horse serum, particularly, serum sickness and sometimes anaphylactic shock, which were always serious and sometimes fatal (Bush and Jansen 1995; Corrigan et al. 1978; Dart and McNally 2001; Jurkovich et al. 1988; Klauber 1972; Loprinzi et al. 1983; Minton 1974; Offerman et al. 2001; Otten and McKimm 1983; Seifert 2006). Patients receiving ACP who experienced hypersensitive reactions responded to treatment with antihistamines and steroids, but severe cases required epinephrine (ovine antivenoms made during the last decade rarely involve the risk of serum sickness, but see below). Anaphylactic shock has also been reported in dogs after treatment with equine antivenom (Conceição et al. 2007). Unfortunately, prospective research data were never available for ACP.

From the start ovine antivenom based on Fab fragments showed promise in controlling symptoms resulting from both hemotoxic and neurotoxic rattlesnake bites (Clark et al. 1997; Dart et al. 1997; Seifert et al. 1997), and only about 5 to 25% of the patients suffered hypersensitivity reactions. Consequently, Fab antivenoms became more popular, and that of ACP declined. The number of acute hypersensitivity cases and general decline in sales caused Wyeth to stop producing ACP in 2001.

A divalent equine antivenom, Antivipmyn (Fab_2H), produced by Instituto Bioclon in Mexico from venoms of *C. durissus* and *B. asper* using rabbits (*Oryctolagus cuniculus*), was tested against the venoms of eight American pitvipers, including those of *Agkistrodon piscivorus, C. adamanteus, C. atrox, C. horridus*, and *C. molossus*, and was found to be very effective in neutralizing their antigen activities (Sánchez et al. 2003a, 2003b). In addition, a polyvalent (Crotalinae) equine antivenom produced by the Institudo Clodomiro Picado, Costa Rica, from venom of *Bothrops asper, Crotalus durissus durissus*, and *Lachesis stenophrys* has effectively neutralized the venoms of *A. contortrix, A. piscivorus, C. adamanteus, C. horridus*, and *C. viridis*, but not that from *C. scutulatus* in tests using laboratory mice (*Mus musculus*) (Arce et al. 2003). It is not known if these antivenoms produce acute hypersensitivity in humans.

Wyeth also produced a quite effective equine divalent (*Micrurus fulvius* and *M. tener*) antivenom that neutralized the toxic antigens contained in the venoms of coralsnakes in the United States (Gennaro et al. 2006). Unfortunately, due to a low demand (very few coralsnake bites occur annually), the company terminated its production in 2008. Today, no specific coralsnake antivenom is available in the United States, and venom from *Micrurus* is not always neutralized by the antivenoms of other elapid species (Minton 1980b).

Recently produced Latin American antivenoms have proven promising for use against North American coralsnake venoms. Anticoral (Elapidae) monovalent antivenom produced by immunizing horses with the venom of *M. nigrocinctus* at the Institudo Clodomiro Picado, Costa Rica, reduced the lethality of venom from *M. fulvius* in the laboratory mice (Arce et al. 2003). In addition, divalent antivenom produced in some Mexican laboratories from the combined venoms of the U.S. coralsnake *M. fulvius* and that of the Mexican species *M. nigrocinctus* has proven effective against the venoms of both *M. fulvius* and *M. tener*. These antivenoms are not readily available, though, and Sánchez and Pérez (2006) and Sánchez et al. (2008) have suggested that it is now crucial to begin considering alternative treatments for North American coralsnake envenomation.

Ovine antivenom—Crotalidae Polyvalent Immune Fab (Ovine) or CroFab antivenom [it and its developmental versions are also referred to in the literature as CroFab™, CroFab2, CroFab®, or CroFav] is manufactured by Protherics Inc., Nashville, Tennessee, and marketed in the United States by Savage Laboratories, New York. It is a sterile, nonpyrogenic, purified, lyophilized preparation of ovine Fab (monovalent) immunoglobulin fragments isolated from proteins produced in the blood of healthy sheep immunized against the venom of a specific North American species of pitviper (either

Agkistrodon piscivorus, C. adamanteus, C. atrox, or *C. scutulatus*). It is also effective against the venom of *A. contortrix* (Lavonas et al. 2004); this species was not included in the original clinical tests of ovine Fab antivenoms because of its usually less severe envenomations, and Caravati (2004) thought it may not be cost effective for the generally milder bites of this snake.

Crofab is white, and a vial consists of Fab with a molecular weight of 50,000 D, >85% total antibody in Fab and <3% in FC, <1.5 g of total protein, and <0.5% albumin (Dart and McNally 2001). Fab antivenoms are resistant to a very wide range of temperature and handling conditions, so CroFab can be stored for a relatively long period at 2–8°C. It is only effective against pitviper venom, and more so against the venoms of some species than others. It does, however, have the advantage of neutralizing the neurotoxic venoms of *C. helleri* and *C. scutulatus* (as does also equine antivipmyn [Fab_2H] serum made from the venoms of *Bothrops asper* and *C. durissus durissus*, and produced by Laboratorios Bioclon/Silanes in Mexico; Galán et al. 2004), but is not very effective against coralsnake venoms.

A monovalent form of the antivenom was first approved by the Food and Drug Administration (FDA) in 2000, and later a divalent and the present polyvalent forms were produced. In contrast to ACP, the prospectives of Fab ovine antivenoms are well known (Dart et al. 1997).

More than 40,000 vials of CroFab are produced annually, but it takes about 8 weeks to complete a batch, and its production is complicated and very labor intensive (see previous discussion). Treatment with CroFab may be very expensive; depending on the severity of the envenomation, as many as 44 (normally 3–18, average 4–8) vials of the antivenom may be required over several days (Camilleri et al. 2005; Offerman et al. 2002).

Use of CroFab is not without problems. Initial positive results are sometimes short-lived. Apparently the small Fab antibody fragments are often absorbed by the kidneys before they have had a chance to neutralize all of the snake's venom. More than 25% of patients receiving an ovine Fab antivenom experience recurrence phenomena, consisting of either local symptoms (cough, hives, labored breathing, rash, severe itching, swelling) or those of coagulopathy (platelet abnormalities [thrombocytopenia], abnormal clotting [prothrombin] time, lower fibrinogen level) within a short time; anaphylaxis is rare, but has occurred (Bogdan et al. 2000; Boyer et al. 1999, 2001; Bush et al. 1999; Dart et al. 2007; Dart and McNally 2001; Fazelat et al. 2007; Jackson and O'Conner 2007; Kirkpatrick 1991; Lavonas et al. 2004; Lintner et al. 2006; Seifert and Boyer 2001; Stanford et al. 2005). These appear to be the result of a pharmacokinetic and pharmodynamic mismatch between the antivenom and target venom antigens; tissue penetration and neutralization by the antivenom is incomplete, and clearance of antivenom not bound to its venom target is significantly faster than clearance of some venom components, allowing signs and symptoms of envenomation to recur (Seifert and Boyer 2001). In such cases, patients usually respond quickly to additional administration of the antivenom. Bebarta and Dart (2005) have suggested that a delay, if possible, in the use of CroFab may help in the avoidance of coagulation recurrence; however, tests to confirm this are required.

It is not known whether or not CroFab is excreted in human breast milk, so caution should be used when it is administered to a woman who is breast-feeding an infant. It is also not conclusively understood if this antivenom can cause fetal damage, and so should only be given to a pregnant patient if clearly required. However, a study by Offerman et al. (2002) has indicated that treatment with CroFab is suitable for envenomated children.

The directions for the administration of the antivenom included within the CroFab packet should be strictly followed. The recommended dosage procedure to avoid the development of hypersensitivity reactions is an initial dose of 4–6 vials of CroFab, with an additional dose of 6 more vials, followed by doses every 6 hours for up to 18 hours if required.

Some foreign laboratories are in the process of developing a new generation of antivenoms that retain the IgG's antibody arms by keeping them attached while eliminating the long allergic tail. A similar antivenom against scorpion stings has been used successfully in Mexico for years. Such improvements in antivenoms could make their use in snakebite cases more effective and cheaper—but, alas, this is in the future.

Seasnake antivenom—Although some seasnakes are capable of causing fatal envenomations of humans, because of their relatively small fangs bites by this group of elapids seldom produce envenomation symptoms and need no treatment with antivenom. Heatwole (1999) recommends that antivenoms should not be used unless absolutely necessary, and then should be administered within two days after the envenomation.

Pickwell (1978) thought that antivenom produced for any particular seasnake could potentially be used to treat the bite of another such species, but this should be confirmed. Experimental treatment with antivenom produced from venom of the Australian tiger snake, *Notechis scutatus*, has had some positive results (Reid 1980b). Unfortunately, specific seasnake antivenoms are usually unavailable in many oceanic areas, and, according to Russell (1983), only the Commonwealth Serum Laboratory, Victoria, Australia, manufactures an antivenom that will neutralize venoms from 12 seasnake species. However, it is only readily available in the Australian region.

Medical facilities should treat bites by *Pelamis platura*, which reaches Mexico and the United States, in the same way as other snakebites are treated (as discussed above).

Vick et al. (1975) reported that all lots of antivenom developed for the hooked-nosed seasnake, *Enhydrina schistosa* (which has the most potent venom of all snakes; Zug and Ernst 2004), were mostly effective in treating mice injected intravenously with the venom of *P. platura*. Mice given the antivenom began to recover, depending on the initial dose of venom, in 2–2.5 hours postinjection. By 4 hours all surviving mice were essentially normal, and none died after 4 hours. So, antisera produced from *E. schistosa* can potentially be used to treat human envenomations by *P. platura*.

Helodermatid lizard antivenom—Experimental antivenoms have been produced for the venom of the two species of *Heloderma* at the Poisonous Animal Research Laboratory, Arizona State University, Tempe, and at the Venom Poisoning Center, Los Angeles County–University of Southern California Medical Center, Los Angeles. So this is possible, but, because of the extremely low incidence of fatal bites by these lizards, there has been no commercial demand for it.

How Does Antivenom Work? Studies were conducted in France by Bon (2007) on the kinetics of specific Fab antibodies during treatment of envenomation by the adder, *Vipera berus*. After intramuscular venom injection, the resorption of venom follows a complex process in which the Fab fragments bind to toxic venom antigens (up to 85%

neutralization of phospholipase A_2 from the venom of *C. atrox*; Price and Sanny 2007). Neutralization is fast during the first 24 hours but then proceeds at a slower rate over the next 72 hours, resulting in a half-life of elimination of about 36 hours.

Bon (2007) also tested the effect of the antivenom therapy. It appeared that the venom detoxification can be explained by a redistribution of the venom from the extravascular compartment to the vascular one, where the antivenom isolates it and its antibodies neutralize the venom's toxic antigens. In the tests involving the venom of *V. berus*, Fab'_2 was more efficient than Fab because of more appropriate pharmacokinetic parameters.

Do Some Plants Inhibit Snake Venoms? The use of plants to treat snakebites is ancient. According to Greek mythology, Hercules used an antidote made from leaves of a poplar tree [*Populus alba*?] when he was bitten by a venomous snake.

Globally, many medicinal plants contribute bioactive compounds that are currently used to treat envenomation by local snakes. Most of this use is based on old folk treatments devised by native peoples. But do they effectively neutralize snake venom or relieve its symptoms?

About 83% of the reports have been from the Americas (Soares et al. 2005). Historically, a number of native and introduced plants were administered in several ways by Native North Americans to treat snakebites, and other folk remedies were brought to North America by European settlers and the Chinese (Duke 1986; Foster and Duke 1990; Klauber 1972; Lewis and Elvin-Lewis 1977; Moerman 1986; Weiss 1988).

Rattlesnake bites were not rare during the famous Hopi Tribe snake dances (see the account of *C. viridis* and Klauber 1972 for descriptions), and they supposedly used an herbal antidote to treat envenomations experienced by their priests during the dances. Coleman (1928) interviewed members of the tribe to determine the specific plant used and how the antidote was prepared. The tribal members refused to identify the plant. It was a secret known only to one member of the tribe, who passed it on to another when dying. However, Coleman was told some things about how it was prepared. The leaves and stems of the plant were boiled, according to 2 tribal members, respectively, for either 10 minutes or 2 to 3 hours. Many of the tribe were said to have a constant supply on hand and that it was effective for 2 to 3 months after it was made. The wound was moistened with the antidote fluid after a tourniquet had been applied and the punctures lanced. Apparently, the antidote was also ingested; a typical dose was approximately 50 mL, and only one dose was swallowed.

Coleman obtained a "pint" of the plant preparation, and upon returning to his laboratory placed the antidote in his ice box. Two months later, he removed it to test its effect on dried rattlesnake [species not given] venom injected into 350 g guinea pigs (*Cavia porcellus*). The liquid was slightly turbid, odorless, and a pale amber color; tasted slightly bitter; and had a pH of 5.09. The antidote was fed to 2 fasting guinea pigs with no adverse effects in 24 hours, and dissection revealed no visual damage to their digestive tracts. The amount fed was 3 to 4 times that said to have been the dose for a 68 kg man. Coleman then injected 2 mL into the leg muscles of a guinea pig, which produced intense swelling and lameness within 24 hours. The area around the injection wound was hardened for 2 to 3 days, but all adverse symptoms disappeared within a week.

Several mixture concentrations of antidote and dried venom dissolved in various

NaCl solutions were then injected into the leg muscles of the rodents at various periods, and another series of the animals was fed the antidote very shortly after a venom injection. The animals administered the plant antidote in both ways became very ill and most died.

Coleman used venom that was old and had not been kept in a vacuum and two different age groups of guinea pigs (the second, younger, group proved more susceptible to the venom). Other experimental problems were also evident, but Coleman concluded that the plant antidote does not neutralize the venom in vitro.

Sheldon (1929) reported the treatment of an apparent timber rattlesnake bite to the leg of a Native American who chewed the root of a plant resembling "flax," and then put the chewed remains on the wound. This caused him to vomit [probably because he swallowed some of the resulting juice]. He repeated the process three times, and this supposedly resulted in a retardation and eventual cessation of swelling at the level of the bite and above.

In the eastern and central United States several plants, many of them common garden or wild species, have been used, mostly by Native Americans, to treat snakebite. Parts of the following species were listed by Foster and Duke (1990) as being ingested (I), chewed (C), or used as either a poultice (P), rub (R), or tincture (T) to be applied to the wound, or as an external wash (W) to cleanse it: *Adiantum capillus veneris* (P), *Allium sativum* (I), *Antennaria plantaginifolia* (P), *Arctium lappa* (seed W), *A. minus* (seed W), *Arisaema triphyllum* (W), *Aristolochia serpentaria* (I, weak root tea), *A. tomentosa* (I, weak root tea), *Botrychium virginianum* (root P), *Calla palustris* (P), *Caltha palustris* (I), *Catalpa bignonioides* (I), *Ceanothus americanus* (I, leaf tea), *Cimicifuga racemosa* (T), *Cunila origanoides* (I, tea), *Diphylleica cymosa* (I), *Echinacea angustifolia* (I, root tea; C, root), *E. pallida* (I, root tea; C, root), *E. purpurea* (I, root tea; C, root), *Eryngium yuccifolium* (P, W), *Eupatorium rugosum* (P), *Euphorbia corollata* (root P), *E. ipecacuanhae* (root P), *Foeniculum vulgare* (seed P), *Fraxinus americana* (I, inner bark tea), *Goodyera pubescens* (I), *Helianthus annuus* (P), *Hieracium venosum* (I, tea), *Hypericum hypericoides* (C), *Juniperus communis* (W), *Lilium canadense* (P, R), *Liriodendron tulipifera* (weak W), *Nicotiana tabacum* (P), *Polemonium reptans* (I, root tea), *Rudbeckia hirta* (I, tea), *Sanicula marilandica* (root P), *Solidago canadensis* (I, flower tea), *Trillium erectum* (W), *Triosteum perfoliatum* (root P), and *Uvularia perfoliata* (I). Soares et al. (2005) noted that compounds from the areal parts of *Eclipta prostata* are effective in counteracting envenomation by *Agkistrodon contortrix*, and *Crotalus viridis*; and Kuppusamy and Das (1993) reported that tannic acid (gallotannin) inhibits hyaluronidase activity and is able to neutralize hemorrhaging in mice envenomated by *Crotalus adamanteus*.

However, whether or not the above listed plants are truly effective in controlling snake envenomations is still to be determined in controlled laboratory studies. If at least some of them are, they may provide an alternate way of relieving envenomation symptoms in areas of the world where antivenom and hospitals are not readily available.

Elsewhere some plants have also been reported to be very effective in counteracting the venoms of some of the world's deadliest snakes. In Thailand, a concoction made from the ground root of turmeric (*Curcima longa*) and alcohol is ingested, and apparently produces remarkable resilience to the venom of the king cobra (*Ophiophagus hannah*) and other neurotoxic snakes. Eric Lattman (*in* Backshall 2008) conducted experiments on the plant's ability to neutralize venoms, and found that it was almost

100% effective against neurotoxins, and certainly more effective than conventional antivenoms. Studies in Nigeria by Asuzu (*in* Raloff 2005) indicate that the bark of the African locust bean tree (*Parkia biglobosa*) when mixed with parts of four unidentified shrubs markedly reduces the effects on laboratory animals of venoms from spitting cobra (*Naja nigricollis*), puff adder (*Bitis arietans*), and West African saw-scaled viper (*Echis ocellatus*). In addition, Soares et al. (2005) listed 79 plants, mostly South American, used to treat snake bites (which see for details).

Conservation of Venomous North American Reptiles

There is a consensus among conservationists and herpetologists that venomous snakes and the two helodermatid lizards, like other North American reptiles, have declined in numbers during the last century. This is difficult to determine accurately, however, as for most species there are no "before" and "after" data on population sizes, especially for those occurring in Mexico. Two other factors that make it difficult to assess population status are behavioral secretiveness and inhospitable and hard-to-reach habitats. Most venomous North American reptiles are difficult to collect in meaningful numbers for mark-recapture studies. Those with the best data are the denning rattlesnakes *Crotalus horridus, C. oreganus,* and *C. viridis*. Nevertheless, these creatures are having a more difficult time surviving in the human modern world. Dodd (1987, 1993) and Greene and Campbell (1992) offer excellent discussions of the need for conservation and/or methods to improve the lot of venomous reptiles.

WHY ARE OUR VENOMOUS REPTILES IN DECLINE?

The problems venomous reptiles face today are the same as those affecting all wild animals. Fortunately, no North American venomous reptile has become extinct in historical times. Unfortunately, however, with the ever-increasing encroachment of humans into wild habitats, many parts of North America have become very dangerous for these animals, and some probably face extinction if conservation methods are not soon applied. The most devastating impacts are habitat loss or alteration, including death on our roads, environmental pollution, and the introduction of invasive species; wanton killing; overcollection for various reasons, including the pet trade and the so-called roundups; and the loss of genetic variability. Venomous reptiles do suffer from parasites and pathogens (see species accounts), but little data concerning the prevalence of disease are available for wild populations. Possibly, global climate change will affect them in the future.

A reptile's geographic distribution and habitat choice may expose it to a greater chance of population decline. Species are usually characterized as either of special concern, threatened, or endangered. However, Mace and Lande (1991) proposed a new system of threats based on species population dynamics and realistic time scales for conservation plans to work. They characterized taxon status as either critical, with 50% probability of extinction within 5 years or 2 generations, whichever is shorter; endangered, 20% probability of extinction within 20 years or 10 generations, whichever is longer; or vulnerable, 10% probability of extinction within 100 years. Using these criteria, Greene and Campbell (1992) identified pitvipers as at least vulnerable to extinction if they have small distributional ranges, have specialized habitats, and/or are subject to high, human-caused mortality over a substantial part of their distribution.

Occurrence on more than one of the following lists indicates an increased vulnerability over those taxa listed only once: (1) North American venomous reptiles with restricted or small distributions and/or highly fragmented ranges (Campbell and Lamar 2004; Conant and Collins 1998, Ernst and Ernst 2003, Stebbins 2003); (2) those with restricted ranges, including endemic island species: *Agkistrodon bilineatus, A. taylori, Crotalus catalinensis, C. enyo cerralvensis, C. e. furvus, C. helleri caliginis, C. mitchellii angelensis, C. m. muertensis, C. molossus estabanensis, C. ruber exsul, C. ruber lorenzoensis, C. tortugensis, C. willardi amabilis, C. w. meridionalis, C. w. obscurus, C. w. silus*, and *C. w. willardi;* (3) North American venomous taxa with small and/or highly fragmented distributions (including some endemic island species): *Agkistrodon contortrix pictigaster, Crotalus enyo cerralvensis, C. e. furvus, C. helleri caliginis, C. lepidus maculosus, C. l. morulus, C. mitchellii angelensis, C. m. muertensis, C. molossus estabanensis, C. m. oaxacus, C. oreganus abyssus, C. o. cerberus, C. o. concolor, C. ruber lorenzoensis, C. r. lucasensis, C. scutulatus salvini, C. tigris, C. totonacus, C. viridis nuntius*, and the five subspecies of *C. willardi;* (4) North American taxa currently experiencing habitat conversion or living in specialized vulnerable habitats: *Agkistrodon piscivorus conanti, A. p. piscivorus, Heloderma horridum. H. suspectum, Crotalus adamanteus, C. atrox, C. cerastes cerastes, C. lepidus klauberi, C. mitchellii pyrrhus, C. molossus, C. oreganus oreganus, C. pricei, C. scutulatus, C. stephensi, C. tigris, C. virdis viridis, C. willardi* (southeastern Arizona), *Sistrurus catenatus* (all subspecies), *S. miliarius* (all subspecies), *Micruroides euryxanthus, Micrurus fulvius*, and *M. tener;* and (5) those venomous taxa that have experienced abnormally high mortality during the last century: *Crotalus adamanteus, C. cerastes, C. horridus, C. oreganus lutosus, C. o. oreganus, C. v. viridis*, and *C. willardi*. From this list, and those above, it can be seen that almost all venomous North American reptiles have been, or are being, adversely affected and are vulnerable. The one possible exception is the seasnake *Pelamis platura*, but if the current rate of pollution of our oceans continues and its prey species are adversely affected, it too could face great population declines.

WHAT ARE THE MAJOR THREATS TO NORTH AMERICAN VENOMOUS REPTILE POPULATIONS?

Habitat Destruction or Alteration

Critical North American habitat is being destroyed or made inhospitable for our reptiles every day. All of this is due to the increasing use of wild areas by humans, and is not a new occurrence. The original European settlers cut down the eastern forests, altered waterways and drained wetlands, and plowed or burned the prairie grasslands and eradicated the American bison (*Bison bison*) and various prairie dog (*Cynomys* sp.) towns, thus further altering the natural ecosystem. These are century-old, recognized examples of habitat destruction on our continent that originally affected such snakes as *Agkistrodon contortrix, A. piscivorus, C. horridus, C. oreganus, C. viridis*, and *Sistrurus catenatus*.

Modern-day lumbering practices have adversely impacted *Crotalus adamanteus, C. oreganus, C. willardi*, and *S. miliarius miliarius* (especially in Hyde and other North Carolina counties). Removal of trees not only directly alters the preferred habitat of woodland venomous snakes, but indirectly affects them by eliminating critical habitat of their major prey. In addition to logging, mining has become a problem for highland populations of *C. lepidus*, *C. pricei*, and *C. willardi* in southeastern Arizona (Johnson and Mills 1982), and for *A. contortrix* and *C. horridus* in the Appalachian Mountains.

Little good can be said of both the changing of waterways and wetlands and the destruction of grasslands. In Florida, the Southwest, and parts of the Midwest and Pacific states, and on islands, overgrazing by introduced domestic herds of cattle, goats, and sheep has allowed the invasion of nonnative plant species, often cacti, that have made the original habitat inhospitable for some prey species and reptiles alike.

In the southern states, introduction of predatory fire ants (*Solenopsis* sp.) has had an adverse effect on small reptiles, especially through destruction of their eggs, and may impact the oviparous coralsnakes *Micrurus fulvius* (Mount 1981) and *M. tener* in this way. Introduction of goats and pigs onto the islands of endemic Mexican rattlesnakes has destroyed the vegetation and threatened those snakes.

Particularly in the Southwest, suburban sprawl around such metropolitan areas as Los Angeles, San Diego, Las Vegas, Phoenix, and Tucson has destroyed critical habitats for several rattlesnakes and the Gila monster. The critical habitat of *C. tigris* west of Tucson is a good example of this impact (Goode and Smith 2005), and the loss of habitat due to urban development is the primary threat to *C. horridus* in southeastern Virginia (Mitchell and Schwab 1991). The same is true in southern Florida, where the human population has greatly increased in recent decades (Wilson and Porras 1983). Do we really need more housing developments and strip malls?

Urban development has further dissected major reptile habitats with newly constructed roads. More roads crossing critical reptile habitat mean more animals killed by motorized traffic. Snakes like roads! They are easier to travel along or use as critical pathways, and at night are warmer, allowing the animal to heat itself. All herpetologists and naturalists know that one of the best ways of collecting these animals is by nighttime road-cruising.

Whenever we traveled to places with snake populations, we took collecting permits, snake-bags, collecting jars, and a supply of formalin to process the reptiles collected, especially on the roads at night, for the senior author's teaching collection. We were often astonished by the numbers of snakes dead on the roads, and believe this to be a major factor in the decline of reptile populations near urban centers and especially at newly dissected, critical, snake habitats.

This problem is not restricted to the Southwest. Venomous snake populations have been adversely affected in Florida and throughout the Southeast, in the Appalachians, and in many areas of the Midwest and West through roadkills (Brown 1993; Dodd et al. 1989; Gibson and Merkle 2004; Jochimsen and Peterson 2004; Mitchell and Schwab 1991; Reinert 1985; Shepard et al. 2008a; Timmerman and Martin 2003; also see the individual species accounts). We could always count on finding some DOR *C. adamanteus* on the roads in Collier County or *A. piscivorus conanti* along the Tamiami Trail in southern Florida, and have seen many dead *A. c. mokasen* and *A. p. leucostoma* on western Kentucky and southern Illinois roads, and *C. v. viridis* on those in the Dakotas and eastern Wyoming. Roadkills are also a problem in parts of Mexico (Lazcano et al. 2009).

Unfortunately, with future incursion of human populations into major venomous reptile habitats will come a corresponding increase in motorized traffic on the roads there and the deaths of more of these animals.

The effects of environmental pollution on North American venomous reptile populations have only been reported for a few species, but negatively impact the survival of other species as well. The impacts may be direct, like poisoning them through skin absorption or drinking contaminated water; but are usually indirect, like the accidental poisoning of individuals through carrion ingestion of rodents previously killed by eating strychnine baits (Campbell 1953a). Pesticides, like DDD, DDE, and DDT (Hall 1980; Stafford et al. 1976), heavy metal pollutants (see the discussion in the account of *A. piscivorus*), or radioactive pollutants (by *C. oreganus* living near nuclear plants; Arthur and Janke 1986) may also be ingested with prey.

Wanton Killing

The killing of snakes on discovery, particularly venomous ones, has always taken place and is a major threat. Humans do not want to live near or with venomous reptiles. This fear, particularly of snakes, probably had its start, in large part, because of the Adam and Eve story in the Bible, where the snake is portrayed as evil. Granted, some North American snakes are very dangerous, but the chance of the average human encountering one is remote. From early on, settlers in the United States raided and killed those species whose numbers congregated at winter dens (see discussions under *C. horridus, C. oreganus*, and *C. viridis*). The pouring of gasoline into gopher tortoise burrows to drive out *C. adamanteus* (prohibited in Florida and Georgia) is very destructive to the snake and other wildlife (Speake and McGlincy 1981; Speake and Mount 1973). In the past, several states, or counties within them, have even offered bounties on various venomous species (Brown 1993; Dodd 1987; Furman 2007; Galligan and Dunson 1979; Klauber 1972).

Overcollection

Venomous reptiles are collected in significant numbers for various reasons; unfortunately, this is one of the most devastating impacts on their populations. All taking of individuals, especially that of adult females, has a deleterious effect on a population's breeding potential.

Few good reasons exist for removing these reptiles from the wild, but the increasing role of their venoms in medicine production and medical research is one. Many are collected to provide venom for the production of antivenoms and for legitimate medical research. It is now known that some enzymes contained in the venom of various venomous snakes and the helodermatid lizards can be used in the treatment of diseases and conditions, like type 2 diabetes, blood-clotting disorders, pain, and possibly cancer (see discussion in the chapter "Venom"). The collection of some reptiles for educational purposes, like legitimate zoo displays, is reasonable, but the numbers should be limited. Also, other valuable behavioral, physiological, and other biological research require the collecting of some individuals to provide data useful in survival plans.

Unfortunately, most collection of these animals causes more damage than good. The pet trade, and the trade in leather hides, souvenirs (mostly mounted individuals, heads, or rattles), and canned meat, are especially harmful (see the discussion in the account of *Crotalus adamanteus* and Snyder 1949). The most prized are those that are considered rare, hard to collect, or protected by law (like *C. willardi obscurus*). Many of the animals taken from the wild adjust poorly to captivity, experience short life spans, and linger away until they die of malnutrition or disease (Murphy and Armstrong 1978). Also, they are a danger (see the chapter "Envenomation by North American Reptiles") and should only be kept by persons who are experienced and know how to properly care for them.

One of the most troublesome practices is that of the so-called organized roundups or hunts held in several states, particularly Kansas, Oklahoma, Pennsylvania, and Texas. These usually involve rattlesnakes, mostly *C. atrox, C. horridus*, and *C. viridis* (Campbell et al. 1989; Fitch 1998, 2002; Fitch and Pisani 1993; Galligan and Dunson 1979; Kilmon and Shelton 1981; Painter and Fitzgerald 1998; Pisani and Fitch 1993; Reinert 1990; Schmidt 2002; Speake and Mount 1973), but copperheads, *A. contortrix mokasen,* have also been the species of a hunt in York County, Pennsylvania (C. Ernst, personal observation). Many of the snakes are slaughtered outright for the above trades, but some are supposedly returned to the place where they were collected. Because the snakes are usually dumped all together into large pens, how can it be told for certain which is which and where a particular snake was collected? Those "returned" are usually released in the nearest likely habitat regardless of their original collecting site. Such transferred individuals do not know the area and probably move about searching for homeward cues. In so doing, they are more vulnerable to predation, use up valuable fat stores needed for reproduction or winter survival, and encounter roads on which they may be killed. As stated above, not only do roundup and hunt events deplete the numbers of individuals in populations (see the above reference list for examples of decreasing numbers of snakes collected over consecutive events), but the removal of gravid or reproductive size/age females reduces the population's ability to recover lost numbers.

Loss of Genetic Variability

Some populations of venomous reptiles in Mexico and the United States, like the mountaintop subspecies of *C. willardi* and the endemic island taxa of *Crotalus*, are already naturally vulnerable to loss of genetic variability. Their isolation makes the introduction of new genetic combinations through gene exchange impossible. Once this happens, a genetic "bottleneck" occurs where the genotype slowly moves toward a single one. If no drastic environmental change occurs, such populations are probably pretty safe. However, if the opposite takes place, their new genetic composition may not contain the adaptations (through mutation and/or genetic recombination) needed to cope fast enough with the environmental stress, and the population becomes extinct.

Some other North American populations of venomous reptiles are also experiencing increasing isolation, as their habitats are subdivided by the spread of civilization, and could eventually experience on the local level the same types of genetic problems as endemic ones. This is particularly true in populations of *Agkistrodon bilineatus* in Latin America (Greene and Campbell 1992; Parkinson et al. 2000), the various desert species of *Crotalus* in southern California, and *C. horridus* in the New Jersey Pine Barrens (Bushar et al. 2005).

Global Climate Change

The possibility that global temperature change is occurring and its effects on humans are of much current debate. Although whether or not global climate change is taking place is still uncertain, we should consider the worst-case scenario that it is fact, and err for the good of all species on earth, which includes the venomous reptiles of North America.

The greatest potential danger to venomous reptiles from global climate change is the adverse effects on their habitats caused by the warming of air and soil and changes in the seasonal rainfall patterns that might cause the habitat to be dry for at least part of the year. Such an event would certainly cause changes in vegetation and most likely an overall drying of some regions (could the southwestern deserts and that of the Great Basin move northward?). This in turn may cause a change for the worse in the potential prey species living in the reptiles' habitats.

Our venomous reptiles do not have temperature-dependent sex determination, as occurs in turtles, so their sex ratios should not become skewed toward one sex or the other; however, neonates will be adversely affected by any change in their potential food supply, as will be female pitvipers that must replenish their energy reserves between breeding events. If the species can adapt to these new conditions, change may be for the better but probably will not be for most species.

A warming of the North American continent, however, could allow some species to increase their ranges northward, but at the same time possibly eliminate populations of other species, like *Agkistrodon piscivorus* and *Sistrurus catenatus* (Lee et al. 2005; Patten et al. 2005; Seigel 1986) that depend on wetland habitats that would probably dry up. The endemic subspecies of *Crotalus willardi* would probably also be in trouble as they lose prey variability. It is possible that, if they survive the change, *Crotalus o. oreganus* could eventually invade Alaska, *C. horridus* and *C. viridis* could move farther north in Canada, and some Mexican venomous species could reach the United States.

WHAT ARE POSSIBLE CONSERVATION METHODS?

Following are possible solutions to some of the major threats previously discussed.

Education

Educating the public about the natural role of our venomous reptiles, their life histories, and possible legitimate uses is the key to conserving them (Carmichael 2008; Dodd 1987, 1993; Sasaki et al. 2008; Snider et al. 2005). It should cause a lessening of malicious killing.

It is hard to convince the public that venomous creatures should also be conserved like birds and bunnies, and this is possibly the major hurdle to overcome if our venomous reptiles are to survive into the future (note how few species of venomous reptiles are protected under the U.S. Endangered Species Act, table 4.1). All animals have their important roles to play in our ecosystems, and the public must be made aware that venomous reptiles have such roles (medicine, rodent control, etc.), and that they are not as dangerous in the wild as supposed (see the chapter "Envenomation by North

American Reptiles"). Such education is best aimed at the young, as adults are pretty much set in their beliefs regarding venomous creatures (Dodd 1993). Federal and state conservation agencies, biologists in our colleges and universities, zoos, conservation societies, and naturalists must lead the way in educating the public to the plight of all animals. When species' survival plans are developed for venomous reptiles, the role of education must be stressed.

Habitat Preservation

Critical habitats of declining species must be preserved by law if necessary, but certainly by private citizens with such sites on their properties. Land use practices must be examined and altered if needed. Any protection must include habitats used in all of the seasons, especially hibernacula protection for reptiles that communally den during the winter (*C. horridus, C. oreganus, C. viridis*) or critical feeding areas like the wetlands inhabited during the summer by *S. catenatus* (Seigel 1986). The creation of protected reserves or conservation easements is helpful in such cases. Of course, these all involve public or private funding that is not always available. Each specific site has its own characteristics and requires a different suite of solutions. An across-the-distributional-range plan may not work everywhere (one plan does not necessarily "fit" all). The plan must be flexible enough to be modified to match the needs at a specific locality. No colony or population is too small to disregard.

The spread of suburbia must also be considered. All developmental encroachment into critical habitat should require an environmental impact study that includes venomous reptiles before permission to proceed is granted. The construction of roadways through or alongside critical areas should be kept at a minimum. A possible solution to the roadkill problem is the construction of culverts and underpasses beneath roads (snakes in northern Virginia are known to use these; C. Ernst, personal observation). Another possible solution is to construct baffles or barriers that prevent snakes or helodermatids from crawling onto roads (Dodd 1993; Southall 1991). These cost money and should be included in the original road construction plans; however, will the public be willing to fund such measures to conserve venomous reptiles?

Critical habitat must be preserved and invasive species, like introduced domestic goats and pigs that kill and eat venomous reptiles, must be removed or strictly controlled on the Mexican islands of endemic rattlesnakes.

Control of Overcollecting

National and state/province protection (tables 4.1 and 4.2) and required collecting permits are a great help in controlling overcollecting for the pet and commercial trades. This protection should include limits on the numbers of individuals to be taken and severe penalties for offenders. A national restriction of venomous reptiles in the pet trade should also be considered (this would greatly reduce the number of annual envenomations; see the chapter "Envenomation by North American Reptiles"); some states do this now, but others need to join the movement. Organized roundups and hunts should be prohibited or severely monitored and controlled by both federal and state authorities.

TABLE 4.1. PROTECTED AND THREATENED STATUS OF VENOMOUS REPTILES OF THE UNITED STATES

Family and Taxon	ESA	States	IUCN	CITES Appendix	Threats
Helodermatidae					
Heloderma suspectum		AZ (L), NM (E), NV (P)	NT	II	1, 2, 3, 4, 5
Heloderma s. cinctum		CA (P), UT (T)	NT (under *Heloderma suspectum*)		7 (CA)
Viperidae					
Agkistrodon contortrix		IA (E), MA (E)	LC		1, 2, 4, 5, 7
Agkistrodon piscivorus		IN (T)	LC		1, 2, 4, 5
Crotalus adamanteus		FL (L), MS (L), NC (L)	LC		1, 2, 3, 4, 5
Crotalus atrox		AR (SC), AZ (L), TX (L)	LC		1, 2, 3, 4, 5
Crotalus c. cerastes		UT (SC)	LC (under *Crotalus cerastes*)		
Crotalus helleri		CA (L)	LC (under *Crotalus oreganus*)		1, 2, 5
Crotalus horridus		CT (E), IL (T), IN (T), KS (N/C), MA (E), MD (L), MN (SC), NH (E), NJ (E), NY (T), OH (E), PA (C), RI (L), TX (T), VT (E)	LC		1, 2, 4, 7, 9
Crotalus horridus (*atricaudatus* morph)		VA (E)	LC (under *Crotalus horridus*)		1, 2, 5
Crotalus lepidus		AZ (L)	LC		1, 4
Crotalus l. lepidus		NM (E)	LC (under *Crotalus lepidus*)		1, 4
Crotalus mitchelli		UT (SC)	LC		1
Crotalus oreganus		CO (E), UT (L)	LC		1, 2, 4, 5
Crotalus pricei		AZ (L)	LC		1, 2, 4, 5
Crotalus ruber		CA	LC		1, 2, 4, 5
Crotalus scutulatus		CA, UT (SC)	LC		1, 2, 4, 5
Crotalus tigris		AZ (L)	LC		1, 2, 3, 5
Crotalus viridis		IA (E)	LC		1, 2, 4, 5, 7
Crotalus willardi		AZ (T)	LC		1, 4, 5, 6
Crotalus w. obscurus	T	AZ (T), NM (E)	LC (under *Crotalus willardi*)		1, 4, 5, 6
Sistrurus catenatus		AZ (E), CO (E), IA (E), IL (E), IN (T), MN (SC), NY (E), PA (E), WI (E)	LC		1, 2, 4, 5, 7
Sistrurus c. catenatus		MI (SC), MO (E)	LC (under *Sistrurus catenatus*)		1, 2, 4, 5, 7
Sistrurus c. tergeminus		MO (E)	LC (under *Sistrurus catenatus*)		1, 2, 4, 5, 7
Sistrurus miliarius		KY (T)	LC		1, 2, 4, 5
Sistrurus m. miliarius (red phase)		NC (L)	LC (under *Sistrurus miliarius*)		1, 2, 4, 5
Sistrurus m. streckeri		TN (T)	LC (under *Sistrurus miliarius*)		1, 2, 4, 5

TABLE 4.1. PROTECTED AND THREATENED STATUS OF VENOMOUS REPTILES OF THE UNITED STATES *(continued)*

Family and Taxon	ESA	States	IUCN	CITES Appendix	Threats
Elapidae					
Micrurus fulvius		AL (L), FL (L), LA (L), MS (L), NC (L), SC (L)	LC		1, 2, 5
Micrurus tener		AR (SC), LA (L), TX (L)	LC		1, 2, 5

Key to Abbreviations. Status: C, candidate for listing; CE, critically endangered; E, endangered; EX, extirpated; L, permit/license; LC, least concern; NT, near threatened; P, protected, but not listed; SC, special concern; SP, special protection; T, threatened. Threats: 1, habitat alteration or destruction; 2, roadkills; 3, leather or meat trade; 4, overcollecting for either the pet or souvenir trade, scientific biological research, or venom research; 5, wanton killing; 6, loss of genetic variability; 7, rare or uncommon; 8, specific threat undetermined; 9, extirpated.

Although not all species do well in captivity, in the future permitted captive breeding may help alleviate the need to take animals from the wild.

Survival Plans and the Need for More Life History Data

Conservation models (Halama et al. 2008; Hamilton 2005; Jenkins and Peterson 2008; Martin et al. 2008; Schofer 2005) and plans (Anton 2004; Possardt et al. 2005) are only as good as the data entered. Unfortunately, due to the secretive nature of most North American venomous reptiles, not enough data concerning critical aspects of their biology are known. This is particularly evident for some of the Mexican snakes, where most information is anecdotal. Fortunately, more field studies have been funded in recent years, but we also need reproductive (Aldridge1975, 1979a, 1979b, 2002; Aldridge and Brown 1995; Aldridge and Duvall 2002; Aldridge et al. 2008), physiological (Beaupre 1993, 1995, 1996, 2008; Beaupre and Duvall 1998a, 1998b; Beaupre and Zaidan 2001), and behavioral studies to provide additional critical data; some of these can more profitably be carried out in the laboratory. The use of landscape modeling and GIS (geographic information system) data is useful in determining habitat needs.

Should We Move Them?

Relocation of "nuisance" individuals from places of possible encounter with humans and those where habitat is to be destroyed has been tried (Brown et al. 2008; McCrystal and Ivanyi 2008; Nowak 1998, 2005; Nowak et al. 2002; Parsons and Sarell 2005), but with limited success. Although the reptiles are moved away from humans, transportation of venomous snakes or lizards out of their home ranges to other distant areas puts them in danger (see previous discussion under "Overcollecting"), and should only be used as a last resort. Either way, the animals will probably survive only a relatively short period in either environment, so the success rate is low.

TABLE 4.2. PROTECTED AND THREATENED STATUS OF VENOMOUS REPTILES OF CANADA AND MEXICO

Family and Taxon	National	Provinces/States	IUCN	CITES Appendix	Threats
		CANADA			
Viperidae					
Crotalus horridus	EX	Ontario (E)	LC		1, 2, 5, 9
Crotalus oreganus		British Columbia	LC		1, 2, 4, 5
Crotalus viridis viridis		Alberta (P) Saskatchewan	LC (under *Crotalus viridis*)		1, 2, 5
Sistrurus catenatus	T	Ontario			1, 2, 5
		MEXICO			
Helodermatidae					
Heloderma horridum	T	Entire Mexican range	LC	II	1, 2, 3, 4, 5
Heloderma suspectum	T	Sonora, Sinaloa	NT	II	1, 2, 3, 4, 5
Viperidae					
Agkistrodon b. bilineatus	SP	Entire Mexican range	NT		1, 2, 5
Agkistrodon contortrix		Chihuahua, Coahuila	LC		1, 2, 4, 5
Agkistrodon taylori	T	Entire Mexican range	LC		1, 2, 4, 5
Crotalus atrox	SP	Entire Mexican range	LC		1, 2, 3, 4, 5
Crotalus basiliscus	SP	Entire Mexican range	LC		1, 2, 5
Crotalus catalinensis	T	Isla Santa Catalina	CE		1, 4
Crotalus cerastes	SP	Baja California Norte, Sonora, Isla Tiburon	LC		1, 2, 4, 5
Crotalus enyo	T	Entire Mexican range	LC		1, 2, 5
Crotalus helleri		Entire Mexican range	LC		1, 2, 5
Crotalus h. caliginis		Isla Coronado	LC		1, 4, 5, 6
Crotalus lepidus	SP	Entire Mexican range	LC		1, 2, 4, 5
Crotalus mitchelli	SP	Entire Mexican range	LC		1, 2, 4, 5
Crotalus m. angelensis		Isla Angel de la Guarda	LC		1, 4, 5, 6
Crotalus m. muertensis		Isla El Muerto	LC		1, 4, 5, 6
Crotalus molossus	SP	Entire Mexican range	LC		1, 2, 4, 5
Crotalus m. estebanensis		Isla San Esteban	LC		1, 4, 5, 6
Crotalus pricei	SP	Entire Mexican range	LC		1, 2, 4, 5
Crotalus ruber	SP	Entire Mexican range	LC		1, 2, 3, 4, 5
Crotalus r. exsul	T	Isla Cedros	LC (under *Crotalus ruber*)		1, 4, 5, 6
Crotalus r. lorenzoensis		Isla San Lorenzo de Sur	LC (under *Crotalus ruber*)		1, 4, 5, 6
Crotalus scutulatus	SP	Entire Mexican range	LC		1, 2, 4, 5
Crotalus tigris	SP	Sonora, Isla Tiburon	LC		1, 2, 3, 5
Crotalus tortugensis	SP	Isla Tortuga	LC		1, 6
Crotalus totonacus (under *Crotalus durissus*)	SP	Entire Mexican range			1, 2, 5
Crotalus viridis	SP	Chihuahua, Coahuila	LC		1, 2, 5
Crotalus willardi	SP	Entire Mexican range	LC		1, 4, 5, 6
Sistrurus catenatus	SP	Entire Mexican range	LC		1, 2, 4, 5

TABLE 4.2. PROTECTED AND THREATENED STATUS OF VENOMOUS REPTILES OF CANADA AND MEXICO *(continued)*

Family and Taxon	National	Provinces/States	IUCN	CITES Appendix	Threats
Elapidae					
Micruroides euryxanthus	T	Chihuahua, Sonora, Sinaloa, Isla Tiburon	LC		Specific threats unknown, but probably 1, 2, 4, 5
Micrurus distans	SP	Entire Mexican range	LC		1, 2, 5
Micrurus tener (under *Micrurus fulvius*)	SP	Entire Mexican range	LC		1, 2, 5

Key to Abbreviations. Status: C, candidate for listing; CE, critically endangered; E, endangered; EX, extirpated; L, permit/license; LC, least concern; NT, near threatened; P, protected, but not listed; SC, special concern; SP, special protection; T, threatened. Threats: 1, habitat alteration or destruction; 2, roadkills; 3, leather or meat trade; 4, overcollecting for either the pet or souvenir trade, scientific biological research, or venom research; 5, wanton killing; 6, loss of genetic variability; 7, rare or uncommon; 8, specific threat undetermined; 9, extirpated.

Gaining Human Cooperation Is the Answer

All of the above possible solutions depend on the public recognizing the need for, and practicing, conservation of North America's venomous reptiles. This will not be easy, and at some localities may be extremely difficult due to learned prejudices toward these animals. But we must not lose hope; it is critical that we continue in our efforts to increase public awareness and appreciation of these creatures.

Identification of the Venomous Reptiles of Canada, the United States, and Northern Mexico

The identification of the individual species of lizard or snake presented in this book is based on a series of dichotomous keys. To identify a venomous reptile, the reader should refer first to the following "Key to the Families of North American Venomous Reptiles." After comparing it to the characters expressed in the key, the family to which it belongs should be evident. Next, go to that family account in the text where a second key to the genera belonging to that family is presented. Use this key to determine the proper genus for your reptile, and then go to the account of that genus in the text. There you will find a third key to identify the various species belonging to that genus (if the genus is monotypic, having only one species assigned to it, only the generic key is needed). Once the species has been determined, its description (see "Recognition") and color photograph(s) should be compared with your animal to finalize the identification. Variations in head scales of elapids and pitvipers are illustrated in figures 2–4; figure 5 illustrates the ventral head scales of snakes.

In the text three standard measurements are used: total body length (TBL), the distance between the most anterior point of the snout and the posterior tip of the tail (to the base of the rattle in rattlesnakes); snout-vent length (SVL), the distance from the most anterior tip of the snout to the posterior border of the vent (because many snakes have lost a portion of their tails for various reasons, this is a more useful measurement than TBL); and tail length (TL), the distance from the posterior border of the vent to the posterior tip of the tail.

Reptiles' integument is covered with dry, keratinous scales. Those on a snake's head may be enlarged (*Agkistrodon, Micruroides, Micrurus, Pelamis, Sistrurus*), consist only of

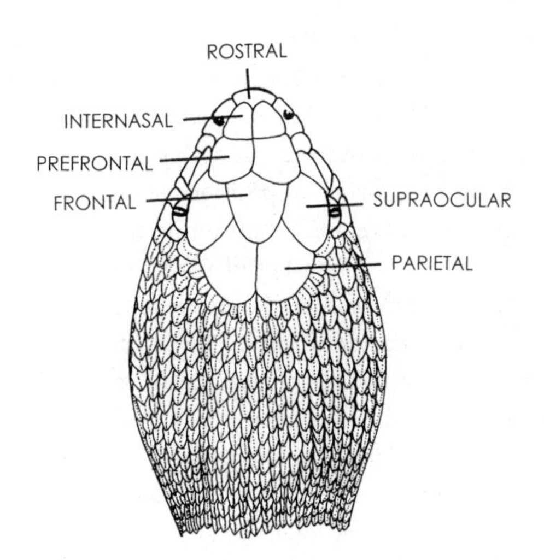

Figure 2. Head scalation of *Micruroides, Micrurus, Agkistrodon,* and *Sistrurus*, dorsal view.

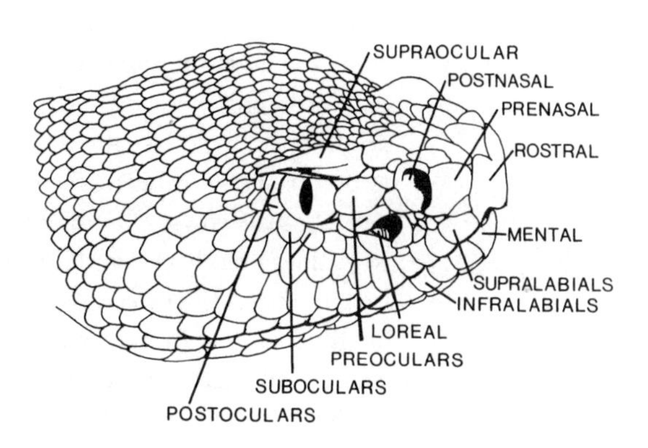

Figure 3. Head scalation of *Crotalus*, dorsal view.

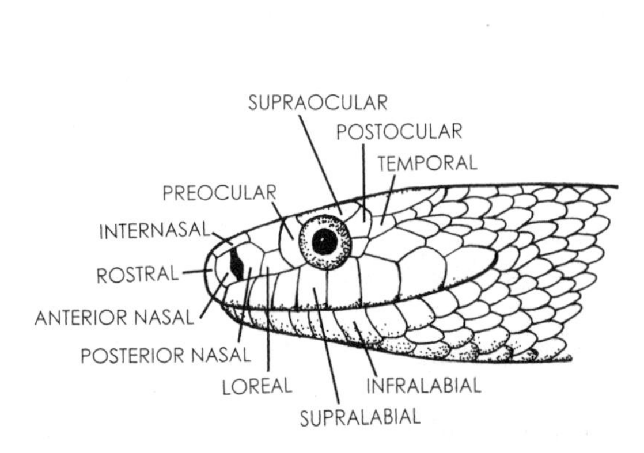

Figure 4. Snake head scalation, lateral view.

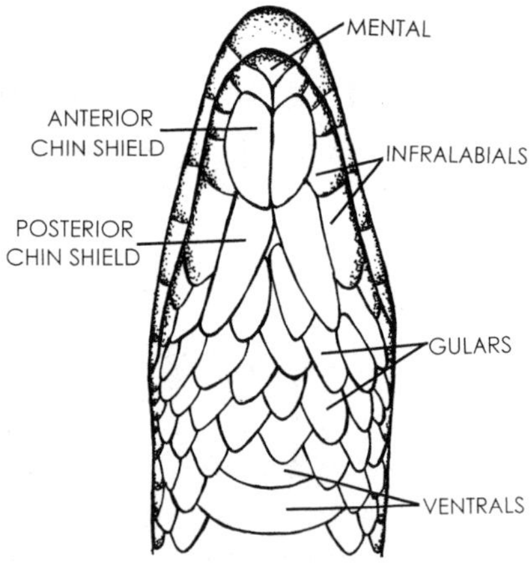

Figure 5. Snake head scalation, ventral view.

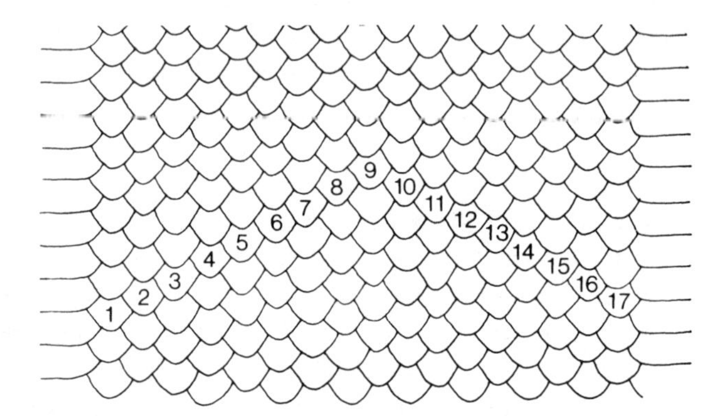

Figure 6. Counting snake scale rows.

small scales (*Crotalus*), or may be bead-like (*Heloderma*) (fig. 7); the terminal scales on a snake's tail may be modified into a rattle (*Crotalus, Sistrurus*). Dorsal body scales may be smooth (*Micruroides, Micrurus, Pelamis*), have a longitudinal keel or carinate appearance (*Agkistrodon, Crotalus, Sistrurus*), or may be bead-like (*Heloderma*). The scales may also be either pitless (*Heloderma, Micruroides, Micrurus, Pelamis*) or contain a pair of apical pits on their surface (*Agkistrodon, Crotalus, Sistrurus*). The scales occur in rows along the body (fig. 6), which vary in number at particular positions along the body (anterior, midbody, posterior just before the anal region), and can be used as taxonomic characters. Usually there is a reduction in the number of rows from anterior to posterior; for instance, the copperhead, *Agkistrodon contortrix*, usually has 23 or 25 anterior scale rows, 23 at midbody, and 19 or 21 posterior rows. The number of scale rows does not change with growth (fig. 6). On the belly of nonmarine snakes are a longitudinal series of enlarged, rectangular ventral plates (scutes) arranged in a single longitudinal row from throat to the anal vent (divided by a midventral groove in *Pelamis*); an enlarged flap, the anal plate, either undivided (*Agkistrodon, Crotalus, Sistrurus*) or divided (*Micruroides, Micrurus, Pelamis*), covering the anal opening (vent); and one or two adjacent rows of small subcaudal scales posterior to the anal vent. Counts of ventrals and subcaudals, and whether or not the anal plate is single or divided, are

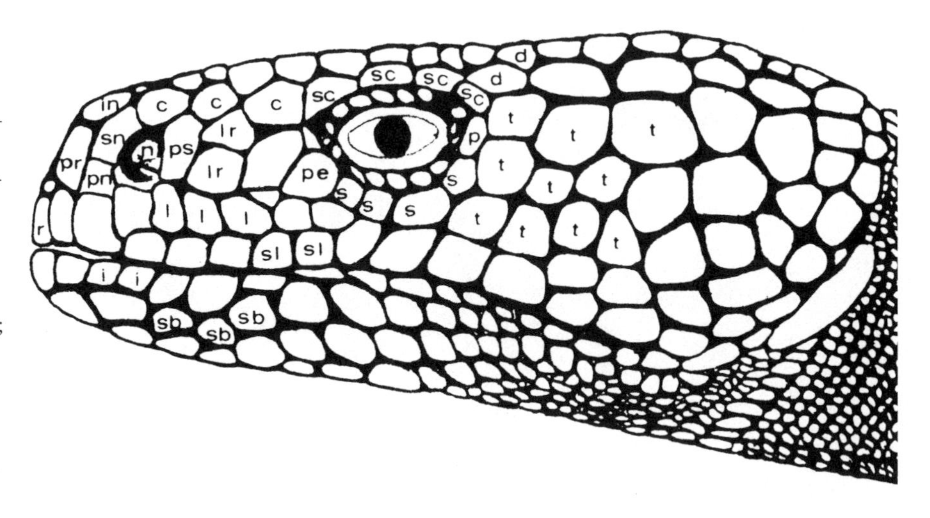

Figure 7. Head scalation of *Heloderma*, lateral view. Scale legend: *c*, canthal; *d*, dorsal; *i*, infralabial; *in*, internasal; *l*, lorilabial; *lr*, loreal; *n*, naris; *p*, postocular; *pe*, preocular; *pn*, prenasal; *pr*, prerostral; *ps*, postnasal; *r*, rostral; *s*, subocular; *sb*, sublabial; *sc*, superciliary; *sl*, supralabial; *sn*, supranasal; and *t*, temporal.

major taxonomic and sexual characters in snakes. *Heloderma* lizards have rows of squarish scales crossing their venters. They lack an anal plate, but instead have either a pair of enlarged preanal scales (*H. suspectum*) or several small ones (*H. horridum*) just before the anal vent. Of course, the lizards have legs, which are lacking in the snakes.

Another useful tool in identification of a venomous reptile is its dental formula, the number of teeth situated on the bones of one half of the skull. A dental formula is presented in the description of each species.

KEY TO THE FAMILIES OF NORTH AMERICAN VENOMOUS REPTILES

1a. Legs are present, the dorsal scales are bead-like projections, the ventral scales are squarish or rectangular and occur in transverse rows . . . . Helodermatidae

1b. No legs present, the dorsal body scales are not bead-like, the ventral scales are elongated plates 2

2a. Tail is laterally compressed and oar-like, the ventral scales are divided by a midventral groove. . . . . .Elapidae (in part)

2b. Tail is not laterally compressed or oar-like, the ventral scales are not divided by a midventral groove 3

3a. Tail ends in a rattle or one or two enlarged dry scales. . . . . Viperidae

3b. Tail does not end in a rattle or one or two enlarged dry scales. 4

4a. A pit-like hole is present on the face in front of the eye . . . . Viperidae

4b. No pit is present on the face in front of the eye . . . . Elapidae

Helodermatidae
Beaded Lizards and Gila Monsters
Escorpiónes

This family is most closely related to the that of Old World monitor lizards, Varanidae, which includes the lanthanotines. The 2 families share almost 40 morphological characters (Bernstein 1999; Pregill et al. 1986).

Helodermatid lizards belong to the clade Platynota, which traditionally includes the Varanoidea (*Varanus, Heloderma, Lanthanotus*) and the fossil genera in the Moasauroidea (Beck 2004a, 2005, 2009; Norell and Keqin 1997; Pregill et al. 1986). *Primaderma nessovi* Nydam, 2000, from the Cretaceous Albian-Cenomanian of Utah, is the oldest known montersaurian fossil. Closely related to it is *Etesia mongoliensis* Norell, McKenna and Novacek, 1992, from the late Cretaceous of Mongolia, so the Platynota may have originated in Asia. Fossils assignable to Helodermatidae first appear in late Cretaceous deposits in Montana and Wyoming in the United States and in Alberta and Saskatchewan in Canada (Keqin and Fox 1996). The late Cretaceous species, *Paraderma bogerti* Estes, 1964, from eastern Wyoming, is probably the ancestor of both extant North American helodermatid species (Pregill et al. 1986). Another early Cretaceous *Heloderma*-like platynotan, most closely comparable to *Paraderma*, has been found in Utah (Cifelli and Nydam 1995). Other important later fossil species are *Eurheloderma gallicum* Hoffstetter, 1957, from the late Eocene or early Oligocene of France and the Miocene of Florida (Bhullar and Smith 2008) (the genus may have also been present in the late Paleocene of Wyoming; Pregill et al. 1986); *Lowesaurus matthewi* (Gilmore, 1928), from the middle Oligocene of Colorado and late Oligocene or early Miocene of Nebraska; and *Heloderma texana* Stevens, 1977, from the Miocene of West Texas. We refer the reader to Yatkola (1976) for morphological comparisons between *Eurheloderma, Heloderma*, and *Lowesaurus*.

The *Heloderma* are relatively large, heavy-bodied lizards with broad heads; short, bluntly rounded snouts; lips that bulge laterally; a protractile, terminally forked tongue; and bead-like scales. A Jacobson's organ is present, as are also ear holes and moveable eyelids. The blunt tail is short and fat and cannot regenerate if severed. The scales have been described by Stewart and Daniel (1975). They are keelless, and rounded on the head, body, and limbs, but square or rectangular in shape on the venter. Most contain an internal, polygonal or subconical, osteoderm (bone deposit). At midbody, the dorsal scales occur in nonoverlapping transverse rows; those on the short, strong limbs may slightly overlap. Preanal scales may be large and paired, or small and several in number; no femoral or preanal pores are present. The skin is shed several times a year; the shedding process is described by Bogert and Martin del Campo (1956). The toes are heavily clawed. A bladder is present (Cope 1900; Davis and De Nardo 2007; Shufeldt 1890), although Gabe and Saint Girons (1965) reported these lizards lack one. Reproduction is oviparous. The electrophoretic protein pattern of the blood serum was described by Zarafonetis and Kalas (1960a).

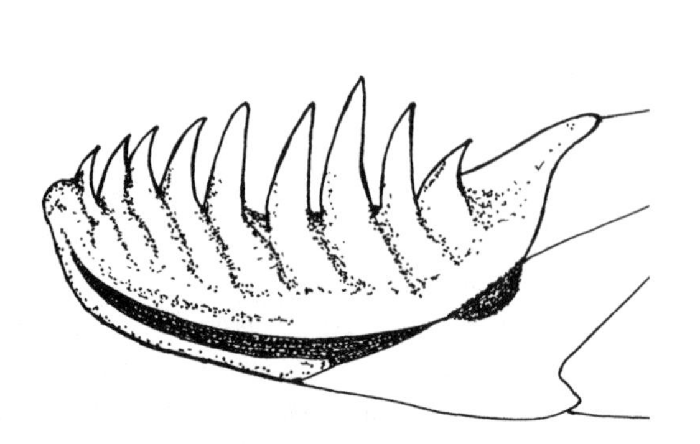

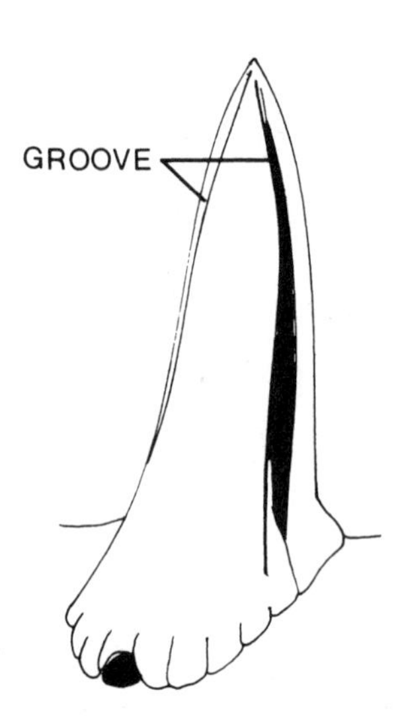

Figure 8. Arrangement of grooved venom delivery teeth on the dentary of helodermatid lizards, and an example of a single tooth.

Osteological characters include a short, broad skull with a well-developed postorbital arch but no supratemporal arch, as the squamosal bone is reduced. The prefrontal and postfrontal bones are in contact in front of the orbit, but do not always prevent the frontal bone from entering the orbit. Lateroventral processes of the paired frontals meet along a suture beneath the olfactory lobe of the brain. The palatine and pterygoid bones are separated, and their teeth are reduced. Only a single parietal bone is present, which lacks either a parietal eye (Gundy and Wurst 1976) or foramen. One premaxilla is present. Nasal and frontal bones are paired, and the narrow vomers are widely separated posteriorly. Subpleurodont teeth are present on the premaxillary, maxillary, pterygoid, palatine, and dentary bones. Vertebrae total 61–76, including 8 cervicals and 30–40 caudals. The shoulder girdle consists of a narrow clavicle, an interclavicle lacking transverse processes, and a longitudinally divided sternum. Other osteological characteristics are discussed in the various papers on fossils mentioned above. The cochlea is described by Schmidt (1964), and the cervical musculature by Herrel and De Vree (1999).

The venom is neurotoxic, and, contrary to that of snakes, the venom delivery system is associated with the lower jaw. The dentary bone bears 7–13 long, recurved, pleurodont teeth (fig. 8) that are grooved at least on the anterior surface, and often on the anterior, medial, and posterior surfaces. The lower infralabial glands are modified to produce venom, which flows from 4 to 9 labial foramina (Bhullar and Smith 2008) up the grooves as the lizard chews. Kwok and Ivanyi (2008) have developed a minimally invasive method for obtaining venom from helodermatid lizards (see for details).

Only two species of *Heloderma* exist on earth today, and both are native to North America: the beaded lizard, *Heloderma horridum*, occurs in western Mexico and southern Guatemala, while the Gila monster, *Heloderma suspectum*, is found in the southwestern United States and northern Mexico.

Heloderma Wiegmann, 1829b
Gila Monsters and Beaded Lizards
Escorpiónes

KEY TO THE SPECIES OF THE GENUS *HELODERMA*

1a. Body pattern yellow or white and black, tongue pink, no enlarged preanal scales, >70 subcaudals, tail <65% of SVL . . . . *horridum*
1b. Body pattern pink or salmon and black, tongue black, 2 enlarged preanal scales, <65 subcaudals, tail <56% of SVL . . . . *suspectum*

Heloderma horridum (Wiegmann, 1829a)
Beaded Lizard
Escorpión, Lagarto Enchaquirado

RECOGNITION: The beaded lizard is large (TBL_{max}, possibly close to 1 m [Campbell and Lamar 2004], but usually adults are 60–70 cm [57–80 cm TBL, 33–47 cm SVL; Beck 2004b]; >70 total vertebrae and about 40 caudal vertebrae are present) and stout, with beaded yellow and black or gray-brown scales and a thick tail. The body is basically black with variable and irregular, scattered, yellow or white spots or small blotches. The pattern of light marks varies between populations (see "Geographic Variation"), and may change and become more complex with age. Some individuals lack light markings. The tail is patterned with 5–7 dark alternating with light bands. The broad head and neck are black. Limbs are mostly black (some lizards may have light markings). The rounded body scales occur in 49–60 rows at midbody. Preanal scales are not enlarged; more than 70 (range, 73–87) subcaudals are present. Dorsal head scales include 1 rostral, 2 (1–4) postrostrals, 1 (2) supranasals, 2 (3) internasals, 3 (2–4) canthals, 3 (1–4) superciliaries (a small series of scales positioned dorsally to the orbit where normally a supraocular would occur), 6–9 dorsal scales lying medially between the posterior superciliaries, and 7–12 total scales lying between internasals and the occiput. Laterally on the head are 1 prenasal, 1 nasal, 1 postnasal, 1–2 (3) loreals, 4–7 lorilabials (lying between the loreals and the supralabials), 2 (1) preoculars, 1 (2) postocular, 2–3 suboculars, 5–6 (4–7) temporals, 11 (10–13) supralabials, 13 (12–15) infralabials (most touch chin shields), and 3–4 rows of sublabials below the infralabials. A shield or wedge-shaped mental scale and 4–8 chin shields are present on the chin. The eyes are black with rounded pupils, the ear opening is narrowly oblique or ovoid, and the tongue is pink with ventral transverse folds or wrinkles. The limbs are stout with heavy claws. Claws are 8–10 mm long, and the length of claw/SVL of the fourth foretoe and fourth hindtoe are, respectively, 2.2–2.7% and 1.7–2.3% (Bogert and Martin del Campo 1956). Dentition consists of 6–11 premaxillary, 6–9 maxillary, 1–7 (8?) pterygoid, 1–5 palatine, and 8–10 dentary teeth (with numbers 4–7 the longest).

The hemipenis is undivided and lacks a terminal horn. The sulcus spermaticus is a simple flap proximally, but forms a more distinct groove distally; the hemipenis twists during eversion so that the sulcus surface is directed dorsally, and the sulcus runs around the proximal surface facing the tail. The sulcus does not reach the apex, but instead ends in a nude pocket on its distal surface. Zones of thickened but soft transverse flounces are present, and a small, domed region of 9–10 transverse flounces occurs at the proximal end of the sulcus. More extensive flounces lie on the rest of the hemipenis and extend onto its apex. Proximally, the flounces are irregular and almost shallowly papillate (Branch 1982).

According to Applegate (1999), males "tend" to have larger, wider heads and narrower bodies than females, but this is purely subjective. Bogert and Martin del Campo (1956) listed the following combination of characters, by subspecies, for use in distinguishing the sexes: TBL_{max} (most adults)—male, 65–77.5 cm; female, 63.5–71.5 cm; SVL_{max} (most adults)—male, 35.5–45.5 cm; female, 36.5–41 cm; mean HW (at eye)/HL—male, 58–61%; female, 61–63%; mean HL/SVL—male, 18–20%; female, 17–18%; mean TL/SVL—male, 76–82%; female, 73–75%; mean hind limb span/SVL—male, 70–76%; female, 65–74%; mean length of fourth finger claw/SVL—male, 2.7–3.0%; female, 2.4–29%; and mean length of fourth toe claw/SVL—male, 2.1–2.4%; female, 1.8–2.1%. Gienger and Beck (2007) reported that male *H. horridum* have significantly longer tails than females, and are larger (SVL) only when the largest individuals were included in the analysis (they thought sexual differences in TL are probably the result of sexual selection acting through male versus male aggressive behaviors; see below). Beck and Lowe (1991) listed the following mean meristic differences between 8 males and 5 females from Jalisco: TBL—male, 67.1 cm; female, 60.9 cm; SVL—male, 38.3 cm; female, 34.6 cm; TL—male, 28.3 cm; female, 26.5 cm; tail circumference—male, 92.5 cm; female, 93 cm; and mass—male, 833 g; female, 667 g. The longer tail of males probably evolved in response to playing a leverage role during their ritualized combat (see below).

A slight bulge ventrally just behind the anal vent in the lateral postanal areas indicates the position of the male hemipenis, and during the breeding season it is sometimes possible to evert the hemipenis by palpating this area. However, it is still very difficult to determine the sex of *H. horridum* using these characters. Probing for the hemipenis is probably the best way to sex the lizard (Laszlo 1975; E. Wagner et al. 1976), but even this is not always successful (Peterson 1982a, 1982b). Morris and Alberts (1996) have used two-dimensional ultrasound imaging to sex the species, and were successful with 7 of the 10 lizards examined (apparently excessive bowel gas prevented correct sexing of the other 3 individuals), but this is impractical for field studies.

GEOGRAPHIC VARIATION: Four subspecies are recognized. *Heloderma horridum horridum* (Wiegmann, 1829a), the Mexican beaded lizard or Lagarto Enchaquirado, is found along the Pacific versant of western Mexico from Sinaloa to Chiapas. In this subspecies, the dorsal body pattern consists of prominent small light spots and blotches, the supranasal scale is separated from the postnasal by the first canthal scale, six or seven scales lie between the superciliaries on the dorsal surface of the head, and the shield-shaped mental scute contains distinct lateral indentations. *Heloderma h. alvarezi* Bogert and Martin del Campo, 1956, the black beaded lizard or *Heloderma* negro, is found only in the Río Grijalva Depression in central Chiapas and extreme western

Heloderma horridum horridum, female (Fort Worth Zoo, John H. Tashjian)

Heloderma horridum horridum (R. D. Bartlett)

Guatemala. Its body pattern is predominantly dark as light markings are few or absent, its supranasal scale is separated from the postnasal by the first canthal, fewer than eight scales lie between the superciliaries on top of the head, and its mental scute is a broad wedge lacking lateral indentations. *Heloderma h. charlesbogerti* Campbell and Vannini, 1988, the Guatemalan beaded lizard, is found in the Río Motagua Valley and adjacent foothills of eastern and southeastern Guatemala. The dorsal body pattern consists of pale light spots on a black ground color and four or five light bands on the tail, the first canthal prevents the supranasal from contacting the postnasal scale, seven or fewer scales lie between the superciliaries on the dorsal surface of the head, the mental scute is shield-shaped and scalloped along its borders, and females may have enlarged pairs of preanal scales (lacking in the other subspecies). *Heloderma h. exasperatum* Bogert and Martin del Campo, 1956, the Río Fuertes beaded lizard, occurs in western Chihuahua, southern Sonora, and northern Sinaloa. It has a more yellow-speckled dorsal body (especially caudally), the supranasal scale in contact with the postnasal, eight scales between the superciliaries on the dorsal surface of the head, and a shield-shaped mental scute with lateral indentations. Excellent color plates of the pattern variations occurring within the four subspecies are in Beck (2005).

Bogert and Martin del Campo (1956) thought a color pattern cline exists between *H. h. horridum* and *H. h. exasperatum* in the northern part of the species' range, and Beck (2005) reported that intergradation occurs between *H. h. horridum* and *H. h. alvarezi* in the area between the Isthmus of Tehuantepec and Cintalapa, Chiapas.

MtDNA base sequence genetic differentiation between the living taxa of *Heloderma* is greater than suspected. The genus exhibits about a 35% sequence divergence from an *Elgaria* outgroup; *H. horridum* is 20% divergent from *H. suspectum*. Within *H. horridum*, populations of *H. h. exasperatum* from the northernmost range are 6.1% divergent from Jalisco *H. h. horridum*, 9.6% divergent from Chiapas *H. h. alvarezi*, and 10.1% divergent from the *H. h. charlesbogerti* of eastern Guatemala. These subspecific differentiations are well beyond the norm, suggesting that the species, as currently defined, may consist of three species instead of one (Douglas et al. 2003).

CONFUSING SPECIES: The only other American lizard with bead-like scales is the Gila monster, *Heloderma suspectum*, which is sympatric with *H. horridum* only at the latter's northernmost distribution. The two species can be separated by using the above key.

KARYOTYPE: The karyotype has not been described, but is probably similar to that of *H. suspectum* (which see for details). Feltoon et al. (2007) isolated six polymorphic DNA microsatellite loci in *H. horridum*. The number of alleles per locus ranged from

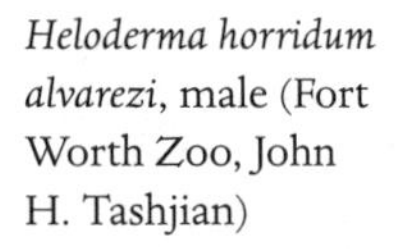

Heloderma horridum alvarezi, male (Fort Worth Zoo, John H. Tashjian)

Distribution of *Heloderma horridum.*

3 to 12, with observed heterozygosity estimates ranging from 0.000 to 0.007 and expected heterozygosity estimates ranging from 0.00 to 0.73.

FOSSIL RECORD: No fossils have been reported.

DISTRIBUTION: The beaded lizard ranges from western Chihuahua (Lemos-Espinal et al. 2003), southern Sonora, and Sinaloa southward along western Pacific Mexico to southern Chiapas and northwestern Guatemala. Its distribution is then interrupted until it is found again in the Departmento de Santa Rosa in southeastern Guatemala. Zoogeography of the species in Middle America is discussed by Savage (1966).

HABITAT: In Mexico, the lizard occurs at elevations of below 100 m to 1,861 m (Monroy-Vilchis et al. 2005), mostly in soft-soiled tropical dry deciduous forests, but it has also been found in pine-oak and scrub as well as thorn thickets and woodlands. It seems to prefer the dry grassy borders of these wooded areas, where it enters hollow, rotting logs or burrows into the plant debris or under the roots of trees. It will enter the riparian habitats of arroyos and rocky slopes in such areas. It avoids the dry cacti habitats on the plains of Tehuantepec, but farther south in Guatemala its habitat is much drier, tropical arid forests. A detailed description of the various habitat types is found in Beck (2005).

Microhabitats used may vary with the season. In Jalisco, during the dry season, most *H. horridum* were observed along gully banks, followed in turn by flat forests and trees (both at the base and arboreal); few were found in rocky crevices. In the wet season, most lizards were again found along gully banks, but with fewer observations; the use of trees increased, as did slightly the utilization of rocky crevices, but observations of flat forests use decreased dramatically (Beck and Lowe 1991).

BEHAVIOR AND ECOLOGY: Beaded lizards are active throughout the year. Most activity in Jalisco takes place from March to November, with a peak in May, possibly in response to the lizard's breeding cycle (Beck 2005), but also with nesting of the spiny-tailed iguana (*Ctenosaura pectinata*), whose eggs are a primary food item. Surface activity drops off sharply in the cool December–February period (Beck and Lowe 1991).

Although diel activity can be either diurnal or nocturnal, *H. horridum* is most likely to be found during the daylight hours. In Jalisco, diurnal surface activity is bimodal, with a strong activity peak between 1600 and 2000 hours and a lesser peak between 0700 and 1000 hours (Beck 2004b; Beck and Lowe 1991). A similar bimodal activity cycle occurs in Michoacán (Hardy and McDiarmid 1969). During Jalisco's wet season, these two activity peaks are more separated than during the dry season, when the lizards become more active earlier in the morning and later in the afternoon. During the hottest times of the day, the lizard seeks shelter. Nocturnal activity there is limited and occurs mostly during the wet season (Beck and Lowe 1991). Elsewhere in Chiapas, the lizard is active toward dusk and shortly thereafter, but not in the morning, and during the wet season it emerges earlier in the day (Álvarez del Toro 1982). Hardy and McDiarmid (1969) found Michoacán *H. horridum* active in the morning, suggesting crepuscular activity. Other records reviewed by Bogert and Martin del Campo (1956) indicate some nocturnal movements, and both Campbell and Lamar (2004) and Smith (1935) reported nocturnal activity.

Beaded lizards try to avoid BTs near 35°C, possibly because their dark bodies absorb heat rapidly as the AT rises, and this helps determine the daily activity cycles mentioned above. No parietal eye is present (Gundy and Wurst 1976), but the animal does display unidirectional monocular optokinetic behavior, during which it orients toward the sun by moving its head and eyes (Ireland and Gans 1977). Whether its occasional basking is, or is not, to warm itself in the morning is questionable, but such optic behavior would help move the lizard into the best position for rapid heat absorption, and could possibly send an alarm of overheating when it is becoming too warm. Beck (2005) has observed basking behavior in January and after summer rain events.

The metabolic rate is also tied to BT. Beck and Lowe (1994) measured the SMR of 12 *H. horridum* at 25°C and 7 individuals at 15°C and found no consistent repeatable daily rhythm. The Q_{10} was 3.0 between the 2 temperatures measured, and the relationship between lizard BM and the SMR was 0.16 $BM^{0.69}$ at 25°C and 0.04 $BM^{0.73}$ at 15°C. The BM exponent of 0.69 differed significantly from the among-species BM exponent of 0.80 for a combination of *H. horridum, H. suspectum*, and 33 other squamate species taken from the literature. However, adult *Heloderma* had a BM exponent of 0.80, and their SMR was lower than that of the other squamates. At 15°C, periods of apnea further reduced the SMR of *H. horridum*. Beck and Lowe concluded that ecology is important in influencing the metabolic rate, and that reclusive lizards, like *Heloderma*, have lower SMRs than do nonreclusive species. During a study of locomotor performance using 12 *H. horridum*, Beck et al. (1995) discovered that at 25°C, the species has a V_{O2max} of 3.08 mL $O_2/(g^{0.73}$/hour)—a high aerobic capacity for a lizard with such a low level of activity and the highest recorded at this BT for any lizard. Its factorial aerobic scope (V_{O2max}/SMR) at 31°C was 30.4—again, the highest ever recorded for a lizard (most reptiles score around 10). Males scored a significantly higher V_{O2} than females, suggesting that sexual selection plays a factor.

Beck and Lowe (1991) never recorded a voluntary BT higher than 37°C in the wild, but thought it well below the lizard's BT_{max}, which has not been determined for *H. horridum*, but is 42–44°C in *H. suspectum*. At the other extreme, *H. horridum* is not very tolerant of cold ETs, as in its tropical habitat it seldom, if ever, experiences them; the coldest BT recorded by Beck (2005) was 17.7°C for one in an underground retreat in January. In Jalisco, BTs during the wet season were 23–32°C, with 25–26°C and 29°C the most frequently recorded (activity during the wet season is often during rain events), but during the dry season, the lizards had BTs of 22.5–37°C, with a decided peak at 31–33°C (Beck 2005). Mean BTs of active Jalisco *H. horridum* during the wet and dry seasons were 27.2°C and 30.2°C, respectively (Beck and Lowe 1991). The preferred BT for activity in the wild is probably around 29.5°C (Beck and Lowe 1991); interestingly, one tested in a laboratory heat chamber by Bogert (1949) maintained its BT within a degree of 29°C. One basking in a tree between 1415 and 1620 hours (AT, 26.8–28.4°C) after a heavy September rainfall kept its BT between 28.2 and 29.4 (mean, 28.8) °C (Beck and Lowe 1991). BT of an active *H. horridum* is generally higher than its corresponding AT, so the lizard's highest daily BTs are achieved when it is surface-active during the day (0900–1800 hours). When the lizard is resting in underground shelters, the BT is rather constant and highly correlated to the temperature at 50 cm depth.

Living in basically dry habitats, *H. horridum* must prevent excessive loss of body water to the environment through the skin and during breathing. This can in part be achieved by remaining underground or in other cool shelters during the hottest times of the day, and becoming surface-active during the cooler portions of the day or during rain events. The lizard will drink if offered water, and captives will soak themselves if water is provided in a large enough container (Beck 2005; C. Ernst, personal observation); it will replenish body water from puddles during or immediately after rains. Substantial weight gain occurred in Jalisco *H. horridum* after rehydration; one 880 g individual gained 52 g of water after a light June rain of 1.5 mm (Beck and Lowe 1991). Some water is also ingested with its prey. Metabolic water does not seem to play an important role as a water source, although possibly some water is stored in the tail fat (Beck 2005).

The beaded lizard sluggishly moves along in a shuffling, undulating manner, moving its limbs in the following order: (1) right front, (2) left rear, (3) left front, and (4) right rear. The body is bent toward the hind foot being raised, and the tail is shifted in the same direction. The body is never in contact with the ground while the lizard is walking (Bogert and Martin del Campo 1956). The rate of movement is correlated to body size; larger lizards move faster than smaller ones. The fastest individual recorded by Beck et al. (1995) achieved a speed of 2.3 km/hour. Each surface activity bout averages 67 minutes, and a mean distance of 236 m is covered during this time (Beck and Lowe 1991). Home range size in Jalisco varies from 12 to 48 (mean, 21.6) ha (Beck 2005; Beck and Lowe 1991).

H. horridum has great endurance. One on a treadmill at submaximal speed walked 0.8 km/hour for more than 3 hours and 40 minutes without tiring before the trial was ended. Its high aerobic capacity is a very important factor in this endurance. In Jalisco, an 800 g active individual had an SMR of 0.94 mL/g/km as it traveled 25 km in 121 hours (Beck and Lowe 1991); at this rate, the calculated energy used per year would be 3,773.7 kJ or 901.9 kCal (Beck and Lowe 1991). In the wild, the lizard forages at a rate less than 33% of that which it can sustain on a treadmill (Beck et al. 1995).

H. horridum is a good climber; its long limbs, strong claws, and semiprehensile tail (Pregill et al. 1986) are all useful adaptations. The lizard frequently ascends to tree branches 5–7 m off the ground (Bogert and Martin del Campo 1956; Campbell and

Lamar 2004; Shaw 1964), possibly in search of bird or lizard nests (see below) or to avoid predators. In Jalisco, tree-climbing mostly occurs during the wet season (Beck and Lowe 1991). Although, in its dry habitats, the beaded lizard has limited occasions to enter deeper waters, it probably is an adequate swimmer, but little data are available concerning its aquatic abilities.

As in some snake species, male *H. horridum* participate in ritualized combat bouts. Such bouts occur during the fall (September–November) when the testes are undergoing spermiogenesis, and may represent a form of reproductive territoriality. Aggression begins when a male crawls onto and straddles the back of another male ("the dorsal straddle" stage). This causes the bottom male to bend its body laterally, putting a front leg and a hind limb above the shoulder and hip of the upper male. The males' bodies are then pressed laterally together as they turn their heads and tails away from each other and push their snouts against the ground. More pushing against each other positions their bodies into a high arch with their venters in contact, and their forelimbs, snouts, and tail tips touching the ground ("the body arch" stage). This continues until pressure causes a collapse of the body arch, with the dominant male achieving the dorsal position. Collapse of the body arch and maintenance of the lower male on its back seem to be the main purposes of the behavior sequence. The superior male then presses his venter against that of the subordinate male, and bites it to keep a hold on its back (*H. horridum* is probably immune to the venom of its species, as occurs in *H. suspectum*). Several such bouts may occur, lasting only a few minutes (in the wild) to more than 16 hours (in captivity). Finally, the subordinate male breaks off contact and shuffles away, leaving the field of battle to the victorious male. Larger males have better leverage, and are normally the winners of combat bouts. Combat may occur simultaneously in several pairs of males (Álvarez del Toro 1982; Beck 2004b, 2005; Beck and Ramírez-Bautista 1991; Campbell and Lamar 2004; Demeter 1986; Ramírez-Velázques and Guichard-Romero 1989).

REPRODUCTION: *H. horridum* probably attains sexual maturity at an age of 2 to 3 years. Size at maturity has not been adequately studied, but according to Campbell (*in* Beck 2005) neonate *H. h. charlesbogerti* reach adult size (>30 cm) in 3 years. Goldberg and Beck (2001) reported that 4 males with SVLs of 32.0–44.4 cm showed evidence of spermatogenesis, and females with TBLs of 61.8 cm and 68.6 cm have laid eggs (Bogert and Martin del Campo 1956; Curtis 1949b).

The following testicular conditions existed in four males examined by Goldberg and Beck (2001): (1) 3 August (Colima)—early recrudescence, with mostly primary spermatocytes and proliferation of cells for upcoming spermiogenesis; (2) 6 August (Sonora)—regressed testes with spermatogonia the predominant cells, although some primary spermatocytes were present, suggesting the next spermatogenic cycle had just begun; the epididymides lacked sperm; (3) 4 September (Sinaloa)—spermiogenesis, with sperm present in enlarged, convoluted epididymides; and (4) 1 October (Jalisco)—testes badly decomposed, but large size of seminiferous tubules, great number of intratubular cells, and vas deferens packed with sperm suggest spermiogenesis was occurring. Similar histological data are missing for females. Smith and Grant (1958) reported that a female from Jalisco collected in mid- to late August contained eight eggs. A female that died in July contained six eggs in its oviducts; the one egg opened was unfertile (Taub 1963).

The above data on male sexual cycles supplied by Goldberg and Beck (2001) support observations of fall combat between males (Beck 2005) and the supposition that males search for females and mate in the fall in the wild (Álvarez del Toro 1982; Ramírez-Bautista 1994); however, captives have copulated or exhibited courtship behavior in April (two records, with oviposition of infertile eggs in late July in one and oviposition of a partially fertile clutch in early July from the other), June (no oviposition), and early August (oviposition of one apparently unfertilized egg in mid-October) (González-Ruiz et al. 1996). Jalisco males return to previously used areas and formerly used burrows in the fall, and Beck (2005) thought this is part of their mate-searching behavior.

Females are probably found more by scent than by sight, as the rate of male tongue-flicks increases as they explore former burrows or areas traversed by females. If one is found, she at first tries to escape, and the male chases her. When he catches her, the male climbs onto her back and performs a dorsal straddle (see male combat above). Next, he uses his chin to rub and stroke the female's head and neck. She may still try to escape, and may carry him on her back for some distance before quieting and accepting copulation. The male then twists his tail under her tail to bring their cloacal vents together. During this behavior, he elevates one hind limb over the dorsal base of her tail while keeping the other hind limb touching the ground. If receptive, the female may raise her tail to admit his hemipenis. The pair then becomes attached, with the male gripping the female with his hind limb, biting the nape of her neck, and stroking her body with his tongue. Mating may take several hours (Álvarez del Toro 1982; Arnett 1976; Beck 2005; North 1996).

Probably a duration of six to eight weeks occurs between copulation and oviposition, as in *H. suspectum* (Beck 2005), with the eggs being laid between October and February (Álvarez del Toro 1982; Beck 2004b; Curtis 1949b; Ramírez-Velázquez and Guichard-Romero 1989). Eggs in clutches are nonadhesive; 15 literature clutches averaged 9 (range, 1–22) eggs. No more than one clutch is laid per year, and not every female oviposits annually (Beck 2005). The white, elongated eggs have rough, leathery shells; 23 that we measured were 53.0–69.0 (mean, 56.7) mm in length and 23.0–40.0 (mean, 29.2) mm in width.

Hatchlings emerge from the eggs during the summer monsoon period in May–August (Beck 2004b; Campbell and Lamar 2004; Gienger et al. 2005; Ramírez-Bautista 1994). Laboratory-incubated eggs have hatched in as little time as 145 days to as much as 226 days (Beck 2005); higher ITs shorten the IP. Contrastingly, Ramírez-Bautista (1994) thought the IP may take 10–12 months in nature, and several reports have suggested 210–270 days as the natural IP (Beck 2005).

Neonates retain a yolk sac at hatching, and have more vividly pronounced yellow body and tail markings (tail bands in *H. h. exasperatum* and *H. h. horridum*) than adults. The light markings usually fade with age, but the juvenile yellow spots may be retained on the tail of *H. h. alvarezi* and the tails of adult *H. h. exasperatum* and *H. h. horridum* may contain some yellow markings after the juvenile tail bands break up. Hatchlings have the following mean measurements: TBL, 20.3 (range, 19.7–20.8) cm; SVL, 12.8 (range, 11.5–14.1) cm; TL, 91.6 (range, 89.0–100.0) cm; and BM, 35.8 (range, 23.0–47.0) g.

GROWTH AND LONGEVITY: Increases in SVL of 15 wild *H. horridum,* over periods of at least 9 months in Jalisco, averaged 10.8 (range, 5.6–15.5) mm per year. SVL was

Heloderma horridum exasparatum, male (Fort Worth Zoo, John H. Tashjian)

closely correlated with TL (r = 0.98), but less so with BM (r = 0.79). Individual BM varied throughout the activity period, and 1 adult had a 30% weight gain (stomach empty) between August 1987 and May 1987. Over the activity season, gain in BM of 5 lizards was about 100 g (Beck and Lowe 1991). Bogert and Martin del Campo (1956) thought that juveniles with 22–30 cm SVLs were in their second year of growth, and those with 30+ cm SVLs were certainly in their third year. Captive hatchlings increased their BM in 1 year from 42 g to 159.6 g, a 3.8% gain (Gonzáles-Ruiz et al. 1996). As occurs in many other reptiles, growth seems to slow after the attainment of maturity.

A wild-caught male *H. h. alvarezi* had lived 33 years, 1 month, and 11 days in captivity at the Institudo de Historia Natural, Tuxtla Guiterrez, Chiapas, Mexico, and was thought to be more than 38 years old, and a wild-caught female *H. h. horridum* had lived for 28 years at the Sacramento Zoo and was thought to be more than 31 years old. Both were still alive when reported by Jennings (1984).

DIET AND FEEDING BEHAVIOR: The following prey have been reported, or are suspected of, being taken by wild *H. horridum*: insects—ants (Formicidae), beetles (*Canthon viridis* and various larvae), cicada (Cicadidae), cockroaches (Blattidae), flies (Diptera), and orthopterans (Orthoptera); small turtles and their eggs (*Kinosternon* sp.); lizard eggs (*Ctenosaura pectinata*); snake eggs (*Drymarchon corais, Drymobius* sp., *Masticophis* sp.); bird eggs (*Cissilopha beecheyi, Lophortyx douglasii, Ortalis poliocephala, Trogon citreolus*); nestlings and other birds (*Leptotila verrauxi, Piaya cayana, Zenaida asiatica*); mice (*Sigmodon* sp.); rabbits (*Sylvilagus* sp.); and possibly other mammals (bats; *Dasypus novemcinctus; Didelphis marsupialis, D. virginiana; Sylvilagus cunicularius*) (Ariano Sánchez 2003; Beck 2004b, 2005; Beck and Lowe 1991; Bogert and Martin del Campo 1956; Bogert and Oliver 1945; Campbell and Lamar 2004; Campbell and Vannini 1988; McDiarmid 1963; Pregill et al. 1986; Sumichrast 1864; Woodson 1949b; Zweifel and Norris 1955). The eggs and nestlings of the western chachalaca (*Ortalis poliocepha-*

lus) and the citreoline trogon (*Trogon citreolus*) are also potential prey in Jalisco (Beck 2004b).

Some of the above vertebrate foods may have been consumed as carrion, which the lizard readily eats, and some of the insects may have been accidentally ingested with other prey. Sumichrast (1864) also reported that earthworms, centipedes, and small anurans are consumed in nature, but there is no collaborative evidence of this. Captives have been successfully fed cockroaches, *Tenebrio* larvae, small frogs, small mice (*Mus musculus*) and juvenile rats (*Rattus norvegicus*), hen's eggs (*Gallus gallus*), and chopped meat (Álvarez del Toro 1982; Ariano Sánchez 2003; Beck 2005; Bogert and Martin del Campo 1956; C. Ernst, personal observation; Woodson 1949b).

The diet consists chiefly of reptile and bird eggs. In Jalisco, 14 fecal samples from 10 lizards yielded reptile egg shells in 10 (71%), bird feathers in 8 (57%), and insects in 5 (36%) (Beck and Lowe 1991). Herrel et al. (1997) identified five distinct stages in the lizard's fixed egg-eating behavior: (1) approach, (2) shell-piercing, (3) uptake, (4) crushing, and (5) swallowing. Piercing and crushing are directly related to the egg-eating behavior. Shell-piercing is accomplished by several bites, during which the anterior teeth are used to puncture, but not crush, the egg. During the piercing, bite forces of 31 N are generated when the force is perpendicular (90°) to the tooth row, but an orientational shift of force away from the perpendicular axis causes an increase in the bite force to 62.4 N at −30° and 61.6 N at −150°. A minimum force of 4.42–13.87 (mean, 6.8) N is required to first crack the egg. Next, the egg is smashed within the oral cavity without additional tooth contact. While crushing the egg, bite forces of 43.4 N are generated when the force is perpendicular to the tooth row, but a shift away from the perpendicular during this stage causes an increase in the force to 87.4 N at −30° and 86.3 N at −150°.

When hunting, *H. horridum* walks slowly along, frequently extending its tongue; an average of 201 tongue-flicks occurs in 30 minutes under natural conditions (Bissinger and Simon 1979). Olfaction seems to be the main sense used to locate prey, especially to find bird and lizard nests. When examining a potential food item, captives increase the rate of tongue-flicking, and, after thoroughly examining the food with their tongues, begin to swallow it (C. Ernst, personal observation). In laboratory tests by Cooper and Arnett (2001), the lizard responded with tongue-flicks to animal odors, but not to plant odors. During additional tests by those authors (Cooper and Arnett 2003), it responded with significantly more tongue-flicks to odors of its normal captive food (*Mus musculus*) than to the odors of animals not offered to it. Further, Hartdegen and Chiszar (2001) tested the ability of *H. horridum* to discriminate between the smell of egg yolk (domestic chicken), rodent blood (*M. musculus*), water, and the irrelevant masking odor of cologne. The lizard had elevated mean tongue-flick scores when exposed to egg yolks (mean, 79.6) and blood (mean, 42.9), as compared to water (approximately 10) and cologne (approximately 28), but a combination of egg yolk and cologne elicited a score of 68. The above indicate that odor recognition, through tongue-flicking, is an important factor in discriminating prey from nonprey items.

Interestingly, the beaded lizard feeds mostly on prey that does not require envenomation. Although it will bite and chew larger live prey, which is probably only occasionally caught (Beck 2005; C. Ernst, personal observation), most observations indicate that venom is not used in subduing prey. Probably, its venom is mostly a defensive adaptation.

Neonates and juveniles may consume meals equal to their BM. After a large meal, fat is stored in the lizard's tail. Tails of helodermatids contain four longitudinal cavities extending the length of the tail, in which fat is stored after large meals. The spaces are separated by a band of muscle, and the two smallest spaces orient toward the vertebral column, while the two larger ones orient ventrally. Other fat is stored around organs in the body cavity (Schneider 1941; Shufeldt 1890).

Espinosa-Avilés et al. (2008) have isolated the following bacteria from the oral cavity (O) and cloaca (C) of *H. horridum*: *Citrobacter brakii*, *C. freundii*, *Citrobacter* sp. (C), *Diphteroides* sp. (O), *Enterobacter aerogenes* (O), *Enterococcus* sp. (O), *Escherichia coli* (O, C), *E. coli* c 1 (O), *E. coli* c 2 (O), *E. coli* c 3 (C), *Klebsiella ozaenae* (O, C), *Neisseria* sp. (O), *Proteus mirabilis* (O, C), *P. penneri* (O), *P. vulgaris* (C), *Pseudomonas aeuroginosa* (O), *Pseudomonas* sp. (O, C), *Staphylococcus coagulase* (O), *Streptococcus viridans* (O), and *Streptococcus* sp. (O).

VENOM DELIVERY SYSTEM: In contrast to the venom apparatus of snakes, that of helodermatid lizards occurs on the lower jaw and serves 8–10 solid, conical, grooved, 2–6 mm teeth, not a pair of hollow fangs on the maxillae (see fig. 8). The middle dentary teeth are longer than those flanking them. The venom is produced in a pair of large, multilobed inferior glands, the glands of Gabe, one located laterally on the side of each jaw (not in a pair of glands on the maxillae as in snakes). The position of the glands can be identified by the swellings on the lower jaw. Each gland is divided by septa into 2–4 lobes, each of which is subdivided into several smaller lobules. These are further subdivided into even smaller lobules containing cavities connected by small tubules. The venom is secreted by granulated columnar cells within the small tubules that discharge the venom into a central storage lumen. The lumen, in turn, is drained by a very short venom duct to the mucous membrane on the outer surface of the gum at the base of a grooved tooth. No muscular action is required to move the venom along, and, in fact, compressor muscles are not present around the glands; instead, the chewing action of the jaws moves the venom along. Venom enters a longitudinal groove on the tooth surface via three openings at its base, and is moved up the groove and into the wound by capillary action. The venom duct is separated from the tooth pedestal by an outer wall of the cup-shaped dental sac, the mucous membrane fold investing the pedestal. The fold is unique in its structure and is thought to play an important role in the discharge of the venom and its transfer into grooves on the larger maxillary teeth. In some instances, the venom may mix with saliva from the upper jaw glands and perhaps be carried into grooves on the maxillary teeth, as well as those on the premaxilla (Beck 2004a, 2005; Bogert and Martin del Campo 1956; Fischer 1882; Fox 1913; Greene 1997; Phisalix 1912, 1917; Russell and Bogert 1981; Shufeldt 1891; Tinkham 1971b). The lizard usually holds live prey very tightly and chews it.

VENOM AND BITES: The venom of *Heloderma* lizards has been studied for more than a century; an entire contributed volume on it was edited by Loeb in 1913. It is highly neurotoxic, and electrophoretically similar to that of cobras (Mebs and Raudonat 1966). It primarily affects the respiratory and cardiac systems (see below).

The main chemical constituents, as in snake venoms, are toxic proteins and peptides, of which the toxicity lies in those that are nondialyzable. Fourteen major components have been identified from *Heloderma* venom. The following list has been adapted from Beck (2005): (1) extendin-3 (in *H. horridum* venom; a peptide)—acts on the pancreas, causing it to release amylase; (2) exendin-4 (exentide, in *H. suspectum* venom; a peptide)—induces insulin release from the pancreas; *Heloderma* extendin-3 and extendin-4 are potential diabetes therapeutics; discussion of the chemical and physiological expression of the extendin peptides of *Heloderma* is beyond the scope of this publication; for this we refer the reader to Y. Chen and Drucker (1997), T. Chen et al. (2006), Christel and De Nardo (2006, 2007), Conlon et al. (2006), and Eng et al. (1992); (3) gilatide (in *H. suspectum* venom; a fragment of the extendin-4 peptide)—improves memory in rodents; (4) gilatoxin (in *H. suspectum* venom; a kallikrein-like, lethal, serine, glucoprotein)—causes rapidly developing hypotension and contraction of the smooth muscles of the uterus (in rats); the reactions are caused by bradykinin polypeptides that influence smooth muscle contraction.; gilotoxin is more toxic in the presence of other venom components than when acting alone (Tu 1991b); (5) helodermin (exendin-2, a peptide)—lowers blood pressure (in dogs); (6) helodermatine (a kallikrein-like protease)—causes a dose-dependent decrease in arterial blood pressure (in rats); (7) helospectin I and II (exendin-1, peptides)—cause amylase release from the pancreas; (8) helothermine (a peptide)—causes lethargy, partial paralysis of hind limbs, intestinal distension, and hypothermia (in rats); (9) *horridum* toxin (in *H. horridum* venom, a kallikrein-like lethal glycoprotein)—causes hypotension, internal hemorrhaging, and hemorrhaging and bulging of the eyes; (10) hyaluronidase (an enzyme that splits hyaluronic acid)—acts to diffuse venom through surrounding connecting tissues at the bite site; (11) nerve growth factor—effects unknown; (12) phospholipase A_2 (enzyme that catalyzes hydrolysis of phospholipid glycerol backbones)—effects unknown for *Heloderma,* but those of snakes act as presynaptic membrane toxins, particularly on the acetylcholine esterases in the synapses of the cardiac and respiratory systems; (13) serotonin (a neurotransmitter hormone)—causes inflammation, excites smooth muscles, vasodilation, and other effects; and (14) an unnamed lethal toxin (a kallikrein-like lethal peptide)—suppresses the contraction of diaphragmatic muscles (in mice). The venoms from both species of *Heloderma* lack NADase and phosphodiesterase activities (Aird 2008a).

More detailed aspects of the venom chemistry and physiology of *Heloderma* are discussed in Aird (2008a), Alagón et al. (1982, 1986), T. Chen et al. (2006), Datta and Tu (1997), Eng et al. (1990), Gomez et al. (1989), Holz and Habener (1998), Loeb (1913), Mebs (1968, 1969, 1978, 1995), Mebs and Raudonat (1966), Mochca-Morales et al. (1990), Morrissette et al. (1995), Nikai et al. (1988c, 1992), Nobile et al. (1994, 1996), Raufman (1996), Russell (1983), Russell and Bogert (1981), Tan and Ponnudurai (1992), Tu (1977, 1991b), Tu and Murdock (1967), Uddman et al. (1999), Utaisincharoen et al. (1993), Vandermeers et al. (1991),Yamazaki and Morita (2004), and Zarafonetis and Kalas (1960a). Undoubtedly, by the time this book is published, additional studies will have been published on venoms of *H. horridum* and *H. suspectum*.

Venom from *H. horridum* has a protein concentration of 80 mg/mL and a pH of 6.9–7.0; its electrophoretic pattern shows at least 18 protein bands (some individual variation occurs), which are constant throughout the year (Alagón et al. 1982). Its gilatoxin has a molecular weight of 33,000, an isoelectric point of 4.25, 245 amino acids, and bradykinin-releasing activity, and is fibrinogenolytic (Komori and Nikai 1998).

A bite by a species of *Heloderma* is usually very painful (subsiding in $<$2 hours); other symptoms recorded in human bites include an increase in white blood cell production;

anxiety; biphasic blood clotting, sometimes prolonged (at first accelerated, but later retarded); breathing difficulties (dyspnea); bruising (ecchymosis) at the bite site; bulging of the eyes and contraction of the pupils; convulsions; decrease in neutrophils; diarrhea; edema and swelling (usually subsiding in <4 hours); cyanosis at the bite site; edema of the intestines; fever and chills; faintness or dizziness; general weakness; hemorrhaging (both internal and from the eyes); hypersensitivity around the bite site; hypotension and vasodilation (often rapid); increase in heart rate (tachycardia) with eventual cardiac failure in diastole (acute myocardial infarction); increased nervous irritability; lethargy; lowered BT; pain (mild or severe); muscle contractions of the intestines and uterus; nausea or vomiting; paralysis (limbs and partial of body); muscle rigidity of abdominal wall; numbness at the bite site; profuse bleeding from bites; reduction in flow, oxygenation, and potassium levels of the blood; ringing sensation in the ears (tinnitus); shock (hypotensive anaphylaxis); slurred speech (dysphonia); swelling of the tongue; tearing from eyes; tenderness of lymph glands; and urination, with eventual renal failure (Albritton et al. 1970; Beck 2005; Bogert and Martin del Campo 1956; Bou-Abboud and Kardassakis 1988; Burnett et al. 1985; Campbell and Lamar 2004; Cantrell 2003; Ernst 1992; Heitschel 1986; Kunkel et al. 1984; Loeb 1913; McNally et al. 2007; Piacentine et al. 1986; Preston 1989; Snow 1906; Woodson 1947; Zarafonetis and Kalas 1960a). The bite may be fatal, with death usually occurring from paralysis of the respiratory system and/or cardiac failure.

Although generally feared by native peoples throughout its range, few reports of bites by *H. horridum* have been chronicled, while many case histories of human bites by *H. suspectum* exist (see that species; Beck 2005; Bogert and Martin del Campo 1956; Brown and Carmony 1999); however, many bites by *H. suspectum* go unreported (Miller 1995). Supposedly fatalities have occurred from beaded lizard bites, but these have not been documented. We have found only three case histories: (1) Dugès (1899) related how a bite in Mexico resulted from a man irritating the lizard so much that it eventually bit him, and another person, on the finger. The wounds bled and were very painful, but healed rapidly. No other effects were mentioned. (2) A Florida biological supply dealer was bitten on his left index finger by a 635 cm TBL [?] *H. horridum*. The lizard maintained its grip for a minute or two and was difficult to remove, leaving deep lacerations. Immediate pain of a burning sensation was experienced that radiated into the arm. Within five minutes, the victim became nauseous and vomited. When admitted to the hospital 30 minutes after the bite, the man was suffering from extreme pain in his left hand and arm, nausea, and vomiting. The pain was so severe (likened to a bone fracture or kidney stone by the attending physician) that he screamed and thrashed about on the examining table. The finger and dorsum of the hand was somewhat swollen and red, but with very little blood oozing from the puncture wounds. The symptoms subsided after treatment (see paper), and by the next morning only the finger and dorsal surface of the hand were swollen. The patient was released from the hospital on the third day after the bite incident (Albritton et al. 1970). (3) A man was bitten on his right hand by a captive *H. horridum*. The lizard was removed 30 seconds later. At that time, the victim experienced severe local pain, dizziness, vomiting, and profuse sweating. He washed the wound and took a single dose each of aspirin and diphenhydramine, but then developed shortness of breath with increased local pain and paramedics were called. Upon arrival in the emergency department, he was lethargic, vomiting, and in severe pain, with significant swelling of his right hand, lips, and tongue. His O_2 saturation decreased to 55% soon after arrival without apparent cyanosis, but he required O_2 via a face mask to maintain a normal saturation rate. He was treated (see paper) and placed in the intensive care unit, where he continued to complain of severe pain. Persistent edema and local swelling were evident, and he continued to require supplemental O_2. His white blood cell count was 18,500 k/mm^3 with 80% regimented polymorphonuclear leukocytes. He gradually improved over the next eight hours; his lips and tongue returned to normal size and his O_2 saturation returned to normal on room air. Swelling at the bite site remained, however, along with red streaking up the forearm. Antibiotics were administered, and he had an uneventful further stay in the hospital and was discharged the following day (Cantrell 2003). No antivenom was administered in these three cases, as none is available. Obviously, this lizard's envenomation capabilities should not be taken lightly.

The reported LD_{50} values (mg/Kg BM) of mice and rats for beaded lizard venom are intraperitoneal, 2.0 (Alagón et al. 1982); intravenous, 2.7 (Hendon and Tu 1981); and subcutaneous, 1.0 (Ariano-Sánchez 2003) and 1.4 (Mebs and Raudonat 1966). The LD_{50} for the protozoan, *Paramecium multimicronucleatum*, is 24.7 (range, 20.8–29.3) mg/mL (Johnson et al. 1966). The LD_{50} for humans has not been calculated, but is probably similar to that of *H. suspectum* (0.38–0.63 mL).

PREDATORS AND DEFENSE: Records of predation are rare. Adult beaded lizards have little to fear except humans, but juveniles are probably preyed on by a suite of vertebrates. The eggs must be occasionally destroyed by predators, but data on nest success are lacking. The only reported wild predators are the snakes *Boa constrictor* and *Drymarchon corais* (A. Ramírez *in* Beck 2005). Other vertebrates known to attack various life stages of reptiles—like snakes (*Coluber constrictor, Drymobius* sp., *Lampropeltis* sp., *Leptodira* sp., *Masticophis* sp.), caracara (*Caracara cheriway*), various large hawks (*Buteo* sp.), vultures (*Cathartes aura, C. burrovianus; Coragyps atratus; Sarcoramphus papa*), opossums (*Didelphis marsupialis, D. virginiana*), armadillo (*Dasypus novemcinctus*), coati (*Nasua narica*), raccoon (*Procyon insularis, P. lotor*), badger (*Taxidea taxus*), weasel (*Mustella frenata*), skunks (*Conepatus mesoleucus; Mephitis macroura, M. mephitis; Spilogale putorius, S. pygmaea*), dogs and coyotes (*Canis familiaris, C. latrans*), foxes (*Urocyon cinereoargenteus*), and various cats (*Felis catus; Leopardus pardalis; Lynx rufus; Panthera onca; Puma concolor, P. jagouaroundi*)—occur within the range of the beaded lizard, and potentially prey on it. In tests by Balderas-Valdivia and Ramírez-Bautista (2005), *H. horridum* reacted adversely by moving quickly away from chemical cues of the sympatric snakes (mean escape latency in seconds) *Boa constrictor* (2.81), *Crotalus basiliscus* (3.04), *Agkistrodon bilineatus* (3.53), *Crotalus molossus* (6.36), *Loxocemus bicolor* (9.41), *Drymarchon corais* (10.77), and *Masticophis mentovarius* (16.05). Odors of several other species of snakes not thought to prey on the beaded lizard were also tested, with resulting latent escape times of 16–60 seconds. Judging by the reactions of *H. horridum* to the first group, those snakes are potential, if not already, predators on it. In captivity, a baby alligator (*Alligator mississipiensis*) bit a beaded lizard that later succumbed (Woodson 1949a).

When confronted with a potential enemy, *H. horridum* often flattens itself against the ground, possibly to avoid detection, and if further approached, shuffles off to avoid conflict. If a burrow or other retreat is nearby, it will enter, and when a tree is available, possibly climbs it (the lizard often seeks shelter in trees, especially during the wet season, perhaps to avoid predation). However, if the disturber continues to rapidly approach, *H. horridum* assesses it through tongue-flicking, and gives an open-mouth threat display while loudly hissing. As a last resort, if the would-be attacker comes too near, the lizard may lunge at it and bite (it certainly tries to bite if seized). It then holds on tightly and chews venom into the resulting bite wounds; its venom is probably a defensive adaptation, not one for feeding. The body color pattern may help to avoid detection if the lizard is shaded in vegetation or a shelter, but it may also serve as a warning of the species' venomous nature. A potential predator that has survived a previous bite by *H. horridum* will probably think twice before attacking one again.

PARASITES AND PATHOGENS: The lizard does not always adjust well to the stresses of confinement, and refuses to eat, causing its general health to deteriorate. This leaves it vulnerable to disease; pneumonia seems to be a particular problem (Woodson 1949a). Wild beaded lizards probably also have diseases; five serotypes of *Salmonella* have been isolated from an unidentified wild *Heloderma* (Hoff and White 1977). A Mexican *H. horridum* examined by Thompson and Huff (1944a, 1944b) was negative after two inoculations with the blood protozoan *Plasmodium mexicanum*. Bogert and Martin del Campo (1956) reported one parasitized by a tick in Tehuantepec; and, undoubtedly, the species also harbors endoparasitic helminths, but there is no literature on this.

POPULATIONS: Few population data are available. In Jalisco, Beck and Lowe (1991) captured 15 individuals within an area of approximately 36 ha (approximately 2 km^2; Beck 2005), a density of 0.42/ha or 3.25/km^2. The lizards captured included 8 adult males, 5 adult females, and 2 juveniles; there was an adult sex ratio of 1.6:1, and a juvenile to adult ratio of 0.15:1. Adult males had a mean TBL of 671 cm (SVL, 383 cm) and BM of 833 g; adult females averaged 609 cm in TBL (SVL, 346 cm) and 667 g in BM. One adult male was 1,600 g; the smallest juvenile was 33 g. However, Beck (2005), based on mark-recapture data, thought that only slightly more than 50% of the individuals living there had been caught, and that densities of 18.6/km^2 (30/$mile^2$) are probably not uncommon in undisturbed habitat.

At the time Beck (2005) wrote his book, 82 males and 71 female *H. horridum* were in captivity in institutions in Canada, Mexico, and the United States, a combined male to female ratio of 1.15:1. Neither this ratio, nor the one given above for the Jalisco lizards, is significantly different from unity.

Populations of beaded lizards in undisturbed areas are probably safe for the present time, but those in contact with humans have experienced a decline in both total available habitat and its quality; tropical dry forest is the most threatened forest type in Mexico and Central America (*H. h. charlesbogerti* is particularly in trouble). The reasons for its decline are several, including forestry practices, expanding agriculture (particularly clearing for ranching), some clearing for domestic use, and invasive plants (especially grasses). Where such activities occur, roads soon follow, and with them mortality of lizards on the roads increases. The devastating effects of motorized vehicles on populations of snakes and turtles in the United States have been well documented. Some pet trade activity may occur, but is probably not of significance unless the population of *Heloderma* is of low numbers to begin with, even though the species commands a high price in Japan and Europe. Commercial use of the lizard for leather products and tourist novelties is much more serious. Uninformed persons kill many

Heloderma horridum exasparatum, hatchling (R. D. Bartlett)

each year on the spot when they are encountered because they think the animal is extremely dangerous.

If the species is to survive, all of these practices must be curbed and conservation plans put into effect. Fortunately, some Latin American countries are now setting aside lands as preserves for this and other sympatric species, and the species is now listed under CITES (Johnson and Ivanyi 2004). Also, the species is currently protected from collection and export throughout its tropical range (Beck 2009). Education is the most important ingredient in the conservation of *H. horridum*, and several of the above negative factors could be alleviated through more knowledge of its biology and a better understanding of the lack of danger to humans. Our knowledge of the important life history parameters of this lizard is poor, and mainly based on studies by Beck and his compatriots at only a few places within the species' range. Other data are mostly anecdotal. Additional studies of its ecology and reproduction throughout the lizard's distribution are needed to form viable conservation plans. *Heloderma horridum* is not lost yet, but we must act soon if it is to prosper in the future. The species is listed in Appendix II of CITES Red List for Threatened Species, and listed as threatened in Mexico.

REMARKS: Beck (2004a, 2004b, 2005, 2009), Bogert and Martin del Campo (1956), and Campbell and Lamar (2004) have reviewed *H. horridum*, and Beaman et al. (2006) have published a bibliography. Espinosa-Avilés et al. (2008) reviewed the hematology and blood chemistry of the species.

Heloderma suspectum Cope, 1869
Gila Monster
Escorpion del Norte

RECOGNITION: The Gila monster is a large, chubby lizard (TBL_{max}, 57 cm [Beck 2005]; but most adults are 30–40 cm; <70 total vertebrae and 25–28 caudal vertebrae are present) with beaded yellow, pink, and black scales, and a short, fat tail. The body is patterned with irregular dark brown or black reticulations on a yellow to orange or salmon-pink background. This pattern changes with age, becoming more complex and consisting of 5 dark saddle-like bands, 1 on the neck and 4 between the front and hind legs, and 4–5 tail bands. Adult *H. s. suspectum* become more melanistic with age, and *H. s. cinctum* from black lava areas may have a darker ground color (Beck 1985a). The broad head is black from the snout back to the orbits and laterally to at least the corner of the mouth. The chin and neck are black, and the legs and feet are usually predominantly black. The black tongue has transverse folds or wrinkles on its underside. Body scales are rounded and occur in 52–62 rows at midbody. Ventrally, the preanal scales are paired and enlarged, and 55 (range, 48–62) subcaudals are present. Dorsal head scalation includes 1 rostral, 2 (3) postrostrals, 2 supranasals, 2 (3) internasals, 3 (1–4) canthals, 3 (4–5) superciliaries (a small series of scales positioned dorsal to the orbit where normally a supraocular would occur), and 6–8 dorsal scales lying medially between the posterior superciliaries; total scales between the internasals and occipit, 12–13 (11–15). Laterally on the head are single prenasal, nasal and postnasal scales, 2–4 (5) loreals, 4–9 lorilabials (lying between the loreals and the supralabials), 2 (1–3) preoculars, 1 postocular, 2–3 suboculars, 5–6 (4–7) temporals, 11–12 (10–14) supralabials (the first is in broad contact with the rostral), and 13–15 (12–16) infralabials (most touch chin shields). Also present are 3–4 rows of sublabials on the lower jaw below the infralabials, a wedge-shaped mental scale, and 4–8 chin shields. The eyes are black with round pupils; the ear openings are narrow oblique or ovoid slits. The limbs are stout with heavy claws. Claws are usually 6–10 mm long, and the length of claw/SVL of the fourth foretoe and fourth hindtoe are, respectively, 1.2–3.0% and 1.0–2.1% (Bogert and Martin del Campo 1956). Dentition consists of 6–9 premaxillary, 7 (8–9) maxillary, 1–4 (0) pterygoid, 1–4 (0) palatine, and 8–9 (7–10) dentary teeth (Bhullar and Smith 2008; Bogert and Martin del Campo 1956; Olson et al. 1986).

The hemipenis is single with an undivided sulcus spermaticus bordered by two folds (proximally it is composed of a single flap). No horns or lateral cups are present. The proximal third of the hemipenis is naked; distally 18–20 fine, soft, transverse flounces encircle the organ, but do not fuse with the folds bordering the sulcus spermaticus. The most distal 6–8 flounces are divided by a bare apical zone that extends as an inverted triangle for a short distance (Branch 1982).

It is almost impossible to sex *H. suspectum* by external characters. Lowe et al. (1986) reported that the male has a larger and wider head, and that there are subtle differences in the abdominal shape of males and females, but these are size related. Males are the larger sex in *H. s. suspectum*, but those of *H. s. cinctum* are slightly smaller than females. Males have slightly longer foreclaws and more swollen and longer tails than females. The lateral postanal areas on the ventral side just behind the vent show evidence of the hidden hemipenis in males, and males have four large, quadrate, median, preanal scales with no interruption of the three rows crossing the preanal area. In females, the smaller scales in the anterior row crossing the preanal area are divided by two large median preanal scales. However, these characters seldom lead to a true sex determination; they are subjective, particularly those regarding the preanal scales, and must be used in combination.

Sexing done by probing during zoo breeding programs has sometimes been successful. Probing is easily done with a long, thin, blunt, lubricated rod inserted into the cloaca toward the base of the tail; it should slip into the uneverted hemipenis of males, and thus penetrate deeper (35–50 mm in males, but not so deep, 21–31 mm, in females) (Laszlo 1975; Wagner et al. 1976), but this too is not always accurate (Peterson 1982a, 1982b). Hemipenes can usually be extruded from adult males in the wild by palpation, especially during the breeding season, but sex determination in nature may have to be accomplished by observation of copulation or egg-laying. The hemipenes can be detected through ultrasound imaging of the underside of the tail, even in juveniles (Morris and Alberts 1996; Morris and Henderson 1998), but this can only be done in the laboratory. Although X-ray studies by Card and Mehaffey (1994) revealed sexual differences in the pelvic girdle, studies by Divers and Lennox (1998) failed to reveal any significant differences between males and females.

GEOGRAPHIC VARIATION: Two subspecies have been described. *Heloderma suspectum suspectum* Cope, 1869, the reticulate Gila monster or Lagarto de Cuentas, ranges from Yuma, Yavapai, and Gila counties in Arizona; and Grant, Hildalgo, Luna, and possibly Doña Ana counties in southwestern New Mexico; south through Sonora to northern Sinaloa in Mexico. Adults have the dark body cross bands obscured by black mottled and blotched pigment that predominates over the lighter pink or orange ground color (juveniles have more sharply defined cross bands). The light bands on the tail also contain dark spots or mottling; four or five dark bands occur on the tail. *Heloderma s. cinctum* Bogert and Martin del Campo, 1956, the banded Gila monster,

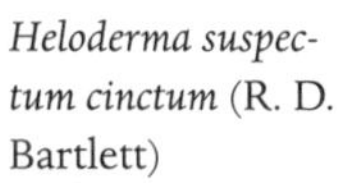

Heloderma suspectum cinctum (R. D. Bartlett)

ranges from Washington County in southwestern Utah; Clark and Lincoln counties in southern Nevada; and Imperial, Inyo, Riverside, and San Bernardino counties in adjacent California (Beck 1985b, 2005; Lovich and Beaman 2007); south, in the Colorado River drainage, into the Grand Canyon, and to Yuma County in southwestern Arizona. It normally retains most of the juvenile pattern, consisting of four well-defined black saddle-like bands across the body between the fore and hind limbs. Light pink or orangish pigmentation predominates. Little dark pigment occurs in the light tail bands, and there are five black tail bands. Beck (1985a) reported that *H. s. cinctum* living in black basaltic areas in southern Utah contain much more dark pigment than those from farther south and may lack light cross bands.

With the exception of a record from the Mojave River, all California records of *H. s. cinctum* occur east of 116° W longitude, where precipitation is highly biphasic, with >24% falling in the warm season, so summer rains may be important in the subspecies' foraging ecology. The California distribution of *H. s. cinctum* suggests that an invasion into the high mountain ranges of the northeastern Mojave Desert during the last interglacial period was along the Colorado River corridor (Lovich and Beaman 2007).

Funk (1966) reported that a Gila monster from Yuma County, Arizona, was an intergrade between the two subspecies, and Daniel Beck (personal communication) has informed us that intergrades are now commonly found.

In contrast to the large mtDNA differences between the subspecies of *H. horridum* (see above), Douglas et al. (2003) found only 0.5% differentiation between *H. s. suspectum* and *H. s. cinctum*.

CONFUSING SPECIES: The only other North American lizard with beaded scales is *Heloderma horridum*, which can be distinguished by using the above key. No other lizard from the southwestern United States and northern Mexico is patterned like *H. suspectum*.

KARYOTYPE: The karyotype is composed of 38 chromosomes, 14 macrochromosomes, and 24 microchromosomes (Matthey 1931a, 1931b).

FOSSIL RECORD: Pleistocene (Rancholabrean) remains of *H. suspectum* have been found in Arizona at Deadman Cave, Pima County (Mead et al. 1984); in Vulture Cave, Mojave County (Mead and Phillips 1981); in the Waterman Mountains, Yuma County (Van Devender 1990); and at Gypsum Cave, Clark County, Nevada (Brattstrom, 1954). In addition, the Pleistocene (late Irvingtonian) remains, identified only as *Heloderma* sp., from Anza-Barrego Desert State Park in Southern California (Gensler 2001) probably belong to this species.

DISTRIBUTION: The Gila monster ranges from extreme southwestern Utah (Washington County), southern Nevada (Clark County), and adjacent San Bernadino County, California, southeastward through western and southern Arizona to southwestern New Mexico (Grant and Hildalgo counties), and south in Mexico through much of Sonora and probably to northwestern Sinaloa. There is also a record from El Dorado in west-central Sinaloa that may represent a disjunct population.

HABITAT: The Gila monster usually lives in arid deserts but not barren areas at elevations of 1,100–2,033 m (Beck 2005), particularly those with scattered cacti, shrubs,

Distribution of *Heloderma suspectum*.

mesquite, and grasses, but may also be found in pine-oak forests, thorn forests, and irrigated lands. Rocky slopes, arroyos, and canyon bottoms (particularly those with streams) may support relatively dense populations in some parts of Arizona and Sonora. The biotic provinces occupied by *H. suspectum* include Sinaloan thornscrub, southwestern desert scrub, Sonoran desert, desert grassland, Mohave Desert, Chihuahuan desert scrub, and Madrean evergreen woodland (Beck 2005).

Underground shelters are important in all habitats. Beck and Jennings (2003) reported the following mean days' duration of shelter use by seasons for *H. suspectum* from southwestern New Mexico: spring, 5.1; dry summer, 4.5; wet summer, 4.2; fall, 13.6; and winter, 88.7. Studies indicate some fidelity to particular shelters. The sheltering lizard can escape the extremes of temperature and weather on the surface above. Using radiotelemetry, Beck and Jennings (2003) monitored the use of such retreats by 8 to 10 *H. suspectum* for 6 years, and recorded the timing, frequency, and duration of visits to more than 250 shelters. They used transects to assess the availability of shelters, recorded structural features of all of them, and used data loggers to monitor seasonal differences in their use. The shelters were not randomly chosen by the lizards. They spent more time in areas where more shelters were available than elsewhere, and chose the retreats on the basis of rockiness (usually dug under boulders or small rock outcrops), slope and entrance characteristics, depth, relative humidity, and temperature. Lizards stayed longer and had greater fidelity to those retreats used during extreme weather periods (like winter). These were paralleled by seasonal changes in the aspects of the shelters chosen. In winter, the retreats tended to be south-facing, deeper, rockier, and warmer than those used at other seasons. In contrast, dry-summer shelters had a social aspect tied to intraspecific behaviors and reproduction, and were those used, often simultaneously, by several *H. suspectum*.

Shelters used by the animal in Utah were in rocky areas (59% in loose Navajo sandstone, 41% on basaltic lava slopes or flows). Desert tortoise (*Gopherus agassizii*) burrows and woodrat (*Neotoma lepida*) mounds were used 8% and 10% of the time, respectively. Sixty-seven percent of the entrances faced east, southwest, or south; the few facing northwest or northeast (2%) were only used in the summer, and all hibernation sites faced south. One lizard dug inside its shelter for a total of seven hours during late April and early May, so the amounts of digging may be significant (Beck 1990). Shelters are co-occupied with desert tortoises only during the nesting season, and the lizards occupied almost exclusively those of female tortoises, probably because they were nesting sites (see "Diet and Feeding Behavior"; Gienger and Tracy 2008b).

Pianka (1967) thought *H. suspectum* is restricted to the Sonoran Desert because it is a secondary carnivore and almost certainly dependent on the summer rains of this desert (see also Lovich and Beaman 2007), with their concomitant predictable burst of warm-season production and breeding of its prey species. It is thus directly dependent upon the length of the growing season and the predictability of warm-season production. Pianka, however, did not consider the Mojave Desert populations of *H. suspectum* that are subject to a more winter rainfall-dominated regime.

BEHAVIOR AND ECOLOGY: The year's activity usually starts as early as March, when a few *H. suspectum* emerge from hibernation, but the first major emergence and/or dispersal period normally occurs in the first two weeks of April (De Nardo 2005; Gates 1957), and some activity continues into October, or rarely November, when hibernation begins. However, some Arizona Gila monsters bask in January and February (Beck 2005; Lowe et al. 1986). Nevada Gila monsters are most active from late May through June, but activity rapidly declines in August to about 50% of that in June and only 33% of that in May (Gienger 2003). In Utah, *H. suspectum* are most active from April into June; 64% of the annual activity and 77% of the total distance moved take place between April and early July (Beck 1990). The maximum days per month that any Utah lizard is surface-active are 14. During the active period as much as 97% of the time is spent in shelters and less than 13% of the energy budget is spent on surface activity. In the spring and fall it is mostly diurnal, but the summer heat forces the lizard to become crepuscular or nocturnal from June through September (contrary to Lowe et al. 1986). In the Mohave area of northwestern Arizona, activity decreases notably in June and July (Jones 1983). In Sonoran Arizona, most are annually surface-active for a total of no more than 190 hours, involving a maximum of approximately 115 days; 98% of the year is spent under cover (Lowe et al. 1986). Juveniles may have a longer period of activity, beginning at the same time as that of adults but peaking in July instead of April–May (Beck 2005).

Depending on weather conditions, *H. suspectum* may be active at any time during the day. In Utah, 68% of daily activity occurs between 0830 and 1230 hours (Beck 1990). Arizona lizards are diurnal mostly in April and May, with a second peak in September and October, but are nocturnal or crepuscular during the hot summer months (De Nardo 2005). In Arizona and New Mexico, the lizard is most surface-active from 0900 to 1100 hours in the spring; activity is bimodal during the dry (0800–0900 hours and 1800–2100 hours) and wet (0800–1000 hours and 1800–2100 hours) summers, and peaks at 1000–1200 hours in the fall and winter (Beck 2005). Beck (1990) noted, however, that activity reports are biased by the time of day investigators devote to their field studies. In laboratory tests conducted by Lowe et al. (1967), the lizard showed essentially the same activity pattern regardless of whether subjected to either constant light or constant darkness.

Beck (2005) listed mean activity BTs of 29.3–31.0°C for wild *H. suspectum* from several studies. BTs of nondormant animals recorded by Beck (1990) show that in Utah *H. suspectum* spends more than 83% of the year at 25°C or lower, and more than 50% of the year at or below 20°C. Mean monthly BTs (excluding those during activity) range from 12.3°C (December) to 28°C (July). Lizards in shelters have their BTs significantly correlated with the surrounding ATs and STs. Activity BTs average 29.3 (range, 17.4-36.8) °C, and activity occurs at ATs of 10–34°C and at STs of 20.5–32.0°C. Basking occasionally occurs, and one Gila monster observed basking in late April and early May maintained a mean BT of 28.5°C; usually baskers maintain BTs of 28–31°C (Beck 2005). *H. suspectum* has a black peritoneum that helps block ultraviolet rays from reaching the internal organs while basking or otherwise surface-active (Hunsaker and Johnson 1959). Bogert and Martin del Campo (1956) reported a mean BT of 28.7°C for 3 captives in a cage with a gradient in heat lamps.

In Arizona, the lizard may emerge to bask from January to March with BTs as low as 12.7°C (Lowe et al. 1986). The mean preferred deep BT range of active Arizona lizards is 22–34 (mean, 29) °C, and active individuals have been taken with BTs as low as 16°C and as high as 37°C, but few allow their bodies to heat above 33.8°C without seeking shelter (Lowe et al. 1986). It is very intolerant of BTs of 37°C or higher.

No obvious correlation exists between the O_2 capacity of Gila monster blood and the lizard's thermal relationships. The partial pressure of O_2 required to produce 50% saturation of its blood at approximately 38°C and 37–40 mm Hg of CO_2 is 60–72 mm Hg, high by mammalian standards (Dawson 1967).

Based on data from Bogert and Martin del Campo (1956) and that supplied to him by Warren, Brattstrom (1965) reported a mean activity BT of 27.2°C and minimum and maximum voluntary BTs of 24.2°C and 33.7°C, respectively. His calculated CT_{min} and CT_{max} were −3.0°C and 48.0°C, but he questioned the latter temperature. Bogert and Martin del Campo (1956) reported that a BT of 44.2°C produced potentially lethal paralysis. Normally the lizard behaviorally maintains its BT within 7°C of the CT_{max}.

Gila monsters have unusually low SMRs. The Q_{10} for the SMR between 20°C and 25°C is 2.9 (Beck 1990). Beck and Lowe (1994) measured the SMRs of 16 Gila monsters at 15°C and 26 individuals at 25°C. Adults had a mean BM of 513 g (range of juveniles and adults, 26.4–802.0 g). No constant repeatable metabolic daily rhythms were noted, and the Q_{10} temperature coefficient for metabolism between 15 and 25°C was 3.0. Metabolic rates of adult *H. suspectum* were lower than those of other squamates tested, and at 15°C periods of apnea contributed to a further reduction in the SMR. A reduced metabolic rate at lower BTs provides a significant energy savings during periods of inactivity. The relationship between BM and SMR is $SMR = 0.05\ BM^{0.65}$ at 15°C and $SMR = 0.17\ BM^{0.67}$ at 25°C. Beck (2005) presents a detailed discussion of metabolism and energetics in *H. suspectum*.

Few are seen above ground in southwestern Utah from late November through February, when cold ETs cause these lizards to hibernate. Typical overwintering sites are animal burrows, rock crevices, or small holes under boulders, preferably on south-facing slopes, where the lizard may bask at the entrance on warm winter days (Beck

1990; Lowe et al. 1986; Repp 1998a; Shaw 1950). A chamber may be excavated at the end of these holes, and several may hibernate in the same burrow. The same site may be used during summer estivation. Gila monsters suffer little from the several months of fasting, living on fats deposited during the active period (Bogert and Martin del Campo 1956). Winter BTs recorded by Beck (1990) were 11.3–15.2°C, while STs at 75, 25, and 5 cm depths were 7.0–15.3°C, 2.0–16.5°C, and −2.5–30.5°C, respectively.

Water availability is critical for *H. suspectum*. Captives languish if not provided free access to water for drinking and soaking. In nature, moisture loss is avoided by spending most of the time in burrows, where the relative humidity is nearly always high. Pianka (1967) even thought the lizard was restricted to areas of periodic but not infrequent rainfall (see above and Lovich and Beaman 2007). Gila monsters are often active after summer rains. Previously, it was thought that most of the moisture obtained by the lizard came from its prey, but it will drink from puddles when they are available.

Gila monsters experience high evaporative water loss (cutaneous, respiratory, and cloacal) relative to their BM. The lizard's beaded skin is permeable to water, and may have relatively high rates of water passage. Cutaneous water loss undergoes a consistent temperature-dependent increase between 20.5 and 40.0°C (Q_{10}, 1.61; with an evaporative water loss of 0.378–0.954 mg/g per hour). Ventilatory water loss does not exhibit a significant temperature-dependent response, but has a range of 0.304–0.663 mg/g per hour. During heat stress, *H. suspectum* may receive some relief from cloacal evaporative cooling, although this may result in a loss of critically needed body water. Evaporative water loss from the cloaca is low and relatively constant between 20.5 and 35.0°C (Q_{10}, 8.3×10^7), but it increases from 0.0008 at 35°C to 7.3 mg/g per hour at 40°C. This rise in cloacal evaporative water loss corresponds to a lowering of the BT relative to the ET. Dehydration to 80% of initial BM leads to a delay in the onset and an attenuation of the dramatic increase in water loss via the cloaca (De Nardo et al. 2004).

During the driest stress periods in Utah (July–September) and other Mojave Desert areas that do not receive significant summer rains, *H. suspectum* may remain in its burrow and estivate or become more crepuscular or nocturnal. While sleeping or estivating it may even lie on its back (Kauffeld 1943).

The bladder plays an osmoregulatory role by acting as a long-term physiological water reservoir. Water stored there serves as a reserve against moderate dehydration. The water buffers increases in plasma osmolality when food and water are unavailable. Water is absorbed from the urinary bladder into circulation, and this absorption and drinking water provide similar osmoregulatory benefits within 24 hours, although drinking water provides a more immediate osmotic benefit. During food and water deprivation, plasma osmolality increases 2.5 times faster in individuals with an empty bladder compared with those with a full one. During rehydration, stereotyped binge drinking increases the lizard's BM almost 22%, resulting in a 24% reduction in plasma osmolality and a substantial increase in bladder water within 24 hours (Davis and De Nardo 2007).

Home ranges of 2 Utah males and 1 female tracked by Beck (1990) were 66.2 ha, 32.6 ha, and 5.6 ha, respectively. The creatures traveled from only a few meters around their shelters to more than 1 km (mean, 213 m). The rate of travel was about 0.25 km per hour, with an average duration of about 51 minutes per movement event. The greatest total distances traveled were 1,190 m (15–30 April), 3,555 m (May), 3,150 m (June), and 2,000 m (1–15 July). In Nevada, the home range of 7 lizards (sex not given) averaged 64.2 (range, 21.4–147.4) ha (Gienger 2003); in New Mexico, 6 Gila monsters had an average home range of 36.5 ha: 3 males, 28.7–82.7 ha; 1 female, 25.2 ha; and 2 subadults, 11.7–15.0 ha (Gallardo et al. 2002); and in New Mexico, 7 individuals had an average home range of 58.1 ha: 4 males, 41.0–88.2 ha; and 3 females, 6.2–104.8 ha (Beck and Jennings 2003).

Heath (1961) attached trailing devices to 9 Gila monsters and followed them for 17 days. When on the move, they averaged 215 m per day, almost identical to that later noted by Beck (1990). The daily distance moved was 27–457 m, and the most rapid sustained movement was 320 m in 3 hours. However, movements while foraging are slow with intermittent periods of exploration and inactivity according to Jones (1983). Jones also reported increases in the mean daily movement in April and May (mean for both sexes, 26.5 m; males, 21.7 m; females, 32.4 m), and hence higher than in June and July (mean for both sexes, 2.8 m; males, 3.1 m; females, 2.4 m). Some movements may be triggered by the lizard trying to find a suitable humidity regime (De Nardo 2005).

Maximal speeds are low; those recorded by John-Alder et al. (1983) were 0.70 km per hour at 25°C and 1.03 km per hour at 35°C, but endurance declines as speed is increased toward the maximum. Beck et al. (1995) reported a maximum sprint speed of 1.7 km per hour. Helodermatid lizards, while not fast and unable to engage in brief periods of high-speed burst locomotion, are capable of sustaining slow movement for long periods (Beck and Lowe 1991). Beck et al. (1995) reported that *H. suspectum* is able to sustain a speed rate of 1 km per hour for up to 16 minutes.

H. suspectum can return when displaced short distances. Sullivan et al. (2004) translocated 25 Gila monsters considered nuisances by Phoenix, Arizona, residents distances of 0–25,000 m from their capture point. Eighteen translocated <1,000 m returned successfully to their original place of capture within 2 to 30 days. None of the 7 lizards translocated >1,000 m successfully homed, and experienced high daily rates of speed, were deprived the use of familiar refuges, and probably suffered greater predation.

Gila monsters rarely climb above ground, but Cross and Rand (1979) saw 2 climb 90 cm and 2.5 m up desert willows on separate occasions, and Campbell and Lamar (1989) have observed them as high as 5–7 m in trees.

Agonistic interactions between male *H. suspectum* have been observed in captivity by Demeter (1986) and Grow and Branham (1996), and in the wild by Lowe et al. (1986) and Beck (1990). Male combat in Arizona is restricted to the period April–July (Lowe et al. 1986), and in captivity is most intense during March–July (Grow and Branham 1996). Beck's (1990) analysis of a three-hour contest between two large males revealed similarities with varanid lizard and pitviper combat. Nine of the 10 major behavioral acts in combat interaction between 2 captive males recorded by Demeter (1986) were observed by Beck, who watched 2 fight in the field: (1) dorsal straddle, (2) frontal head nudge, (3) lateral head shove, (4) neck arch, (5) tail wrap, (6) lateral head bite, (7) lateral tail thrash, (8) dorsal head pin, and (9) roll. Seven additional major behavioral acts were also identified: (1) head raise (raising of head and stiffening of front limbs by inferior lizard following a neck arch; done in response to head nudge and shove by the other lizard); (2) circling (moving in a semicircular path around another lizard); (3) body twist (while in dorsal straddle, a twisting of body by the inferior lizard and placing its gular region against the neck of the superior lizard so that the bodies entwine; the superior lizard usually responds with a neck arch); (4) lateral rocking (rocking motion from side to side while in dorsal straddle, often resulting in a roll; initiated by the inferior lizard, this apparently serves to force the separation of the superior lizard); (5) dorsal body press (turning and pressing the dorsal body surface of the superior lizard against the back of the inferior lizard; usually performed under boulders, where the superior lizard presses the inferior lizard downward by pushing against a boulder with the forelimbs); (6) high stand (standing side by side, each lizard performs a head raise); and (7) scoop (pressing the snout under another lizard, scooping it upward). Upon approach, the lizards perform head nudge, shove, neck arch, and head raise, often switching roles; circling sometimes precedes these actions. The aggressor then mounts the other male in a dorsal straddle. The lizards repeatedly intertwine tails, untwining them at intervals. The inferior lizard typically responds to a dorsal straddle with a neck arch, while the superior lizard performs a dorsal head pin. The superior lizard may also thrash its tail. Considerable struggling may take place during the next phase, and the inferior lizard often walks with the superior lizard clinging to its back. Lateral rocking sometimes separates the lizards but, if not separated, they usually proceed into a body twist. Finally, the inferior lizard initiates a body twist, and two males remain in this position until one gains the superior position in a dorsal straddle or sometimes they break contact, from the force exerted during twisting. A typical bout lasts an average of 10 (range, 4–15) minutes, but several encounters may be included, extending more than 3 hours (Beck 1990). After fighting continuously for 3 hours, the lizard does not repeat the behavior for more than 10 hours (Beck 2009).

The objective during each bout is apparently to gain and maintain a superior position during a dorsal straddle. The inferiorly positioned lizard either rolls or twists its body in an effort to break the dorsal straddle and gain the top position.

Male combat suggests that Gila monsters have a definite social system; these usually solitary animals are not roaming aimlessly when searching for potential mates or foraging sites. Beck (1990) thought that common shelter use and seasonal movements that bring individuals back to communal areas, establishment of dominance through male-male combat, and scent-marking are all elements of a structured social system. During combat, a lizard may approach the limit of his physical endurance, especially while performing dorsal straddles, tail thrashes, and body twists (Beck 1990).

Captive females may become aggressive toward other females as oviposition approaches (Grow and Branham 1996).

REPRODUCTION: The exact size and age of sexual maturity in the wild is unknown, but those copulating and ovipositing in captivity have all been more than 40 cm in TBL (SVL, 27 cm; Peterson 1982a) and at least 3 years old (Seward *in* Beck 2005). SVL of females examined by Goldberg and Lowe (1997) that were vitellogenic or contained oviductal eggs was 23.9–29.5 cm. Stebbins (2003) gives the mature SVL range for *H. suspectum* as 22.8–35.5 cm.

The best descriptions of the sexual cycles are by Goldberg and Lowe (1997). They examined 112 Gila monsters, 100 from the vicinity of Tucson, Arizona; the others were from other sites in Arizona and Sonora, Mexico. They found the following sequential four-stage male cycle: (1) spermiogenesis (May into early June): testes at maximum size, with spermatids metamorphosing into sperm, the lumen of seminiferous tubules lined with sperm, and masses of sperm in the epididymides; (2) regression (late June): testes of minimum size, with depleted germinal epithelium and a few residual sperm; (3) regression (late June into August): predominantly spermatogonia and Sertoli cells in the seminiferous tubules; and (4) recrudescence (September–April): renewed germinal divisional epithelium containing primary and/or secondary spermatocytes, and empty epididymides.

The ovarian cycle reported by Goldberg and Lowe (1997) contained three stages: (1) inactivity (March–July): enlarged (>8 mm) follicles; (2) vitellogenesis (April–May): yolk deposition and follicle enlargement occurring (possibly yolking began previously during the winter retreat period; Lowe et al. 1986), with large follicles common and present into June; and (3) ovulation (May, August): eggs present in the oviducts. All of the females that they examined were approaching oviposition, so possibly annual reproduction is the female norm; however, observations on New Mexico females suggests that all mature females do not lay eggs each year (Beck 2005). Annual reproduction is probably tied to energy resources, which vary yearly in the wild. In captivity, well-kept females often reproduce annually; a frequently fed female maintained in a Tucson greenhouse produced clutches containing up to seven fertile eggs in each of five consecutive years (Martin *in* Beck 2005).

Hensley (1950) examined a female preserved in October that contained 5 large (diameter, 35–37 mm) and a series of 25 additional smaller (1.0–7.5 mm) ovarian eggs. Because the female was collected in May and had not been allowed to breed, he thought the larger eggs represented those that would normally have been fertilized and laid that year, and that the larger (6.8–7.5 mm) of the smaller groups of eggs presumably would have matured and been laid the following year. Three females collected in April and May by Gates (1956b) contained 40–52 undeveloped eggs, the largest being 5.5 mm long. A female at the Seattle Zoo laid eggs in 2 successive years (Wagner et. al. 1976).

Ortenburger (1924) reported a July copulation in the field; however, based on the above sexual cycles, this date is doubted by Goldberg and Lowe (1997). Lowe et al. (1986) concluded that most mating by wild Arizona *H. suspectum* mainly takes place in May, which seems more likely, and elsewhere mating in the wild occurs during the period April–June (Beck 1990, 1993; Degenhardt et al. 1996). Captives have mated in January, March–June, September, and December (Gates 1956b; Peterson 1982a; Shaw 1964, 1968; Wagner et al. 1976). Copulation lasts from 30 minutes to more than an hour, but most pairings take 40–60 minutes. The male initiates courtship by moving about with much tongue-flicking, apparently seeking the female's scent, and rubbing his cloaca on the ground in a sideways motion (Wagner et al. 1976). When a female is located, the male lies beside her and rubs his chin against her neck and back while holding her with his hind leg. She may resist and try to bite him while crawling out from underneath, but the male does not bite her. If receptive, she raises her tail and exposes her vent. The male then moves his tail beneath hers, brings their vents together, and inserts his hemipenis. Copulatory movements are slow and convulsive, occurring every 3 to 4 seconds (the mating posture is illustrated in Gates 1956b). Mating is a much subdued activity when compared to male combat, and some probably occurs underground during the time shelters are shared by the sexes.

Clutches of 1–13 (mean, 5.98; n = 45) eggs have been reported. The eggs are elongate, with white, rough, leathery shells; mean dimensions of 30 eggs we examined were length, 66.5 (range, 55.0–75.0) mm; diameter, 33.7 (range, 29.0–39.0) mm; and weight, 39.6 (range, 31.1–48.4) g. When first laid, they are shiny and moist, but with slow drying become soft and leathery in 15–20 minutes (Lowe et al. 1986).

Oviposition in the field occurs in July and August (Goldberg and Lowe 1997; Lowe et al. 1986; Stebbins 2003), but data concerning the duration of incubation are confusing (see Beck 2005). Captive IPs have lasted from 30 to 142 days (Eidenmüller 1993; Eidenmüller and Wicker 1992; Shaw 1964; Visser 2002), but average about 125 days (Peterson 1982a). However, Lowe et al. (1986) reported that in Arizona the eggs overwinter underground and hatch the following late April into early June, with most neonates appearing in May, after a natural IP of about 10 months. This would indicate that the eggs may undergo diapause, like that which occurs in the *Kinosternon* turtles of the region (see Ernst and Lovich 2009). Instead, however, neonates probably hatch, but remain in the nest cavity over winter and emerge the following spring, another behavioral pattern practiced by many other reptiles. By Lowe et al.'s (1986) calculations this would mean an approximately one-year period between copulation and hatching of the eggs (or emergence of the young), which seems very long in regard to the considerably shorter IPs recorded in captivity. More study of the ITs that occur during incubation in wild nests is needed to explain the differences between the IP in captivity compared to that in the wild.

Mean TBL and BM of 33 neonates taken from the literature were, respectively, 15.8 (range, 8.8–19.1) cm and 36.0 (range, 21.0–47.0) g; Beck (2005) reported SVLs of 13.5–14.1 cm and BMs of 23.1–33.4 g. Young Gila monsters have an extremely large amount of yolk in their abdominal cavities at hatching. Several days may be required between the pipping of the egg and the final emergence of the small lizard. The ground color of neonates is usually lighter in color than that of adults so that the dark patterns of the young are more pronounced.

GROWTH AND LONGEVITY: Wagner (*in* Beck 2005) reported that captive hatchling *H. suspectum* can show a 2.4-fold increase in TBL (16.0–37.8 cm) and an 11.7-fold increase in BM (34.5–403.5 g) within their first year. Woodin (*in* Shaw 1968) reported young brought to the Arizona-Sonora Desert Museum were 18.3–18.5 cm long in late spring (5 May–4 June), 17.2–18.1 cm on 20 July, and 20.5 cm on 19 August. Since he never saw newly hatched Gila monsters in the fall, Woodin thought these TBLs represented first-year growth since emergence from the nest. Eidenmüller and Wicker (1992) reported that a captive 16.8 cm TBL, 38.2 g hatchling grew to 22.1 cm and 71.5 g in approximately 6 months. Bogert and Martin del Campo (1956) and Tinkham (1971a) reported captive growth rates of 7–10 mm per year for adults. Growth in the wild is also slow in adults. Beck (1990) recorded average growth rates of 4.8 mm per year in lizards initially with <30.0 cm SVLs, and 2.1 mm per year in larger individuals. Beck (2005) gave the following mean annual SVL growth rates: year 1 (107–220 mm), 38.8 mm; year 2 (221–255 mm), 34.7 mm; year 3 (256–290 mm), 14.1 mm; year 4 (291–325 mm), 7.2 mm; year 5 (326–360 mm), 2.6 mm; and year 6 (361+ mm).

The Gila monster is relatively long-lived. One survived 32 years and 2 months in captivity at Ball State University (Cooper and List 1979); another (sex unknown) lived 27 years, 10 months, and 29 days after its initial capture (Snider and Bowler 1992); and

2 others survived more than 24 years in captivity (Crosman 1956). Beck (2005) reported minimum adjusted ages for wild-studied individuals of 15.0–23.5 years.

DIET AND FEEDING BEHAVIOR: A number of foods have been listed for the Gila monster, but there have been relatively few actual observations of natural predation by it. Reported prey include the eggs of desert tortoises (*Gopherus agassizii*), lizards' eggs (*Cnemidophorus tigris*[?], *Holbrookia maculata, Phrynosoma solare*) and young (*Cnemidophorus gularis*), snake eggs, eggs (and young?) of ground- and tree-nesting birds (*Callipepla gambelii, Toxostoma curvirostre, Zenaida macroura*), cottontail rabbits (*Sylvilagus audubonii*), jack rabbits (*Lepus californicus*), the young of ground squirrels (*Ammospermophilus leucurus; Spermophilus tereticaudus, S. variegatus*) and chipmunks (*Tamias* sp.), kangaroo rats (*Dipodomys deserti, D. merriami* [carrion]; *Dipodomys* sp.), cotton rats (*Sigmodon* sp.), and woodrats (*Neotoma albigula*) (Arnberger 1948; Barrett and Humphrey 1986; Beck 1986, 1990, 2005; Bogert and Martin del Campo 1956; Campbell and Lamar 2004; Coombs 1977; De Nardo 2005; C. Ernst, personal observation; Gallardo et al. 2002; Gienger and Tracy 2008a, 2008b; Hensley 1949, 1950; Jones 1983; Lowe et al. 1986; Ortenburger 1924; Ortenburger and Ortenburger 1927; Shaw 1948a; Spiess 1998; Stahnke 1950, 1952; Stebbins 1954; Stitt et al. 2003; Tinkham 1971a; Woodson 1949b; Zylstra et al. 2005). The above prey data were collected from adults. McGurty (2002) removed three reptile eggs and a small amount of mammal hair from an Arizona individual of neonate length (SVL, 13 cm), and 3 reptile eggs from another juvenile (SVL, 17.5 cm), so the egg-eating habit is there from the beginning.

Lowe et al. (1986) thought Arizona Gila monsters are carnivore specialists on the newborn young of rodents and lagomorphs, and that these basic foods are opportunistically augmented with birds and lizards and the eggs of birds, lizards, snakes, turtles, and tortoises. They also noted that young Gila monsters can consume more than 50% of their own BM at a single feeding, and that adults consume 35%. If an adult can eat three or four meals in the spring, it can probably store enough fat to supply it until the following spring.

The diet in Utah (based on 4 direct observations and 20 fecal samples) consisted of young cottontails, 42%; desert tortoise eggs, 29%; young ground squirrels, 16%; mourning dove eggs, 8%; and the carrion of young kangaroo rats, 4% (Beck 1990).

Nine Gila monsters observed during 1 April–13 July by Jones (1983) foraged mostly in April and May in areas where eggs of ground-nesting birds were abundant, but there was a decrease in egg availability in June and July. Thereafter, the lizards shifted their diets to small mammals and moved to areas where these were fairly plentiful. The Gila monsters rarely backtracked, and consumed an average of 46 (range, 36–83) % of the eggs found in each nest. Eggs were either swallowed whole or the contents were partially eaten after the egg shells were broken (captives also consume eggs in this way). No lizard foraged at the same nest twice, and only once did one consume all of the eggs available in a nest. The lizards ate an average of 1.3 eggs per day, foraged on an average of 3.3 eggs per nest, and raided an average of 0.4 nest per day. They spent an average of 11.4 minutes eating eggs, and males foraged on a greater number of nests (mean, 0.47) per day than females (mean, 0.32). The distance these *H. suspectum* traveled each day was positively correlated with the percentage of eggs eaten per nest in males, and there was also a positive relationship between lizard BM and the number of eggs consumed each day.

In contrast to the observations by Jones (1983), Beck (1990) reported that his Gila monsters spent 5–15 minutes excavating the nests of the birds incubating their eggs or protecting nestlings, and that all of the young or eggs in a nest were consumed.

Not all prey is taken at ground level; Feldner (*in* Beck 2005) watched a Gila monster consume the nestlings of a curved-billed thrasher (*Toxostoma curvirostre*) in a nest 3.6 m up a palo verde tree. In addition, the Gila monster's egg predation is not always easy, as female desert tortoises have been observed defending their eggs against it (Barrett and Humphrey 1986; Zylstra et al. 2005); however, the more agile lizard wins these bouts.

Those we have kept readily consumed chicken (*Gallus gallus*) eggs, canned dog food mixed with raw eggs, a dead hamster (*Mesocricetus auratus*), and small mice (*Mus musculus*) and rats (*Rattus norvegicus*). Captives have also eaten collared lizards (*Crotaphytus collaris*; Stebbins 1954).

The lizard feeds infrequently on large quantities of food, weighing up to 33% of its BM at one time (Beck 1990; Christel et al. 2007; Lowe et al. 1986), probably in response to the difficulties in finding prey at regular intervals in its severe habitat. Prey is rarely if ever envenomated, and, in fact, is commonly just overpowered and crushed in the jaws when seized and promptly swallowed. This supports the statements by Bogert and Martin del Campo (1956) and the suggestion by Beck (1990) that the role of the venom delivery system is primarily defensive.

The venom of the lizard does contain the digestive enzyme extendin-4 (Chen et al. 2006; Christel et al. 2007), but it only comprises <5% of venom dry weight. Christel et al. (2007) measured metabolic and intestinal responses to feeding in the presence and absence of circulating extendin-4. Following the ingestion of rodent or egg meals of up to 10% of the lizard's BM, metabolic rates peaked at 4.0- to 4.9-fold of SMRs and remained elevated for 5–6 days. Specific dynamic action of these meals (43–60 kJ) was 13–18% of total meal energy. Feeding caused significant increases in mucosal mass, enterocyte width and volume, and regulation of D-glucose uptake roles and aminopeptidase-N activity. Total intestinal uptake capacity for L-leucine, L-proline, and D-glucose became significantly elevated 1–3 days after a meal. However, absence of circulating extendin-4 had no impact on metabolism following a meal or the postmeal response of intestinal structure and nutrient uptake, but it did significantly increase activity of aminopeptidase-N in the intestines.

The absence of extendin-4 reduces blood levels of glucose and triglycerides, but its presence has a slow opposite effect after feeding (Christel and De Nardo 2007; Christel et al. 2007).

Prey is detected by hearing, sight, and, probably, olfaction. Wild, foraging Gila monsters frequently flick their tongues, particularly when nearing prey, and captives thoroughly examine the food given to them with their tongues. Bissinger and Simon (1979) reported a mean normal tongue-flick rate of 235.4 per 30 minutes for *H. suspectum*. The rate of tongue-flicking increases after a strike, and leads to olfactory searching by the lizard (Cooper and Arnett 1995). This strike-induced chemosensory searching lasts at least 10 minutes—rather long for a squamate reptile.

Cooper (1989) tested the use of chemical senses by *H. suspectum* while foraging. The lizard flicked its tongue more often in response to house mouse odors than to the odor of two control substances, but the number of tongue-flicks elicited was no greater for the mouse odor than for the control odors. Nevertheless, the lizard bit in a significantly greater proportion of tests with prey odors than it did with control stimuli. In additional tests, the lizard was able to distinguish with a greater rate of tongue-flicking between mice, water, and the nonprey items earthworms and fish (Cooper and Arnett 2003). It also exhibited greater mean tongue-flick scores when introduced to mouse blood (42.9) and egg yolk (79.6) than to cologne (approximately 28) or water (approximately 10), but tongue-flicked more (68) at a mixture of egg yolk and cologne (Hartdegen and Chiszar 2001), so odor strength may play a role.

VENOM DELIVERY SYSTEM: The venom apparatus of *H. suspectum* is essentially described above under *H. horridum*. It differs chiefly from that of *H. horridum* in having only one entrance to a groove on the dentary tooth. Each dentary tooth has an anterior groove, and most contain grooves on both the anterior and posterior surfaces. The 3 anterior-most dentary teeth are 1–3 mm long, while those following are longer, 4–5 mm, with teeth 4–7 the longest.

VENOM AND BITES: The Gila monster does not strike like a snake, but can lunge quickly forward or turn its head sideways and bite if one gets too close or it is held. Once it has sunk in its teeth, the lizard holds on with a bulldog grip (Arrington 1930) and "chews" in the venom. Bogert and Martin del Campo (1956) reported one clinging to an automobile door handle for 15 minutes. If one remains distant from a Gila monster, and does not foolishly reach into crevices or under rocks where it may be hidden, or avoids handling the lizard, no bite should occur (most recorded bites have occurred while the lizard was being held). *H. suspectum* are not aggressive unless provoked, but they are dangerous and their venom apparatus most likely evolved as a protective device (see above).

An adult Gila monster may yield 0.75–1.25 mL of venom (Brown and Lowe 1955; Minton and Minton 1969). The LD_{50} for a 20 g mouse is 80 mL (subcutaneous; Minton and Minton 1969) or 3.0 (2.1–4.2) mg/kg (Johnson et al. 1966), and 1.35 mg/kg BM (intracardial) for rats (Patterson 1967a). It is 24.7 (20.8–29.3) mg/mL of medium for *Paramecium multimicronucleatum* (Johnson et al. 1966).

Mice injected with sublethal doses of venom experience bulging and hemorrhaging of the eyes, hemorrhaging of the intestines and kidneys, and some bleeding in the lungs. After 2–3 days the kidneys are swollen with tubular necrosis or interstitial bleeding around the tubules and Bowman's capsule, resulting in kidney failure (Mebs 1972, 1978).

Van Denburgh and Wright (1900) reported that 5–6 drops of fresh venom will kill a dog in 15–18 minutes. Dogs and rats injected intravenously with venom from the Gila monster experienced a reduction in carotid artery flow, rapidly developing hypotension, changes in intrathoracic and postcaval blood pressures, tachycardia, and irregular breathing. The venom stimulated smooth muscles, causing noncholinergic and nonhistaminic contractions in rats and guinea pigs and temporary inhibition in muscles of the ascending colon in the rats. It does not affect clotting and prothrombin times in rabbits or cats (Patterson 1967a, 1967b; Patterson and Lee 1969).

The lethal dose for a human adult is probably about half the total yield (0.38–0.63 mL) from an adult lizard, but this quantity is rarely released in natural bites (Minton and Minton 1969). A bite results in severe pain (sometimes lasting for days; Caravati and Hartsell 1994), swelling, flushing at and around the site, nausea and vomiting, weakness, profuse sweating, dizziness, paresthesias, tachycardia, hypotension, swelling of lymph nodes near the site, and breathing difficulties. Even hatchlings can produce a bite that can be painful for several hours (Lowe et al. 1986).

While human envenomation is rarely severe, some deaths have occurred. According to Storer (1931), humans have died in as little time as 52 minutes after envenomation, but, if true, this must be exceptional. Woodson (1947) uncritically reported 29 (21%) fatalities in 136 cases, and Grant and Henderson (1957) reported 7 (29%) fatalities in 24 cases. Human fatalities from Gila monster bites are reviewed in Beck (2005), Bogert and Martin del Campo (1956), Brennan (1924), Brown and Carmony (1999), and Stahnke et al. (1970). Other nonfatal case reports are related by Arrington (1930), Beck (2005), Bogert and Martin del Campo (1956), Bou-Abboud and Kardassakis (1988), Brown and Carmony (1999), Caravati and Hartsell (1994), Duellman (1950), Engelhardt (1914), Grant and Henderson (1957), Heitschel (1986), Hooker and Caravati (1994), Miller (1995), Norris and Bush (2007), Piacentine et al. (1986), Preston (1989), Roller (1977), Russell and Bogert (1981), Shannon (1953), Shufeldt (1882), Stahnke et al. (1970), Streiffer (1986), Strimple et al. (1997), and Tay Zavala et al. (1979, 1980).

When a bite occurs, the lizard should be removed as soon as possible to limit the amount of venom flowing into the wound, and all rings or other jewelry at the bite site should be removed at once. The bitten part should be immobilized and mild pressure applied to control bleeding, but use of a tourniquet or constriction band, heat, ice, or shock with a stun gun should be avoided. Never pack the wound in ice or try to increase or decrease blood flow in the injured part. The victim should then be taken immediately to a hospital. There, her or his vital signs should be monitored every 15 minutes, and the diameter of the bitten part, and a point proximal to it, recorded every 15–30 minutes during the first 6–8 hours. The wound should be totally cleaned and sterilized, and any teeth removed. An antitetanus agent (antitoxin/toxoid) and broad-spectrum antibiotic should be administered. Symptoms should be treated as they occur, and an infusion of electrolyte solutions, antishock drugs, or pain killers administered as needed. Necessary blood tests should be administered to monitor blood cell levels, and, if indicated, an electrocardiogram taken. The bitten part should be immobilized in a functional position at, or near, heart level. Except on the first day, the bite site should be soaked for at least 15 minutes 3 times a day in 1:20 aluminum acetate solution (Burow's solution), during which time the fingers (or toes) should be exercised. Puncture wounds should be covered with a light, sterile dressing (Beck 2005; Russell and Bogert 1981).

There has been no urgency in producing a commercial antivenom since bites seldom occur (and most come from handling captive lizards). Because human envenomation by *H. suspectum* is rare, there is currently no commercially available antivenom to counteract its venom; but two have been produced on an experimental basis (Russell and Bogert 1981): (1) Poisonous Animal Research Laboratory, Arizona State University, and (2) Venom Poisoning Center, Los Angeles County, University of Southern California Medical Center.

Some research has been conducted on the effect the species' venom has on other animals. Patterson (1967a) reported that in dogs and rats, Gila monster venom causes an immediate reduction in carotid artery blood flow to the brain, and that these animals suffer hypotension, increased heart rates, and breathing difficulties. The venom may also stimulate smooth muscle tissue (Patterson 1967b), but has little or no effect on blood coagulation (Patterson and Lee 1969).

Components of the venom of *H. suspectum* stimulate insulin production and slow the release of glucose. A drug, Byetta® (exenatide), manufactured by Amylin Pharmaceuticals and Eli Lilly and Company, is now being used to treat type 2 diabetes.

PREDATORS AND DEFENSE: Known or potential predators of *H. suspectum* are snakes (*Crotalus mollosus, Lampropeltis getula*); large predatory birds—golden eagles (*Aquila chrysaetos*), hawks (*Buteo jamaicensis, Parabuteo unicinctus*), vultures (*Cathartes aura, Coragyps atratus*), owls (*Bubo virginianus*), ravens (*Corvus corax, C. cryorileucus*), and roadrunners (*Geococcyx californianus*); and carnivorous mammals—armadillos (*Dasypus novemcinctus*), badgers (*Taxidea taxus*), canids (*Canis familiaris, C. latrans; Urocyon cinereoargenteus; Vulpes macrotus*), cats (*Felis catus, Leopardus pardalis, Lynx rufus, Panthera onca, Puma concolor*), coatis and raccoons (*Nasua narica, Procyon lotor*), opossums (*Diadelphis virginiana*), and skunks (*Conepatus mesoleucus; Mephitis macroura, M. mephitis*) (Beck 2005; Bogert and Martin del Campo 1956; Brown and Carmony 1999; Cashman et al. 1992; Funk 1964a; Hensley 1949; Russell and Bogert 1981; Sullivan et al. 2002). Some of these animals may only prey on lizard carrion when the opportunity arises.

Actual observations of predatory attacks on the lizard are rare. Most predation probably occurs on the egg and juvenile stages. Hensley (1949) mentions an attack by a domestic dog, and Funk (1964a) reported that a black-tailed rattlesnake (*Crotalus molossus*) purportedly ate one. Ivanyi (*in* Beck 2005) reported a small *H. suspectum* was killed by coyotes when it inadvertently entered their outdoor enclosure at the Arizona-Sonora Desert Museum. Lowe (*in* Bogert and Martin del Campo 1956) saw a badger expel a Gila monster from its burrow, but the badger did not attack it.

Bogert and Martin del Campo (1956) listed three defensive behaviors by *H. suspectum*. First, it will try to escape by slowly crawling backward, and if possible, seek refuge in an inaccessible burrow or other site. If it is prevented from fleeing, the Gila monster will try to intimidate its attacker. This involves hissing and open-mouth threats; if the enemy approaches too close, the lizard lunges toward it. Finally, as a last resort, should it have an opportunity, it will bite ferociously. If handled, wild Gila monsters will squirm, try to bite, and defecate.

In the past, investigators considered the venom apparatus as primarily a prey-subduing adaptation (Pregill et al. 1986), but Beck (2005), Bogert and Martin del Campo (1956), and Lowe et al. (1986) thought it more likely to have evolved only as a protective mechanism. Because a large part of the lizard's food consists of eggs or nestlings that can be swallowed without previous envenomation, and the rodents it eats are swallowed following the first crunch in the jaws without being envenomated (Beck 1990), there is no reason to doubt that venom evolved as part of their defensive mechanism.

The body color and pattern may also serve two defensive purposes. Against the right-colored soils and vegetation Gila monsters tend to be cryptic, particularly at twilight or at night. Color and pattern may also serve as a warning device (Cloudsley-Thompson 2006), supplementing a venom apparatus that evolved primarily as a defensive device capable of causing excessive pain rather than death. Greene (1988) thought the black-and-yellow pattern of the Mexican *H. horridum* to be primitively cryptic, whereas the black-and-pink coloration of *H. suspectum* is derived and probably intended to warn.

PARASITES AND PATHOGENS: More data are available on the parasites and pathogens of *H. suspectum* than on those of *H. horridum*. Several types of parasites have been identified from the Gila monster. The coccidian protozoan *Eimeria becki* has been isolated from fecal samples (Upton et al. 1993), and Mahrt (1979) found unidentified hemogregarines in the blood. Several endoparasitic helminths have also been reported. The cestodes *Oscheristica whitentoni* (Goldberg and Bursey 1991) and unidentified tapeworms (Bogert and Martin del Campo 1956); unidentified encysted larval acanthocephalans (Bogert and Martin del Campo 1956); and the nematodes *Hexamita* sp., microfilaria of *MacDonaldius andersoni*, and *M. seetae* [?], *Oswaldocruzia pipiens*, *Piratuba mitchelli*, *Skrjabinoptera phrynosoma*, *Splendididofilaria cerophila*, *Trichomonas* sp., and unidentified microfilaria have all been found in the species (Baker 1987; Chabaud and Frank 1961a, 1961b; Goldberg and Bursey 1990, 1991; Griner 1983; Mahrt 1979; Ryerson 1949; Smith 1910; Stabler and Schmittner 1958; Yamaguti 1961). Additional reported ectoparasites include lice [?] (Landry *in* Woodson 1949a), mites (Hoffman and Sánchez 1980), and ticks (Campbell and Lamar 2004; Smith 1910).

Like other reptiles, the Gila monster suffers from diseases. Captives have experienced gout-like urate deposits (Appleby and Siller 1960). Others have died of pneumonia (*Pseudomonas reptilivorus*; Cardwell and Ryerson 1940; Jacobson 1984; Landry and Wiley *in* Woodson 1949a), or developed *Salmonella enteritis* (Dolensek and Cook 1987). Canker sores, melanomas, neoplasms, and tumors are not uncommon in captives (Cooper 1968; Hatkin 1984; Jacobson 1980, 1981, 1984; Machotka 1984; Tinkham 1971a; Woodson 1949a). Hatkin (1984) reported a hepatic thrombus associated with fungal hyphae, and a *H. suspectum* at the University of Kentucky developed a cyst-like tumor between the base of its tail and left kidney (C. Ernst, personal observation). Hanel et al. (1999) isolated the bacterium *Clostridium innocuum* and an additional unidentified species of the genus from the blood of a zoo captive *H. suspectum*.

POPULATIONS: Few data have been published on the population structure and dynamics of the Gila monster. Lowe et al. (1986) thought it not rare or uncommon, but infrequently seen in Arizona. It is assumed that Gila monster populations seldom reach high population densities. In southwestern New Mexico, a population had an estimated density of 5/km^2 (Beck 1994), similar to that recorded for a southern Arizona population, 15–31/4 km^2, by Martin and Lowe (*in* Beck 2005).

Beck (1990) marked 27 lizards at a 2 km^2 site in southwestern Utah; 14 (52%) were adults with SVLs of at least 32 cm, 13 (48%) were more than 32.5 cm, but only 1 (3.7%) immature individual was caught. The total BM of 22 of these lizards was 10.52 kg (mean, 479 g; range, 145–880 g). In comparison, the SVL size classes of 27 wild Arizona Gila monsters were (loosely interpreted from a bar graph) <22 cm, 8%; 22.1–25.5 cm, 8%; 25.6–29 cm, 15%; 29.1–32.5 cm, 52%; and 32.6+ cm, 22%; similarly, the same SVL classes for 25 wild New Mexico lizards were 9, 9, 12, 42, and 33%, respectively (Beck 2005). At a different site in southern Arizona, Martin and Lowe (*in* Beck 2005) found the majority (70%) of the lizards to be large adults more than 29 cm SVL; subadults composed less than 10% of the sample. Beck (2005) measured 206 *H. suspectum* in the University of Arizona preserved reptile collection and recorded the following numbers of individuals in each SVL class: 10.7–14.9 cm, 28; 15.0–22.0 cm, 28; 22.1–22.5 cm, 40; 25.6–29.0 cm, 67; 29.1–32.5 cm, 41; and 32.6+ cm, 2. Apparently, populations of *H. suspectum* are skewed toward larger, mature individuals. A table generated by Beck (2005) to show the annual mortality and survival rates of adults indicated, respectively, means of 15.0 (range, 6.5–27.5) % and 85.0 (range, 72.5–93.5) % for these two dynamic factors. Recruitment of individuals into the population was low, 4–15 individuals per year.

It is very difficult to determine the sex of a living Gila monster (see "Recognition"), so sex ratios of natural populations are uncertain. However, Bogert and Martin del Campo (1956) presented data based on 92 preserved *H. suspectum* greater than 22.5 cm SVL. Their total sample included 45 males and 47 females, a 1:1 ratio, but the individual sex ratios of the two subspecies were skewed toward one sex or the other: *H. s. cinctum*, 2.7:1 (16:6) and *H. s. suspectum*, 1:1.4 (29:41). The sex ratio of the 196 (those that could be sexed) of the 206 preserved *H. suspectum* examined by Beck (2005) was 103 males to 93 females (1.11:1), not significantly different from 1:1.

Human beings present the most danger to the Gila monster. In the past, they often tried to kill these lizards on sight. Tinkham (1971a) reported that some are destroyed by persons who intentionally run their automobiles over them, and claims that he saw a driver swerve out of his way to hit one. Many were previously killed to stuff as ornaments for the tourist trade, and others were collected for the pet trade.

Today, however, habitat destruction plays a larger deteriorating role, with the continual spread of agriculture and urbanization, more and more road construction, natural and manmade fires, and the introduction of foreign plants. Although *H. suspectum* is somewhat resilient, these practices often degrade or fragment the habitat so much as to make it unsupportable for the lizard.

Heloderma suspectum was one of the first legally protected reptiles (Grant 1952), and is now protected everywhere it occurs in the United States and Mexico. The pet and tourist trades should no longer be major factors, and it should never be molested or killed. But how effective these laws are in preventing habitat destruction is questionable. It has been our experience that developers usually gain the upper hand over conservation efforts; this must be reversed if *H. suspectum* is to survive anywhere but in the most inhospitable habitats to humans. Translocation of urban Gila monsters, as with other reptiles, has been only partially successful (Sullivan et al. 2004).

The species is currently protected in Arizona, New Mexico, and Nevada, and the subspecies *H. s. cinctum* has state protection in California and Utah.

REMARKS: Beck (2005), Bogert and Martin del Campo (1956), and Russell and Bogert (1981) listed several myths regarding the Gila monster: (1) the tongue is a stinger (no, it is part of an olfactory mechanism); (2) the breath is venomous (the venom, sometimes mixed with saliva, must be introduced into a puncture to have an effect); (3) the venom may be spat (the venom is not purposely spat at an attacker; however, hissing lizards may sometimes accidentally expel droplets of saliva); (4) it has no anus and this is what makes it so mean and venomous (no, it has a cloaca and an anal vent like other lizards); (5) it jumps at its tormentor (Gila monsters have no jumping ability); (6) the lizard turns on its back so that the venom produced in glands in the lower jaw can better be injected (they seldom, if ever, turn over onto their backs when defending themselves); (7) they will bite out pieces of flesh (the teeth are not adapted for cutting, only chewing); (8) Gila monsters are not immune to the venom of their species (they are immune to their own venom [Brown and Lowe 1954]; venom is secreted when they bite each other); and (9) Gila monsters are the product of hybridization between rattlesnakes and nonvenomous snakes, especially those of the genus *Pituophis* (this is ridiculous).

The species has been reviewed by Beck (2004a, 2004c, 2004d, 2005, 2009), Bogert and Martin del Campo (1956), and Campbell and Lamar (2004), and Beaman et al. (2006) presented a bibliography. Brown and Carmony (1999) have published a generalized book on the lizard's natural history.

Elapidae
Elapid Snakes

This family contains 60 genera and 297 species of advanced, dangerous snakes (David and Ineich 1999) that have potent neurotoxic venom, and includes the coralsnakes, cobras, mambas, and seasnakes. The family is well represented in Australia, southeastern Asia, Africa, and South America (3 genera, 72 species; Campbell and Lamar 2004), but only 5 species occur in the area covered in this book.

Within the Colubroidea, the Elapidae has genetic links to the Atraspididae; some African, Malagasy, and Asian colubrid genera; and *Homoroselaps* (Slowinski and Lawson 2005). It is not closely related to the Homalopsine and Pareatine colubrids, or to the family Viperidae. The only major feature separating elapids from colubrids is their hollow fangs at the anterior end of the maxillae.

The Elapidae is probably the most derived of all snake families. It is best characterized by its proteroglyphous fang structure, and most of our knowledge of its morphology is based on the study by Bogert (1943). One or more permanently erect, hollow fangs are present near the front of each shortened maxillary bone (fig. 9). Unlike the maxillary of the Viperidae, that of the Elapidae has very limited mobility (*Acanthophis* sp.; Fairley 1929). Because the fangs are permanently erect and the maxillary bone is basically rigid, an elapid cannot stab its prey as do solenoglyphous viperids. It must bite its victim, and because its fangs are short, normally has to chew its venom into the wound.

The prefrontal bone is rigidly attached to the frontal bones, and the maxillary is not hinged where it is attached to the prefrontal, although there are muscle attachments to the ectopterygoid bone and to the posterior end of the maxilla. Except in *Micruroides* and the seasnakes, the fangs are the only teeth on the maxillae and are longer than the other teeth on that bone. All maxillary teeth are essentially the same size in the North American seasnake, *Pelamis platura*. The fangs are permanently erect and incapable of folding upward against the roof of the mouth when it is closed, as occurs in the Viperidae. Instead, the fangs fit into pockets on the inside of the mandibular gums when the mouth is closed. The fangs are shorter compared to those of snakes belonging to the family Viperidae (see the measurements presented in the individual species accounts of both families), but fang length, as in the Viperidae, is positively

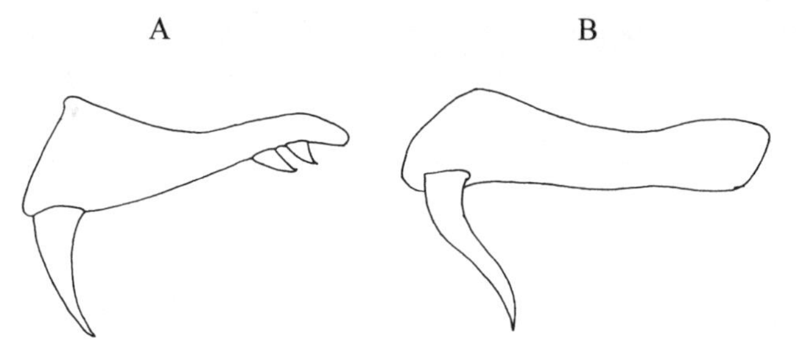

Figure 9. The proteroglyphous maxillary fang arrangement of North American terrestrial elapids: A, *Micruroides*; B, *Micrurus*.

correlated with the individual snake's body length. Correspondingly, fang width, thickness, and the lengths of the entrance lumen, discharge orifice, and points distal to the discharge orifice are correlated to fang length. Elapid fangs are short because large, permanently erect fangs would require a deepening of the mouth cavity to prevent the fangs from perforating the floor of the mouth.

Elapid fangs are broad at their base but gradually taper to needle-like points, and are only slightly curved as compared to the fangs of viperids (Klauber 1939). Each fits into a socket in the maxillary bone and has an opening adjacent to the distal end of the venom duct. The venom flows from the duct through an entrance lumen into the fang's central canal (or, in some elapids, down an incompletely enclosed groove on the outer surface of the fang). The canal extends downward through the fang to a variably long discharge orifice on the anterior surface of the fang, just above a solid tip. Both the wall of the venom canal and the outer surface of the fang are covered with enamel.

Fangs become worn over time or are sometimes broken, and are replaced by a process that is essentially identical in both the Elapidae and Viperidae (Bogert 1943; Burkett 1966; Ernst 1962, 1982a, 1982b; Fitch 1960; Klauber 1939, 1972). A series of 5–6 replacement fangs is normally present under the gum on each maxillary bone in both families of snakes (Bogert 1943; Ernst 1982a; Klauber 1939, 1972). These are located behind and above the functional fang in alternating right and left sockets in the bone, and replacement is from front to back. The teeth in either the right or left series are shed in separate successions; one series is replaced before shedding is initiated in the second series. The fangs in the replacement series are approximately perpendicular to the anterior curved portion of the maxillary bone to which they will eventually become attached. The functional and first replacement fangs lie essentially side by side for a while before the functional one to be replaced is shed. The functional fang is alternately found attached to the inner and outer adjacent sockets. Of these two sockets, the inner is usually located slightly anterior to the outer one.

The venom is produced in a pair of large serous, tubular, venom glands. One is located on each upper jaw below and behind the orbit and above the rear corner of the jaw. The gland usually bends downward around the corner of the mouth to the lower jaw, particularly when filled with venom. When the gland is empty, its posterior corner is pushed upward, eliminating the curve. The gland's structure and discharge mechanism are essentially those described in the Viperidae account.

The venom duct is completely surrounded, at least proximally, by masses of a narrow serous accessory gland containing epithelial mucous-secreting cells that may act as valves to regulate the flow of venom to the fang. Venom drawn from the lumen of the venom gland is less toxic than venom taken from the fang, so the accessory gland's secretions, although nontoxic, may activate some venom components. As in the Viperidae, the venom duct ends adjacent to the fang, within a sheath of connective tissue surrounding the fang's base. This sheath helps direct the flow of venom into the fang's canal and outward into the prey. Also, the oral mucosa of the lower jaw contains small, tubular, infralabial glands that produce small amounts of a secretion similar to the venom produced in the true venom glands and whose ducts open near the bases of slightly grooved mandibular teeth. The premaxillae are toothless, but teeth occur on the pterygoids, palatines, and dentary bones.

Some elapid species are capable of erecting their palatine teeth to aid in moving prey into the esophagus (Deufel and Cundall 2003). The dentary lacks a coronoid bone. Postfrontal, coronoid, and pelvic bones are absent. The hyoid is *Y*- or *U*-shaped with two superficially placed, parallel arms. Body vertebrae have short, recurved hypapophyses. The right lung is present, but the left lung is absent or greatly reduced; seasnakes may have an additional tracheal lung (Wallach 1998). The spiny hemipenis has a bifurcate centripetal sulcus spermaticus. Dorsally, the head is covered with enlarged plates, but loreal scales are absent. The pupils are round. Body scales are usually smooth; photomicrographs of the micro-ornamentation of body scales of various species are presented in Roze (1996). Ventral scutes are well developed only in arboreal and fossorial species. Reproduction is oviparous or ovoviviparous.

Fossil elapids date from the Miocene of France (*Palaeonaja* Hoffstetter, 1939). Indeterminate Miocene species of *Micrurus* have been found at a Barstovian site in Webster County, Nebraska (Holman 1977), and at a Hemphillian site in Alachua County, Florida (Auffenberg 1963), so the family had already reached the Americas by this epoch. Two major clades exist; one, the Elapinae, consists mostly of terrestrial African, American, and Asian species, and another, the Hydrophiinae, is composed of Australian seasnakes (Slowinski et al. 1997). Species from both clades are present in North America—the coralsnakes (*Micruroides euryxanthus, Micrurus distans, M. fulvius*, and *M. tener*) and the seasnake (*Pelamis platura*). Molecular studies by Keogh (1998) have indicated that the Elapinae is paraphyletic, with the kraits (*Bungarus*) and seasnakes a sister group to all other elapids. Also, immunological assessment of serum albumins by Cadle and Sarich (1981) shows that the American *Micruroides* and *Micrurus* are close allies with other elapids instead of derivatives of South American colubrids, and that a late Oligocene–early Miocene separation between the New and Old World elapid lineages may have occurred.

The origin of seasnakes is not clear. There seem to be two marine clades, the seasnakes (Hydrophiinae) and the seakraits (Laticaudinae), which evolved separately from Australian terrestrial elapids. Rasmussen (1997) even suggested the possibility of three independent origins of seasnakes, but refined this to two clades in 2002. The fish-egg-eating hydrophiid *Aipysurus eydouxii* seems to have its own venom evolutionary trajectory (Li et al. 2005). In addition to its loss of fangs and greatly atrophied venom glands, its venom's only neurotoxin gene, a three-fingered one, has a dinucleotide deletion, resulting in loss of neurotoxic activity. The venom also contains 16 unique phospholipase A_2 clones and cloning of reverse transcription-polymerase chain reaction products. Apparently less diversification of the phospholipase A_2 has occurred in this species than in other seasnakes. These changes are a consequence of the snake's shift to a different ecological feeding niche.

American coralsnakes exhibit highest species richness in the equatorial region, becoming scarcer at higher latitudes. Neither TBL nor range size increases with latitude, so there is little support for Bergmann's Rule. Range area and median range latitude are, however, positively correlated above 15°N, indicating a possible Rapoport Rule effect at high northern latitudes. Available continental area strongly influences geographic range size, and range size is positively correlated to TBL (Reed 2003).

One of the most interesting aspects of elapid biology is the coralsnake mimicry question. Whether or not coralsnakes and other red, yellow, and black banded colubrid snakes form a mimicry complex has been discussed for years (see Savage and Slowinski 1990 and 1992 for a detailed classification of all coralsnake color patterns), and several opposing theories exist. As many as 115 species (approximately 18%) of harmless or mildly toxic species of all American snakes are regarded as coralsnake mimics (Savage and Slowinski 1992).

Some mimics apparently migrated from areas of sympatry with coralsnakes to allopatric areas, but kept the DNA phenotypes for mimicry (Harper and Pfennig 2007, 2008). Mimics living on the edge of their model's range, where the model was rare, resemble the model more closely than mimics in the center of the model's range, where it was common. When free-ranging natural predators on the edge of the model's range were given a choice of attacking replicas of good or poor mimics, they avoided only the good mimics. In contrast, predators from the center of the model's range attacked both good and poor mimics equally. Generally, although poor mimics may persist in areas where their model is common, only the best mimics should occur in areas where their model is rare. So only the best mimics may occur at the edge of the model's range (Harper and Pfennig 2007).

Past objections to the mimicry theory have been based largely on the supposition that once bitten by a coralsnake, a predator will die from the bite (Rijst 1990); however, the apparent poor vision and relatively poor venom delivery system of coralsnakes preclude that every bite will be fatal (Lindner 1962; McNally et al. 2007; Russell 1967a).

Another view is that coralsnake patterns are primarily aposematic (Gehlbach 1972; Hecht and Marien 1956; Mertens 1956, 1957; Pough 1988; Savage and Slowinski 1992), particularly since there is an absence of countershading with the continuation of the dorsal bands onto the venter. Also, the patterns may be cryptic, presenting a visual illusion of disruption as the snake crawls. Can coralsnake patterns be aposematic, cryptic, and mimic at the same time? Greene and Pyburn (1973) thought this concept perfectly feasible. Grobman (1978) suggested that similar color patterns have arisen independently of natural selection in unrelated sympatric species occupying similar habitats (pseudomimicry), while Wickler (1968) suggested that the dangerously venomous coralsnakes are the mimics of mildly venomous colubrid snakes, not the models for these species.

In review discussions, Greene and McDiarmid (1981) and Pough (1988) concluded that field observations and experimental evidence refute previous objections to the coralsnake serving as the Batesian model in the mimicry hypothesis, and that their bright colors serve as warning signals to predators. Supporting this, Smith (1975) showed that the naive young of some tropical reptile-eating birds instinctively avoid the red, yellow, and black banded pattern of coralsnakes. Coralsnakes seem to exhibit near-infrared reflectance that renders the snake extremely visible against a leaf litter substrate (Krempels 1984), and this may advertise a warning to potential avian predators.

Hallwachs and Janzen (*in* Pough 1988) reported that in Costa Rica, scavengers (mostly birds) that quickly consume road-killed snakes of other species typically leave dead coralsnakes alone, and in a laboratory study by Brodie (1993), color-banded models were attacked by birds less often than were brown ones. In tests using banded models of snakes sympatric or allopatric to predators and a plain brown control, models of allopatric banded snakes were attacked more often than were sympatric ones. In addition, during Costa Rican field tests using tricolor (red-yellow-black) plasticine models, no birds attacked the models (Brodie and Janzen 1995). From these observations of avoidance, Pfennig et al. (2001) concluded that predators avoid coralsnakes in sympatry with them. However, their data also indicate that avoidance of banded patterns should weaken with increasing latitude and elevation as coralsnakes become rarer. Apparently, similar colored and patterned harmless snakes, or mildly venomous ones, would gain some protection from bird predation with resemblance to sympatric coralsnakes. This may not be true with mammalian predators. Caged, wild coatimundis (*Nasua narica*) confronted with sympatric live snakes, including two species of coralsnakes, did not avoid the coralsnakes or other snakes resembling them, but they did avoid coralsnake models, showing that tests with abstract snake models cannot unconditionally serve as evidence of an aposematic function of coralsnake coloration (Beckers et al. 1996). Pfennig et al. (2007) reported evidence that variable levels of predation on good mimics probably reflect frequency-dependent (apostatic) predation.

It has been supposed that coralsnakes are nocturnal, and that their bright colors would be meaningless at night, but data from Greene and McDiarmid (1981), Jackson and Franz (1981), and Neill (1957) show *Micrurus fulvius* to be mostly diurnal. Studies of concordant geographic pattern variation by Greene and McDiarmid (1981) strongly suggest that some colubrid species of *Atractus, Erythrolamprus, Lampropeltis*, and *Pliocerus* are involved in mimicry systems with sympatric coralsnakes.

Parent birds may respond to the sight of a nearby snake by actions that reveal the location of their nest. If bright, banded snake patterns elicit this behavior more than do cryptic patterns, then banded snakes that eat nestlings should have a hunting advantage over cryptic snakes. Smith and Mostrom (1985) examined this theory in field tests with American robins (*Turdus migratorius*) and a red, yellow, and black banded coralsnake model. The model elicited no more response than did a plain brown snake model.

Based on current evidence, the most likely advantage of a bright, banded pattern to snakes is to confer protection against predators, either as camouflage or as warning coloration in mimicry systems.

KEY TO THE GENERA OF NORTH AMERICAN ELAPIDAE

1a. Marine, tail oar-like, nostrils valved, olive dorsally, yellow ventrally .... *Pelamis*

1b. Terrestrial, tail not oar-like, nostrils not valved; body pattern of red, yellow (white), and black bands.. 2

2a. Most of head black, face black to level of angle of lower jaws; first broad body band behind the light collar red; usually 1–2 teeth on maxilla behind the fang .... *Micruroides*

2b. Only front of head black, face black to just behind eyes; first broad body band behind the light collar black; no teeth on maxilla behind the fang..... *Micrurus*

Micruroides Schmidt, 1928
Sonoran Coralsnakes
Coralillo de Occidente

Micruroides is monotypic, containing only one species, *M. euryxanthus*. Cadle and Sarich (1981), based on immunological comparisons of serum albumins, showed that the New World coralsnake genera *Micruroides* and *Micrurus* are more closely related to other elapids, rather than derived from a lineage of South American colubrids. Their results suggest a late Oligocene–early Miocene separation between the New and Old World elapid lines. Murphy (1988) discovered that *Micruroides* and the seasnake, *Pelamis platura,* share five L-lactate dehydrogenase heterotetramer isozymes, while *Micrurus fulvius* has only two of the isozymes. He proposed the common ancestor of coralsnakes had a five-banded isozyme pattern. Further studies based on allozymic and morphological characters show the two North American coralsnake genera are not closely related even though they occur together at the northern extent of the range of North American coralsnakes (Slowinski 1995), with *Micruroides* more closely related to Asian kraits (*Bungarus*) and *Micrurus fulvius* and *M. tener* most closely related to several tropical *Micrurus*.

Micruroides euryxanthus (Kennicott, 1860)
Western Coralsnake
Coralillo Occidental

RECOGNITION: This small (TBL_{max}, 66 cm; Babb *in* Rossi and Rossi 1995), red, white (or yellow), and black banded snake has most of its face black posteriorly to the level of the angle of the lower jaws, and the first broad body band behind the light collar is red. The red and white (or yellow) bands touch, but both the red and black bands are usually wider than the shorter light bands, which may be restricted to only a few scale rows in width in some individuals. The banding extends onto the venter, and follows a red-yellow-black-yellow-red pattern (the "tricolor monad" banding pattern of Savage and Slowinski 1990). Eleven to 13 black bands cross the body; the tail has 1–2 black bands separated by white or yellow. Some black pigment may be present in the red bands. Dorsal body scales are smooth and lie in 17 anterior rows and 15 rows at midbody and immediately in front of the vent. There are 205–250 ventrals and 19–32 subcaudals present; the anal plate is divided. Dorsal head scalation includes 1 wider than long rostral, 2 internasals, 2 prefrontals, 1 frontal, 2 supraoculars, and 2 parietals. Lateral head scalation includes 2 nasals (the prenasal touches the first supralabial), 1 preocular (which touches the postnasal), 2 (3) postoculars, 1 + 2 temporals, 7 supralabials, and 6–7 infralabials. Loreal and subocular scales are absent. On the lower jaw is a mental scale that is separated from the 2 small chin shields by a pair of infralabial scales. Following these are several rows of gular scales before the ventral scutes.

The 7–8 subcaudals long, bifurcate hemipenis has a divided sulcus spermaticus; it approaches a capitate condition with the spines expanded in the distal region but gradually becoming shorter toward the apex, which bears a small papilla-like projection. A longitudinal naked fold begins at the base and extends almost parallel to the sulcus spermaticus. A few scattered small spinules may also be present up to the zone of larger spines. It is illustrated in Dowling (1975).

The dental formula consists of 4 palatine and 7 dentary, but no pterygoidal, teeth; the hollow fang on the maxilla is followed, after a diastoma, by 1–2 solid teeth.

Males have 11 (range, 9–13) black body bands, 224 (range, 205–236; 250 in 1 individual from Isla Tiburon; Zweifel and Norris 1955) ventrals, 25–27 (range, 23–32) subcaudals, and TLs 8–9% of TBL; females have 12–13 (range, 9–16) black body bands, 233 (range, 219–245) ventrals, 24 (range, 19–27) subcaudals, and TLs 6–7% of TBL.

GEOGRAPHIC VARIATION: Three subspecies are currently recognized. *Micruroides euryxanthus euryxanthus* (Kennicott, 1860), the Arizona Coralsnake or Corallilo Occidental, ranges from south-central Arizona and southwestern New Mexico, south to northern Chihuahua and Sonora, Mexico; an isolated population occurs on Isla Tiburon, Mexico. It has fewer than 95 rows of mid-dorsal red body and tail scales, varying amounts of black pigment in the red bands, 225–245 ventrals in females, and 214–236 ventrals in males. *Micruroides e. australis* Zweifel and Norris, 1955, the Sonoran Coralsnake or Coralillo de Sonora, occurs in southwestern Chihuahua and southern Sonora, Mexico. It has yellow bands 2.5–4.0 scales long, 100+ rows of red mid-dorsal body and tail scales, no black pigment in the red bands, 225–228 ventrals in females, and 213–226 (250; Zweifel and Norris 1955) ventrals in males. *Micruroides e. neglectus* Roze, 1967, the Mazatlan Coralsnake or Coralillo de Mazatlán, has only been found in Sinaloa, Mexico, at Cosaliá and in the vicinity of Mazatlan. Only 3 specimens (2 males and 1 female) have been collected (Hardy and McDiarmid 1969; Meik et al. 2007; Roze

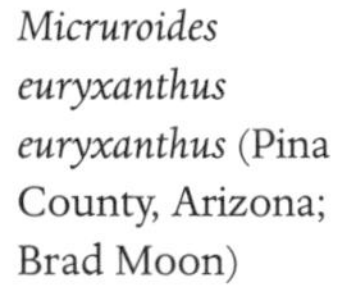

Micruroides euryxanthus euryxanthus (Pina County, Arizona; Brad Moon)

Micruroides euryxanthus euryxanthus (Cochise County, Arizona; Paul Hampton)

1967). It has yellow bands 0.5–2.0 scales long, 100+ rows of red mid-dorsal and tail scales, no black pigment in the red bands, 206–208 ventrals in the two males, 226 ventrals in the only female, and totally black parietal scales.

CONFUSING SPECIES: *Micrurus distans*, *M. fulvius*, and *M. tener* have a broad, black body band behind the light neck band, 6–25 black body bands, and 4 chin shields. Sympatric harmless red, black, and yellow banded colubrid snakes do not have the bands crossing the venter, and usually have pale snouts and the red bands touching black bands. Shovel-nosed snakes (*Chionactis*) have white bands touching red saddle-like blotches, but the snout is not black and the supralabials are normally light colored.

KARYOTYPE AND FOSSIL RECORD: Neither karyotype nor fossil record is yet reported.

DISTRIBUTION: *M. euryxanthus* ranges from south-central Arizona and southwestern New Mexico southward in Mexico to about Mazatlan, Sinaloa (although it has yet to be found in the central part of that state), and east to western Chihuahua. A relict population occurs on the Isla Tiburon, Mexico.

HABITAT: *Micruroides* has been found in a variety of dry habitats with rocky or gravelly soils. It is most often associated with riparian zones or arroyos, but also inhabits scrub and brushy areas, grasslands, and cultivated fields. This snake sometimes enters buildings, perhaps to escape the heat. Being a secretive burrower, it is seldom seen above ground, preferring to remain in some subterranean chamber beneath rocks or logs or within old stumps. This coralsnake's ecological distribution is often correlated with that of its chief prey, blind snakes (*Leptotyphlops*), and it has been found in the channeled burrow systems of these small snakes (Lowe et al. 1986). On Isla Tiburon the snake occurs near sea level, and in Arizona it ranges to elevations of more than 1,900 m.

Distribution of *Micruroides euryxanthus*.

The northern *M. e. euryxanthus* is associated with the Sonoran Desert, where it occupies areas with relatively high amounts of summer rainfall, and is primarily found in the subtropical thornscrub biome and habitats containing palo verde cacti (*Cercidium* sp.). In contrast, the more southern *M. e. neglectus* lives in warm, humid deciduous forests (Meik et al. 2007).

BEHAVIOR AND ECOLOGY: Most details of the behavior and ecology of *M. euryxanthus* remain unknown. Since the review of its biology by Shaw (1971) the ensuing years have added little to our knowledge of this snake, and most new data are from the northernmost subspecies, *M. e. euryxanthus*. Its secretive, burrowing nature makes it difficult to both find and research, but nevertheless, an extensive ecological study is needed.

The snake probably first becomes seasonally active in March or April in Arizona and starts to hibernate there in October or early November; Gates (1957) found one on 30 November hibernating more than 1 m deep below the soil surface. Collection records seem to indicate that it is most active from July to mid-September, with a peak in late June and early July in Arizona (Lowe et al. 1986); more than 80% of New Mexico sightings occur in July–September (Degenhardt et al. 1996). No observations of basking have been reported, but at night, the snake heats itself by crawling onto warm road surfaces. Other than that, nothing has been published on its thermal requirements.

It is mostly crepuscular or nocturnal in the summer, although sometimes abroad during overcast days or immediately after rains. Meik et al. (2007) found one crossing a road at 2219 hours during warm, humid conditions on 17 July. Spring (March–May) and fall (September–October) activity may be more diurnal, possibly during the morning, as in the more eastern coralsnakes, *Micrurus fulvius* and *M. tener*.

REPRODUCTION: Stebbins (1985, 2003) surmised a range of 28–33 cm as the shortest TBL of an adult, and Goldberg (1997) reported the SVLs of mature males and females to be 32.0 cm and 35.6 cm, respectively, but the age at which these lengths are achieved is unknown.

Knowledge of the reproductive cycles of *Micruroides* comes chiefly from Goldberg (1997). Males with regressed testes containing spermatogonia and Sertoli cells were found in June and August. No males were in recrudescence. Spermiogenesis with metamorphosing spermatids occurred from April through November, and mature sperm were present in the epididymides and vas deferens all year long.

A female examined by Goldberg (1997) contained yolking eggs on 31 May, and another had oviductal eggs on 23 June; Funk (1964b) had previously reported oviductal eggs on 20 July. Early vitellogenic females were found on 30 May and 13 July. Females examined on 1 May and 1 June were reproductive, indicating that not all females reproduce in a given year.

Courtship and mating behavior have not been described. It has been assumed that mating takes place in the spring, but mature sperm can be found in the epididymides and vas deferens throughout the year, so some mating activity can occur at other times, especially in the fall. Clutches may contain two to six eggs, but usually only two or three are laid (Lowe et al. 1986; Stebbins 2003). On 27 July, Funk (1964b) dissected a 41.6 cm female and removed the 2 oviductal eggs mentioned above; their measurements were 34.6 mm × 6.1 mm and 39.3 mm × 6.2 mm. Both eggs were thin-shelled with no apparent embryonic development. Oviposition is correlated to summer rains in July and August (Lowe et al. 1986), and the eggs may be laid in rotting wood (Fowlie 1965), but most are probably oviposited under rocks or in underground burrows. Hatching occurs in the early fall (Lowe et al. 1986; Shaw 1971). Newly hatched young have TBLs of 19.0–20.3 mm (Lowe et al. 1986).

GROWTH AND LONGEVITY: No growth data have been reported, and nothing is known of its longevity in the wild. Most *Micruroides* do not adapt well to captivity, and so do not live long. Shaw (1971) reported that one survived 3 years and 8 months at the San Diego Zoo, but Rossi and Rossi (1995) reported a captive longevity of more than 10 years.

DIET AND FEEDING BEHAVIOR: In nature, *M. euryxanthus* is definitely a predator of reptiles, particularly those with smooth scales. The leading prey is the western threadsnake, *Leptotyphlops humilis* (Lowe et al. 1986; Vitt and Hulse 1973; Woodin 1953). Other snakes, including *Chilomeniscus cinctus, Chionactis occipitalis, Coluber constrictor, Diadophis punctatus, Gyalopion canum, G. quadrangulare, Hypsiglena torquata, Phyllorhynchus browni, Sonora semiannulata,* and *Tantilla* sp., and the lizards *Cnemidophorus* sp., *Elgaria kingii, Eumeces* sp., and *Xantusia vigilis* have been consumed in the wild (Campbell and Lamar 2004; Finnigan *in* Rossi and Rossi 1995; Lowe et al. 1986; Meik et al. 2007; Roze 1996; Vitt and Hulse 1973). Captives have eaten or tried to eat the snakes *Chilomeniscus cinctus, Chionactis occipitalis, Diadophis punctatus, Hypsiglena torquata, Leptotyphlops dulcis, L. humilis, Sonora semiannulata,* and *Tantilla planiceps*, and the lizards *Anniella pulchra* and *Scincella lateralis* (Campbell 1977; Gates 1957; Lindner 1962; Lowe 1948; Rossi and Rossi 1995; Veer et al. 1997; Vitt and Hulse 1973; Vorhies 1929; Woodin 1953). Captives have refused the following prey during feeding tests: snakes—*Arizona elegans; Phyllorhynchus browni, P. decurtatus; Rhinocheilus lecontei;* and *Thamno-*

phis cyrtopsis, and lizards—*Anniella pulchra; Cnemidophorus tigris, C. velox; Coleonyx variegates; Sceloporus undulatus; Urosaurus ornatus; Uta stansburiana;* and *Xantusia vigilis* (Vitt and Hulse 1973); captive conditions may have played a role in some rejections. *Micruroides* seem to prefer smooth-scaled reptiles. Shaw (1971) reported his long-lived captive fed regularly on five species of small local lizards and four species of small snakes, but did not name them. Fowlie (1965) thought the snake may also eat insect larvae, but this has not been proven. Of 16 preserved museum specimens examined by Vitt and Hulse (1973), only 1 (6%) contained food.

The venom delivery apparatus of *Micruroides* is relatively poor (see below), and reptilian prey is not always immediately incapacitated (Lindner 1962; Lowe 1948; Vitt and Hulse 1973; Vorhies 1929; Woodin 1953). As little time as 7 minutes (Lowe 1948) or as much as 30–40 minutes (Lindner 1962; Vorhies 1929) may elapse before the prey is weak enough to be swallowed, and often the prey puts up a spirited fight during this period (Veer et al. 1997).

Feeding behavior generally follows a consistent pattern (Vitt and Hulse 1973). *Micruroides* apparently locates prey by its odor, first responding to it with increased tongue-flicking. The snake then approaches the prey with a series of jerky movements. When close, the snake touches the prey a number of times with its tongue, then bites it with the mouth opened at approximately 30°. Usually the bite is toward the animal's posterior end. When bitten, the prey usually attempts to crawl away. Then *Micruroides* uses a chewing motion to move anteriorly toward the head of its victim. In most cases, envenomation does not take place (judging from the absence of tooth puncture marks and behavior of prey after being bitten). The snake continues chewing until it reaches the anterior end of the animal's head and then begins ingesting the prey. Swallowing is accomplished by a series of chewing motions separated by irregular short pauses. After the prey has been swallowed, a shelter is usually sought where the snake can digest its meal in leisure. A predation of a *Leptotyphlops humilis* observed by Vitt and Hulse (1973) took 22 minutes from the time the snake first showed interest (increased tongue-flicks) to when ingestion was complete. The predator and prey were 47.0 cm and 25.5 cm long, respectively. Usually, when *L. humilis* is taken, the initial bite is in the anterior one-third of the prey's body, whereas, in other prey species, the bite is normally more posteriorly oriented. Not all prey exhibit symptoms of envenomation by the time they are swallowed, but some *L. humilis* do.

Behavior is different when prey is refused. When potential prey is located, the frequency of tongue-flicks by *Micruroides* increases initially, then stops, and the snake resumes resting. If touched by the potential prey, the snake assumes a defense posture and its movements become jerky. Usually the head is buried under an anterior coil while the tail is raised, exposing the ventral side. Occasionally the cloaca is everted and gas is expelled with a popping noise. During the tail display the snake usually attempts to escape by burrowing or crawling under some object. No attempt is made to bite the prey.

VENOM DELIVERY SYSTEM: Eight *Micruroides* with 21.5–51.5 cm TBLs we examined had FLs of 0.1–1.0 (mean, 0.6) mm; Bogert (1943) reported 4 *M. euryxanthus* 40.1–45.6 cm long had 0.7–1.0 mm FLs. A positive correlation between FL and TBL is evident. Occasionally, a fang will be shed and remain in the puncture wound during an envenomation (McNally et al. 2007).

VENOM AND BITES: This small snake has relatively potent neurotoxic venom, and should be considered dangerous, especially to children. It should not be handled unless necessary (see below), and then protective gloves should be used.

The maximum dry venom yield is 6 mg, and the lethal dose for a human is probably 6–8 mg; the average protein content, 1.2 mg, is the highest of U.S. coralsnakes (Roze 1996). The snake's short length, along with a correspondingly small mouth and short fangs, makes it very difficult for *Micruroides* to deliver a penetrating bite to a human, but human evenomations have occurred.

Russell (1967a) reported three nonlethal cases, all resulting from handling the snake. Common symptoms included immediate, but not severe, pain at the site of the bite, with the pain persisting from 15 minutes to several hours. Nausea, weakness, and drowsiness occurred several hours later. Paresthesia (abnormal sensations) was also experienced; this was limited to the finger bitten in one case, but spread to the hand and wrist in the other two bites. Two victims experienced no symptoms after 7 to 24 hours; but symptoms persisted for 4 days in the third person, the one bitten by the largest snake (55 cm). One of those bitten was a physician who took detailed notes concerning the bite and its symptoms. He pulled the snake from his finger almost as soon as the bite occurred. About three hours later, paresthesia developed involving the finger and hand. This was followed by a progressive deterioration of his handwriting ability over a period of about 6 hours, at the end of which his handwriting looked like that of a "5-year-old child." The doctor experienced headache, nausea, difficulty in focusing his eyes, and some slight drooping of the upper eyelids. Inability to focus properly resulted in his walking into doors on several occasions. He also thought it possible he experienced some photophobia (sensitivity to light) and increased lacrimation (production of tears). Weakness and drowsiness occurred, and his memory was vague when he tried to remember the incident. There was no change in heart rate or difficulty breathing. In another case described by McNally et al. (2007), a 40-year-old male experienced swelling and numbness at the bite site on his left hand, some facial numbness, and a metallic taste in his mouth; he was admitted to a hospital for observation and discharged 13 hours later without complications. If the snake is small, the fangs may not penetrate human skin (Oliver 1958). Only 12 human envenomations were recorded between 2001 and 2005 (Seifert 2007a).

The venom's effects on a night lizard (*Xantusia vigilis*) were described by Lowe (1948). The envenomation took place at 2232 hours, and by 2234 hours slight paralysis of hind limbs was apparent. The hind limbs were completely paralyzed, and the front limbs showed slight paralysis by 2235 hours. Fifteen seconds later, the front limbs were nearly completely paralyzed, and the lizard experienced undulatory convulsions over its entire body. Total loss of righting ability occurred at 2237 hours, and at this time the lizard's breathing was deep and rapid. It had apparently died at 2238 hours, as breathing ceased (total time elapsed since bite: 5 minutes, 15 seconds), but when Lowe dissected the lizard at 2239 hours, he found the heart beating slowly. He applied normal saline solution to the heart, but at 2239 hours the heart stopped beating (total time elapsed since bite: 7 minutes, 30 seconds).

PREDATORS AND DEFENSE: The rate of predation on *Micruroides* is unknown as few observations have been reported. However, due to its small size it probably is at least occasionally the victim of carnivorous mammals (skunks [*Conepatus*, *Mephitis*,

Spilogale], badgers [*Taxidea taxus*], foxes [*Urocyon*, *Vulpes*], coyotes [*Canis latrans*], raccoons [*Procyon lotor*]), hawks, kestrels, owls, roadrunners, ophiophagous snakes (*Coluber constrictor*, *Lampropeltis* sp., *Masticophis* sp.), and possibly also tarantulas. It is cannibalized by its own species (Lowe et al. 1986).

The particular colored banding pattern of *Micruroides* serves as a warning to potential attackers (Cloudsley-Thompson 2006; see "Elapidae"). In tests on the reaction of naive roadrunners (*Geoccyx californicus*) to venomous *Micruroides* and *Crotalus atrox*, and the nonvenomous bullsnake, *Pituophis catenifer*, live coralsnakes and models of them did not elicit overt avoidance responses from the birds, but models elicited fewer pecks than did crawling snakes. The bullsnake and rattlesnake were also not overtly avoided, but did bring on a significantly higher number of leaps and wing flips by the birds, thus reacting as if both snakes were dangerous prey (Sherbrooke and Westphal 2006).

Micruroides is shy and gentle if not severely disturbed, but if startled, it coils, tucks the head under its body, and elevates and turns the tail so that the ventral surface faces toward the disturbance, and waves it about. The moving, banded tail may draw predators away from the tucked-under head in a decoying behavior. Vitt and Hulse (1973) found that three (14%) of the animals they studied had scars on the tail that possibly indicated a predatory attack there.

Another common defensive behavior of this snake is that of discharging air and fecal matter from its cloaca vent, producing a moderately loud popping sound (amplitude, 50.0–53.5 dB; frequency range, 0.42–5.52 kHz), and distinct temporal patterning and harmonics (Bogert 1960; Young et al. 1999). Air is apparently drawn into the cloaca through the vent, and then expelled through the same opening. The popping sound is caused primarily by the M. sphinctor cloacae, but may also involve other extrinsic muscles. The sounds last about 0.2 second and are repeated at 0.3- to 0.5-second intervals.

If neither of the above defensive behaviors is successful in discouraging its disturber, and the snake is touched or is approached too closely, it will strike quickly and viciously in a sidesweeping motion. If its mouth makes contact, a bite with continuous chewing may occur.

PARASITES AND PATHOGENS: Acanthocephalans parasitize the species (Ernst and Ernst 2006). Unidentified oligancanthorhynchid larval cystocanths were reported from this snake by Goldberg and Bursey (2000).

POPULATIONS: Nothing has been reported regarding population size, density, or dynamics. The snake's seemingly rareness is due more to its secretive, nocturnal, and burrowing habits than to actual scarcity. It is probably much more common in some areas than expected.

A conservation plan for such a poorly known species is hard to formulate. Certain things, though, should be considered. Most individuals are found at night on roads, and a high percentage of these are DOR. Areas of road where the snake is more commonly found should be targeted for drift fence construction that leads to culverts through which the snake can cross beneath the road (of course, this will not stop it from seeking the warmth of the road surface; see above). The continuous destruction of its habitat as civilization encroaches onto the Sonoran Desert must also be checked.

Micruroides euryxanthus australis (Between Navajoa and Alamos, Sonora; John H. Tashjian)

Most important, however, is the education of the public about the virtues and beauty of this little creature and the general lack of danger from it.

REMARKS: *Micruroides euryxanthus* is the northernmost of the coralsnakes in western and tropical America. Due to its distribution and several characters considered primitive elapid features (like additional solid teeth behind the fang on the maxilla), it has usually been thought to be near the origins of coralsnake evolution (Bogert 1943; Schmidt 1928). The species was reviewed by Campbell and Lamar (2004), Davidson and Eisner (1996), Ernst and Ernst (2003), and Roze (1982, 1974, 1996).

Micrurus Wagler, 1824
American Coralsnakes
Corales

Approximately 67 species of *Micrurus* occur in the Americas (Campbell and Lamar 2004), but only 3 in the North American distribution presented in this book. The cranial osteology of 13 species is described in Scrocchi (1992).

KEY TO *MICRURUS*

1a. Red body bands that lack black pigment, yellow body bands very narrow (0.5–2.0 scales wide) or absent, western mainland Mexico . . . . *distans*
1b. Red body bands with black pigment, yellow body bands 3–4 scales wide, southern United States 2
2a. Red body bands with no or only a few small black spots, black neck band does not touch the parietal scales, east of Mississippi River. . . . . *fulvius*
2b. Red body bands with widely scattered black spots or blotches, black neck band touches the parietal scales, west of Mississippi River. . . . . *tener*

Micrurus distans (Kennicott, 1861)
West Mexican Coralsnake
Coral del Oeste Mexicano

RECOGNITION: This black-snouted, red, yellow, and black banded snake (TBL_{max}, 107.5 cm) has very narrow (0.5–2.0 scales wide) to absent yellow bands, and very wide red bands that lack black pigment. The red bands are 6–25 scales wide anteriorly, but narrow to 5–10 scales wide at midbody. The 6–20 black body bands are 3–6 scales wide, usually include 2–3 ventrals, and are 3 to 4 times wider than the light bands that separate them. Three to 6 tail bands are present. A bright yellow band crosses the back of the head, and is followed by a black, 6–8 scales wide, nuchal band starting behind the parietals. On the head, the black dorsal pigment covers the frontals and preoculars, and extends posteriorly to cover the anterior 30–50% of the parietals and the anterior dorsal portion of the temporals. Some light pigment may occur on the anterior supralabials, and a light spot may be present on the rostral and anterior portion of the internasals. The light lower jaw is dark mottled. Body scales are smooth and pitless, and lie in 13–14 rows anteriorly, 15–16 rows at midbody, and 15 rows near the vent. The venter has 197–242 ventrals, 38–55 subcaudals, a divided anal plate, and no supraanal keels (tubercles). Lateral head scalation includes 2 nasals, 1 preocular, 2 postoculars, 1 + 1 temporals, 7 supralabials, and 7 infralabials (the first 4 touch the anterior chin shields); no loreal scales are present. On the lower jaw the mental does not touch the chin shields, and the posterior pair of chin shields is the largest. The pupils are round.

The long bifurcated hemipenis has a divided sulcus spermaticus extending from its base onto each lobe. The organ is basally naked, but covered distally by first a region of small spines, and then longer spines that become shorter toward the apex of the lobe. The lip of the sulcus is naked for its entire length. A large longitudinal naked fold begins almost at the base of the hemipenis and runs approximately parallel to the sulcus, ending before the large lobe spines begin (the hemipenis is illustrated in Dowling 1975 and Roze 1996).

Some Sonoran *M. distans* have one or two solid teeth on the maxilla behind the hollow fang (Bogert and Oliver 1945), a trait uncommon in the genus *Micrurus*.

Males (TBL_{max}, 73 cm) have 6–15 black body bands and 4–7 black tail bands, red and black body bands 13–18 and 2–6 scales wide, respectively, 197–217 ventrals, 46–55 subcaudals, TLs 13.3–15.9% of TBL; females (TBL_{max}, 107.5 cm) have 7–17 black body bands and 3–4 black tail bands, red and black body bands 6–13 and 3–6 scales wide, respectively, 216–242 ventrals, 38–44 subcaudals, and TLs 9.5–11.8% of TBL. (See "Geographic Variation" for subspecific differences in band and scale counts.)

GEOGRAPHIC VARIATION: *Micrurus distans* has four subspecies (characteristics and distributions mostly taken from Campbell and Lamar 2004). *Micrurus distans distans* (Kennicott, 1861), the West Mexican Coralsnake or Coral del Oeste Mexicano, is recognized by having a nuchal band 4–7 scales wide, the first red body band 18–25 scales wide, 11–17 black body bands, and yellow body bands 0.5–2.0 scales wide; males have other red body bands 13–18 scales wide, black body bands 2–4 scales wide, 4–6 tail bands, 208–214 ventrals, and 46–52 subcaudals; females have other red bands 6–13 scales wide, black body bands 3–6 scales wide, 3–4 tail bands, 222–235 ventrals, and 38–41 subcaudals. The subspecies is found from southwestern Chihuahua through southern Sonora and northwestern Nayarit. *Micrurus d. michoacanensis* (Dugès, 1891), the Michoacan Coralsnake or Coral de Michoacán, has a nuchal band 4–8 scales wide, red body bands 5–10 scales wide, 6–11 black body bands (each 3–4 scales wide), yellow body bands 0.5 scale wide or absent, and 3 tail bands; males have 208–213 ventrals and 47–50 subcaudals; females have 224–230 ventrals and 38–39 subcaudals. This snake occurs in the Río de las Balsas Basin of Michoacan and Guerrero and the Pacific coastal region of Guerrero. *Micrurus d. oliveri* Roze, 1967, Oliver's Coralsnake or Coral de Oliver, has a nuchal band 4–6 scales wide, red body bands 5–10 scales wide, 11–14 black body bands (each 3–4 scales wide), and 0.5–1.0 scale wide yellow body bands; males have 3–6 tail bands, 197–209 ventrals, and 50–55 subcaudals; females have 4–5 tail bands, 216–218 ventrals, and 43–44 subcaudals. It is found in Colima, adjacent Jalisco, and Michoacan (Reyes-Velasco et al. 2008). *Micrurus d. zweifeli* Roze, 1967, Zweifel's Coralsnake or Coral de Zweifel, has red body bands about 5–10 scales wide, 19–20 black body bands (each 4–6 scales wide), and 1–2 scales wide yellow body bands; the only known male has 6 tail bands, 217 ventrals, and 48 subcaudals; females have 4 tail bands, 237–242 ventrals, and 39–41 subcaudals. It is only found in central and southern Nayarit and adjacent Jalisco in the Río Grande de Santiago watershed and in Aguascalientes (Quintero-Diaz et al. 2000). *M. d. distans* and *M. d. zweifeli* intergrade in southern Nayarit (Roze 1996).

CONFUSING SPECIES: Within the Mexican range covered in this book, *M. distans* may be confused with *Micruroides euryxanthus,* which has almost its entire head black and a red body band immediately following the light nuchal band. Of the possible

Micrurus distans distans (California Academy of Sciences, John H. Tashjian)

Micrurus distans distans (Alamos, Sonora; Eric Dugan)

sympatric nonvenomous mimics, *Lampropeltis triangulum* has red and yellow body bands separated by a black band and the black bands occurring in pairs separated only by a narrow yellow band, and both *Sonora michoacanensis* and *Sympholis lippiens* possess a loreal scale. Differences from other confusing extralimital species of Mexican *Micrurus* are discussed in Campbell and Lamar (2004), and the use of red and yellow bands separated by a black band is only accurate north of Mexico City.

KARYOTYPE: The karyotype is undescribed, but probably contains 32 chromosomes (16 macrochromosomes, 16 microchromosomes) as in reported karyotypes of *Micrurus*.

Distribution of *Micrurus distans*.

FOSSIL RECORD: No fossils have been reported.

DISTRIBUTION: *Micrurus distans* ranges, in western mainland Mexico, from Municipio de Yécora, Sonora (Van Devender and Enderson 2007) and southwestern Chihuahua southward along the Pacific coastal plain to Guerrero.

HABITAT: Ecosystems used by the species include dry subtropical scrub and thorn woodlands and tropical deciduous forests. It occurs at elevations ranging from sea level to 1,415 m (Campbell and Lamar 2004; Van Devender and Enderson 2007).

BEHAVIOR AND ECOLOGY: Practically nothing is known about the biology of this animal. A good ecological study is needed.

M. distans is probably active in every month, and collection data indicate it is mostly crepuscular or nocturnal. The snake is often active after rain events. Hardy and McDiarmid (1969) collected them on several occasions at night after light rains at ATs of 24.2 and 26.3 °C, and Suazo-Ortuño et al. (2004) reported one active at an AT of 29.1°C and relative humidity of 84.6%. Activity is not restricted to ground level; the one found by Suazo-Ortuño et al. (2004) had climbed a tree to a height of about 4.5 m.

REPRODUCTION: Like other coralsnakes, *M. distans* is oviparous, but nothing else is known about its reproductive biology.

GROWTH AND LONGEVITY: Growth and longevity are unreported.

DIET AND FEEDING BEHAVIOR: Roze (1996) stated that the diet consists of colubrid snakes, but suitable-sized, smooth-scaled lizards are also probably eaten. The climbing individual observed by Suazo-Ortuño et al. (2004) may have ascended the tree in search of sleeping lizards, or possibly even nestling birds.

VENOM DELIVERY SYSTEM: The fangs of a 56 cm SVL (TBL, 65.2 cm) *M. distans* were 2.4 mm long, and those of a second 44.3 SVL (TBL, 51.2 cm) individual were 1.6 mm long.

VENOM AND BITES: Like other *Micrurus*, *M. distans* has potent neurotoxic venom. If touched, it will strike from side to side in an attempt to bite its attacker (Zweifel and Norris 1955). However, human envenomations rarely occur in the wild; this is probably due to the snake's secretiveness, habitat, and possible nocturnal activity when most humans are not about. Between 2001 and 2005, only one bite by *M. distans* was recorded (Seifert 2007a), but fatalities have occurred (Dart et al. 1992a). The venom can be neutralized by antivenom from *M. fulvius* (Dart et al. 1992a), but this specific antivenom is no longer produced by Wyeth Laboratories; however, venom from *M. distans* should be counteracted by the newly produced Coralymm MR antivenom.

PREDATORS AND DEFENSE: No predators have been reported, but the species must have its share of ophiophagous snake, bird, and mammal enemies. Gadow (1911) reported that wild turkeys (*Meleagris gallopavo*) and peafowl (*Pavo cristatus*) feed on coralsnakes, so possibly some *M. distans* are taken.

Its tricolored warning pattern gives it some protection. When disturbed, *M. distans* may at first strike, but after some time, calms down, hides its head beneath its body, coils the tip of its tail, raises the tail above the body, and moves it about, apparently to divert attention from its vulnerable head (Zweifel and Norris 1955); this is reminiscent of the defense behavior of *Micruroides euryxanthus*, described above.

PARASITES AND PATHOGENS: No parasites or pathogens have been reported.

POPULATIONS: No data are available. The snake is almost certainly more secretive than uncommon.

REMARKS: Smith and Chrapliwy (1958) proposed that *Micrurus proximans*, from Nayarit, is ancestral to the adjacent subspecies of *M. diastema* and *M. distans*, and that its close similarity to the eastern Mexican *M. diastema affinis* may be significant in this connection. *M. distans* has been reviewed by Campbell and Lamar (2004) and Roze (1967, 1982, 1996).

Micrurus fulvius (Linnaeus, 1766)
Harlequin Coralsnake

RECOGNITION: This snake has a TBL_{max} of 129.5 cm (McCollough and Gennaro 1963a), but most adults are shorter than 80 cm. It has a red, yellow, and black banded body (the "tricolor monad" pattern of Savage and Slowinski 1990). Black body bands total 11–19, those on the tail 3–5; 12–24 red bands cross the body. Banding continues onto the venter, and there may be some black pigment on the red bands. Occasionally body pattern variants occur (Meachem and Myers 1961). A bright yellow band occurs on the occiput, separating the black neck band from the parietal scales. The snout is black. The body scales are smooth and occur in 15 rows throughout; the anal plate is divided. Ventrals total 197–233, and subcaudals 30–47; no keels (tubercles) are present on the supraanal scales. Head scalation usually consists dorsally of 1 rostral, 2 internasals, 2 prefrontals, 1 frontal, 2 supraoculars, and 2 parietals; laterally are 2 nasals, 1 preocular, 2 postoculars, 1 + 1 (2) temporals, and 7 supralabials; there are no loreal or subocular scales. On the lower jaw are a mental, which does not touch the chin shields, 2 pairs of chin shields (the posterior pair being the larger), and usually 7 pairs of infralabials (the first 4 pairs touch the anterior chin shields). The pupils are round. The hemipenis has a bifurcated sulcus spermaticus extending from the base to nearly the apex of each fork. Each fork tapers gradually toward the apex and ends in a spine-like papilla. The base of the organ is naked, after which small spines and scattered spinules cover it up to the bifurcation, where larger spines begin. The sulcus lip is naked for its entire length, but is covered on both sides with small spines. Large spines occur before the bifurcation of the organ and gradually diminish in size toward the apex; the bifurcation region is without spines. A large longitudinal naked fold begins almost at the base of the organ and runs approximately parallel to the sulcus, ending shortly before the bifurcation, where the large spines begin. The dental formula consists of the fang on the maxilla, 4–6 palatine teeth, no pterygoidal teeth, and 6–7 teeth on the dentary.

Males (TBL_{max}, 81.2 cm) have 197–217 ventrals and 40–47 pairs of subcaudals; females (TBL_{max}, 129.5 cm) have 219–233 ventrals and 30–38 pairs of subcaudals. Males have longer tails: means, 13.8% of SVL in males and 9.3% in females (Jackson and Franz 1981). Males lack keels on the body scales anterior to and above the anal plate.

GEOGRAPHIC VARIATION: No subspecies are currently recognized. Coralsnakes from southern Florida were designated *Micrurus fulvius barbouri* by Schmidt (1928) on the basis of a few specimens lacking black spots on their red bands. Subsequently, Duellman and Schwartz (1958) reviewed this character in Floridian *fulvius*, and found that those from southern Florida often have black spotted red bands and that Schmidt's designation was invalid.

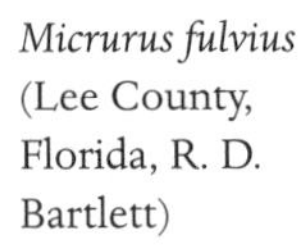

Micrurus fulvius (Lee County, Florida, R. D. Bartlett)

Micrurus fulvius
(R. D. Bartlett)

CONFUSING SPECIES: Within the species' southeastern range, the colubrid snakes *Lampropeltis triangulum elapsoides* and *Cemophora coccinea* are considered mimics. Both have a red, yellow, and black banded pattern, but their red and yellow bands are separated by black bands, and their snouts are red instead of black. In addition, *Cemophora* has no bands crossing its white or cream-colored venter. See the above genera key for comparisons with other North American coralsnakes.

KARYOTYPE: The karyotype is unreported, but considered similar to that described for *M. tener*.

FOSSIL RECORD: Several Pleistocene records exist for Florida: Irvingtonian (Meylan 1982) and Rancholabrean (Auffenberg 1963; Gut and Ray 1963; Hirschfeld 1968; Holman 1958, 1959b, 1996; Martin 1974; Weigel 1962).

DISTRIBUTION: *Micrurus fulvius* ranges from southeastern North Carolina southward, mostly on the coastal plain, through Florida, and west through central and southern Georgia, Alabama, and southern Mississippi to southeastern Louisiana.

HABITAT: The species is found in a variety of habitats. It seems to prefer dry, open, or brushy areas, and has been taken in xerophytic rosemary scrub, seasonally flooded pine flatwoods, xerophytic and mesophytic hardwood hammocks, and in Florida, occasionally in marshy areas (Dodd and Franz 1995; Dundee and Rossman 1989; Jackson and Franz 1981; Neill 1957), where it spends most of the time buried in the soil, leaf litter, logs, or stumps. It may also hide in gopher tortoise (*Gopherus polyphemus*) burrows, where the two reptiles are sympatric.

In north-central Florida, *M. fulvius* is the third most often trapped snake in upland habitats, but does not seem to occupy ones in lowland. Its upland niche overlaps that of *Cemophora coccinea, Coluber constrictor, Masticophis flagellum,* and *Sistrurus miliarius*, and its niche breadth ranks third among the four most abundant snakes trapped there by Dodd and Franz (1995).

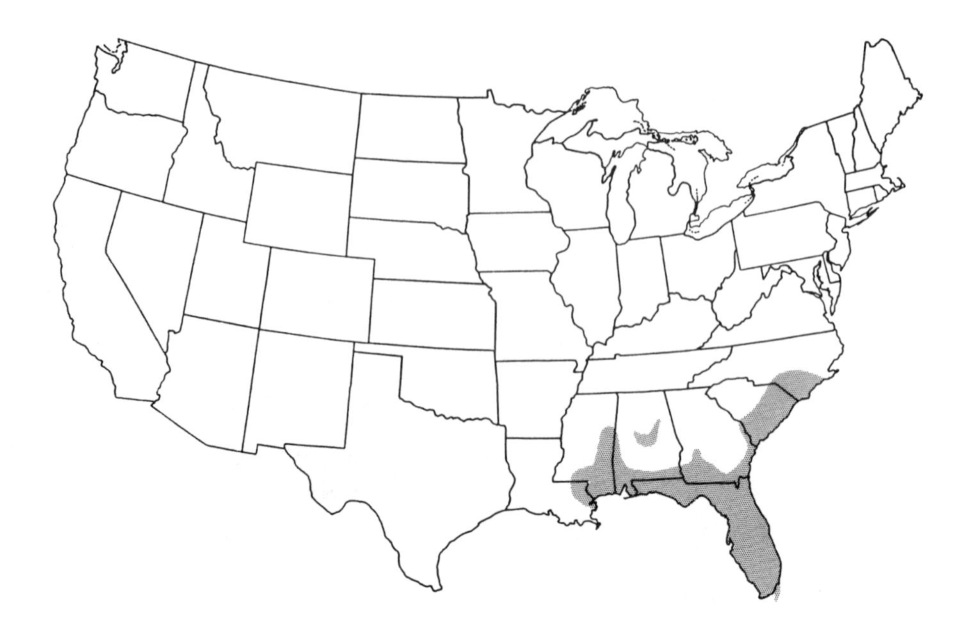

Distribution of *Micrurus fulvius*.

BEHAVIOR AND ECOLOGY: In Florida, *M. fulvius* is active in every month, but it has a distinct bimodal activity pattern, with more activity from March to May and, again, from August to November; the snake is least active in December–February (Dalrymple et al. 1991a; Jackson and Franz 1981). Gravid females are not very active in June and July, but become more active in August–December after oviposition. North of Florida the species is forced to hibernate, normally underground (Neill 1948 reported no winter records for this species in Georgia).

There is a misconception that coralsnakes are nocturnal, but observations on various North, Central, and South American species show that these snakes often are active during the day (Greene and McDiarmid 1981), and Neill (1957) reported that only 1 of 121 active *M. fulvius* was taken at night. He also summarized the few literature records for nocturnal captures, and concluded that this snake is largely diurnal in both Georgia and Florida, prowling in the early morning shortly after sunrise to about 0900 hours. It is most often seen on bright, sunny mornings, but occasionally prowls in the late afternoon or early evening. During April to August, Florida *M. fulvius* are surface-active from 0700 to 0900 hours, remain under cover for much of the day, and resume activity during the late afternoon at 1600–1730 hours (Jackson and Franz 1981). In March and September–November, it emerges in mid- or late morning (0900–1000 hours), and with the exception of a midafternoon quiescent period from 1330 to 1600 hours, remains active most of the day. Carr (1940) reported night activity, and has observed it in the water.

Captive *M. fulvius* have been kept successfully at ATs of 22–32°C (Rossi 1992; Vaeth 1984), but thermal data are lacking for wild individuals. Deckert (1918) reported that *M. fulvius* from around Jacksonville, Florida, became active and may have basked when the next day AT reached about 14.5°C after a night AT of −3°C.

REPRODUCTION: Jackson and Franz (1981) reported that most male *M. fulvius* 45 cm SVL or longer contain sperm, whereas smaller ones do not. This length is reached in 11 to 21 months.

During the male sexual cycle in Florida, recrudescence occurs in the fall and regression during the spring (Jackson and Franz 1981). Adult testes mass was <0.26 g in January–August and >0.26 g in September–November; the right (anterior) testis is usually larger than the left (posterior). The cycle is essentially synchronous with that described for *M. tener* by Quinn (1979a). Males contain less storage fat than females; small amounts of fat are present in the fall (particularly November), and may correspond either to testes recrudescence or to an increased level of surface activity. Liver mass seems to be homogeneous throughout the year. Austin (1965) described the sperm.

Most females mature at an SVL of about 50–56 cm at about 26 months of age (Jackson and Franz 1981; Roze 1996). In Florida, vitellogenesis occurs in late winter and early spring (March–May), with follicles reaching preovulatory size by early June; oviposition occurs from mid-June into July (Jackson and Franz 1981; Telford 1955). Oviductal hypertrophy is already underway by mid-April (Jackson and Franz 1981). A Florida female collected by Jackson and Franz on 9 June contained shelled eggs and distinct corpora lutea 5.0–6.5 mm in diameter, suggesting that they are of short duration. Gravid and spent females contain several size classes of follicles, which will become later clutches. Georgia females have oviposited in mid-July (Zegel 1975). Data gathered by Jackson and Franz (1981) suggest an inverse relationship between lipid and reproductive cycles in females; fat bodies are largest in February into early May and smallest in June. The timing of adult female liver masses corresponds to that occurring in storage fat bodies, with the smallest liver masses in gravid females.

Mating occurs in the spring (April–May) and possibly also in the early fall (late August–early October). Courtship is similar to that described for *M. tener*. Although Zegel (1975) described the courtship behavior of *M. fulvius*, he did not witness a successful copulation; Vaeth (1984), however, described a successful mating between a male *M. tener* and a female *M. fulvius*. The male stopped moving when he had approached within 2 cm of the female's head and his vent was in apposition at the left

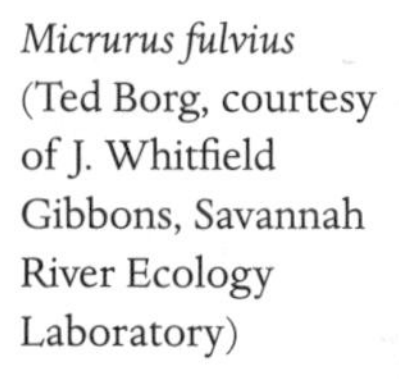

Micrurus fulvius (Ted Borg, courtesy of J. Whitfield Gibbons, Savannah River Ecology Laboratory)

side of her vent. He then elevated his tail and tried to wrap it under and around that of the female. She responded by slightly elevating her tail and gaping her cloaca. Intromission was accomplished rapidly with the right hemipenis. The elapsed time to intromission was about 25 minutes. No attempt was made to completely entwine tails nor was there any cloacal rubbing prior to copulation. Shortly after hemipenis insertion, the male moved the anterior third of his body off the female. After approximately 10 minutes of coital activity, localized rhythmic contractions occurred in the male's cloacal region and continued intermittently for 12 minutes. In this particular mating, ejaculation was not completely efficient, as a pool of seminal fluid accumulated beneath the two snakes soon after the contractions were observed. Presumably, this seminal fluid was leaking from the exposed part of the sulcus on the basal portion of the hemipenis. Neither snake made any other body movements while mating. Only an occasional tongue-flick occurred, and these may have been in response to Vaeth's movements. After 61 minutes of copulation, the male withdrew his hemipenis and crawled away.

Oviposition normally occurs from mid-June into July, but early clutches have been laid in late May. Clutches contain 1–13 eggs, but 4–7 are most common. The white to cream-colored eggs are very elongate (20–47 mm × 6–14 mm, 3–6 g), and are usually laid underground or beneath leaf litter. C. Ernst and his students found four eggs in a hollow depression beneath an old wooden tie on an abandoned railroad embankment in Collier County, Florida. Several eggs in the clutch may adhere to each other.

The young hatch in August and September after an IP of 70–90 days, depending on the IT, and have 17.8–20.3 cm TBLs (Allen and Neill 1950c). Jackson and Franz (1981) reported that a recent hatchling had an SVL of 26.6 cm, and that individuals with SVLs of 24.9–30.3 cm were collected in October and November.

GROWTH AND LONGEVITY: *M. fulvius* doubles its size in less than 2 years, and by 3 years of age may grow to nearly 60 cm SVL. Growth is negligible in January–February of any year. After hatching, neonates show some growth in October and November; beginning in August, a spurt in growth occurs in 40–50 cm SVL males, and yearlings grow during the entire annual active period (Jackson and Franz 1981). North-central Florida *M. fulvius* trapped by Dodd and Franz (1995) had an average SVL of 65.2 (range, 39.2–71.5) cm and a mean BM of 31.5 (range, 15.0–62.0) g.

Natural longevity is unknown, but a wild-caught male lived an additional 10 years, 8 months, and 8 days at the Brookfield Zoo (Snider and Bowler 1992).

DIET AND FEEDING BEHAVIOR: Although Florida *M. fulvius* may feed nearly year round, except possibly during January–February, feeding activity is most intense in September–November with a lesser peak in April–May (Jackson and Franz 1981).

During foraging, *M. fulvius* usually crawls slowly and pokes or probes with its head beneath leaf litter (C. Ernst, personal observation). These head probes are stereotyped (Greene 1984). Poking involves repeated forward and lateral head movements, and is accompanied by frequent tongue-flick clusters. Apparently both visual and chemical stimuli elicit attacks. Rapid prey movements seem to bring on a quicker attack by the snake and sometimes override aversive chemical cues. Approach is slow if the prey is stationary or moving slowly, but rapid if the prey is crawling quickly away.

Prey is usually seized with a quick forward movement of the anterior or entire body of the coralsnake, but it is occasionally seized with a quick sidewise jerk of the head and neck. Neither *M. fulvius* nor *M. tener* has an efficient strike; Greene (1984) thought this is perhaps due to poor vision resulting from their relatively small eyes. The snake typically holds its prey, sometimes chewing it, until it is immobilized by the venom.

The prey is not usually released before it is swallowed, and is almost always ingested head first. Preswallowing maneuvers suggest that either tactile and/or chemical cues are used to recognize the anterior end of the prey. Scale overlap on the prey is the major cue for locating its head (Greene 1976). Swallowing is accomplished, as in other snakes, by alternating jaw movements as the coralsnake literally crawls over its prey. Ingestion is almost always followed by much tongue-flicking and opening and closing of the jaws to work them back into position.

M. fulvius feeds almost exclusively on elongated, smooth-scaled reptiles: amphisbaenians (*Rhineura floridana*), lizards (*Eumeces fasciatus, E. inexpectatus; Neoseps reynoldsi; Ophisaurus ventralis, Ophisaurus* sp.; *Scincella lateralis*), and snakes (*Cemophora coccinea; Carphophis amoenus; Coluber constrictor; Diadophis punctatus; Farancia abacura; Lampropeltis triangulum; Micrurus fulvius; Opheodrys aestivus; Elpahe guttata; Seminatrix pygaea; Stilosoma extenuatum; Storeria dekayi, S. occipitomaculata; Tantilla coronata, T. relicta; Thamnophis sauritis, T. sirtalis; Virginia valeriae*) (Chance 1970; C. Ernst, personal observation; Greene 1984; Heinrich 1996; Jackson and Franz 1981; Krysko and Abdelfattah 2002; Loveridge 1938, 1944; Neill 1968; Obrecht 1946; Palmer and Braswell 1995; Schmidt 1932). Conant (1975) listed anurans as a food item, but these are probably rarely eaten, if at all. Zegel (1975) reported that two captive hatchlings ate a *Sceloporus undulatus* and a centipede, and Greer (1998) reported a captive readily accepted dead *Agkistrodon contortrix, Cemophora coccinea, Opheodrys aestivus, Elaphe guttata,* and *Storeria dekayi*.

The snake is opportunistic, taking suitable prey that crosses its path when it is hungry. However, the skinks (*Eumeces, Neoseps, Scincella*) and the smaller snakes it takes may not be cost-effective in the amount of energy gained compared to that needed to find, attack, and swallow them (Greene 1984).

Kirkwood and Gili (1994) reported that a juvenile and an adult *M. fulvius* actually weighed 0.2 g and 2.5 g less, respectively, a week after having eaten laboratory mice (*Mus musculus*) at the London Zoo; mice are not a normal food of the snake.

VENOM DELIVERY SYSTEM: Fangs of 12 *M. fulvius* (TBL, 27–88 cm) that we measured were 0.6–2.7 mm in length; adults (TBL, 65–88 cm) had FLs of 1.6–2.7 mm. Bogert (1943) reported that 3 70.0–81.7 cm individuals had 2.1–2.5 mm fangs.

VENOM AND BITES: The venom attacks the nervous system, primarily the respiratory center, resulting in loss of muscle strength, difficulty in breathing, and death in extreme cases (Weis and McIsaac 1971). The main toxins in elapid venoms are post- and presynaptic neuorotoxins and cardiotoxins; depolarization of the membranes of muscle fibers (particularly the bulbar ones) is common. Average total protein content is 0.89 mg (Roze 1996). Proteolysis activity is absent or negligible, although L-leucine aminopeptidase activity may occur. Ramsey et al. (1972), however, identified six homogeneous and one heterogeneous protein fractions in lyophilized venom from *M. fulvius;* three possessed anticoagulant activity, and all seven fractions exhibited strong phospholipase A activity and hemolyzed red blood cells. Anticholinesterase activity normally occurs in bites (Kumar et al. 1973). Phospholipase activity is strong; the venom contains a very potent phospholipase A_2 with myonecrotic or cardiotoxin-like properties (Aird and da Silva 1991; Alape-Girón et al. 1994; Roze 1996; Vital Brazil 1987; Weis and McIsaac 1971). When injected into mice, venom from *M. fulvius* produces a characteristic stiffening and erection of the tail, a positive Straub reaction as striking as, or even greater than, that produced by morphine (Macht 1947). Khole (1991) reported the subcutaneous, intraperitoneal, and intravenous LD_{50} values of mice as 1.30 mg/kg, 3.6–9.0 μg/16–18 g mouse, and 0.2–0.4 mg/kg, respectively. The intravenous ratio of toxicity of mice is 0.6; that of dogs is 0.5 (Brown 1973; Cohen and Seligmann 1966).

The maximum recorded dry venom yield is 38 mg (Roze 1996). Normal wet and dry venom yields from *M. fulvius* are 2–12 and 3–5 mg, respectively (Ernst and Zug 1996; Minton and Minton 1969; Norris 2004; Roze 1996; Russell and Puffer 1970). Fix (1980) and Fix and Minton (1976) reported that 8 of 14 adult *M. fulvius* from which they extracted venom gave dry yields in excess of 6 mg, and that 4 produced dry yields of 12 or more mg. The lethal dose for a human is probably 4–7 mg (Bolaños et al. 1978; Minton and Minton 1969; Roze 1996). Arce et al. (2003) reported mean interperitoneal and intravenous mouse LD_{50} values of 0.70 (0.6)–0.82 μg/g and 0.34 (0.28–0.40) μg/g, respectively. Other LD_{50} values reported in micrograms for a 16–20 g mouse (*Mus musculus*) are 9–26 (Bolaños et al. 1978; Minton and Minton 1969); 0.74–0.97 mg/kg (Russell and Puffer 1970; Sánchez and Pérez 2006); for a 70 kg human, the subcutaneous LD_{50} is 91 mg/kg, and the number of lethal doses per bite is about 0.02 (Brown 1973). A positive correlation exists between snake TBL and venom yield; longer snakes have greater venom yields.

In laboratory tests by Halter (1923), a bitten green frog (*Rana* sp.?) died after paralysis of its legs. A large domestic cat (*Felis catus*) yowled continually, dragged its envenomated leg, and tried to escape after being bitten at 1000 hours. By late afternoon, the cat remained quietly in a corner with its head lowered, wheezing audibly as it breathed. If urged to walk, it yowled with each step, and crouched again, as if to sleep, even though its eyes were open. In the evening, its breathing was much labored, and both hind limbs were paralyzed. The cat crouched all the next day in one spot with its head lowered. By evening its breathing was even more labored; it heaved convulsively, and its pupils did not respond to bright light. The cat was found dead the next morning, 48 hours after being bitten. It was in rigor when discovered; a pool of serous fluid had issued from its mouth and nostrils, and its pupils were widely dilated. The bite site showed no perceptible swelling or discoloration. When moved, large amounts of uncoagulated blood issued from its mouth and nostrils. Two days after the cat was bitten, Halter (1923) reported that an eastern hog-nosed snake, *Heterodon platirhinos*, and a brown watersnake, *Nerodia taxispilota*, were also envenomated by the same coralsnake; they died within a few hours, demonstrating that *M. fulvius* can produce several fatal bites within a very short time.

Vital Brazil (1987) and Weis and McIsaac (1971) reported that cats injected with crude venom experience a rapid and temporary drop in blood pressure, and that neuromuscular and respiratory depression occur in an hour; they eventually die, presumably from hypotension. The venom acted similarly in some respects to the cardiotoxin from cobra venom, and Vital Brazil presumed it contained a fraction that blocked the acetylcholine receptors. However, Snyder et al. (1973) isolated a neurotoxin from *M. fulvius* venom that produces an irreversible neuromuscular blockage in chicken (*Gallus domesticus*) biventer cervicis nerve-muscle preparation, inhibits acetylcholine-induced contraction in a dose-dependent manner, and does not show the depolarizing activity of crude venom.

Hemolytic symptoms involving the washed red cells of various laboratory animals occurred in 30 minutes (guinea pigs) and 2 hours (dog, mouse, chicken) after introduction of *M. fulvius* venom; however, the blood cells of sheep, rabbits, monkeys, and humans were not affected (Cohen and Seligmann 1966).

Coralsnakes are responsible for only about 1% of the annual envenomations by snakes in the United States. Symptoms occur in 75% of persons bitten by *M. fulvius*, and neurologic symptoms may not begin until 10 to 12 hours later, giving a false sense of security (Kitchens and Van Mierop 1987; Norris 2008; Norris and Bush 2007). Fortunately, less than 20 human envenomations are recorded in the United States each year, and most of these occur while the snake is handled. Seifert (2007a) reported only 82 envenomations by *M. fulvius* in the period 2001–2005, with no deaths involved. One Florida hospital treated only 39 bite cases during a 12-year period (Kitchens and Van Mierop 1987). Although these bites occurred throughout the year, most were in the spring and fall, corresponding to the snake's most active periods. Parrish (1966) reported that only 0.3% of 2,836 persons hospitalized in the United States for snakebite in 1958–1959 were bitten by coralsnakes (all 3 species combined). The annual fatality rate of envenomations by *M. fulvius* and *M. tener* in the United States is less than 1% of all fatal snakebites (Parrish 1963). More than 30% of envenomations by *M. fulvius* and *M. tener* occur when the snake is confused with a harmless mimic.

Serious, and in some cases fatal, bites (potentially up to 10% of total bites) do occur (Neill 1957; Norris 2008; Scheppegrell 1928; Willson 1908). Pain, usually not severe, may occur at the bite site, and if the bite is on a limb, may slowly extend up that extremity. Other reported symptoms include abdominal pain, abnormal reflexes, apprehension, blurred or double vision, breathing difficulties with eventually bulbar respiratory paralysis, contracted pupils, contusions and swelling at the bite site, convulsions, delirium, dizziness, drooping eyelids, drowsiness, dyspnea, euphoria, facial spasms, fasciculations, fluttering eyelids and pupil contraction, fullness of the throat, general weakness, headache, heavy perspiration, lethargy, muscle soreness, nausea and vomiting, pain (mild and transient), paresthesia, personality changes, salivation, swallowing difficulty, swelling of the tongue, and a weakened pulse. Treatment may be very expensive (a vial of antivenom may cost up to $150) and several days in the hospital is usually required. Stickel (1952) estimated the human fatality rate from bites of *M. fulvius* to be 20–75%; fortunately, the percentage is lower than the former figure, especially if the bite is treated with antivenom (Neill 1957). In Florida, 2–5% of all snakebites are caused by *M. fulvius* (McCollough and Gennaro 1963b), and in Alabama only 2% of all snakebites (Parrish and Donovan 1964). Case histories of bites are related by Andrews and Pollard (1953), Clark (1949), Gennaro et al. (2006), Greer (1998), Kitchens and Van Mierop (1987), Neill (1957), Parrish and Donovan (1964), Parrish and Kahn (1967a, 1967b), Stejneger (1898), True (1883), Werler and Darling (1950), and Willson (1908).

Currently, due to the relatively few cases of envenomation by coralsnakes, no specific antivenom against the venom of *M. fulvius* and *M. tener* is manufactured in the United States, but use of the Mexican (Coralymm MR) antivenom, produced from the venoms of tropical *Micrurus*, shows promise in the treatment of envenomations by our North American species (Bolaños et al. 1978; Sánchez et al. 2008; Sánchez and

Pérez 2006). In an observational study of bites by *M. fulvius* and *M. tener* by Morgan et al. (2007), seven poison centers in Florida and Texas were queried about cases reported to them. Of 109 victims, 51.4% received antivenom and 48.6% did not. No deaths occurred in either group. In spite of published recommendations, some providers chose not to administer antivenom to suspected coralsnake victims. Presence or absence of systemic effects and/or severe clinical findings may have played an important part in their decisions. Many victims with only local symptoms did not receive antivenom and never developed systemic effects.

Because *M. fulvius* has relatively short fangs, the chance of a human receiving a bite in the wild when not handling one is slim. Nevertheless, Carr (1940) has reported instances of large *M. fulvius* actually attacking humans, but such behavior must be rare; heavy shoes or boots and thick trousers should be sufficient protection for the hiker. The species is apparently immune to the venom of its own kind (Peterson 1990).

PREDATORS AND DEFENSE: Recorded observations of predation on *M. fulvius* include those by the imported fire ant (*Solenopsis invicta*), which preys upon the eggs and young; ophiophagous snakes (*Drymarchon corais, Lampropeltis getula, Micrurus fulvius*); predatory birds (*Buteo jamaicensis, B. lineatus; Falco sparverius; Lanius ludovicianus*); and domestic cats (*Felis catus*) (Belson 2000; Brugger 1989; Chance 1970; Greene 1984; Jackson and Franz 1981; Loveridge 1938, 1944; Roze 1996; Stoddard 1978). Note that larger *M. fulvius* do not hesitate to eat smaller members of their own species.

Would-be predators of *M. fulvius* do not always fare well. Brugger (1989) observed an adult male red-tailed hawk die with a partially eaten Florida coralsnake in its talons. Six punctures (presumably by fangs) occurred on the left tarsus of the bird and nine more in the skin over the left tarsus gastrocnemius muscle. The muscle was swollen and the tissue surrounding two punctures discolored orange. The bird died of flaccid paralysis typical of the neurotoxic effects of elapid venoms.

When approached, *M. fulvius* will often flatten the posterior portion of the body, tuck its head under an anterior coil, ball up its tail, and wave it about, thus drawing the predator's attention away from the head (Greene 1973b); Gehlbach (1972) has demonstrated that this behavior will deter some potential mammalian predators. When pinned down, however, the snake will strike and chew on the restraining implement. These strikes are usually sideways, rapid, and often vicious. Its banded color pattern may give it some protection from predators (see above).

PARASITES AND PATHOGENS: The trematode *Ochetosoma ellipticum* has been found in *M. fulvius* (Franz 1974). One we had at the University of Kentucky developed respiratory difficulties, stopped feeding, became emaciated, and died. *M. fulvius* is also plagued by intestinal nematodes (Ernst and Ernst 2006).

POPULATIONS: Its secretive fossorial habits render the species not readily observable; it often lives in heavily populated urban areas without being detected. This gives a false impression of rarity, when the snakes may be quite plentiful. For instance, Beck (*in* Shaw 1971) reported that during a 39-month period 1,958 *M. fulvius* were turned in for bounties in Pinellas County, Florida, and the species comprised 14.3% of 230 snakes of 10 species trapped in upland north-central Florida habitats during 1989 and 1990 (Dodd and Franz 1995). Both Carr (1940) and Deckert (1918) thought it common in Florida.

The most serious threats to current populations of *M. fulvius* are the destruction of its habitat and highway deaths. The population in Alabama has declined since the introduction of aggressive fire ants (Mount 1981). Loss to collection for the pet trade or venom industry seems negligible. In addition to the species being extremely dangerous, it does not usually adapt well to captivity, and should not be kept as a pet.

REMARKS: The species was reviewed by Campbell and Lamar (2004), Davidson and Eisner (1996), Roze (1967, 1982, 1996), and Roze and Tilger (1983).

Micrurus tener (Baird and Girard, 1853)
Texas Coralsnake
Coral Texano

RECOGNITION: *M. tener* (TBL_{max}, 121.3 cm; Tennant 1984) has a tricolored nomad banding pattern similar to that of *M. fulvius*, but melanistic individuals have been encountered (sometimes with serious results; Gloyd 1938). Black body bands total 10–30, and tail bands 3–7. Twelve to 24 red body bands are present. Some body bands may extend onto the venter, and scattered black pigment is present in the red bands. A bright yellow band crosses the occiput, but does not prevent the black neck band from contacting the parietals. Body scales are smooth and pitless, and lie in 15 rows throughout. The venter has 197–233 ventrals, 30–47 subcaudals, and a divided anal plate. Keels are absent from the supraanal scales. Cephalic dorsal scales consist of 1 rostral, 2 internasals, 2 prefrontals, 1 frontal, 2 supraoculars, and a pair of parietals. On the side of the head are 2 nasals, 1 preocular, 2 postoculars, 1 + 1 (2) temporals, 7 supralabials, and no loreal or subocular scales. The lower jaw contains the mental separated from the 2 pairs of chin shields (the posterior pair are the largest), and 7 infralabials (the first 4 touch the anterior chin shields). The pupils are round. The hemipenis is essentially like that described for *M. fulvius*.

Males (TBL_{max}, 80 cm) have 181–216 ventrals, 38–46 subcaudals, and TLs >12% of SVL. Females (TBL_{max}, 121.3 cm) have 202–232 ventrals, 26–39 subcaudals, and TLs <10% of SVL. Mean SVLs of males and females are 53.9 cm and 62.0 cm, respectively (Quinn 1979a).

GEOGRAPHIC VARIATION: Four subspecies are recognized (Campbell and Lamar 2004). *Micrurus tener tener* (Baird and Girard, 1853), the Texas Coralsnake or Coral Texano, ranges from southwestern Arkansas and northern and central Louisiana southwestward through Texas to Coahuila, Nuevo Leon, and central Tamaulipas. It has 200–211 ventrals, 38–46 subcaudals, and 3–4 black tail bands in males; 219–227 ventrals, 26–34 subcaudals, and 2–3 black tail bands in females; a 7–10 scales wide black nuchal band; and 10–15 black body bands. *Micurus t. fitzingeri* (Jan, 1858), Fitzinger's Coralsnake or Coral de Fitzinger, is found in Colima (Reyes-Velasco et al. 2009), from Guanajuato and Querétaro to Morelos, and in Zacatecas. It has 208–216 ventrals, 41–44 subcaudals, and 5–6 black tail bands in males; 222–232 ventrals, 31–35 subcaudals, and 3–5 black tail bands in females; the black nuchal band 5–6 scales wide; and 19–26 black body bands. *Micrurus t. maculatus* Roze, 1967, the Tampico Coralsnake or Coral de Tampico, found only in the vicinity of Tampico, Tamaulipas, has 181–185 ventrals, 43–45 subcaudals, and 5–7 black tail bands in males; 202–208 ventrals, approximately 31 subcaudals, and 4–5 black tail bands in females; a black nuchal band 6–7 scales wide; and 13–17 black body bands. *Micrurus t. microgalbineus* Brown and Smith, 1942, the Potosí Coralsnake or Coral Potosíno, ranges southwestward from southern and southwestern Tamaulipas and central and eastern San Luis Potosí to central Guanajuato. It is characterized by having 198–204 ventrals, 41–45 subcaudals, and 5–7 black tail bands in males; 216–225 ventrals, 32–39 subcaudals, and 4–5 black tail bands in females; a black nuchal band 6–8 scales wide; and 17–25 black body bands.

CONFUSING SPECIES: *Micrurus distans* and *M. fulvius* can be differentiated by using the above key. The Sonoran coralsnake, *Micruroides euryxanthus*, has a wide red band behind the light collar instead of a black one.

KARYOTYPE: Thirty-two diploid chromosomes are present; 16 macrochromosomes and 16 microchromosomes (Graham 1977). All chromosome pairs are homomorphic except pair 6, which is heteromorphic (ZW) in females; males are ZZ.

FOSSIL RECORD: Pleistocene fossils have been discovered at Irvingtonian and Rancholabrean sites in Kendall and Travis counties, Texas (Hill 1971; Holman 1969a; Holman and Winkler 1987).

DISTRIBUTION: *Micrurus tener* occurs from central and northern Louisiana and southwestern Arkansas southwestward through southern Texas, and southward in Mexico through eastern Coahuila, northeastern Nuevo Leon, and Tamaulipas to San Luis Potosí, eastern Guanajuato, Queretaro, and Hidalgo. It has also been recently found in northeastern Colima (Reyes-Velasco et al. 2009).

HABITAT: Habitats occupied include mixed deciduous hardwood and pine forests, subtropical dry thorn scrub woodlands, riparian zones in dry areas, tallgrass prairie, mesquite-grassland, and even the debris of urban areas. It apparently does not occur in cloud forest habitats (Martin 1958). In eastern Texas, it seems to prefer oak-hickory woodlands with thick plant litter where it can hide and also find the semifossorial reptiles upon which it preys. In western Texas, this snake is found in oak-juniper canyons with scattered rocks and rock crevices for cover. Over its range, the species has been recorded from elevations near sea level to more than 2,000 m.

BEHAVIOR AND ECOLOGY: Seasonally, *M. tener* is most active in the spring (March–May) and fall in Texas (Price 1998; Vermersch and Kuntz 1986). Farther south, in Mexico, it probably can be found surface-active in all months. In the United States, it is forced underground during the colder periods of winter.

Werler and Dixon (2000) suspected that in Texas *M. tener* is active, depending on the ET, at any time of the day or night, but mostly it moves in the morning with lesser activity in late afternoon. It is particularly active after rain events. In arid southern Texas, it has been seen crossing roads early in the night when diel ATs were very high and the humidity was low. Price (1998) reported that the snake is highly fossorial, spending most of the time underground, but that it is surface-active during the day in spring and after a rainfall, and that it does not seem to be very active at night. Price thought females are most active in the fall when they are replenishing their lost fat supplies.

Distribution of *Micrurus tener.*

Ruick (1948) reported the snake may perform a type of sidewinding locomotion while trying to escape across a road surface, and Mitchell (1903) reported that it will return continuously to the same retreat. The species is a good swimmer (C. Ernst, personal observation).

In Mexico, *M. tener* seems to be mostly crepuscular or nocturnal; Werler and Dixon (2000) usually encountered the snake at 2000–2400 hours, and Ruick (1948) found them crossing roads on moonlit nights. However, it is also active in the early morning or on overcast days (Campbell and Lamar 1989).

Ruick (1948) collected 10 *M. tener* while road cruising when the AT was about 24°C. It seems to have little tolerance for high temperatures. Werler and Dixon (2000) reported that captives are best kept at ATs not exceeding 30°C, and that a range of 23–26°C is best; individuals kept at temperatures above 30°C languish and die.

REPRODUCTION: Price (1998) thought males attain sexual maturity in 12–21 months at an SVL of about 40 cm; the smallest male *M. tener* undergoing spermiogenesis found by Quinn (1979a) was 40.2 cm SVL. In Texas males, a complete regression with spermatogonia and Sertoli cells occurs in May through August, with a peak in June. Early recrudescence with spermatogonial divisions and primary spermatocytes occurs from June through October, with a peak in July. In August and September, males have late recrudescence with secondary spermatocytes and undifferentiated spermatids. Active spermiogenesis with mature sperm in the lumen takes place from August to April. Seminiferous tubule diameter is greatest from November to March, and testes weigh the most from December to February. Sperm resides in the ductus deferens from February to December, and in the epididymis from April to December (Quinn 1979a).

Females mature in 12–21 months at an SVL of 50–55 cm (Ford et al. 1990; Price 1998; Quinn 1979a; Roze 1996). Ovary weights increase from March through April, decline slightly in May, and then more rapidly in June; follicle lengths show the same pattern (Quinn 1979a).

According to Vermersch and Kuntz (1986), mating takes place in April–May and again in August–September in south-central Texas, but in that state, the snake may breed at any time during the annual activity cycle. Late August and September matings result in the female storing the sperm within her oviducts until ovulation the following summer (Vermersch and Kuntz 1986; Werler and Dixon 2000).

Females probably lay down pheromone trails that males may use to find them. A courting male *M. tener* crawls to the female and flicks his tongue several times over her back at midbody. He then raises his head and neck at about a 45° angle and, leaving his neck at that angle, tilts his head down and touches his nose to the female's back. In this position he quickly and smoothly runs his nose along the female's back to about 5 cm behind her head. In about 60% of courtships, the male moves along the female's back from rear to front; if he moves from front to back, he reverses direction when he reaches the area of the female's vent. The male usually does not flick his tongue during this advance, but does align his body over the female's body. His body and tail are dipped laterally at his vent region along the female's side at her vent. Immediately anterior to the vent, the male's body is positioned on the female's back, and immediately posterior to her vent his tail is projected upward at about a 30° angle, but does not touch the female's tail, which is normally rested flat on the ground. Several vent thrusts are then made by the male in an attempt to copulate, and his hemipenes may be partially everted. If intromission is at first unsuccessful, the male will move his entire body back and forth on the female's dorsum in about 2.5 cm strokes at a rate of a stroke per second. This sequence may end in several rapid strokes by only his snout. Nose strokes are about 1.25 cm long and occur at a rate of about 2 per second. An unreceptive female does not gape her vent, and tries to crawl away. The male then pursues her and repeats the sequence (Quinn 1979a). The observed courtship sequence occurred 5 times in 40 minutes, but copulation was not successful (Quinn 1979a).

Clutches contain 2–13 eggs (although 3–6 eggs are probably the norm), which are laid in late May to July in leaf litter, sawdust piles, rotting logs, or stumps, and probably also in animal burrows. Tryon and McCrystal (1982) recorded a period of 37 days between copulation and oviposition in captive females. Reported measurements of the elongated, white eggs are length, 20.0–47.0 mm; width, 6–14 mm; and weight, 4.2–5.6 g (Campbell 1973; C. Ernst, personal observation; Sabath and Worthington 1959).

The IP covers about 60–65 days, and the young hatch in late August to early October (C. Ernst, personal observation). Hatchlings emerge from slits in the egg shell about 10 mm long; the total time needed for emergence is approximately 4 hours (Campbell 1973). Hatchlings have 15.0–20.5 cm TBLs and BMs of 2–3 g (Campbell 1973; C. Ernst, personal observation).

GROWTH AND LONGEVITY: No growth data are available. A female of unknown age when captured survived 18 years, 3 months, and 30 days at the Fort Worth Zoo (Snider and Bowler 1992).

DIET AND FEEDING BEHAVIOR: *Micrurus tener* has been reported to prey on anurans (unidentified), lizards (*Cnemidophorus gularis; Eumeces fasciatus, E. tetragrammus; Ophisaurus* sp.; *Sceloporus undulatus; Scincella lateralis*), and snakes (*Agkistrodon contortrix, A. piscivorus; Arizona elegans; Diadophis punctatus; Ficimia olivacea; Lampropeltis calligaster; Leptotyphlops dulcis; Masticophis flagellum; Micrurus tener; Opheodrys aestivus; Pantherophis guttata, P. obsoleta; Salvadora grahamiae; Sonora semmiannulata; Storeria dekayi, S. occipitomaculata; Tantilla gracilis, T. planiceps, T. rubra, Tantilla* sp.; *Thamnophis marcianus, T. proximus; Tropidoclonion lineatum; Tropidodipsas sartorii; Virginia striatula, V. valeriae*) (Clark 1949; Curtis 1952; Fisher 1973; Greene 1973a, 1984; Kennedy 1964; Klauber 1946; Mitchell 1903; Price 1998; Reams et al. 1999; Roze 1996; Ruick 1948; Schmidt 1932; Strecker 1908; Swannack and Forstner 2003; Tennant 1998; Vermersch and Kuntz 1986; Werler and Dixon 2000). Vermersch and Kuntz (1986) reported that juveniles may prey on arthropods, but this has not been proven, and Greene (1984) thought that anurans and small mammals are only rarely taken.

Feeding behavior of *M. tener* is similar to that of *M. fulvius* (which see for details). Prey is usually ingested head first, and ventral scale overlap is used to locate the head prior to swallowing (Greene 1976).

VENOM DELIVERY SYSTEM: Adults have FLs of 3.0–3.3 mm (n = 10); other data concerning the venom delivery system are lacking.

VENOM AND BITES: As with other coralsnakes, the venom of *M. tener* is highly neurotoxic, acting primarily on the respiratory and cardiovascular systems by blocking impulse transmission across the synaptic junctions between nerves and muscles. Probably the venom chemistry is similar to that reported for *M. fulvius*, but unfortunately, most data for *M. tener* cannot be separated from that of *M. fulvius*, under which name it was reported. Possani et al. (1979), however, did describe the phospholipase A_2 of venom from *M. t. microgalbineus*. It contains 0.49 mg of protein in contrast to the 0.87 mg of *M. fulvius* (Roze 1996).

Venom of *M. tener* is less toxic than that of *M. fulvius*. The LD_{50} for laboratory mice is 1.15 mg/kg BM (Sánchez and Pérez 2006), and the average dry venom yield and human lethal dose are, respectively, 10–12 mg and 4–7 mg (Minton and Minton 1969; Roze 1996). There are ample case reports on envenomations, some fatal (Clark 1949; Gloyd 1938; Neill 1957; Parrish and Kahn 1967a; Schwartzwelder 1950; Shaw 1971; Stejneger 1898; Stimson and Engelhardt 1960; True 1883; Werler and Darling 1950; Wright and Wright 1957). Reported envenomation symptoms in humans, some in contrast, are breathing difficulty, BT elevation, burning sensation or pain (sometimes cyclic) at bite site and limb of bite, cold skin, conjunctivitis of eyes, difficulty in swallowing, dizziness, drowsiness, feeling of swollenness in throat, inflammation at bite site, leg pains (although bite was to hand), nausea with periodic vomiting, negligible or no swelling (edema) and pain at the bite site, no hemorrhaging or bruising at bite site, no sloughing of skin, profuse perspiration, rapid and elevated heartbeat, soreness about face (especially jaw bone, teeth, and temples) and throat, soreness of eyes and skin, swelling (edema) at bite site, swelling of neck glands, stiff joints, and warm, dry skin. Some of these symptoms may have been the results of the treatments applied. Most serious symptoms do not become evident until several hours after the initial envenomation. Fortunately, bites by this species are rare (Parrish 1966), probably much less than 3% of all annual snake envenomations in Texas; Seifert (2007a) reported a total of only 82 recorded cases between 2001 and 2005. Not all envenomations display puncture wounds (Norris and Dart 1989). Bites can be treated with the antivenom Coralymm produced from the venom of tropical coralsnakes (Sánchez et al. 2008).

PREDATORS AND DEFENSE: American bullfrogs (*Rana catesbeiana*), common kingsnakes (*Lampropeltis getula*), adult Texas coralsnakes (*M. tener*), and, possibly, Virginia opossums (*Didelphis virginiana*) are known predators (Clark 1949; Curtis 1952; Minton 1949; Tennant 1998).

In our experience, when first discovered, *M. tener* may either flee or remain still, but if further disturbed, it will perform the same defensive behaviors described for *M. fulvius*. It thrashes about and bites, often with a sidewise strike, if handled.

PARASITES AND PATHOGENS: The snake is a host of the nematodes *Cosmocercoides dukae* and *Kallicephalus inermis* (Harwood 1932; Schad 1962).

POPULATIONS: Of 2,083 snakes collected by Clark (1949) in the Hill Parishes of Louisiana, only 9 were *M. tener*, and of 85 snakes observed during more than 8,800 km of road cruising in western Louisiana by Fitch (1949), only 2 were this species. Because the snake is fossorial and secretive, reports concerning its abundance are normally understatements. In contrast, it can be locally common. Tennant (1998) reported that it was a common Texas snake. Ruick (1948) considered *M. tener* common south of Corpus Christi, where he collected 10 during 4 collecting trips along the same 16 km mile stretch of road; Vermersch and Kuntz (1986) thought its populations in the 8 south-central counties of Texas to be moderate in size.

M. tener faces the same survival challenges that *M. fulvius* does: habitat destruction, highway deaths (3 of 10 DOR snakes recorded by Ruick [1948] were this species), and fire ant predation. Collection for venom extraction and the pet trade is not a serious threat.

REMARKS: *M. tener* is closely related to the more southern species, *M. tamaulipensis* (Lavín-Murcio and Dixon 2004).

Formerly, the trivial name *tenere* was used for this species, but Frost and Collins (1988) have shown that the proper spelling is *tener*. The species was reviewed by Campbell and Lamar (2004), Davidson and Eisner (1996), Roze and Tilger (1983), and Werler and Dixon (2000).

Pelamis Daudin, 1803
Yellow-bellied Seasnakes
Serpiente del Mar

The genus is monotypic, containing only the world's most widely distributed seasnake, *P. platura*.

Electrophoretic studies of serum and tissue enzymes and gene-sequencing studies of rRNA and mtDNA have shown *Pelamis* to be most closely related to the other hydrophiid seasnake genera *Aipysurus, Emydocephalus, Enhydrina, Hydrophis*, and *Thalassophina* (Cadle and Sarich 1981; Keogh 1998; Mao et al. 1983; Murphy 1988). This generally substantiates the morphological relationships proposed by McDowell (1972) and Voris (1977).

Pelamis platura (Linnaeus, 1766)
Yellow-bellied Seasnake
Vibora del Mar

RECOGNITION: This marine snake (TBL_{max}, 114.3 cm) is easily recognized by its unique oar-like tail, which is flattened from side to side (about 10 times higher than wide), sharply demarcated two-toned body coloration (dark back, yellow belly), and dorsally placed valved nostrils.

The dark dorsal body coloring is either black, olive, olive-brown, or brown; the

Pelamis platura (Costa Rica; Dallas Zoo, John H. Tashjian)

Pelamis platura
(R. D. Bartlett)

venter is yellow, cream, gray, or pale brown; juveniles are more brightly colored than adults. Body pigmentation tends to be in the form of broad longitudinal stripes, but some individuals have undulating stripes, bars, or spots (see below). The laterally compressed tail has a variable pattern of dark and light reticulations or, in a few snakes, large spots. Except in entirely yellow individuals, the head is counter-shaded, dark on top, light beneath.

In his monograph on seasnakes, Smith (1926) defined seven body patterns in *Pelamis*: (1) black above and yellow or brown below, with the two colors sharply defined, head black above, usually with a yellow upper lip; (2) black above and brown below separated by a yellow stripe, head black above; (3) a black dorsal stripe that is either wavy throughout or broken into spots posteriorly, yellow laterally and ventrally, head black above with yellow on the lips more pronounced and extending onto the snout; (4) black above and yellow or brown below, with a lateral series of black spots that may fuse into a stripe, head black above; (5) black above and yellow below, with a ventral series of black spots or bars, head black; (6) yellow, with a black dorsal stripe anteriorly and transverse dorsal spots and bars on the sides and venter posteriorly, head with black variegations; and (7) yellow, with dorsal black or brown-bordered cross bars, bars on the venter alternating with the dorsal ones, head with black variegations. Color patterns of Costa Rican *P. platura* fall into the first four of Smith's seven color varieties (Bolaños et al. 1974).

According to Tu (1976), the most common pigment pattern in the eastern Pacific population is a tricolored one of black (olive)-yellow-brown. The next most common is bicolored black (olive)-yellow. A rare unicolor yellow phase also occurs in less than 1% (4 of 3,077) of the snakes Tu examined (but in 3% of the snakes at Golfo de Dulce, Costa Rica; Kropach 1971b).

The body scales are smooth and nonoverlapping, and in 39–47 anterior rows, 44–67 midbody rows, and 33–46 rows at the tail. The 260–465 ventral scutes are divided by a midventral groove, and are almost as small as the dorsal scales; the anal plate is subdivided. Subcaudals total 39–66. Picado (1931) provides a photomicrograph of the body scales of *Pelamis*, and Voris and Jayne (1976) described the costocutaneous muscles.

The head is angular, narrowing considerably in front of the eyes. The top of the head is covered with enlarged scales: 1 rostral, 2 nasals (in contact dorsally), no internasal scales, 2 prefrontals, 1 frontal, 2 supraoculars, and 2 parietals. Some individuals may have 2 or more small interparietal scales. Lateral head scalation includes 1 (2) preocular, 1–2 suboculars, 2 (3) postoculars, 2 + 2 + (2–3) + (3–4) temporals, 7–11 supralabials (number 2 touches the prefrontal, numbers 4–5 are below the orbit and usually separated from it by suboculars), and 10–13 infralabials. Chin shields are small with the anterior pair separated by small scales, and the mental is ungrooved. The tongue is very short, barely capable of protruding from the mouth. On the maxilla, the short anterior fang is followed by a diastema that separates it from 7 (5–6)–10 solid posterior teeth of almost the same size as the fang. Other teeth present are 7 (5–6) palatines, 22–28 pterygoidals, and 15–18 dentaries.

The hemipenis is narrow and feebly bilobed, and has small spines over most of its length, but papillae at the extreme tips. The sulcus spermaticus is forked near the tip, and lips of the sulcus are prominent and fleshy, but not spiny (McDowell 1972). It is illustrated in Mao and Chen (1980) and Wright and Wright (1957).

Adult males are shorter (TBL_{max}, approximately 70 cm; mean SVL, 45.2–52.5 cm), with longer tails (mean, 6.8% of SVL, 13% of TBL), shorter heads (mean, 31.6% of SVL), and narrower heads (mean, 14.9% of SVL) than females (TBL_{max}, 114.3 cm; mean SVL, 48.1–58.1 cm; mean TL 6.2% of SVL, 11% of TBL; HL, 34.8% of SVL; and mean HW, 15.2% of SVL; respectively), and their dorsal scales have 1–3 tubercles while those of females are smooth (Huang 1996; Kropach 1975; Mao and Chen 1980). Unfortunately, none of these is sufficient by itself for sex determination. Huang (1996) presents regression coefficients for HL, HW, and HD on SVL for both sexes.

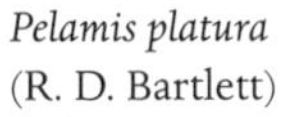
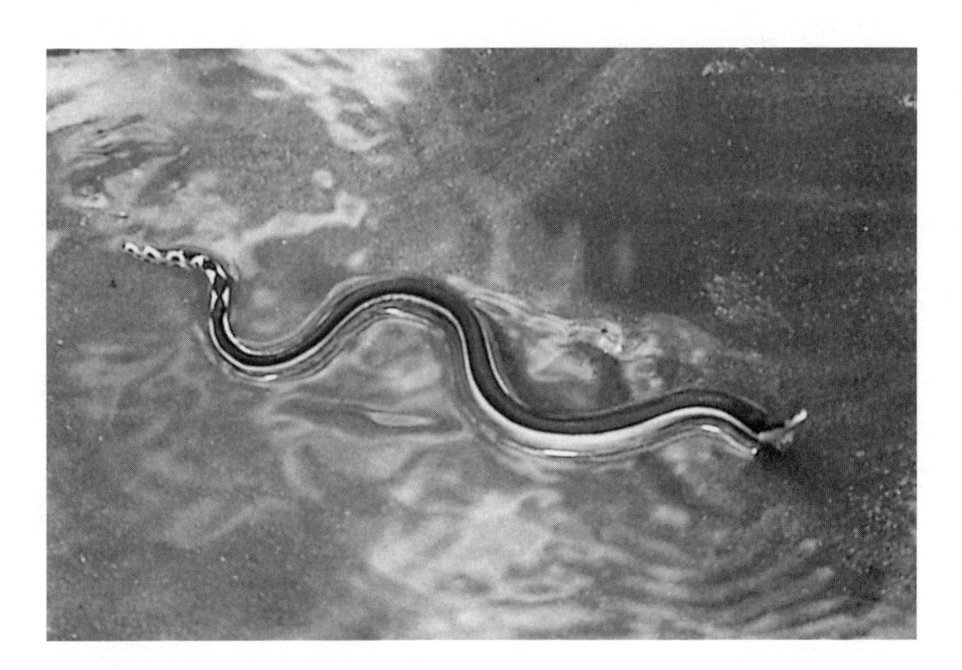

Pelamis platura
(R. D. Bartlett)

GEOGRAPHICAL VARIATION: No serious quantitative study of possible subspeciation in *Pelamis* has been published, but, according to Smith (1926), different patterns predominate in widely separated oceanic populations.

CONFUSING SPECIES: No other snake in the Americas has the color pattern or flattened, oar-like tail of *P. platura*.

KARYOTYPE: Each body cell contains 19 pairs of chromosomes consisting of 20 macrochromosomes and 18 microchromosomes. Macrochromosome pairs 1–2 are metacentric, pair 3 is subtelocentric, and pairs 4–9 are telocentric. The pair of sex chromosomes is metacentric; females are heteromorphic ZW, males are homomorphic ZZ (Bolaños 1983; Gutiérrez and Bolaños 1980).

FOSSIL RECORD: No fossils are known. The adaptive zone of hydrophiid snakes, like *Pelamis*, is of relatively recent origin and was probably first occupied by the snake no earlier than the late Oligocene Epoch in response to the marine thermal regimes that were present then. Hecht et al. (1974) discussed the possible evolution of *Pelamis* to these marine thermal regimes, and we refer the reader to their paper for details.

DISTRIBUTION: *Pelamis* is mostly restricted to the warm tropical and subtropical oceans, where its resident and breeding populations are probably circumscribed by the 25°C isotherm and the 100 m isobath (Hecht et al. 1974). It ranges in the Indian and Pacific oceans from the Seychelles (Lawrence 2007) and eastern Africa, Madagascar, Arabia, and India, throughout coastal southeastern Asia, Indonesia, southern Russia (Kharin 2007), Japan, the eastern coast of Australia, New Zealand, and the Pacific Islands to the western coast of the Americas from Ecuador, and the Galapagos Islands, north to Baja California and the Gulf of California (see Hecht et al. 1974 for a detailed

Distribution of *Pelamis platura*.

discussion of its Indian and Pacific oceans distribution). It has also recently been found on the Caribbean coast of Colombia (Hernandez-Camacho et al. 2005 [2006]).

P. platura is the only seasnake recorded from the Hawaiian Islands (McKeown 1996). In California, waifs have been seen in waters off San Diego (Stebbins 2003), and it has been collected at San Clemente, Orange County (Pickwell et al. 1983), and Los Angeles Bay (Shaw 1961). The snake probably only reaches California during El Niño years, when the waters are warmer.

P. platura could possibly invade the Caribbean Sea (Dunson 1971). Kropach (1972) collected as many as 60 *P. platura* in an hour at the Pacific entrance of the Panama Canal, so the potential exists for the snake to make its way into the Caribbean Sea via that waterway. Although it does not seem to voluntarily enter the tidal waters of bays or river mouths, and may thus naturally avoid freshwater, the snake can tolerate freshwater for extended periods (see below). Cozzi (1980) and Priede (1990) speculated that, with the heating of the oceans by global warming, the snake may colonize the Atlantic Ocean and eventually even reach British waters in the future.

HABITAT: *P. platura* is totally marine. Although usually considered pelagic, it is most often found within 1 to 20 km of the coast in the Pacific Ocean off Central America, and seems to prefer the more shallow inshore waters. Normally it occupies waters of 22–30°C, where it spends most of the time drifting at the surface (it dives during periods of rough water; Tu 1976), often in large numbers among the flotsam in surface slicks formed at the interface of two currents. Such habitats probably provide good foraging.

BEHAVIOR AND ECOLOGY: The natural history of *P. platura* is probably the best known of the seasnakes. Annually, it has been collected from December to June along the Pacific Coast of Colombia (Alvarez-León and Hernández-Camacho 1998). In the Galapagos Islands the snake has been recorded from February to November (Reynolds and Pickwell 1984). Farther north off Panama and Costa Rica it is present in every month, with noticeable increases in numbers during the drier months. Seventy-nine percent of the 635 juveniles collected by Vallarino and Weldon (1996) in the Gulf of Panama were taken in July and September–December. Mexican records show that it is present off the coast of Michoacan from November to April, but is most often observed in March and April (Duellman 1961). Two-thirds of the records between Jalisco and Baja California are for the months of January to April (Casas-Andreu 1997), and Hardy and McDiarmid (1969) collected the snake on 15 February and 1 March and again in October and June in Sinaloa.

Nocturnal activity during 1930–2400 hours has been recorded at Bahia Honda, Panama, by Myers (1945). The 22 snakes collected were 220–311 mm long, and were probably juveniles. All made their appearance in the same way by leisurely swimming upward toward the surface. Upon reaching the surface, each snake protruded its head above the water and floated upward until nearly in a horizontal position while breathing in air. They then slowly swam downward toward the bottom, and none seemed interested in feeding. Myers suspected *P. platura* feeds during the day and spends the night on the bottom, occasionally rising to the surface to breathe, and this has been corroborated by Tu (1976). From 14 January to 3 February, Tu collected 3,077 individuals during 0700–1300 hours, but none from 1500–1800 hours. Perhaps surface waters become too warm at that time, or their fish prey less active. The Panamanian juveniles collected by Vallarino and Weldon (1996) were taken between 0700 and 1300 hours. All sightings of the species in the Gulf of Carpentaria, Australia, by Limpus (2001) occurred at 0936–1730 hours.

When surfaced, the snake often drifts with the current. Surface swimming is by sideward undulations aided by the laterally compressed tail, which may act as a paddle. The snake has the rare ability to swim both forward and backward. When it desires to do so, it can move rapidly through the water. However, such activity is only used for local movements (usually while foraging), and it is doubtful if it actively swims for long distances. Instead, long-distance dispersal is probably passive, with the snake being moved while floating in ocean currents. On occasion they may be blown or drift to the extremes of their range in the eastern Pacific, and this is most likely how they reach Southern California (though rarely).

While it is a graceful swimmer, *P. platura* is poorly adapted for crawling on land. The laterally compressed tail is a hindrance and the lack of elongated ventral scutes prevents the gripping of the ground for crawling. When washed onto a beach it is almost helpless, and soon dies of heat exhaustion or dehydration. In addition, it may not be able to breathe normally when out of water (Minton 1966).

Subsurface swimming is slower than that at the surface (2–4 cm per second), and the snakes usually assume a posture in which the tail is elevated and the posterior position of the body is in a nearly vertical position. Undulatory movements when in this posture involve torsional and rolling motions of the body which, through changes in the camber of the tail keel and body, may contribute to the thrust (Graham et al. 1987b). The total thrust power generated by a 51 cm *P. platura* is 3.641×10^{-4} J at 15 cm per second and 29.877×10^{-4} J at 32 cm per second.

Surfacing depends on the condition of the surface water; fewer snakes surface during periods of rough water (Rubinoff et al. 1986; Tu 1976). Dives to maximum depths of 6.8 m in the dry season and 15.1 m during the wet season occur in the Gulf of Panama (Rubinoff et al. 1986). Pinney (1994) reported the snake is capable of diving to a maximum depth of 152 m.

Most dives are characterized by a four-phased pattern: (1) a nearly vertical descent, with undulatory motions involving rolling or torsional changes in the curve of the body and tail keel; (2) a more bouncy descent, with bobs occurring at 1.7 per minute until reaching a depth where there is no tendency to sink or rise; (3) a gradual descent at about 0.11 m per minute; and (4) a rapid final descent at about 3–4 m per minute (Graham et al. 1987a; Priede 1990). The gradual ascent phase accounts for about 82% of the total underwater duration of each dive and may reflect a period when the snake gradually ascends, and, by Boyle's Law, compensates for buoyancy lost because of the decline in lung volume. An intracardiac blood shunt also exists that assures management of lung O_2 reserves in a manner that augments cutaneous breathing and establishes favorable transcutaneous diffusion gradients that help the removal of built-up nitrogen gases in the blood (to avert the "bends") and the uptake of O_2 (Graham et al. 1987a; Seymour 1974).

Before diving, *P. platura* overinflates its lung, possibly to as much as 20% of body volume (Graham et al. 1975; Priede 1990). It retains enough air in its large lung to keep itself positively buoyant. However, as it swims deeper water pressure squeezes the air in the lung and the snake's overall density becomes greater until it becomes neutral or negatively buoyant at that depth (Graham et al. 1975). As the O_2 in the lung is used, the resulting CO_2 is excreted through the skin rather than diffusing into the lung. Elevation of the head, like during ascent, collapses the posterior saccular portion of the lung (Seymour et al. 1981).

P. platura may remain submerged for long periods of time. Although Dunson (1971) noted that seasnakes are capable of staying underwater for anywhere from 2 to 8 hours (species not given), Rubinoff et al. (1986) observed that the maximum voluntary submergence time for *P. platura* was 213 minutes; but, in 202 complete dives, only 19 exceeded 90 minutes. Diving snakes may be able to avoid anaerobiosis by having a reduced metabolic rate, an enhanced rate of cutaneous O_2 uptake, or both. The recorded O_2 capacity of *Pelamis*'s blood (expressed as the average volume of O_2 that can be held as a percent of blood volume) is rather high, 10.2%, but the O_2 capacity of its blood changes with BT (Pough and Lillywhite 1984). It increases when the BT is raised from 10 to 20°C, but drops if the BT is raised further to 40°C; however, in nature, the snake's BT probably varies little during a 24–hour period. Cutaneous breathing is one of the main physiological adaptations of *Pelamis*; it can remove O_2 from water at rates up to 33% of total standard O_2 uptake, and can excrete CO_2 into the water at rates up to 94% of standard O_2 consumption (Graham 1974a). As much as 18 mL of O_2 per gram of BM can be absorbed through the skin (Heatwole 1999).

Another serious problem confronting the snake is that of osmoregulation: balancing its salt and water content with that of seawater. While marine mammals have kidneys efficient enough to rid the body of excess sodium ions (Na^+), the reptilian kidney is weak and cannot excrete higher concentrations of Na^+ than are found in the blood; the snake's urine is always hypoosmotic to the blood plasma. Therefore, additional structures must supplement the kidneys to prevent excess Na^+ from building up in the body fluids. Although birds and marine turtles have well-developed salt-secreting glands in their nasal passages for this purpose, no such discrete compact gland has been found in *P. platura* (Schmidt-Nielsen and Fange 1958; Taub and Dunson 1967); however, serial sections of the head prepared by Burns and Pickwell (1972) have confirmed its presence. Well-developed nasal glands are found in some other seasnakes (Burns and Pickwell 1972). Also, mucoid cell types are absent from the labial glands of *Pelamis*, so these may not aid in salt secretion. The snake may lack development in the above glands, but it does have a very well-developed, posterior, sublingual gland that acts as a salt gland, excreting NaCl into the mouth for expulsion (Dunson 1971; Dunson et al. 1971). This gland usually comprises about 0.04% of the total BM (Dunson and Dunson 1975), and the fluid excreted has a higher NaCl content than seawater. During periods of salt loading, more Na^+, Cl^-, and K^+ are excreted via the mouth than by the cloaca (Dunson 1968). *P. platura* kept in freshwater for 48 days have shown no decrease in salt gland Na-K ATPase activity or in gland weight, even though Na^+ concentration dropped markedly. Na-K ATPase has been consistently found in high concentrations in tissues specialized for active ion transport (Dunson and Dunson 1975). *Pelamis* can survive in freshwater for periods exceeding six months (Dunson and Ehlert 1971).

The snake's skin is permeable to water but not to Na^+ (Dunson and Robinson 1976); *P. platura* has a very low rate of exchange of Na^+ with seawater. Influx and efflux of Na^+ are balanced, but water is not, and there is a net loss of water amounting to about 0.4% BM per day that occurs primarily through the skin. So the major osmotic problem of *Pelamis* in seawater is water balance, not salt balance (Dunson and Robinson 1976; Dunson and Stokes 1983).

The latitudinal distribution of *P. platura* in the eastern Pacific Ocean lies between the northern and southern 18°C surface isotherms, and its CT_{max} and CT_{min} are 36°C and 11.7°C, respectively (Graham et al. 1971). It generally avoids WTs above 33°C; those observed in Australia by Limpus (2001) were at a mean WT of 24.7 (24.7–24.8) °C. With rapid cooling this snake will stop feeding at 16 to 18°C, but it has a high resistance to cold and can withstand 5°C for about an hour. In laboratory tests, it did not acclimate to 17°C, and, thus, was not able to survive for long periods at such WTs (Graham et al. 1971). There is a general avoidance of surface WTs cooler than 19°C (Rubinoff et al. 1986). BTs of *P. platura* caught in surface waters along the Pacific coast of Middle America have been 26.9–31.0°C (Dunson and Ehlert 1971) and in the Gulf of Panama 28–29°C (Hecht et al. 1974). Feeding slows at 26°C, and effectively stops at 23°C (Hecht et al. 1974). Brattstrom (1965) gave the BT of one as 24.9°C (AT, 25°C), and Pickwell et al. (1983) found a dying *P. platura* on a Southern California beach after the WT had dropped to 16°C. It loses efficient motor control of swimming and floating at 16°C, but slowly acclimated individuals can survive for at least a week at 33–35°C (Hecht et al. 1974). The optimal WT range for *Pelamis* seems to be 28–32°C.

BT is mostly determined by the surrounding water, but Graham (1974b) reported that its dark dorsal body surface absorbs solar radiation, and consistently elevates the BT slightly above the WT when the snake is basking at the surface in a calm sea. There seems to be no positive evidence that the snake moves up or down the water column to thermoregulate, and Graham's laboratory experiments indicate *P. platura* neither seeks nor avoids heat when given thermal choices.

REPRODUCTION: *Pelamis* is ovoviviparous, and the young are born at sea, so relatively little has been observed of its breeding habits. A serious study of the animal's reproductive behavior is needed.

In the Gulf of Panama, males mature at TBLs of at least 50 cm, but females must grow to about 64 cm (Kropach 1975). Goldberg (2004) reported Costa Rican males were mature at 52.2 cm SVL and females at 53.1 cm SVL. Vallarino and Weldon (1996) noted that Panamanian females with 60–74.2 cm TBLs produce young. The above lengths are probably attained in 2 to 3 years.

The reproductive cycles of both sexes in Costa Rica were described by Goldberg (2004). He examined 4 SVL 52.2–61.0 cm males to determine the testicular cycle; the histology was similar to that reported for *Micruroides euryxanthus* (Goldberg 1997). Only two stages were present: (1) recrudescence (two February males) with renewal of the cycle characterized by a proliferation of germinal cells, and spermatogonia and primary spermatocytes predominating; and (2) spermiogenesis (two May males) with forming sperm, the lumina of seminiferous tubules lined with mature spermatozoa, and metamorphosing spermatids. In February, 2 females had inactive ovaries and no yolking, but another 53.1 cm female contained 3 oviductal eggs. In May, 1 female had inactive ovaries, 3 females contained oviductal eggs (1, 63.3 cm, had 4), and another 58.3 cm female contained 2 well-developed embryos. A 61.4 cm July female had 4 embryos, and another July female was undergoing early vitellogenesis. In the Philippines, most gravid females are found in June–August (Alcala 1986).

Breeding populations apparently occur only where the mean monthly WTs are above 20°C (Dunson and Ehlert 1971). Kropach (1975) expected to find seasonal reproduction in the Gulf of Panama because of ecological seasonality of the region, but instead found young with newborn TBLs in every month. McCoy and Hahn (1979) and Visser (1967) obtained similar results in the Philippines and South Africa, so breeding may take place throughout the year.

Courtship behavior has not been fully described. Dunson (1971) presented a photograph of two individuals "knotted" together in mating activity, and McKeown (1996) and Vallarino and Weldon (1996) reported intertwined snake couples. In the copulation observed by Vallarino and Weldon, the male had the posterior 10–15 cm of his body wrapped around the female in 3 coils. The couple remained attached for about 1.5 hours after capture, but frequently tried to separate and swim away from one another. All mating occurs in the water, possibly near the surface in the slicks.

Although the young are thought to be born at sea, Minton (1966) found a 23 cm juvenile in a Pakistan mangrove swamp on 22 March, and Solórzano (2004) reported that Costa Rican females are particularly abundant in the shallow waters of bays during the time of the year when parturition occurs. So probably some females enter such habitats at this time to have their young. The female may surround and protect her newborns during their first few days of life (Rose 1950).

Brood size averages 4.5 young (n = 16). In the eastern Pacific, clutches contain 2–6 young (Goldberg 2004; Kropach 1975), but elsewhere as many as 1–10 young may be born (Branch 1979; Rose 1950; Vanderduys and Hobson 2004; Visser 1967). Fitz Simons (1978) reported that southern African females produce up to 18 young at a time, but this is doubtful. Mass averages 57.8 (12–79) g. Mean RCM is 28.8 (12.0–34.5) % of female before parturition BM, and 42.2 (13.6–52.7) % of female after parturition BM. There is a slight correlation between female length and clutch size (Vallarino and Weldon 1996).

The GP is unknown, but Kropach (1975) observed gravid females in the laboratory and thought it is at least five months and more likely six or more. Panamanian births reported by Vallarino and Weldon (1996) took place between 10 and 23 September. In Queensland, Australia, females birthed on 25 May (1 birth) and 13 June (78 cm SVL female, 2 births and 4 other fully formed young in the oviducts) (Vanderduys and Hobson 2004). During the 25 May birth, within 1 minute of being returned to the water, the 90 cm SVL female lowered the posterior third of her body and began parturition. The neonate emerged head-first, and the total birth experience lasted 3–4 minutes. McKeown (1996) reported tail-first births.

Neonates are more brightly colored than adults; they have a mean TBL of 25.3 (22–29) cm, and a mean BM of 8.8 (6.0–14.3) g.

GROWTH AND LONGEVITY: Growth of young *P. platura* seems rapid, at least during the first year, but the rate is unknown. Male growth slows when they reach 50 cm and mature, but females maintain their rate of growth. Individuals shorter than 30 cm are first-year juveniles; those 30–50 cm are mostly juveniles and subadults, but may include a few adult males; individuals 50–60 cm are adult males and subadult females; and all snakes longer than 60 cm are adults (Kropach 1975; Vallarino and Weldon 1996).

Pelamis apparently does not do well in captivity. Snider and Bowler (1992) reported that an unsexed *P. platura*, of unknown age when wild-caught, lived only 3 years, 5 months, and 10 days at the Houston Zoo, and Shaw (1962) mentioned another captive that survived only 2 years and 4 months.

DIET AND FEEDING BEHAVIOR: Foraging seems mostly to occur only at the surface during the day, and is primarily associated with current slicks (Dunson 1975; Klauber 1935; Klawe 1964; Kropach 1975; Paulson 1967; Voris 1972). *P. platura* practices a type of ambush behavior to catch its fish prey (Heatwole 1999; Klauber 1935). It floats motionless among the debris trapped in the slick, where small fish seek out the debris to feed or to hide and may congregate among the floating materials in large numbers. As a fish swims by, the snake strikes rapidly sideward with its head and seizes it in its mouth. This is usually followed by the snake swimming backward dragging its prey. The fish is immobilized quickly by the snake's venom, which is chewed into the bite wound, and is usually swallowed head first very quickly to avoid impalement by the fish's fin spines. Sensitivity to movement in the water and possibly olfaction seem to be more important in prey detection than vision; Heatwole (1999) reported that finely chopped fish dropped into an aquarium holding these snakes caused a "frenzy" in which an object encountered, even another snake, was bitten. Most snakes collected in slicks contain food, while those caught elsewhere usually do not (Kropach 1971a).

Pelamis is strictly piscivorus, feeding on a broad variety of fish. Voris and Voris (1983) examined the digestive tracts of 235 snakes, and found remains of the following families of fish (number of digestive tracts in parentheses): Acanthuridae (4), Atherinidae (4), Blennidae (2), Carangidae (16), Chaetodontidae (1), Clupeidae (1), Coryphaenidae (11), Engraulidae (8), Fistularidae (2), Kyphosidae (1), Lutjanidae (1), Mugilidae (29), Mullidae (44), Nomeidae (1), Polynemidae (83), Pomacentridae (8), Scombridae (5), Serranidae (11), Sphyraenidae (1), Stromateidae (1), and Tetraodontidae (1). No preference for fish size has been found for any size class of *Pelamis* (Kropach 1975), but larval and juvenile fish are primarily taken. Reported fish prey (natural and in captivity) include *Abudefduf troschelli, Acanthurus xanthopterus; Anchoviella* sp.; *Auxis* sp.; *Blenniolus brevipinnis; Caranx caballus, C. hippos, C. marginatus; Carassius auratus; Chaetodon humeralis; Chloroscombrus orqueta; Coryphaena hippurus*; Cyprinidae; *Decapterus lajang; Diodon histrix; Engraulis mordax; Fistularia cornea; Fundulus parvipinnis; Gillichthys mirabilis; Heterostichus rostratus; Hypsoblennius* sp.; *Kyphosus* sp.; *Lobotes pacificus; Lutjanus* sp.; *Melanorhinus cyanellus; Mugil cephalus, M. durema; Mulloidichthys rathbuni; Peprilus medius; Polynemus approximans; Pesenes whiteleggi; Pseudupeneus grandisquamis; Selar crumenophthalmus; Sphoeroides* sp.; *Sphyraena* sp.; *Spirinchus starksi*; *Thunnus albacares*; and *Vomer declivifrons* (Campbell and Lamar 2004; Klawe 1964; Kropach 1975; Peterson and Smith 1973; Pickwell et al. 1972; Shaw 1962; Visser 1967; Voris and Voris 1983). In addition to fish, a captive has accepted a green tree frog (*Hyla cinerea*) and a newly metamorphosed American bullfrog (*Rana catesbeiana*) (Switak 2002).

VENOM DELIVERY SYSTEM: The venom delivery apparatus of *P. platura* is poorly developed when compared to that of terrestrial elapids. More adaptive emphasis seems to have occurred in the development of numerous small teeth to hold prey (see "Recognition") than on well-developed fangs. Indeed, the fang situated at the front of the maxilla is hardly larger than the other teeth on this bone. Twenty-five (TBL, 43.8–75.5 cm) *Pelamis* we examined had a mean FL of 1.68 (0.9–2.8) mm, and a positive correlation between FL and TBL was evident. Heatwole (1999) gives the average FL as 1.5 mm.

The venom gland is compartmentalized and contains cuboidal, columnar, and mucoid cells. Accessory venom glands may be present that circumscribe the main duct in the suborbital region (Burns and Pickwell 1972). The duct draining the venom gland empties through a primary opening into a crescent-shaped cavity. This cavity is in turn drained by a secondary opening passing through a ventral hardened wall into the basal orifice of the fang. The venom leaves the fang through a slit-like primary orifice near its tip. Replacement fangs may be present, and are attached or not attached to the venom duct, depending on their stage of development. The entire apparatus and basal portion of the functional fang are ensheathed in a cutaneous membrane. A perspective drawing of the venom delivery system is presented in Schaefer (1976).

VENOM AND BITES: The venom is neurotoxic and very deadly to fish. Its action is postsynaptic, causing muscle paralysis that leads to respiratory failure and death, usually within a minute after being bitten (Tu 1991a).

Crystallized venom yields reported by Bolaños et al. (1974) and White (1995) averaged 0.28–0.38 mg; for those snakes measuring more than 60 cm, the average was about 0.87 mg. In contrast, 13 adult (TBL, 62–83 cm) *P. platura* examined by Pickwell et al. (1972) produced an average dry venom yield of 2.8 mg. Dry yields have ranged from 0.9 to 5.0 mg, with the greatest amounts generally associated with the longest snakes (Bolaños et al. 1974; Pickwell et al. 1972; Tu 1976; Zug and Ernst 2004).

Tu et al. (1975) isolated a major neurotoxin from eastern Pacific *P. platura* that contained 55 amino acids (low for potent neurotoxins) with 4 disulfide linkages that seem essential for toxic action. This toxin (*Pelamis* toxin α) comprised 4.5% of the venom (it has an intravenous LD_{50} for mice of 0.31 μg/g; Khole 1991). Two other toxins, b and c, made up 0.95% and 1.6%, respectively. Their intravenous LD_{50} values for mice are 0.044–8.40 μg/g and 0.93–1.00 μg/g, respectively (Khole 1991; Mackessy and Tu 1993; Tu et al. 1975). Liu et al. (1975) isolated and partially characterized a toxin from the venom of Taiwanese *Pelamis* that contained 60 amino acids and they named pelamitoxin a; it is possible that this toxin is the same as *Pelamis* toxin b. The toxin discovered by Liu et al. was very similar to hydrophitoxin b from *Hydrophis* seasnakes and *schistosa* 5 toxin from the venom of *Enhydrina schistosa*. The venom of these latter snakes is known to be capable of producing human fatalities, and Rogers (1903) thought the venom of *P. platura* is as toxic as that of *E. schistosa*. Amino acid sequences of *Pelamis* toxins are given in Mackessy and Tu (1993), Mori et al. (1989), Takasaki (1998), and Tu (1991a). Electrophoretic variants of phospholipase A_2 are also present in the venom (Durkin et al. 1981). The molecular weights of two toxic fractions isolated by Pickwell et al. (1972) were 11,700 and 6,000, respectively, agreeing with the findings of Bolaños et al. (1974). Other chemical and physical properties of the venom are reported in Shipman and Pickwell (1973), and Gawade (2004) composed a pharmacological classification of snake venom neurotoxins.

Venom doses in laboratory studies have been more potent when the venom is administered intravenously (LD_{50}, 0.18–0.44 μg/g) than via the subcutaneous route (LD_{50}, 0.67 mg/kg) to mice (Khole 1991). Vick et al. (1975) reported a mouse intravenous LD_{50} of 0.11 μg/kg, and the minimum dose to kill 99% of the mice as 0.15 μg/kg. Injected mice developed rapid and labored breathing, abnormal dilation of the pupils, increased heart rate, prostration, and loss of the body righting reflex, all symptoms of neurotoxic envenomation.

Fayrer (1872) reported that "a small fowl" died within 3 hours and 24 minutes of being bitten (apparently on the thigh). The minimum lethal venom dose to a pigeon is 0.075 mg/kg, and to a mudfish (*Saccobranchus fossilis*) 0.25 mg/kg (Rogers 1903). Other lethal venom dosages include guinea pig, 1 mg/kg (Nauck 1929); *Rhesus* monkey, 0.16 mg/kg (Pickwell et al. 1972); dog, 0.05 mg/kg (Pickwell et al. 1972); and sheep, 10 mg/60 kg (Bolaños et al. 1974). Lethal data for laboratory mice include an LD_{50} of 0.09–0.44 intravenous mg/kg (Mackessy and Tu 1993; White 1995), a venom dose of 2.2 mg for a 16–18 g mouse (Bolaños et al. 1974), and a dose of 0.5 mg/kg for a 20 g mouse (Barme 1968). The *Pelamis* toxin α produced respiratory paralysis in rabbits preceded by an initial increase in the breathing rate and a drop in blood pressure (Mori et al. 1989).

The LD_{25} of *Pelamis* venom for humans by body mass (BM) has been estimated as 3.7 mg for a 45 kg human, 4.4 mg for one with a BM of 54 kg, 5.9 mg for a 72 kg person, and 7.5 mg for a human weighing 91 kg (Pickwell et al. 1972). So, it appears that this snake, with its small mouth and fangs and relatively low venom yield, poses little hazard to most humans, if it is not handled.

However, danger from the venom of *P. platura* was reported as long ago as 1693 (Ravenau de Lussan *in* Taylor 1953), and this early report stated that the bite was mortal. Even though the snake may be common, envenomations by it are rare, probably due to the mild disposition of the snake and the reasons stated above. Those bites that have been documented have usually been very mild, and often produced no effects. Symptoms of such bites include drooping eyelids (ptosis), hematological abnormalities, inflammation or swelling at bite site, muscle tenderness or pain, and paralysis of any muscles (Kropach 1972; Senanayake et al. 2005). However, Mackessy (*in* Mackessy and Tu 1993) witnessed an envenomation in Mexico where the victim was bitten in the chest by a small *P. platura* and experienced blurred vision; a metallic taste; numbness of the hands, lips, and throat; and shortness of breath. Also, after a hand bite in Costa Rica, a victim experienced slight edema (within the first hour), mild discoloration (within two hours), extreme sensitivity to touch (lasting 36 hours) at the bite site, and deep pain and slight stiffness of the wrist and thumb (within an hour and continuing for 28 hours), but no systemic effects were noted (Solórzano 1995). All symptoms, except slight sensitivity at the bite site, disappeared within 58 hours, and no medical treatment was sought. In another Colombia bite, a worker, although given *Micrurus* antivenom, became sick, vomited, and was unable to eat (Campbell and Lamar 2004).

Circumstantial evidence that serious bites can occur resides in the fear of the snake by fishermen and coastal residents in many parts of its range. Fatal bites have been reported, although mostly as anecdotes. Becke (1909) recounted that an Australian diver bitten on the finger by an apparent *P. platura* suffered convulsions and died 48 hours later despite amputation of the finger within an hour of the bite. Fitz Simons (1919) recorded several fatal bites resulting from beached snakes in South Africa, Halstead (1970) listed four mortal envenomations, Kinghorn (1956) noted that one death had definitely been recorded in India, Swaroop and Grab (1954) stated that fatal bites had occurred in Central America and Mozambique, Wall (1921) reported a death in Sri Lanka, and White (1995) suggested that, although bite cases were usually mild, fatalities had been recorded or suspected.

PREDATORS AND DEFENSE: Few accounts of predation on this snake have been published. Known or potential predators include octopi (*Octopus* sp.), crabs, fish (*Lut-*

janus aratus, L. argentiventris, L. guttata; Sphoeroides cf. *annulatus* [*S. lobatus*?]), birds (*Fregata magnificens; Haliaeetus leucogaster; Larus fuliginosus, L. pacificus; Pelecanus occidentalis*), and mammals (*Hydruga leptonyx, Zalophus californicus*) (Alvarez-León and Hernández-Camacho 199; Branch 1979; Duellman 1961; Hawkins 2005; Heatwole and Finnie 1980; Pickwell 1972; Pickwell et al. 1983; Reynolds and Pickwell 1984; Smith 1926; Van Bruggen 1961; Weldon 1988; Wetmore 1965).

Some of these records may represent carrion eating, rather than predation, since studies by Caldwell and Rubinoff (1983) and Rubinoff and Kropach (1970) showed that a variety of potential bird and fish predators from the eastern Pacific, when tested, made no attempts to attack live snakes.

The snake is usually mild mannered, and divers have swam among them without being attacked (Pickwell 1972; Pickwell et al. 1972). Some, however, may be aggressive; if handled, the snake will turn and bite to defend itself. Minton (1966) reported that one tried to bite repeatedly when picked up with a forceps.

The bright contrasting yellow belly may be a warning device to would-be predators, or the contrasting dark back and bright venter may act aposematically so that there is no selection of one particular pattern by a predator. The contrastingly colored tail may also act as a warning to some potential predators (Greene 1973b), and many *P. platura* have wounds on their tails (Weldon and Vallarino 1988), possibly indicating that the multicolored tail is used as a decoy to direct predators away from the head. In addition, the knotting behavior described below may help retard predation.

There is some circumstantial evidence that either the flesh or the skin of *Pelamis* may be poisonous, and thus sicken predators that bite or ingest the snake (Reynolds and Pickwell 1984; Weldon 1988).

PARASITES AND PATHOGENS: Data are sparse on the parasites of *Pelamis*, and we could discover no information on its diseases.

It is parasitized by the trematode *Torticaecum nipponicum* and the nematode *Paraheterotyphlum australe* (Ernst and Ernst 2006; Fischthal and Kuntz 1975; Johnston and Mawson 1948; Jones 2003; Sprent 1978).

A problem faced by *P. platura* that does not occur in terrestrial snakes is the fouling of the body surface by encrusting marine invertebrates. These include barnacles (*Conchoderma virgatum; Dichelapis warwicki; Dosima fascicularis; Lepas anserifera, L. tenuivalvata; Platylepis ophiophilus*) and bryozoan ectoprocts (*Electra angulata*?; *Membranipora tuberculata*), which form commensal relationships with the snake when they attach to its skin (Alvarez and Celis 2004; Cuffey 1971; Dean 1938; Deraniyagala 1955; Kropach and Soule 1973; Minton 1966; Reynolds and Pickwell 1984; Schulz 2003; Wall 1921; Zann et al. 1975). Fouling may retard swimming and interfere with courtship or mating behavior. Frequent skin shedding (Shaw 1962; Zeiller 1969) helps remove these unwanted guests, as does also the practice of knotting and tight coiling during which the commensals, and possibly ectoparasites, may be scraped off between the coils (Pickwell 1971).

POPULATIONS: *Pelamis* sometimes occur in huge aggregations. Belcher (*in* Smith 1926) reported having seen thousands swimming on top of the water in the Mindoro and Sulu seas; Tu (1976) collected more than 3,000 in less than a month off the coast of Costa Rica; more than 3,000 were also captured there in 3 weeks by Bolaños (1983); and Kropach (1971a) estimated thousands to be drifting in surface slicks off Panama.

Campbell and Lamar (2004) reported that "assemblages of millions may form," but this must be an exaggeration. In contrast, Vallarino and Weldon (1996) only collected 635 in the Gulf of Chiriquí, Panama, in 19 months. Elsewhere, Tu (1974) reported that only 7 (0.05%) of the total 13,927 seasnakes captured in the Gulf of Thailand in 1969 and 1972 were *P. platura*. Neither does the snake seem so plentiful off the coast of Australia; Limpus (2001) observed only 84 (75% of all seasnakes recorded) in 1 day along a 99.4 km ocean line in the Gulf of Carpentaria, off northern Australia, and Schulz (2003) counted only 13 strandings of *P. platura* along 1,500 km of coastline in northeastern New South Wales.

Several size classes are usually present in *Pelamis* populations: 20–30 cm neonates and juveniles, 31–50 cm subadults or adults, and 51+ cm adults. The size class distribution of a group of 73 Philippine *P. platura* collected by McCoy and Hahn (1979) was strongly bimodal; 48 (46 females, 2 males) had TBLs of 34.3–43.7 cm, while the remaining 25 (18 females, 7 males) were 47.6–69.3 cm long. The frequency of size-class distribution may vary during the year (Kropach 1975), particularly as neonates enter the population. Myers (1945) reported that on 1 March at Bahia Honda, Panama, the seasnakes observed were largely juveniles with only a smaller proportion of adults, and Kropach (1975) noted that of 278 *P. platura* collected in September in Golfo Dulce, Costa Rica, 119 (43%) were neonates. The 32 Australian *P. platura* that passed close enough to the boat to have their TBL estimated by Limpus (2001) were 25–60 cm, with 88% longer than 45 cm.

The sex ratio is usually about 1:1; Kropach (1975) reported 340 males out of a sample of 712 snakes. However, the above sample of Philippine *P. platura* collected by McCoy and Hahn (1979) contained only 12% males.

Populations of *P. platura* do not currently seem threatened, but this may change with an increase in global warming and further pollution of the oceans. Some are killed by accidents. Individuals often show wounds caused by boat props, and those washed ashore, for whatever reason, usually do not find their way back to the water, overheat and die, or are predated.

REMARKS: Lanza and Boscherini (2000) have shown that the trivial name should be spelled *platura* to match the feminine generic name *Pelamis*.

According to Murphy (1988), isozymes show *P. platura* to be most closely related to *Emydocephalus annulatus*. The species was reviewed by Mackessy and Tu (1993), Pickwell and Culotta (1980), and Smith (1926); also, Culotta and Pickwell (1991) published a bibliography.

Heloderma horridum horridum, male (Fort Worth Zoo, John H. Tashjian)

Heloderma horridum alvarezi, female (Fort Worth Zoo, John H. Tashjian)

Heloderma horridum exasparatum (R. D. Bartlett)

Heloderma suspectum suspectum (R. D. Bartlett)

Heloderma suspectum cinctum (R. D. Bartlett)

Micruroides euryxanthus euryxanthus (Portal, Arizona; Steve W. Gotte)

Micruroides euryxanthus euryxanthus (Robert E. Lovich)

Micrurus distans distans (R. D. Bartlett)

Micrurus distans distans (Alamos, Sonora; Eric Dugan)

Micrurus fulvius (Aiken County, South Carolina; Carl H. Ernst)

Micrurus fulvius, Key Largo morph (R. D. Bartlett)

Micrurus tener tener (R. D. Bartlett)

Micrurus tener tener (Fort Worth Zoo, John H. Tashjian)

Pelamis platura (Costa Rica; Dallas Zoo, John H. Tashjian)

Agkistrodon bilineatus bilineatus (R. D. Bartlett)

Agkistrodon bilineatus bilineatus (Alamos, Sonora; Eric Dugan)

Agkistrodon contortrix contortrix (Natchitoches Parish, Louisiana; Brad Moon)

Agkistrodon contortrix laticinctus (California Academy of Sciences, John H. Tashjian)

Agkistrodon contortrix mokasen, green morph (Menifee County, Kentucky; Roger W. Barbour)

Agkistrodon contortrix phaeogaster (R. D. Bartlett)

Agkistrodon contortrix pictigaster (Western Texas, R. D. Bartlett)

Agkistrodon piscivorus piscivorus (Beaufort County, North Carolina; Carl H. Ernst)

Agkistrodon piscivorus conanti (Lakeland, Florida; John H. Tashjian)

Agkistrodon piscivorus leucostoma (Trigg County, Kentucky; Roger W. Barbour)

Agkistrodon taylori, juvenile (Tamaulipas; Houston Zoo, John H. Tashjian)

Sistrurus catenatus catenatus (Ontario; Robert E. Lovich)

Sistrurus catenatus edwardsi (R. D. Bartlett)

Sistrurus catenatus tergeminus (R. D. Bartlett)

Sistrurus miliarius miliarius, gray morph (Hyde County, North Carolina; Ray Folsom, Hermosa Beach, California; John H. Tashjian)

Sistrurus miliarius miliarius, red morph (Hyde County, North Carolina; R. D. Bartlett)

Sistrurus miliarius barbouri (Tamiami Trail, Florida; Ray Folsom, Hermosa Beach, California; John H. Tashjian)

Sistrurus miliarius streckeri (R. D. Bartlett)

Viperidae
Viperid Snakes

The vipers are venomous snakes that evolved from colubrid ancestors, probably different from those that may have given rise to the elapids. Differences between the vertebrae of colubrids and viperids are greater than those between elapids and viperids (Johnson 1956). The fossil history of snakes dates from the early Cretaceous (Barremian) more than 120 million YBP; that of true vipers (Viperinae) dates from the early Miocene about 23.8–22.8 YBP (Szyndlar and Rage 2002). The earliest fossil examples of North American Viperidae are from early Miocene (Arikareean) deposits in Nebraska; pitvipers (Crotalinae, see below) date from the early Miocene (Hemingfordian) of Delaware (Holman 1979a, 1995).

The family is found in Asia, Europe, Africa, and the Americas. McDiarmid et al. (1999) proposed that it consists of about 32 genera and 223 species; however, David and Ineich (1999) recognized 33 genera and 236 species. Lawson et al. (2005), based on the study of mitochondrial and nuclear DNA, listed 34 genera. The species are assigned to 4 subfamilies: Azemiopinae, Causiinae, Crotalinae, and Viperinae. Only pitvipers of the subfamily Crotalinae (about 19 genera, 158 species; David and Ineich 1999) occur in the Americas, where, according to Campbell and Lamar (2004), 11 genera and 115 species occur. Species richness is greatest in Central America (Reed 2003).

American pitviper species are extremely variable in both TBL and geographic range size, and range size is positively correlated with TBL. The TBL of rattlesnakes is measured from the tip of the rostrum to the base of the tail; the rattle length is excluded. Available continental area strongly influences individual species' range size, and trends in range sizes may have been structured more by historical biogeography than by macroecological biotic factors. Little support exists for Bergmann's Rule, as body size does not increase significantly with either latitude or elevation. Range area and median range latitude are positively correlated above 15°N, indicating a possible Rapoport Rule effect at high northern latitudes (Reed 2003).

Pitvipers evolved from true vipers (Viperinae) in the Old World (Brattstrom 1964; Darlington 1957), and these ancestral pitvipers apparently reached the Americas by crossing the Bering Land Bridge from Asia. However, the lineage of Old World pitvipers most closely related to American ones has not been determined (Castoe and Parkinson 2006). Today, 27 species in the genera, *Agkistrodon* (4), *Crotalus* (21), and *Sistrurus* (2), occur in the portion of North America covered in this book.

The families of venomous snakes are differentiated by their venom delivery systems. The Viperidae has evolved an advanced solenoglyphous dentition (fig. 10) that allows the snake to strike, envenomate, and withdraw from struggling prey to avoid injury.

The maxillae have become shortened horizontally while becoming deep vertically, and are capable of movement on the prefrontal and ectopterygoid bones, thus allowing the fangs on the maxillae to rotate until they lie against the palate when not in use

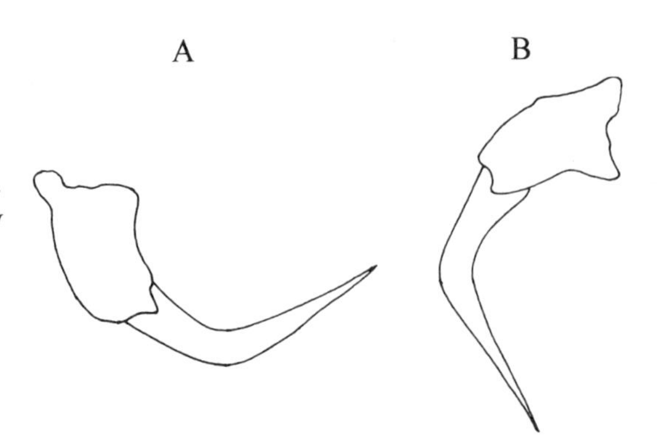

Figure 10. The solenoglyphous fang arrangement of viperid snakes; note the shortened, rotational maxillary bone: A, fang at rest against roof of mouth; B, fang rotated into striking (stabbing) position.

and the mouth is closed (Cundall 2002; Klauber 1972). During the strike the maxillae of adults rotate 70–110° (Cundall 2002), causing the fangs to move downward in a vertical plane and forward into stabbing positions. The fang itself is not capable of movement, but is rigidly fixed within an individual socket located in the maxillary bone; no other teeth are present on the bone. The combination of relatively long, curved fangs (in regard to the shorter, rigid, proteroglyphous fangs of the family Elapidae; see individual species accounts in both families for comparisons) and the ability to rotate them forward constitute the basis of a very efficient biting mechanism (Kardong and Bels 1998). The opposite fangs can be rotated separately. The ability to rotate the fangs present in vipers is almost entirely absent in all other snakes; Fairley (1929) noted that some Australian elapid death adders (*Acanthophis* sp.) also possess to a variable degree the power of elevating and rotating forward the fangs, but the mechanism of rotation differs essentially from that of the vipers. The African *Atractaspis* sp. of the moleviper family Atractaspidae are also solenoglyphous and have the ability to rotate their fangs.

To understand this process it is necessary to remember that the cranial bones of nearly all snakes are loosely joined together and allow a great amount of movement and distention that enables them to swallow thicker-bodied prey. Elastic ligaments connect the bones and are responsible for the great ability of distention.

Several other bones help to transmit the tilting motion to the maxillary bone. The quadrate bone drops posteriorly and causes the pterygoid bone to move forward. This in turn causes an anterior motion of the palatine bone which, being connected to the maxilla below the lacrimal hinge, pushes it forward.

Contraction of several muscles actually puts the mechanism into motion. The sphenopterygoid muscles are the elevator muscles of the fangs. They arise along the median ridge of the base of the skull, and, extending backward, are inserted upon the enlarged posterior terminus of the pterygoid bone. When these muscles contract, the pterygoids are pulled anteriorly. This pushes the ventral side of the maxillary bone forward, while its dorsum is held in place by the lacrimal hinge, resulting in the tip of the fang pointing downward instead of backward. The external pterygoid muscle opposes the sphenopterygoid muscle. It originates from a joint between the quadrate bone and the lower jaw, and extends anteriorly to insert on the outer surface of the maxillary bone. Contraction by it pulls the maxilla rearward, resulting in a backward and upward movement of the point of the fang. During the strike, the mouth opens nearly 180°, and the fangs point forward in a stabbing position. Vipers do not bite, but instead thrust the fangs like a sword into their prey (Kardong and Bels 1998).

Groombridge (1986) and Young and Jackson (2008) reported that in the genus *Crotalus*, the connective tissue link between the fascia of the pterygoideus and the anterior body of the M. compressor glandulae muscle is developed into a prominent tendinous band. A more or less prominent longitudinal plane of separation within the pterygoideus forms a distinct glandulae portion. Few or many lateral fibers of the pterygoideus do not pass posteriorly with the main body of the muscle, but divert laterally toward the venom gland, where they attach to the tendinous band that frequently extends to form an additional aponeurosis (a thicker and denser, deep fascia that covers, invests, and forms the terminus and attachment point of muscles) on the lateral surface of the glandulae portion. This band becomes confluent with the capsule of the venom gland and also, to a variable extent, with the transversoglandulare ligament. The arrangement of the M. pterygoideus glandulae, therefore, is a specialization that increases venom expulsion.

Normally in the pitvipers (Crotalinae), the pterygoideus glandulae muscle passes between the ectopterygoid bone and the venom gland; however, in the species of *Agkistrodon*, the connection between the venom gland and this unmodified, poorly developed muscle is either absent or consists of only a few fibers (Groombridge 1986; Young and Jackson 2008). In addition, the skull of these snakes has a single, robust medioventral basioccipital process consisting of contributions from both the basioccipital and basisphenoid bones. This process serves as a major site for insertion of the rectus capitus muscles. Ruben and Geddes (1983) thought that *Agkistrodon* fang length may be correlated with this structural arrangement, and that it facilitates prey envenomation. Once the fangs engage the prey, the cranium and upper jaw move forward and downward on the prey as the posterior end of the cranium is elevated, pushing the fangs deeper into the prey (Kardong 1975). The rectus capitus are the major muscles involved in these movements. Other myology associated with the jaw and strike apparatus is reviewed in Kardong (1990).

The viper fang is an elongated, very pointed, curved tooth (Ernst 1982a; Klauber 1939, 1972). It contains two cavities: the pulp cavity, which is located on the concave side, and the venom canal, which lies on the convex side of the tooth. The venom canal has an opening at either end. At its beginning is a short, relatively wide, anterior slit, the entrance lumen, which receives venom from the venom duct. Another narrower, more elongated, slit-like opening, the discharge orifice, through which venom leaves the fang to flow into the object bitten, is located on the anterior side of the fang near its tip. Both openings are produced from gaps in the suture formed when the embryonic tooth folded from a flat one to a narrow, tubular one. It is often possible to trace a more or less evident depressed line representing the suture between the two openings.

Microscopic inspection of a cross section of a fang reveals that the venom canal is only a deep groove that became enclosed when the walls crossed over it anteriorly. Some opisthoglyphous (fixed rear-fanged) colubrid snakes still portray this primitive (less derived) condition. The depressed line indicates where the walls met. Because the surface of the venom canal is composed of enamel, as is also the outer surface of the fang, the inner lining is actually the anterior surface and the outer layer is the original posterior surface of the embryonic flat tooth.

The functional fangs of all venomous snakes are replaced periodically (Burkett 1966; Ernst 1962, 1982a; Fitch 1960; Klauber 1939, 1972; Smith 1952; Tomes 1877). This happens whether or not the fang is damaged or worn, but it is not known if the replacement process is more rapid when the fang has been broken before a change would normally occur. Because the fangs are rather delicate, such a provision for replacement appears necessary. It also provides for longer fangs as the snake grows. Some reported average replacement rates for vipers are *Agkistrodon bilineatus*, 20.7%; *A. contortrix*, 19.7%, 20.1%; *A. piscivorus*, 19%, 33.3%; *Calloselasma rhodostoma*, 12.8%; *Daboia russelii*, 10.1%; and *Deinagkistrodon acutus*, 12.3% (Burkett 1966; Ernst 1982a, 1982b; Fitch 1960).

On each maxillary bone is a pair of fang sockets positioned side by side with the inner a little forward of the outer. These sockets are used alternately to hold the functional fang. A section through the jaw shows that the future fangs lie in two alternating rows of sockets behind the fangs, a series leading to each of the two anterior sockets. These future fangs are not fully formed, but are in various stages of development from back to front. The most mature of the 5–6 reserve fangs always lies behind the vacant socket (Ernst 1982a; Klauber 1939, 1972). The first replacement fang of *Daboia russelii* is only 0.1–0.3 mm distant from the functional fang (Ernst 1982a). When the time for a change arrives, this reserve fang moves into the vacant socket anterior to it and becomes fastened in it and attached to the venom canal. The old fang drops out, leaving a vacant socket, which will be occupied, in due course, by the next replacement fang. The active fangs, therefore, alternate between the inner and outer positional sockets.

Separating the two rows of sockets is a heavy, protecting wall of connective tissue. This wall is of greater thickness than that which separates each succeeding fang from its fellow anterior fang on the same side. It is impossible for a fang of the inner row to move into the outer row, and vice versa.

The reserve fangs can be found in all degrees of advancement. The fangs are replaced so frequently that measurements do not ordinarily indicate a larger size of the first reserve fang as compared with the functional fang that it will succeed, but this is known in some rattlesnakes of the genus *Crotalus*. Ernst (1982a) reported the first reserve fang of *Daboia russelii* was only slightly longer than that of the functional one, but never more than 0.1 mm longer. He also reported that the teeth in the replacement series in that snake range in graduated lengths from that of the most anterior reserve fang backward to a short posterior spike about 0.2 mm long.

During embryonic development, the growing fangs are developed not as a unit, but from the point upward (Ernst 1962, 1982a). The point is fully formed and hardened long before the upper part takes shape. In addition, while the morphology of the fang reveals derivation from a flat plate rolled into a tube, the actual growth does not proceed in this manner. Instead, the tubular shape is evident from the earliest period of development. Klauber (1939, 1972) described a typical series in *Crotalus atrox*, which proceeded from a functional fang through six reserve fangs; the last (distal) reserve fang was only a small conical object attached to the connective tissue wall.

Barton (1950) reported that neonate *Crotalus horridus* are fully equipped with replacement fangs, as were several newborn *Agkistrodon contortrix* examined by Ernst (1962). The fact that the reserve fang in one of Barton's rattlesnakes was already assuming the functional position suggests that the first fang-shedding must take place at a very early age.

In the United States, 99% of envenomations by snakes are caused by pitvipers. Venoms of viperid snakes are predominantly hemotoxic, but neurotoxic components are present in many species. The venom is produced in a pair of serous, tubular, glands, one located on each upper jaw below and behind the orbit and above the corner of the mouth. The gland is elongated and typically almond- or pear-shaped, with its broadest portion posterior. When filled with venom, it bends downward posteriorly around the corner of the mouth.

The venom gland in species of *Crotalus* is composed of two discrete secretory regions: a small anterior accessory gland and a large posterior main gland that are joined by a short duct. Within the main gland, the venom is produced in several, normally 4–5, lobes of secretory cells that may make up as much as 80% of the gland's total cell count. These cells are similar to those found in the Duvernoy's gland of the Colubridae. The main gland has at least four distinct cell-types: secretory cells (the dominant cell-type), mitochondria-rich cells, horizontal (secretory stem) cells, and "dark" (myoepithelial) cells. The anterior accessory gland contains six cell-types, including mucosecretory cells and several types of mitochondria-rich cells. Release of venom into the tubules draining into the lumen of the main gland is by exocytosis of granules and by release of intact membrane-bound vesicles (Mackessy 1991). At least in the South American pitviper *Bothrops jararaca,* α_1-adrenoreceptors trigger the production of venom in the secretory cells by activating phosphatidylinositol 4, 5-biphosphate hydrolysis and extracellular signal-related kinase phosphorylation (Kerchove et al. 2008).

The venom is drained from the cells through small tubules into a hollow central lumen. The lumen in turn joins the venom duct, through which the venom flows anteriorly to the base of the fang. The venom duct is completely surrounded, at least proximally, by masses of secreting cells that may act as valves to regulate the flow of venom to the fang. Venom drawn from the lumen of the venom gland is less toxic than that drawn from the fang, so the accessory gland's secretions, although nontoxic, may activate some venom components. The venom duct does not enter the fang, but instead opens adjacent to it within a sheath of connective tissue surrounding the base of the fang. The tissue sheath acts as a seal around the fang that directs the flow of venom into the fang's central venom canal and outward into the object bitten.

The muscle M. levator anguliorís (or M. adductor externus superficialis; McDowell 1986) surrounds the venom gland dorsally and posteriorly, and is the only muscle involved in the discharge of venom from the gland. The dorsal portion of the muscle originates from the skull or dorsal head muscles via fasciae. The most important fibers originate in the parietal portion of the skull and insert in the posteroventral bend of the venom gland. A large ligament is also present that attaches the anterior-dorsal portion of the gland to the postorbital-parietal region of the skull. When emptying the venom gland, the muscle compresses and pulls it dorsally and medially and creates lateral and posterior pressure to force the venom through tubules into the central lumen and then anteriorly into the venom duct.

In both the Viperidae and Elapidae, fangs work in conjunction with muscles and other structures to form a complete venom delivery system that functions like a hypodermic syringe and needle (a case of human ingenuity mimicking nature). The body of the syringe represents the venom gland and venom duct to its throat; the plunger, the muscles surrounding the venom gland; and the needle, the fang.

Postfrontal, coronoid, and pelvic bones are absent, as also are teeth on the premaxillae. The prefrontal bone does not contact the nasal bone, and the ectopterygoid is elongated. A scale-like supratemporal bone suspends the quadrate. The hyoid is either *Y*-shaped or *U*-shaped, with two long superficially placed, parallel arms. All body vertebrae contain elongated hypapophyses. The left lung is absent or vestigial; a tracheal lung is usually present (Wallach 1998). The hemipenis is bilobed or double, with proximal spines and distal calyces, and a bifurcate or semicentrifugal sulcus spermaticus. Both left and right oviducts are well developed. Head scalation is similar to that of colubrids, but some, like most *Crotalus* rattlesnakes, have the dorsal surface covered with small scales instead of enlarged plates. Body scales are keeled. Pupils are vertical slits (Allen and Neill 1950a). Reproduction is either oviparous or ovoviviparous (ovoviviparous in North American species). Detailed discussions of the various characteristics of vipers are presented in Ineich et al. (2006), Marx et al. (1988), and Marx and Rabb (1972); we refer the reader to these publications.

The species in Crotalinae are called pitvipers because of the pair of small holes in their face that open on each side between the eye and nostril. The maxilla is hollowed out dorsally to accommodate the pit, and the membrane at the base of the hole is extremely sensitive to infrared radiations. It allows the snake to detect modest temperature fluctuations within its surroundings, especially those emitted from warm-blooded prey, but it is also used to direct behavioral thermoregulation. This suggests that the pits might be general purpose organs used to drive a suite of behaviors (Krochmal et al. 2004).

The pit organ is a sensitive infrared receptor that records changes from the normal background heat radiation, detects warm prey, and helps direct the snake's strike. It is usually located in the loreal scale of those crotalids having such a scale, or, in those without, in the position on the side of the face corresponding to the region of this scale. Because of this location, the cavities are sometimes referred to as loreal pits. In rattlesnakes (*Crotalus, Sistrurus*) the opening lies below a line from the nostril to the orbit and slightly closer to the former. Each crotaline pit is subdivided into an outer chamber and a smaller inner chamber by a cornified epidermal membrane about 0.025 mm thick. The principal component of the membrane is a single layer of specialized parenchyma cells with osmiophil reticular cytoplasm that lies between two layers of extremely attenuated epidermis (Bullock and Fox 1957; Lynn 1931; Noble and Schmidt 1937). The membrane is innervated by the trigeminal cranial nerve (V), particularly by its ophthalmic and maxillary branches. These pits are similar to, but not identical to, the elaborate lip pits present between the labial scales of pythons and some boas, but 5 to 10 times more sensitive. Noble and Schmidt (1937) thought that crotaline pit organs may have evolved from those on the upper jaw of boids, but this has not been proven. Instead, it is more probable the two types of snakes evolved the pit organs via convergence. Many axons enter the membrane and the innervation is rather dense. The axons lose their myelin, taper to about 1 μm, then expand into flattened palmate structures that bear many branched processes terminating freely over an average area of about 1,500 μm^2, overlapping only slightly with adjacent units but leaving virtually no area unsupplied (Bullock and Fox 1957). *Crotalus* have about 500 to 1,500 axon endings per mm^2.

Bullock and Fox (1957) reported the transmission spectrum of the fresh membrane in *Crotalus* rattlesnakes shares broad absorption peaks at 3 and 6 μm and about 50% is transmitted in other regions out to 16 μm. The visible light spectrum is at least 50% transmitted; however, much is probably lost through reflection. The strongest absorption takes place at wavelengths shorter than 490 μm. A continual transmission of impulses occurs from the pit membrane to the brain (Bullock and Cowles 1952). The rate of this continual message is independent of the snake's BT, but is dependent upon the average radiation from all objects in the receptive field. The membranes are highly sensitive to infrared environmental wavelengths of 15,000–40,000 Å, and any warm or cold object causes a temporary change in the rate of impulse transmission, the response being correlated to sudden ET changes. So the pit organ serves to recognize the presence of any object that is warmer or colder than its surroundings; during tests by Roelke and Childress (2007), three species of pitvipers distinguished between warm and cool targets, while four species of true vipers and two species of colubrid snakes did not exhibit this ability. The field is determined to include a cone extending horizontally from 10° across the midline to a point approximately at right angles to the rattlesnake's body from 45° above to 35° below the horizontal. This allows the receptive fields of the two pits to overlap in front of the snake, and together they survey a 180° field anterior to it. Sensitivity varies with the wavelength, but is generally greater to infrared emissions in the range of 2-3 μm than to shorter or longer wavelengths. The snakes seem able to detect and respond to temperature contrasts of as little as 0.001°C or less (Bakken and Krochmal 2007). Possibly the snakes can detect an ET change of about 0.003°C in 0.1 second, but this estimation may be false, as, according to Krochmal and Bakken (2005), it has been wrongly applied. Either way, the pit membrane is not as sensitive as it might be; its sensitivity is not higher than that calculated for human thermal receptors (Bullock and Fox 1957). Still, Bullock and Diecke (1956) have shown it is certainly sensitive enough to detect objects with STs differing by only 0.1°C, and there is probably no adaptive advantage to having a more sensitive receptor. Several data inputs are needed, like the signal strength (prey size and distance are important) and the quantity and quality of the background "noise" (which must be filtered out). Ideally, prey should be within 0.75 to 1.0 m, while predators should be within 0.75 to 1.5 m away (Krochmal and Bakken 2003).

By visualizing the temperature contrast images formed on the facial pit membrane using optical and heat transfer analysis (including heat transfer through the air in the pit chambers as well as by thermal infrared radiation), Bakken and Krochmal (2007) found the image on the membrane to be poorly focused and of very low temperature contrast. Heat flow through air in the pit chambers severely retards sensitivity, particularly for small snakes with small facial pit chambers. The opening of the facial pit seems to be larger than optimal for detecting small prey at 0.5 m. Angular sharpness (resolution) and image strength and contrast vary greatly with the size of the pit opening, resulting in the patterns of natural background temperatures obscuring prey and other environmental characteristics, creating false patterns. It appears important for snakes to select ambush sites with uniform, noncomplex backgrounds and strong thermal contrasts.

Interestingly, certain other membranes within the mouths of pitvipers may also be sensitive to thermal stimulation (Chiszar et al. 1986a; Dickman et al. 1987). The infrared data received via the pit organs are integrated with visual data in the optic tectum of rattlesnakes (Hartline et al. 1978; Newman and Hartline 1981). De Cock Bunning (1983) thought that, depending on the influence of ecological demands, visual or chemical cues are the main information in the behavioral phases before the strike, but in situations with little input (i.e., at night or in a rodent's burrow, etc.), hunting behavior is guided primarily by radiation of warm objects (see Ernst and Zug 1996; Ford and Burghardt 1993; Krochmal et al. 2004; Molenaar 1992, Newman and Hartline 1982; or Zug and Ernst 2004 for additional information on pit organs).

KEY TO NORTHERN AMERICAN GENERA OF VIPERIDAE

1a. No scaly rattle or button at end of tail . . . . *Agkistrodon* (vol. 1)
1b. Scaly rattle or button present at tip of tail 2
2a. Dorsal surface of head covered with nine enlarged plate-like scales . . . . *Sistrurus* (vol. 1)
2b. Dorsal surface of head covered with small scales, or less than nine enlarged plates . . . . *Crotalus* (vol. 2)

Agkistrodon Palisot de Beauvois, 1799
American Moccasins

Molecular (DNA) studies by Castoe and Parkinson (2006) and Knight et al. (1992) indicate that the genus *Agkistrodon* is monophyletic, and more closely related to the New World genera *Bothriechis, Crotalus,* and *Sistrurus* than to the Asian genera of pitvipers (Kraus et al. 1996). In contrast, studies of the immunoelectrophoretic patterns of venom proteins of the *Agkistrodon* indicate that the genus is most closely related to the Japanese species *Gloydius blomhoffii* (Tu and Adams 1968), and that the *Agkistrodon* share more slow-evolving proteins with the Asian *Hypnale* and the polyphyletic (Castoe and Parkinson 2006) *Trimeresurus* complex (Dowling et al. 1996). Minton (1990b, 1992), using serum immunologic relationships, concluded that the American *Agkistrodon* and the Asian species of *Gloydius* are closely related, but that *Calloselasma, Deinagkistrodon,* and *Hypnale* are about equally related to *A. piscivorus* as to *Vipera palaestinae,* but are not closely related to the other species of American *Agkistrodon.* A multigene phylogenetic analysis by Parkinson et al. (2002) indicates that *Agkistrodon* is most closely related to *Gloydius* and the New World *Crotalus* and *Sistrurus.*

The intrageneric relationships are confusing, depending on what characters are examined and by which method. A molecular study by Parkinson et al. (2000) using mtDNA and tRNA sequences showed strong evidence that *A. contortrix* is basal to its three congeners, and that *A. piscivorus* is basal to the cantils, *A. bilineatus,* and *A. taylori.* A later mtDNA study by Castoe and Parkinson (2006) confirmed these results and, in 2002, Parkinson et al. reported that *A. bilineatus* and *A. taylori* are sister species more closely related to *A. piscivorus* than to *A. contortrix,* agreeing with their previous results (Parkinson et al. 1997). MtDNA studies by Knight et al. (1992) showed *A. bilineatus* and *A. piscivorus* closely related. Conant (1986), on morphological grounds, thought *A. contortrix* closely related to *A. bilineatus* [including *A. taylori*], and *A. piscivorus* less so because it had become adapted to an aquatic existence. Similarly, Van Devender and Conant (1990) proposed a phylogeny of *Agkistrodon* that had *A. bilineatus* and *A. taylori* most closely related, followed by a close relationship of the two to *A. contortrix* and a farther relationship with *A. piscivorus.* Immunological distances reported by Cadle (1992) confirmed this. In an immunological study using blood plasma, Minton (1990b) thought *A. bilineatus* and *A. piscivorus* slightly closer than *A. piscivorus* to *A. contortrix.* Contrastingly, Tu and Adams (1968), in an immunological study using venom proteins, reported that *A. bilineatus* was quite distinct from its two more northern congeners.

Further comparisons of venom from the three species of American *Agkistrodon* by Jones (1976) revealed both geographical differences between and within the species (but see criticism by Gibson 1977).

Study of the scale surface contours by Chiasson et al. (1989) indicated that *A. contortrix* is more closely related to the tropical *A. bilineatus* than to *A. piscivorus.* The surface contours of the body scales of *A. contortrix* and *A. bilineatus* are caniculate, while those on *A. piscivorus* are caniculate / cristate. How this relates to the evolution of the various species of *Agkistrodon* is unclear, as Price (1989) in another study of scale microdermatoglyphics thought the species of *Agkistrodon* form a closely related group, since they have a similar basal scale pattern of 1–2 μm punctuations without any other visible small elements. A neonatal basal scale pattern consisting of polygonal cells with diameters of approximately 25 μm is shared by all the species, as is also a similar apical pattern. However, comparisons of the chemical makeup of skin keratin suggest that *A. contortrix* is not so close to *A. bilineatus*, and that the genus may not be monophyletic (Campbell and Whitmore 1989).

Klauber (1956) and Gloyd and Conant (1990) have evaluated the use of the names *Agkistrodon* Palisot de Beauvois, 1799, and *Ancistrodon* Wagler, 1830, for the genus, and have presented convincing legalistic arguments for the use of the former.

The genus was reviewed by Campbell and Lamar (2004) and Gloyd and Conant (1990), and various osteological and myological characters are described by Kardong (1990).

KEY TO AMERICAN SPECIES OF *AGKISTRODON*

1a. Loreal scale present, normally 25 midbody scale rows 2
1b. No loreal scale present, midbody scale rows normally 23 . . . . *piscivorus*
2a. Two yellow, cream, or white stripes present on each side of the face, one extending posteriorly over the orbit to the temporal region, and the other running backward along the upper jaw to the corner of the mouth; a medial light stripe extending vertically across the rostrum onto the lower jaw; body color dark brown or black; body pattern of variable width, light cross bars 3
2b. No facial stripes present; body color pinkish, reddish, gray-brown, or chestnut; body pattern usually consisting of dumbbell-shaped dark cross bars (or, in *A. c. laticinctus*, broad ones) . . . . *contortrix*
3a. Dorsal cross bands distinct in adults; lower facial stripe not bordered below by a dark line posterior to the second supralabial; 45–56 subcaudals in males; 40–47 subcaudals in females . . . . *taylori*
3b. Dorsal bands usually obscured in adults; lower facial stripe bordered by a dark line posterior to the second supralabial; 55–71 subcaudals in males; 46–67 subcaudals in females . . . . *bilineatus*

Agkistrodon bilineatus (Günther, 1863)
Cantil
Cantil de Agua

RECOGNITION: The cantil is a thick-bodied snake that grows to a TBL_{max} of 138 cm (sex not given; Cuesta-Terrón 1930), but most adults are smaller than 80 cm. Ground color is quite variable (dark gray, dark olive or brown, tan, reddish-brown, or black), and some individuals also show shades of lavender or contain yellow or bronze hues. Body color changes with age, and melanism may occur. About 12–15 (10–19) brown to blackish body bands cross the back and sides. They are irregularly bordered with white, cream, or yellow, and lighter than the body color bands. The end of the rattleless tail is normally pale or greenish-gray. The venter is gray to grayish-brown or reddish-brown, and paler along the midline. It is patterned, particularly laterally, with irregular pale marks, and bronze ventrolateral spots may be present. The face is patterned laterally with two light stripes: a dorsal one that runs from the snout backward over the orbit, and a broader lower one that starts on the rostrum and extends downward to the supralabials and then backward to about the corner of the mouth (see "Geographic Variation"). A medial, variable-width, light bar extends downward from the rostrum to the upper lip. The dorsal body scales are pitted and keeled. Body scale surface structure is canaliculate with stigmata present (Chiasson et al. 1989); the basal morphotype is acellular (punctate), and the apical morphotype is cristate (punctate) (Price and Kelly 1989). The body scales lie in 25 (23–24) anterior rows, 23 (21–25) rows at midbody, and 19 (18–21) rows near the vent. On the underside are 134–137 (127–144) ventrals and 65 (46–71) subcaudals (4–38 may be undivided), and an undivided anal plate. Dorsally on the head are 9 plates: 2 internasals, 2 prefrontals, 1 frontal (sometimes subdivided), 2 supraoculars, and 2 parietals (subdivided into several scales in some individuals). Pertinent lateral head scales include 2 nasals, 1 loreal, 2–3 preoculars, 2–3 suboculars, 2–4 postoculars, several rows of temporals, 8 (7–9) supralabials, and 10–11 (8–13) infralabials. On the chin are a mental scale and 2 elongated chin shields. The pupils are vertical ellipses, and the irises are paler dorsally but darker ventrally. The tip of the tail is noticeably downturned.

The bilobed hemipenis has a central sulcus spermaticus that divides to send a branch to the apex of each attenuated lobe. The sulcus lips are smooth near the base, but calyculate over most of the lobes. The base of each lobe is circled with 25–40 large spines, and the distal portion contains calyces that continue down the sulcus to about the dividing point of the sulcus. The basal stalk of the hemipenis is almost naked (Malnate 1990). Additional data are presented in the account of *A. taylori*.

Teeth present, excluding the maxillary fang, consist of 3–4 (5) palatines, 15 (14–17) pterygoids, and 15 (11–17) dentaries (Kardong 1990).

Males (TBL_{max}, 109 cm) have 13 (10–18) cross bands, 127–142 ventrals, 55–71 subcaudals, and average TLs of 21% of TBL. Females (TBL_{max}, 103.5 cm) have 14–15 (11–19) cross bands, 128–144 ventrals, 46–67 subcaudals, and average TLs of 19% of TBL.

GEOGRAPHIC VARIATION: Currently, three subspecies are recognized (Campbell and Lamar 2004). *Agkistrodon bilineatus bilineatus* (Günther, 1863), the Mexican Cantil or Gamarilla, lives in the Pacific versant of Mexico from extreme southern Sonora and western Chihuahua (Lemos-Espinal et al. 2006) to Guerrero and Pueblo, in the Río Grijalva Valley of Chiapas, and in Guatemala and El Salvador. It formerly occurred on the Tres Marías Islands (Zweifel 1960). Both the dorsal and ventral facial stripes are complete; the body color is dark gray to brown or black, with little differences in the cross bands; the few white cross bars narrow and have no central pattern; the dark venter is mottled with light marks; and the chin and throat are well patterned. *Agkistrodon b. howardgloydi* Conant, 1984, the Costa Rican Cantil, Castellana, or Víbora Castellana, ranges along the Pacific coast of Honduras, Nicaragua, and Costa Rica. The dorsal facial stripe barely extends beyond the orbit; the lower face stripe is normally interrupted; the body color varies from chestnut brown to purplish black, with well-marked light areas dorsally separating the dark cross bands (this pattern becomes more melanistic with age); the medial posterior venter is pale and mostly unmarked; and the chin and throat are orangish in contrast to the anterior venter. *Agkistrodon b.*

Agkistrodon bilineatus bilineatus (San Diego Zoo, John H. Tashjian)

Agkistrodon bilineatus bilineatus (Guatemala City, Guatemala; San Antonio Zoo, John H. Tashjian)

russeolus Gloyd, 1972, the Yucatán Cantil or Víbora de Freno, is found in the Yucatán Peninsula (northeastern Campeche, Yucatán, Quintana Roo), and northern Belize. Its upper facial stripe is narrow and may be interrupted behind the orbit; the lower facial stripe is broader than the dorsal, and becomes interrupted at the corner of the mouth; the body color is dark chestnut brown with narrow white borders along the conspicuous, broad, dark body bands (increasing melanism may occur with age); the anterior venter is dark, the posterior medially light with a few mottles; the chin and throat are usually dark.

Cantils from the Río Chixoy and central Petén of Guatemala are apparently intergrades that contain character overlap with all three subspecies.

Formerly, four subspecies were assigned to *A. bilineatus* (Gloyd and Conant 1990), but the results of a molecular study using both mtDNA and tRNA, and its allopatry along the Gulf Coast of Mexico, were used by Parkinson et al. (2000) to elevate *A. b. taylori* Burger and Robertson, 1951 to the full species, *A. taylori*.

CONFUSING SPECIES: Other species of *Agkistrodon* can be identified using the above key. The *Crotalus* and *Sistrurus* rattlesnakes usually have a tail rattle.

KARYOTYPE: The species' karyotype consists of 36 chromosomes; 16 macrochromosomes (8 metacentric, 4 submetacentric, 2 submetacentric-subtelocentric, and 2 subtelocentric-telocentric), and 20 microchromosomes. The sex chromosomes have not been described, but are probably ZZ in males and ZW in females, as in other American *Agkistrodon* (Baker et al. 1972; Cole 1990).

FOSSIL RECORD: No fossils of *A. bilineatus* have been reported.

DISTRIBUTION: The cantil ranges southward on the Pacific versant of continental Mexico from southern Sonora (Babb and Dugan 2008; Bogert and Oliver 1945) and western Chihuahua (Lemos-Espinal et al. 2006) to Guerrero (Dixon et al. 1962) and Morelos (Davis and Smith 1953). Its distribution becomes disjunct south of there, where it is found again in southern Oaxaca (Canseco-Márquez and Nolasco-Vélez 2008), the Yucatán Peninsula, Guatemala, and El Salvador; along the Pacific coast of Honduras (Cruz et al. 1979), Nicaragua, and Costa Rica; and in northeastern Campeche, Yucatán, Quintana Roo, and northern Belize (Campbell and Lamar 2004; Gloyd and Conant 1990). It formerly occurred on the Tres Marías Islands (Zweifel 1960) in the Gulf of Mexico, but has not been collected there since the late 1800s and is probably extirpated (Gloyd and Conant 1990).

Crotalinae pitvipers, including the ancestors of *A. bilineatus* and *A. taylori*, invaded tropical America from the north, perhaps as early as the Eocene (Darlington 1957; Savage 1966).

Distribution of *Agkistrodon bilineatus*.

HABITAT: *A. bilineatus* is not aquatic/semiaquatic like the *A. piscivorus*, as has been suspected in the past (Ditmars 1931b). Instead, most live in a variety of mesic habitats: subhumid tropical deciduous or semideciduous forests, around permanent ponds, flooded fields, moist cultivated areas (corn, rice, sugar cane), grassy meadows and cattle pastures, bromeliad hedges, and permanent coastal marshes. The wettest region occupied covers the coast of southeastern Chiapas, Mexico, and southwestern Guatemala, which experiences about 200 cm of precipitation during the wet season. However, other populations are found in more xeric situations, like arid scrub forests, near temporary ponds only filled during the rainy season, rock fences, rock crevices on slopes devoid of water, and well-drained hillsides or dry arroyos far distant from water (Campbell and Lamar 2004; Duellman 1961, 1965a, 1965b; Gloyd and Conant 1990; Hardy and McDiarmid 1969; Oliver 1937; Schmidt and Shannon 1947). Reported elevations range from sea level to 1,500 m.

The original habitat was probably dry forest or savannas, which explains its occupation of isolated xeric areas, like the savannas in the Río Negro (Chixoy) Valley of Guatemala, where it was possibly trapped when intermediate stretches of land dried due to climate change (Campbell and Vannini 1988).

BEHAVIOR AND ECOLOGY: Many gaps exist in our knowledge of the life history of *A. bilineatus*, and most information has been anecdotal. Comprehensive ecological studies are needed, especially as the three subspecies occupy different habitats.

The snake is most active during the rainy season, and remains inactive or possibly estivates during the hot, dry season, but, over its range, some are probably active in any month. A summary of the monthly collection dates given in Gloyd and Conant (1990) is as follows: May (15), June (13, 25, 30), July (21), August (6–8, 15, 22, 25, 30), October (1), November (8), January (2), and February (13). The lack of spring records is interesting, but these dates may represent collection bias favoring the wet season. During the dry season, it is rare to find more than one dormant individual at a site, so

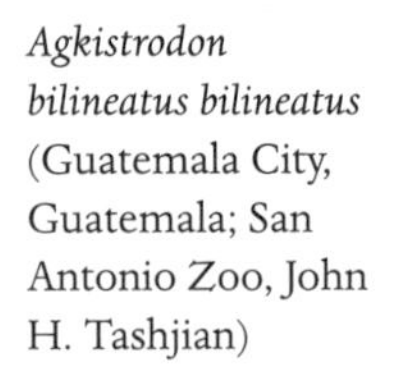

Agkistrodon bilineatus bilineatus (Guatemala City, Guatemala; San Antonio Zoo, John H. Tashjian)

A. bilineatus does not normally den up as occurs in some populations of *A. contortrix* and *A. piscivorus*.

Daily, *A. bilineatus* is usually crepuscular or nocturnal, particularly during the dry season, although some moving individuals have been collected during the day in the wet season, and it may be more diurnal in Costa Rica (Campbell and Lamar 2004; Conant 1990). It is often active, particularly on the roads, after rain events, and the snake is a good swimmer.

The only natural temperature datum is of one collected on 21 July at an AT of 26°C; however, the species is known to at least occasionally bask (Boerema 1989; Gloyd and Conant 1990). Captives will breed if kept at day ATs of 25–30°C and 18–23°C at night (Boerema 1989; Drent 1991; Furiani 1989; Macchiavelli and Mascagni 1990; Peters 1979; Smetsers 1993). Captives have been induced to hibernate by reducing the daytime ATs to 17–22°C and then to a final 12–18°C AT (Macchiavelli and Mascagni 1990).

Although the behavior has not been described, male cantils probably engage in dominance bouts during the breeding season like other pitvipers.

REPRODUCTION: Most of the reproductive data concerning this snake are based on captive observations. From the limited information available, both sexes become reproductive at 2–3 years of age (Boerema 1989; Drent 1991; Macchiavelli and Mascagni 1990; Peters 1979), females at a minimum TBL of 53.3 cm (although most reports of mating or parturition are for those 60.0–87.9 cm and 445–519 g) and males at 60 cm (Boerema 1989; Dugan and Meyer 2007; West 1981). Smetsers (1993) reported that a 1.5-year-old captive male courted a similar age female after she had shed, but that no copulation occurred.

Data on the reproductive cycles of both sexes are few. Females are ovoviviparous. A dead Sonoran female contained three early developed embryos on 10 July (Dugan and Meyer 2007), and Boerema (1989) reported a case of twinning from one egg.

It is thought that the species mates in Costa Rica during December–April (Solórzano 2004). Captives have displayed courtship behavior or have copulated on 10 June, 5 July, 1 and 20 August, 22 October, 10 November, 21 and 25 January, and 25 February (see above reports). Females apparently release pheromones from skin glands that attract and stimulate males to reproductive activity (Smetsers 1993). Captives have been stimulated to breed after a gradual cooling down (Boerema 1989; Macchiavelli and Mascagni 1990; Smetsers 1993), although Furiani (1989) reported that neither a cooling period nor a particular photoperiod is necessary to stimulate mating, but that separation of the sexes for a time, especially during the female's ecdysal period, is more important, as is also the extra administration of vitamins to the female. The female may drag the mating male with her as she crawls about. Copulations have lasted from 50 minutes to 9.0 hours, but most are 2 to 3 hours in duration (see above reports).

In the wild, neonates appear from June through August (Gloyd and Conant 1990; Solórzano 2004; West 1981). In captivity, births have occurred on 6, 25, and 29 April, 19 and 28 May, 30 June, 16 July, and 8 February (Boerema 1989; Drent 1991; Furiani 1989; Macchiavelli and Mascagni 1990). Time elapsed between observed captive matings and parturition has varied from 176 days (Peters 1979) to 395 days (West 1981), with several others lasting more than 200 days. It is doubtful that these long periods represent the actual neonatal developmental period. Instead, sperm stored from a mating and used in a later fertilization was probably involved (see Schuett 1992).

Litters contain 3 (Drent 1991; Dugan and Meyer 2007) to 20 (Álvarez del Toro 1960) young, but most have 8–12 young; the average number of young for 25 litters in the literature was 9.3. Measurements of 54 neonates from the literature were TBL, 15–32 (mean, 25.2) cm; TL, 5.6–6.4 (mean, 3.3) cm; and BM, 7.7–13.0 (mean, 9.3) g. TL is relatively longer in juveniles than adults. Neonates are more brightly colored than adults, and closely resemble the young of *A. contortrix* (those of *A. b. howardgloydi* may be quite red) and *A. piscivorus*. The chestnut to coppery, light-bordered cross bands are darker than the ground color separating them, and the tip of the tail is gray to yellowish-green (see "Diet and Feeding Behavior"). Neonate facial stripes are not as prominent as those of adults.

GROWTH AND LONGEVITY: Available growth data are from captivity. Neonate *A. bilineatus*, 23 cm long at birth, grew to 34–36 cm in 3 months (Peters 1979), and Smetsers (1993) reported that a male and female, both about 6 months old, 25 cm, and 18 g when obtained in October 1990, grew to a TBL of about 75 cm by 1993. A wild-caught *A. b. bilineatus* of unrecorded sex survived 24 years, 4 months, and 19 days at the Institudo Historia Natural, Chiapas, Mexico (Snider and Bowler 1992).

DIET AND FEEDING BEHAVIOR: Wild cantils prey on a variety of animals. Juveniles feed on smaller ones, like insects, anurans, and small lizards, while adults prefer larger prey, like rodents, larger lizards, and snakes (Campbell and Lamar 2004; Gloyd and Conant 1990). Gravid females normally continue to feed during gestation.

The following vertebrate prey have been reported from wild *A. bilineatus*: anurans (*Hypopachus variolosus, Leptodactylus poecilochilus*), lizards (*Ameiva undulata, Ctenosaura similis, Mabuya unimarginata*), snakes (*Imantodes gemmistratus*), small birds, spiny pocket mice (*Liomys pictus, L. irroratus, L. salvinii*), white-footed mice (*Peromyscus* sp.), cotton rats (*Sigmodon hispidus*), and rice rats (*Oryzomys palustris*) (Bogert and Oliver 1945; Campbell and Lamar 2004; Duellman 1954; Gloyd and Conant 1990; Henderson 1970; Solórzano et al. 1999; Villa 1962). Captives have eaten crickets (Gryllidae) and other insects, fish (*Tilapia mossambica*), anurans (*Acris crepitans; Hyla squirella, H. versicolor; Leptodactylus melanonotus; Similisca baudinii*), lizards (*Anolis carolinensis, Eumeces fasciatus, Podarcis muralis, Sceloperus undulatus*), small snakes (*Agkistrodon bilineatus, Regina septemvittata, Thamnophis* sp.), birds (*Turdus grayi*), mice (*Mus musculus*), and brown rats (*Rattus norvegicus*) (Allen 1949; Boerema 1989; Gloyd and Conant 1990; Macchiavelli and Mascagni 1990; Peters 1979; Strimple 1995b; West 1981), but oak toads (*Bufo quercicus*) were ignored by captives (Allen 1949). The small *A. bilineatus* may have been carrion.

Prey is detected by vision, thermal sensing, and odor. Adult cantils actively forage as well as hunt via ambush. Snakes are held after being struck (Gloyd and Conant 1990). Juveniles use their light-colored tail tip to lure smaller prey, waving the tip about to gain the animal's attention (Allen 1949; Álvarez del Toro 1960; Carpenter and Gillingham 1990; Heatwole and Davison 1976; Henderson 1970; Neill 1960; Strimple 1988, 1992, 1995b; West 1981; Wharton 1960). The coiled snake raises its tail into an upright position with the tip bent at an angle, and moves the tip in an intermittent fashion when prey approaches. This action resembles the wriggling of a worm or caterpillar. Once within striking distance the prey is bitten, released, and swallowed later after it has died. Juveniles normally release struggling prey. After feeding and at night the tail is not elevated.

The adult strike is released by visual and thermal cues, but odor plays a role in detecting prey. After the prey has been bitten, the rate of tongue-flicking increases and the injured animal is later trailed and swallowed once it has died or become incapacitated. Chiszar et al. (1979b) reported that during experiments the odor of a mouse (*Mus musculus*) illicited tongue-flicking in previously still *A. bilineatus*, particularly a sustained response after a successful strike.

VENOM DELIVERY SYSTEM: Snakes of the genus *Agkistrodon* have the same typical solenoglyphous fangs as other vipers (Ernst 1982b; Kardong 1979; Klauber 1972). Fifty-eight *A. bilineatus* (TBLs, 18.0–107.6 cm; HLs, 1.1–5.1 cm) measured by Ernst (1962, 1965, 1982b) had FLs of 1.8–12.0 (mean, 6.6) mm; the means TBL/FL and HL/FL were 95.1% and 5.4%, respectively. No sexual dimorphism not related to body size was evident (Ernst 1964). The species' FL_{max} of 12.0 mm placed it third in line behind *Dienagkistrodon acutus* (17.0 mm) and *Calloselasma rhodostoma* (15.1 mm) among total species in the *Agkistrodon* complex, and ahead of both the American species *A. piscivorus* (11.0 mm) and *A. contortrix* (7.2 mm). He calculated straight-line regression coefficients for *A. bilineatus* of $FL = 0.009TBL + 0.699$ ($r = 0.96$) and $FL = 0.206HL - 0.023$ ($r = 0.99$). The curvature of the fang gradually increases from juvenile to adult; the range in the estimated degree of curvature angle was 58–73°, but due to the short length of juvenile fangs, it was extremely difficult to measure their curvature, so a greater error in the estimates exists for those of juveniles. The maximum degree of curvature was only surpassed by *C. rhodostoma* (75°), *A. piscivorus* (74°), and *D. acutus* (74°). A series of replacement fangs are situated along the jaw; total replacement was 20.7%. Allen (1949) reported that a 25.7 cm juvenile had 4 mm fangs that were 7 mm apart at the base.

VENOM AND BITES: *Agkistrodon bilineatus* has a more toxic venom than its North American relatives. Minton et al. (1968) noted that the approximate dry venom yield is 50–95 mg, and Bolaños (1972) reported that 2 milked for the first time yielded an average of 105 mg and a maximum of 182 mg. Extraction from one yielded 0.65 mL of liquid venom (George 1930).

Juveniles and adults have a similar, but not identical, electrophoretic pattern of venom proteins. Contained in the venom of both age classes are 9 bands with molecular weights of 95, 90, 82, 67, 42, 28, 23, 20, and 16 kDa. Adults exhibit an additional 26 kDa band, and their 28 kDa band is brighter than that of juveniles (Solórzano et al. 1999). When dried, the venom is 62% protein (Brown 1973).

Venom of *A. bilineatus* contains a variety of proteinaceous toxins, mostly purine and pyrimidine nucleosides. A novel undecapeptide, arginase ester hydrolase, ATPase, benzoyl-DL-arginine-*p*-nitroaniline, bilineobin, esterases (unidentified), hemorrhagic toxin, hyaluronidase, L-amino acid oxidase, myotoxin-like proteins, NAD nucleosidase, N-benzoyl-L-arginine ethylesterase, 5′-nucleotidase, phosphodiesterase, phospholipase A_2, phosphomonoesterase, *p*-tosyl-*L*-arginine methylester, protease (casein), p-toluenesulfonyl-L-arginine methylesterase, protein C activator, and thrombin-like enzymes have been extracted or have shown activity in venom studies (Aird 2005; Bajwa et al. 1982; Brunson et al. 1978; Denson et al. 1972; Graham et al. 2005; Henderson and Bieber 1985; Huang and Perez 1980; Imai et al. 1989; Jones 1976; Kocholaty et al. 1971; Komori et al. 1993; Landberg et al. 1980; Lomonte et al. 1987; Mebs 1970b; Mebs and Samejima 1986; Nikai et al. 1993, 1995; Ruff et al. 1980; Sifford and Johnson

1978; Stocker et al. 1987; Takahashi and Mashiko 1998; Tan and Ponnudurai 1990; Tu et al. 1967; Weinstein et al. 1985). The virulent hemorrhagic toxin causes lethal hemorrhagic and proteolytic activity; its molecular weight is approximately 48 kDa and its isoelectric point is 4.2 (Imai et al. 1989). Bilineobin has a molecular weight of 26.48 kDa, and consists of 235 amino acids (Nikai et al. 1995). The phospholipase A_2 isolated from the snake's venom has a molecular weight of 14 kDa and an isoelectric point of pH 8.77, and is composed of 123 amino acids (Nikai et al. 1993). Venom from the cantil hydrolyzes DNA (de Roodt et al. 2003).

Bolaños (1972) and Bolaños and Montero (1968) reported crystallized venom LD_{50} values of 21 μg [mcg] intravenous and 29 μg intraperitoneal for 16–18 g mice. Minton et al. (1968) noted an intravenous one of 2.4 mg/kg of mouse, and Solóranzo et al. (1999) recorded an intraperitoneal LD_{50} of 1.25 μg/g of mouse. A *Hyla cinerea* struck by a juvenile cantil became stiff and died in 70 minutes, and an *Anolis carolinensis* died 90 minutes after being struck; both displayed hemorrhagic symptoms (Allen 1949). Horses (*Equus caballus*) bitten have died in less than an hour (Loveridge 1928; Rodas 1938).

Although *A. bilineatus* is the third most frequent source of snakebite on the Yucatán Peninsula (Rosado-López and Laviada-Arrigunaga 1977), data on cantil bites are poor. No complete case reports of human envenomations are available, but some data are presented by Rosado-López and Laviada-Arrigunaga (1977). Most information is anecdotal, but some symptoms have been reported, involving mostly elevated hemorrhagic, hemolytic, and myotoxic activities. Tissue necrosis, blood vessel destruction, and its resulting hemorrhage may be extensive to the point of requiring the amputation of bitten limbs. The venom may cause "spontaneous amputation" if the treatment is careless (Álvarez del Toro 1972 [1973]). The limb's tissues may become so gangrenous that fragments of flesh peel off, eventually exposing the underlying bone. Although Minton et al. (1968) reported that serious local lesions occur but seldom death, the snake's bite will kill a human, sometimes within a few hours (Bogert and Oliver 1945; Gaige 1936).

Hardy (1990) presents a detailed venom bibliography covering the snakes of the *Agkistrodon* complex.

PREDATORS AND DEFENSE: Reports on the aggressive nature of *A. bilineatus* abound, but observations of predatory attacks on it have not been published. Surely at least its young are preyed on by ophiphagous snakes (*Clelia clelia, Drymarchon corais, Drymobius* sp., *Lampropeltis* sp., *Micrurus* sp.), golden eagles (*Aquila chrysaetos*), large hawks (*Buteo* sp.), owls (*Bubo virginianus*), and carnivorous mammals (*Canis familiarius, C. latrans; Dasypus novemcinctus; Didelphis virginiana; Felis* sp.; *Lynx rufus*; *Mephitis mephitis; Nasua narica; Procyon lotor; Taxidea taxus; Urocyon cinereoargenteus; Vulpes macrotus*) within its range.

Upon detection of a potential predator, *A. bilineatus* either remains coiled in position, sometimes nervously vibrating its tail, or tries to crawl to some nearby refuge. If this does not work, and the snake is further disturbed, it displays a vicious temper, striking fiercely and rapidly. It may even advance toward the perceived threat. It has been reported to strike the entire length of its body, but this is most likely an exaggeration. If pinned down or held, the malodorous contents of the anal glands are dispelled. Because of its aggressive behavior and potent venom, most natives fear this

Agkistrodon bilineatus bilineatus (Alamos, Sonora; Eric Dugan)

snake more than the local rattlesnakes. When older and more melanistic, the snake's body pattern does not necessarily camouflage it, but the facial stripes and light-bordered cross bands of younger individuals may possibly serve to warn an attacker of its venomous nature, as may also the bright body colors of the neonates. When confronted with the predatory snake, *Lampropeltis getula*, a cantil performed body bridges (Weldon and Burghardt 1979).

PARASITES AND PATHOGENS: *A. bilineatus* hosts the parasitic nematodes *Rhabdias vellardi* and *Kallicephalus macrovolvus* in its digestive tract and lungs, respectively (Caballero 1954; Caballero y Caballero 1954).

POPULATIONS: Conflicting reports on its abundance exist (Gloyd and Conant 1990). It may be abundant in some areas (Colima, possibly Costa Rica) and seemingly scarce in others. This may, however, be a false impression, as the snake is normally solitary and also quiescent during a large part of the year. It is also more nocturnal during the dry season, so the sighting depends on the observer's time in the field.

Unfortunately, due to habitat destruction for agricultural purposes (Parkinson et al. 2000), the snake's numbers have dwindled over much of its range and it has been extirpated in some areas where it was once common. A few are killed on the roads or by natives, or are taken for the pet trade, but the total loss from these is negligible when compared to that resulting from habitat destruction. Today, it is considered one of the most endangered snakes in Latin America (Greene and Campbell 1992).

REMARKS: The common name cantil apparently derives from the word *kantiil* of the indigenous Tzeltal people of Chiapas, meaning yellow lips, and refers to the lower facial stripe of the snake (Conant 1982).

Agkistrodon contortrix (Linnaeus, 1766)
Copperhead
Zolcuata

RECOGNITION: The copperhead is a pinkish to grayish-brown, or even greenish, stout-bodied snake (TBL_{max}, 134.6 cm [sex not given]; Ditmars 1931b, but most are shorter than 90 cm) with an orange to copper or rust-red, unpatterned dorsal surface of the head, and a series of 10–21 (mean, 15) brown to reddish-brown, saddle-shaped bands on the body. These dorsal bands are usually broader along the sides of the body and narrower across the dorsum, forming a dumbbell-like shape, and small dark spots may occur in the light spaces between the bands (see "Geographic Variation"). Haynie and Knight (1998), Livezey (1949), McDuffie (1963), and Meshaka et al. (1989) describe variation in the banding pattern, and patternless individuals have been reported (Fitch 1959). The tail lacks a rattle, is often faintly banded, and has a yellow, brown, or green tip. Below the eye, the head pigmentation is usually lighter than above it. No light stripes are present, but a dark postocular stripe is often present. A pair of small dark spots occurs on the parietal scales. The venter is cream, pink, or light brown, with dark blotches along the sides of the ventrals. The dorsal body scales are as described for *A. bilineatus*, and lie in 23–25 (24–29) rows anteriorly, 23 (21–27) rows at midbody, and 19 or 21 (18–23) rows posteriorly. According to Stabler (1939), ecdysis occurs about 2 to 6 times a year at a rate of every 1.8 to 6.0 months; we have observed that the process is dependent on food intake and varies annually. The venter has 138–157 (mean, 148) ventrals (with no sexual dimorphism), 37–63 (mean, 45.5) subcaudals, and an undivided anal plate. Pertinent lateral head scales include 2 nasals, 1 loreal, 2–3 preoculars, 2–3 suboculars, 3–4 postoculars, several rows of temporals, 8 (6–10) supralabials, and 10 (8–13) infralabials. The eyes have vertically elliptical pupils (Munro 1949b). Bicephaly has been reported (Mitchell and Fieg 1996).

The hemipenis is deeply bifurcate, with a forked sulcus spermaticus that extends to the tip of each lobe. Approximately 35 large spines lie on the basal 33% of each lobe (most are straight but some are slightly hooked); no spines occur in the crotch between the lobes. The distal 67% of each lobe is covered with small, irregularly placed, flattened papillae, each with a terminal spine. The hemipenis is figured in Dowling (1975) and Malnate (1990). Reese (1947) described the embryological development of the hemipenis.

The shortened maxilla contains only the elongated fang. The rest of the dental formula is as follows: 5 (3-5) palatine teeth, 15–16 (13–20) pterygoid teeth, and 15 (12–18) dentary teeth (Kardong 1990).

The longer, more slender males (TBL_{max}, 115.5 cm; Gloyd and Conant 1990) have TLs 10–19 (mean, 14.5) % of TBL and 38–63 (mean, 47) subcaudals. Females are shorter and stouter (TBL_{max}, 106 cm; Gloyd and Conant 1990) with TLs 10–17 (mean, 13.7) % of TBL and 37–57 (mean, 44) subcaudals. Quinn (1979b) reported an 87 to 100% overlap in the number of tail bands of the sexes in *A. c. laticinctus* and a 42 to 100% overlap in *A. c. mokasen*.

GEOGRAPHIC VARIATION: Five subspecies are recognized (Campbell and Lamar 2004; Conant and Collins 1998; Ernst 1992; Ernst and Ernst 2003; Gloyd and Conant 1990). *Agkistrodon contortrix contortrix* (Linnaeus, 1766), the Southern Copperhead, is pale gray to pinkish in ground color with prominent cross bands that are very narrow across the back and often medially separated, dark lateral spots on the venter, and a pinkish or greenish-yellow tail tip. It ranges from southeastern Virginia southward along the coastal plain to Gadsden, Liberty, Santa Rosa, and Escambia counties, Florida, and west to eastern Texas and northward in the Mississippi Valley to southern Missouri and southwestern Illinois. *Agkistrodon c. laticinctus* Gloyd and Conant, 1934, the Broad-banded Copperhead, has deep, little constricted, reddish-brown crossbands across the back, the tip of the tail is greenish-gray, and the venter is cream-colored with small reddish-brown or black spots and larger irregularly shaped dark blotches. It is found from extreme south-central Kansas (Chautauqua and Cowley counties) southward through central Oklahoma and central Texas to Medina and Atascosa counties. *Agkistrodon c. mokasen* (Palisot de Beauvois, 1799), the Northern Copperhead, is more reddish-brown (some individuals may be greenish), and has a bright coppery head; darker, wider, cross bands, dark round spots on the side of the venter, and often a greenish tail tip. It is distributed from Massachusetts and Connecticut southward on the piedmont and highlands to Georgia, Alabama, and northeastern Mississippi, and west through southern Pennsylvania and the Ohio Valley to Illinois. *Agkistrodon c. phaeogaster* Gloyd, 1969, the Osage Copperhead, resembles the Northern Copperhead, but has a paler ground color with more pronounced dark dorsal bands, no small dark spots between the dark bands, a yellowish-green tail tip, and a grayish, mottled venter. It is only found in northern and central Missouri, eastern Kansas, and the northeastern corner of Oklahoma. *Agkistrodon c. pictigaster* Gloyd and Conant, 1943, the Trans-Pecos Copperhead or Zolcuata de Trans-Pecos, resembles *A. c. laticinctus* in its dorsal pattern, but has a dark reddish-brown to black mottled venter in sharp contrast to cream-colored areas extending onto the belly from the sides, and the reddish-brown midbody cross bands often contain a pair of dark, rounded, lateral spots. It is found only in western Texas from Crockett and Val Verde counties westward through the Big Bend region and Davis Mountains. It barely enters Mexico in extreme northern Coahuila and northeastern Chihuahua.

Agkistrodon contortrix contortrix (Lake Waccamaw, North Carolina; Carl H. Ernst)

Agkistrodon contortrix contortrix (Alabama, Carl H. Ernst)

Intergradation is common where the ranges of the various subspecies meet (Gloyd and Conant 1990). *A. c. contortrix* and *A. c. mokasen* interbreed in northern Georgia (Rubio et al. 2003), on North Carolina's coastal plain (Palmer 1965), and possibly as far north as Fairfax and Prince William counties, Virginia (Ernst et al. 1997), and intergrades between *A. c. phaeogaster* and *A. c. laticinctus* are known from southern Kansas and Oklahoma (Fitch and Collins 1985; Webb 1970).

Soto et al. (2006) found genetic differences in the venom nonenzymatic proteins (disintegrins) of the five subspecies. Two clades were identified: one including *A. c. mokasen* from Kentucky and a second including populations from the other four subspecies in Louisiana, Missouri, and Texas.

CONFUSING SPECIES: Other species of *Agkistrodon* can be distinguished by using the above key. The *Crotalus* and *Sistrurus* rattlesnakes usually have a rattle or at least a button at the tip of their tail. Colubrid snakes belonging to the genera *Lampropeltis, Nerodia, Heterodon,* and *Pantherophis* lack a facial pit, have rounded pupils, and may have patterned heads. The lyresnake (*Trimorphodon lambda*) has elliptical pupils and a patterned head.

KARYOTYPE: Diploid chromosomes total 36: 16 macrochromosomes (4 metacentric, 8 submetacentric, 2 subtelocentric) and 20 microchromosomes. Sex determination is ZW in females and ZZ in males; the smaller female W chromosome is acrocentric or subtelocentric to telocentric (Baker et al. 1972; Cole 1990; De Smet 1978; Zimmerman and Kilpatrick 1973). The NOR is on the long arm of telocentric pair 4 (Camper and Hanks 1995).

FOSSIL RECORD: *A. contortrix* has a rich and widely distributed fossil history (Holman 1995, 2000). Its earliest remains are from Miocene (Hemphillian-Clarendonian) deposits in Nebraska (Brattstrom 1954, 1967; Holman 1979a, 2000). Later, Pliocene (terminal Hemphillian or Blancan) fossils have been unearthed in Kansas, Nebraska, and Texas (Brattstrom 1967; Conant 1990; Holman 1979b; Holman and Schloeder 1991; Rogers 1976, 1984). However, the richest fossil assemblage is from the Pleistocene: the Blancan of Kansas (Brattstrom 1967); the Irvingtonian of Kansas (Holman 1995; Rogers 1982) and West Virginia (Holman 1982); and the Rancholabrean of Georgia (Holman 1967), Kansas (Brattstrom 1967; Preston 1979), Pennsylvania (Brattstrom 1954; Guilday et al. 1964; Holman 1995), Texas (Hill 1971; Holman 1964, 1968, 1969b, 1980, 1995; Johnson 1974; Parmley 1988), Virginia (Guilday 1962; Holman 1968), and West Virginia (Holman and Grady 1987).

In addition, several fossil vertebrae assigned only to the genus *Agkistrodon* could belong to *A. contortrix*: Miocene (mid-Hemphillian)—Nebraska (Parmley and Holman 1995); Pleistocene (Irvingtonian)—Nebraska (Ford 1992); and Pleistocene (Rancholabrean)—Alabama (Holman et al. 1990), Kansas (Johnson 1975), New Mexico (Wiley 1972), and Texas (Hill 1971; Holman 1968, 1969a, 1995; Parmley 1986). The Nebraska records listed above indicate that the genus, and possibly *A. contortrix*, had a greater distribution in the past.

DISTRIBUTION: The copperhead occurs from western Massachusetts and Connecticut and southeastern New York west through the southern two-thirds of Pennsylvania, southern Ohio, Indiana, Illinois, to Missouri and eastern Kansas, and south to Georgia and the panhandle of Florida, Mississippi, Louisiana, eastern and central Texas, and western Texas, and the extreme portions of northern Coahuila and northeastern Chihuahua along the Rio Grande (Río Bravo del Norte) River in Mexico. Except in the northeastern part of the range, the snake is mainly limited by the southern

Distribution of *Agkistrodon contortrix.*

boundary of the Wisconsinan Glacier (Gloyd and Conant 1990). Over its entire range, it occurs at elevations from near sea level to more than 1,500 m; however, it is generally rare at elevations above 760 m in the Appalachians (Huheey and Stupka 1967).

HABITAT: Over its northeastern and upland piedmont and Appalachian range, *A. c. mokasen* is most often found in open canopy, in oak-hickory woodlands, particularly on hillsides with rock crevices and slides, or in areas with downed trees, hollow logs, or brush piles. There, it utilizes relatively open areas with a higher rock density and less surface vegetation than does the sympatric timber rattlesnake, *Crotalus horridus* (Reinert 1984a, 1984b, 1992). Prominent plants are blackberry (*Rubus* sp.), American chestnut (*Castanea dentata*), greenbrier (*Smilax* sp.), hackberry (*Celtis occidentalis*), maples (*Acer* sp.), oaks (*Quercus* sp.), poison ivy and sumac (*Rhus* sp.), pokeweed (*Phytalacca americana*), red cedar (*Juniperus virginiana*), sweetgum (*Liquidamber styraciflua*), tulip poplar (*Liriodendron tulifera*), and Virginia creeper (*Parthenocissus* sp.).

On the southern coastal plain, at least from southeastern Virginia to eastern Texas, and on the piedmont in Texas, *A. c. contortrix* and *A. c. laticinctus* live along the edges of swamps or in mesic woodlands and where American beech (*Fagus grandifolia*), American cane (*Arundinaria gigantea*), American elm (*Ulmus americana*), bald cyprus (*Taxodium distichum*), hackberry (*Celtis lavigata*), holly (*Ilex* sp.), honeysuckle (*Lonicera japonica*), myrtle (*Myrica cerifera*), oaks, pecans (*Carya aquatica, C. illinoensis*), pines (*Pinus* sp.), and tulip poplar predominate.

During a microhabitat modeling study in southeastern Virginia, Cross and Petersen (2001) found the snake significantly more often in areas with a greater number of cover logs, greater understory tree widths, and less leaf and vegetation cover, and at greater distances from overstory trees than at randomly selected sites.

Fitch (1960) reported the most preferred microhabitats of Kansas *A. c. phaeogaster* are in the vicinity of intermittent streams, with mixed groves of cottonwood (*Populus deltoides*), elm (*Ulmus* sp.), locust (*Gleditsia triacanthos*), and blackberry (*Rubus argutus*); fence rows lined with brush and saplings of crabapple (*Pyrus ioensis*), elm, locust, Osage orange (*Maclura pomifera*), and plum (*Prunus americanus*); the vicinity of a pond with a grove of willow (*Salix* sp.), and with a dense ground cover of day flower (*Commelina communis*), rice cut grass (*Leersia oryzoides*), and smartweed (*Polygonum* sp.); and mixed upland thickets of crabapple, elm, locust, oak (*Quercus prinoides*), Osage orange, plum, and sumac (*Rhus glabra*) at the edges of grassland dominated by brome (*Bromus inermis*) or blue-stem (*Andropogon* sp.). Lesser preferred microhabitats were mesic and xeric woodlands of various types, weedy pastures, and fallow fields dominated by sedges and various weeds. The least preferred microhabitats were mesic or more xeric fallow fields and cultivated fields.

In western Texas and northern Mexico, *A. c. pictigaster* occurs in dry woodlands, or even in arroyos with sandy soils and rocky outcrops and cliffs along the edges of deserts. In these xeric areas, the snake is more commonly found in riparian woodlands near permanent or semipermanent waterways. Common plants in the eastern parts are black walnut (*Juglans nigra*), grape (*Vitis* sp.), hackberry (*Celtis laevigata*), Mexican buckeye (*Ungnadia speciosa*), persimmon (*Diospyros texana*), sumac (*Rhus* sp.), willows (*Chilopsis sp., Salix* sp.), and, in the drier western areas, cottonwoods (*Populus* sp.), catclaw (*Acacia greggii*), and prickly pear cacti (*Opuntia* sp.).

Gravid females prefer microhabitats that are clearly separated from those occupied

Agkistrodon contortrix contortrix (R. D. Bartlett)

by males and nonreproductive females. During his more than half century study of Osage Copperheads at the University of Kansas Fitch Natural History Reservation, Fitch captured 335 gravid females in July–early September, of which 275 (82%) were found singly and 55 others in associations of 2–5 females (Fitch and Clarke 2002; but see also "Populations"). Female *A. c. mokasen* use rocky, open, sparsely forested sites with warmer STs until parturition (Reinert 1984b). Such sites are usually close to the hibernaculum, but may be at some distance. The snake is known to enter caves and old mine shafts (McAllister et al. 1995a), and, with the sprawl of suburbia, more copperheads are found in and about buildings, where they hide in gardens, shrubbery, and fringe woods, and under porches or debris.

BEHAVIOR AND ECOLOGY: Annually, *A. contortrix* is active from March–April to late October or early November in the northern parts of its range, with most being seen from May to September, but the copperhead may be active from March until early December. It occasionally emerges on warm days during January and February (Abbuhl 2008) throughout its range. In northern Virginia, we have observed active copperheads from 14 April to 28 October; Palmer and Braswell (1995) found them active from February to December in North Carolina. Kansas *A. c. phaeogaster* have been active from 5 April to 23 November (Fitch 1960).

In the fall copperheads become gregarious and may crawl some distance to a communal hibernaculum, which is sometimes shared with *Crotalus horridus* and smaller numbers of *Coluber constrictor* and *Pantherophis obsoletus* (C. Ernst, personal observation; Neill 1948; Seibert 1965). They usually return to the same area each winter (Fitch 1960). Weathered outcrops with crevices extending below the frost line are often used, as are also caves, gravel banks, stone walls, building foundations, animal burrows, hollow logs and stumps, and sawdust piles. Some copperheads may bask on warm winter days.

At the Mason Neck National Wildlife Refuge, Fairfax County, Virginia, 15–25 copperheads, most about 50 cm but some to 88 cm in TBL, hibernate each year beneath

Agkistrodon contortrix contortrix, melanistic morph (R. D. Bartlett)

a broken concrete sidewalk. Formerly, this population used the broken foundation of an abandoned house as their hibernaculum, but when the house was torn down most of the snakes moved about 50 m to the sidewalk site. The others then overwintered in groups of 2–3 in old fallen trees and stumps.

Along the coastal plain from Virginia to Georgia, although many share hibernacula, some copperheads hibernate singly (C. Ernst, personal observation; Neill 1948). In Virginia it is usually the juveniles that hibernate singly, but we have found some adults overwintering alone. Usually hollow logs or stumps are used as individual hibernacula.

Drda (1968) observed several overwintering in a Missouri cave that were quite active most of the winter; AT in the cave was relatively constant and considerably above freezing. Juveniles crawled deeper into the cave than adults.

Although it is active during daylight in the spring and fall, particularly in the late morning into the afternoon, *A. contortrix* becomes crepuscular or nocturnal during the hot summer months (Guidry 1953; Wright 1987), and is often more active after an evening shower.

The preferred BT range is probably 23–31°C. Twenty-five active Kentucky *A. c. mokasen* had BTs of 12–28 (mean, 20.9) °C at ATs of 8–28 (mean, 21.9) °C and STs of 8–31 (mean, 23.5) °C (C. Ernst, personal observation). The snakes with both the lowest (12°C, 16 October) and the highest (28°C, 10 June) BTs were basking. Abbuhl (2008) found 2 adult *A. c. mokasen* basking in New York on 6 January; the high AT that day was 11.1°C. In Kansas, BTs recorded by Clarke (1958) for active *A. c. phaeogaster* were 7.5–32.0°C. Fitch (1956) determined the preferred BTs of other *A. c. phaeogaster* in eastern Kansas to be 26–28°C, and the preferred activity ranges probably 23–31°C. Brattstrom (1965) reported the minimum and maximum voluntary BTs for 61 individuals to be 17.5°C and 34.5°C, respectively, but Fitch (1960) reported BTs as low as 12.4°C for those under rocks in the early spring. The CT_{min} and CT_{max} are 4–9°C and 41°C, respectively.

Most basking we observed occurred during the morning in either the spring or the fall. Sanders and Jacob (1981) monitored BTs of 20 *A. c. mokasen* by telemetry. Some snakes basking on clear winter days achieved BTs of 10°C or higher. The BTs recorded in the summer varied among snakes of different SVLs, and a significant negative correlation occurred between SVL and the CT_{min}. Fitch (1960) found that gravid female *A. c. phaeogaster* basked more often and seemed to prefer warmer BTs.

Heart rate and breathing rate increase with rising BT (Jacob and Carroll 1982), but BT has no effect on heart rate-breathing response over the range at which copperheads are normally active.

Evaporative water loss rates of South Carolina *A. contortrix* were calculated by Moen et al. (2005). Copperheads had lower rates of water loss in both dry and wet tests than did cottonmouths, *A. piscivorus*, tested under similar conditions. Copperheads in the wet treatment had a least squares mean evaporative water loss of 97.1 (range, 73.4–121.3) mg/hour, while those in the dry treatment lost a mean of 70.5 (range, 46.6–104.9) mg/hour. The rate of loss was positively correlated with the snake's BM.

Three types of movements are made by copperheads (Fitch 1960): (1) movements within the home (or activity) range, (2) abandonment of one home range and occupancy of a second, and (3) migrations to and from hibernacula. Gravid female *A. c. mokasen* and *A. c. phaeogaster* ordinarily do not move far from their hibernacula (Finneran 1953; Fitch 2005). In midsummer they move back to rocky outcrops near the hibernacula, and congregate in small groups at some sheltering hole, hollow log, or rock crevice, presumably to avoid predators and unfavorable weather. Neonates find themselves in a birth microhabitat different from that of adult males and nonreproductive females. They hibernate in this area, and do not move far from it the following spring as the adult males and nonreproductive females migrate to the summer feeding grounds. Ten young recaptured by Fitch moved an average of 42.5 m in an average of 136 days. Average movement of adult males was 70.5 m and 77.5 m for adult females.

Home ranges of Kansas *A. c. phaeogaster* varied between 3.4 ha for females and 9.8 ha for males (Fitch 1958, 1960). The home range diameter of males averaged 354 m and that of females 210 m. Fitch (1960) reported movements on the summer range of 1.5–378 m for individuals that remained within their home range and 442–762 m for those that had apparently shifted home ranges. He also recorded distances of spring or fall captures at hibernacula from points of capture during the summer of 232–1,183 m. Males traveled longer average distances from hibernacula (656 m) than did females (406 m). Radio-equipped Kansas *A. c. phaeogaster* had an average displacement of 11 m/day (including days not moved; Fitch and Shirer 1971). Mean displacement distances for only those days when movement occurred were 18 m for a male, 12 m for a nongravid female, and 12 m for a gravid female.

A marked *A. c. mokasen* in the Shenandoah National Park, Virginia, exhibited homing ability when it returned within 2 days to the place of original capture after having been displaced 0.8 km to a known den site (Martin 1990).

Copperheads climb into low bushes or trees after prey or to bask. Swanson (1952) observed several juveniles resting on laurel bushes 5 cm or more above ground, Engelhardt (1932) captured one in a fork of a tree 1.2 m above ground, Beaupre and Roberts (2001) found another 1.3 m above the ground ascending a tree, and Johnson (1948) saw one more than 2 m high in a tree. We have found the snakes entwined or lying on the branches of small shrubs on several occasions, and have regularly seen them

Agkistrodon contortrix laticinctus (R. D. Bartlett)

crawling or basking on the sides of wind-tilted trees at heights of 2–5 m. Fitch (1960) also reported climbing by Kansas copperheads.

Although this snake has no special affinity for water, it voluntarily enters water bodies, and has been observed swimming on numerous occasions (C. Ernst, personal observation; Fitch 1960; Groves 1977; Smith and Sanders 1952).

Males engage in combat dances mostly in the spring, but sometimes during the summer months (see Aldridge and Duvall 2002 for a list of literature reports). After meeting, the two males make body contact and then almost immediately raise up, facing each other to a height of 30–40% of their TBLs. Much tongue-flicking occurs at this time, and the males seem to try to outstare each other. Then, both sway back and forth in unison with heads bent at a sharp angle. Sometimes their elevated bodies are parallel to each other, or one snake may actually turn its back toward its opponent. One leans over and tries to push the other's head and neck to the ground (topping behavior), while his opponent responds by entwining its aggressor and tries to pin it by pushing with its anterior body and neck. This pushing may continue for some time (usually 20–30 minutes, but sometimes for more than 2 hours), but eventually one male (usually the shorter) is pinned to the ground, breaks off contact, and crawls away (C. Ernst, personal observation; Gloyd 1947; Schuett and Gillingham 1989; Shaw 1948b; Shively and Mitchell 1994; Stewart 1984).

Studies by Schuett (1986) indicate that male copperheads participate in these combat bouts only during the breeding season. Female defense seems to be involved, so he thought the major function of these encounters is mate competition. Although this is probably the stimulus in the wild, male-male combat has taken place in captivity with no females present and at no particular elevated testosterone or corticosterone levels (Schuett et al. 1996). Perhaps this behavior extends to food competition or territoriality involving a food source or critical space, or, possibly, some males just do not like each other.

After the combat, losers, but not winners, are sexually inhibited (Schuett 1996) and have significantly elevated plasma corticosterone levels. Schuett and Grober (2000) suggest that this results from psychoneuroendocrine factors [probably indicating greater stress] rather than from simple exercise. Such higher levels retard the loser's metabolic recovery and result in higher plasma lactate levels in the vanquished male. The corticosterone response has a net effect on metabolic recovery and may be implicated in the protracted suppression of aggressive behavior in losers.

REPRODUCTION: Some female *A. c. mokasen* may be mature at an SVL of 37.5 cm (Mitchell 1994) at an age of about 3 years. In Kansas, about 50% of 3-year-old female *A. c. phaeogaster* are of small adult size (SVL, 50 cm) and sexually mature (Fitch 1960, 1999; Fogell et al. 2002a). The mean SVLs of gravid female *A. c. phaeogaster* at 3 Kansas study sites were Clinton State Park, 63.58 cm; the University of Kansas Succession Area, 64.48 cm; and the University of Kansas Natural History Reservation, 60.66 cm (Fitch and Clarke 2002). Female *A. c. contortrix* in northeastern Texas with SVLs of 47–60 cm are capable of reproducing (Ford et al. 1990). In captivity, female *A. c. contortrix* (TBL, 85–90 cm; age, 3.0–3.5 years), *A. c. laticinctus* (TBL, 70 cm; age, unknown), and *A. c. mokasen* (TBL, 80 cm; age, unknown; and TBL, 85 cm; age, 3.5 years) have produced young (Boerema 1990a, 1990b, 1991a, 1991b; v. d. Velde 1990).

Few data exist concerning the age and size of maturity of wild males. Mitchell (1994) reported Virginia male *A. c. mokasen* to be mature at an SVL of 47.5 cm, and Fogell et al. (2002) reported mating by a wild 73.2 cm SVL *A. c. phaeogaster*. Captive male *A. c. contortrix* with TBLs of 80 and 84 cm and 3.5 years old and a 3.5-year-old 81 cm male *A. c. mokasen* have successfully mated (Boerema 1990a, 1990b, 1991a). The normal size and age of maturity are probably similar to that of the females listed above.

At the time of spring emergence in Kansas, the ova are small (1–9 mm) and occur in several size groups, suggesting that they may mature at different times (Fitch 1960).

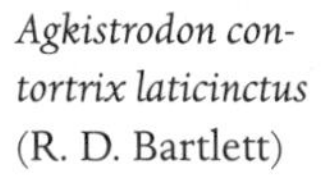

Agkistrodon contortrix laticinctus (R. D. Bartlett)

In May the ova grow rapidly and ovulation occurs in the latter part of the month. Hurter (1911) found 13 76 mm embryos in a Missouri female (presumably *A. c. phaeogaster*) that he dissected on 27 July. Observations that only about 60% of females breed each season seem to indicate a biennial female reproductive cycle (Fitch 1960; McDuffie 1961). Fitch (2003) reported that none of 13 female Kansas *A. c. phaeogaster* captured the year following parturition was gravid and also 10 females recaptured in the third year after birthing were not gravid. However, 4 females were gravid again in the second year and again in the fourth year. A single, large (SVL, 82.5 cm; mass, 572 g) female from an area of high vole production was gravid in its third year, and Fitch (2003) thought she thrived to the extent of possibly being able to reproduce annually. Vermersch and Kuntz (1986) reported that female *A. c. lacticintus* from southern Texas produce young each year, and Allen (1955) reported that a captive female of the same subspecies delivered young in two consecutive years.

The histology and timing of events of the male reproductive cycle have not been described, but are probably similar to those of other North American pitvipers reported by Aldridge and Duvall (2002). Fitch (1960) thought male *A. c. phaeogaster* contained active sperm year round and were capable of mating at any time during the annual activity period. A well-defined seasonal pattern exists in the seasonal testosterone levels of males. Testosterone levels of captive males studied by Schuett et al. (1997) were lowest in April–May, increased in June, were highest in August, declined thereafter up to the time of hibernation in early November, changed little during hibernation (November–January), increased sharply from emergence from hibernation in late winter (February) to March, and then decreased from March to April.

A. contortrix mates in both the spring and summer (Aldridge and Duvall 2002). Observations of matings have been reported by various biologists and these tend to indicate that, although copulation has taken place from April to October, the prime breeding periods are April and May and late August to October. Fitch (1960) found active sperm in the cloaca of a wild-caught female *A. c. phaeogaster* on 19 May. On 24 June a second examination showed still abundant motile sperm, but about 75% of the sperm were dead or very slow in their movements. It seems from these data that sperm is stored in the cloaca by the female for only a relatively short period. Howarth (1974) reported that sperm storage in female copperheads lasts only about 11 days; however, Schuett and Gillingham (1986) reported overwinter viable sperm storage, and Allen (1955) noted a case of a female *A. c. laticinctus* that gave birth to five young at least a year after separation from any male. Sperm apparently survive much longer in the upper end of the oviducts in vascular tissues specialized as seminal receptacles (Fitch 1960; Fox 1956). Kansas female *A. c. phaeogaster* contained active sperm in April–June and October, but other females examined in April, June–August, and October–November were negative (Fitch 1960). Twenty-one (35.6%) of 59 sexually mature female *A. c. phaeogaster* obtained in April and May by Gloyd (1934) contained active sperm. He examined the vas deferens of males in April–August and October and found more or less active sperm in all. Schuett (1982) reported that an October copulation resulted in the birth of young on 3 August.

Courtship is always initiated and performed by the male. Males use their forked tongues to transfer odors to the vomeronasal organ, and follow female pheromone trails with much tongue-flicking during the mating seasons. Males have more deeply forked tongues with longer tines than females. Such an adaptation enhances the odor reception of males needed to find both mates and prey (C. F. Smith et al. 2008).

Seven distinct behaviors are involved: (1) touch-mounting (contact with the female is made with the snout and the head and neck are elevated and placed on her back); (2) chin-rubbing; (3) dorsal-advancing with chin-rubbing (simultaneous advancing and chin-rubbing while mounted on the female); (4) tail-searching (the entire tail is oriented beside that of the female and quivers; while this takes place the male's tail and vent region are pushed beneath her tail, forming a loop; the quivering stops and her tail and vent region are stroked 1–3 times; the male's tail remains looped if intromission is not achieved); (5) no moving while mounted (all motion ceases, with the male's head and neck held close to the female's body); (6) stopping (all body motion ceases except tongue-flicking); and (7) dismounting (Schuett and Gillingham 1988). These behaviors may not always take place in sequence.

The female's response to male courtship involves 10 behavioral acts: (1) advancing (crawling slowly or rapidly forward); (2) remaining stationary (in an outstretched or coiled position); (3) tail-waving (the elevated tip, up to 30°, is moved slowly side to side); (4) whipping of elevated tail; (5) tail-vibrating (rapid side-to-side movements that may produce a tapping or buzzing sound); (6) body-flattening (the body is dorsoventrally compressed); (7) waste elimination from the vent; (8) cloacal-gaping (the elevated tail, up to 45°, is arched and the anal plate lowered; intromission is achieved only after this act); (9) head-raising (1–3 cm above substrate); and (10) tongue-flicking (Gloyd 1947; Munro 1950; Schuett and Gillingham 1988).

Victorious males and those subjugated during combat bouts respond differently to a female's raised head. Males that have lost bouts are repelled by females with raised heads, while the winners of bouts frequently challenge the females that head-lift (Schuett and Duvall 1996). In cases where the female is challenged, combat never occurs and the males resume courtship. This led Schuett and Duvall to propose that female head-lifting is a female choice mechanism involving intraspecific sexual mimicry of male-male combat.

Copulation involves eight different male behaviors: (1) intromission; (2) dismounting until only posterior contact is maintained; (3) stationary (no movements, not even tongue-flicking); (4) cloacal contracting (sporadic contractions of the cloaca and cloacal region); (5) hemipenial enlargement; (6) insemination; (7) backward-crawling (moving backward when the female crawls forward); and (8) hemipenial retraction (withdrawal of the hemipenis and its return in a relaxed state within the male's tail). Females react in three ways during mating: (1) remaining stationary (no movements, including tongue-flicking); (2) advancing (crawling slowly forward pulling the male along); and (3) tail-undulating (the elevated tail, >45°, moving similar to caudal luring movements) (Schuett and Gillingham 1988).

In contrast to the report by Dolley (1939) that the embryos are attached to and develop within the female's oviduct, *A. contortrix* is ovoviviparous (Chenowith 1948). Parturition usually takes place in August or September, but some birthing has also taken place in April, July, and August, and in October and November. Females often aggregate near the time of birthing (Finneran 1953; Fitch and Clarke 2002). The GP lasts 83–150 (mean, 110) days; females may aggregate about the time of parturition, particularly near den sites. While giving birth, the female extrudes the neonates with a series of muscular contractions of her posterior abdominal region, often while lying in a semicircular coil. Neonates are enclosed in a membranous, transparent sac, and emergence from it may take as long as 15 minutes. The head is first pushed through the membranes and the young snake then takes a deep breath or a series of deep, open-mouthed gasps before continuing its struggles to free itself. Chenowith (1948) observed a parturition in which 2 young were born within 10 minutes. The RCMs of 5 females from northeastern Texas averaged 31.1%; 33.7% (4 young), 32.4% (7 young), 41.9% (7 young), 27.3% (10 young), and 27.8% (7 young), respectively (Ford et al. 1990). In the Kansas *A. c. phaeogaster* studied by Fitch (1999), the RCM was 36.1% of gravid weight, or RCM = 0.279 − 0.375 SVL (Seigel et al. 1986).

A single brood may contain 1–21 young (Carpenter 1958; Gloyd and Conant 1990; White 1979; Wright and Wright 1957), but 4–8 are most common. Litter size increases with both age and SVL of the female; larger females of each subspecies produce the larger broods (C. Ernst, personal observation; Fitch 1960; Ford et al. 1990). Specific records from published literature and from Fitch's (1960, 2003) Kansas field studies of *A. c. phaeogaster* describe a total of more than 1,000 eggs or young from more than 200 females with an average brood size of 5.75 young. Fitch and Clarke (2002) reported mean embryo counts of 5.37, 6.65, and 7.13, respectively, for gravid females at 3 Kansas sites. Occasionally, some unfertilized eggs are passed at parturition. The longest 2 subspecies, *A. c. contortrix* and *A. c. mokasen*, produce broods of up to 18 and 21 young, respectively, while those of *A. c. laticinctus* and *A. c. phaeogaster* may have up to 11 and 13 young, respectively; the shortest race, *A. c. pictigaster*, usually births broods with no more than 3–4 young. The number of young produced by a female may vary from year to year, apparently due to different environmental conditions (Seigel and Fitch 1985). Neonate length is also positively correlated to female body length (Ford et al. 1990).

Newborns are patterned like the adults, but are paler in ground color, and their tail tip is yellow. The TBL of neonates is usually 20.0-25.5 (mean, 20.6; n = 158) cm, but may range from 15 to 30 cm; their BM is about 4.5–17.0 (mean, 10.6; n = 52) g. Neonate size and mass depend on the female's food intake mostly prior to ovulation, and varies according to prey availability; entire litters of stunted young have been produced following poor feeding years. The newborn snakes may remain with their mother, at least to their first ecdysis (Anderson 1942; Fitch 1960; Kennedy 1964; Smith 1940).

In laboratory tests, Schuett and Gillingham (1986) found that sperm resulting from fall copulations retains its fertilizing ability until spring, and that fall and spring ejaculates of different males can overlap prior to spring ovulation, producing litters of multiple parentage.

Captive *A. contortrix* have hybridized with both *A. piscivorus* (Mount and Cecil 1982) and *Crotalus horridus* (Smith and Page 1972).

GROWTH AND LONGEVITY: In Indiana, yearling *A. c. mokasen* have 38–43 cm TBLs and 2-year-olds 53–59 cm (Minton 1972) in length. Yearling *A. c. phaeogaster* in Kansas have SVLs of 30–40 cm; those 2 years old, 40–57.5 cm; 3 years old, 45–62.5 cm; 4 years old, 53–73 cm; and 5 years old, 55–75 cm (estimated from Fitch 1960). Males apparently grow faster than females. Mean SVLs in cms by age and sex of Kansas copperheads were 1 year (both sexes, 32.0), 2 years (males, 46.0; females, 44.0), 3 years (males, 55.0; females, 51.0), 4 years (males, 62.5; females, 57.3), 5 years (males, 67.5; females,

Agkistrodon contortrix mokasen (Illinois, Carl H. Ernst)

61.0), 6 years (males, 71.5; females, 64.0), 7 years (males, 74.8; females, 66.0), 8 years (males, 77.0; females, 67.5), and 9 years (males, 79.0; females, 68.5) (Fitch 1999).

Under the right conditions, the species may have a long life. Snider and Bowler (1992) list the following subspecific longevities of individuals kept in captivity: *A. c. contortrix* (male, wild-caught when adult)—23 years, 2 months, 23 days; *A. c. laticinctus* (male, wild-caught, age not reported)—21 years, 6 months, 9 days; *A. c. mokasen* (female, wild-caught when adult)—29 years, 10 months, 6 days; *A. c. phaeogaster* (male, origin unknown, still alive on 31 May 1990)—21 years, 28 days; and *A. c. pictigaster* (male, wild-caught when adult)—20 years, 8 months, 10 days). Fitch (1999) reported an 18-year-old wild *A. c. phaeogaster*, but it is doubtful that many wild copperheads of any subspecies survive 20 years.

DIET AND FEEDING BEHAVIOR: *A. contortrix* has a very broad diet, and, although containing a heat-detecting facial pit, is not restricted to warm-blooded prey. It is known to eat several kinds of invertebrates: millipedes (*Narceus* sp.), spiders (Arachnida), various insects—beetles (Coleoptera), bush katydids (*Scuddaria* sp.), cicadas (*Magicicada tredecim* [captivity]; *Tibicen canicularis, T. linnei, T. pruinosa*), dragonflies (Odonata), grasshoppers and locust (Orthoptera), lepidopteran larvae (Ceratocampidae, Citheroniidae: *Anisota senatoria, A. virginensis, Automeris io, Citheronia regalis*; Saturnidae: *Actias luna, Telea polyphemus* [captivity]; Sphingidae: *Ceratomia catalpae, Manduca sexta* [captivity]), and mantids (*Stegomantis* sp.), as well as poikilothermic vertebrates—salamanders (*Ambystoma opacum; Plethodon cinereus, P. cylindraceus, P. glutinosus; Pseudotriton ruber*), anurans (*Acris crepitans* [captivity], *A. gryllus* [captivity]; *Gastrophryne olivacea; Hyla chrysocelis* [captivity]; *Pseudacris crucifer, P. triseriata; Rana blairi, R. catesbeiana, R. clamitans, R. pipiens* [captivity], *R. sphenocephala, R. sylvatica*), small turtles (*Sternotherus odoratus, Terrapene carolina*), lizards (*Anolis carolinensis; Cnemidophorus sexlineatus; Crotaphytus collaris; Eumeces fasciatus, E. obsoletus; Ophiosaurus attenuatus; Sceloporus undulates; Scincella lateralis*), and snakes (*Agkistrodon contortrix* [captivity]; *Carphophis amoenus, C. vermus; Coluber constrictor; Crotalus lepidus; Diadophis punctatus;*

Lampropeltis triangulum; Masticophis taeniatus; Nerodia rhombifera [captivity]; *Pantherophis obsoletus; Rhinocheilus lecontei; Storeria dekayi, S. occipitomaculata; Tantilla coronata; Thamnophis proximus, T. sirtalis; Virginia striatula*). Reported homoiothermic prey include small birds (*Aegelaius phoeniceus, Archilochus colubris, Cardinalis cardinalis, Carduelis tristis, Catharus guttatus, Dendroica* sp., *Guiraca caerulea, Molothrus ater, Petrochelidon pyrrhonota, Pipilo erythropthalmus* [?], *Seiurus motacilla, Sturnus vulgaris, Zonotrichia albicollis*), young opossums (*Didelphis virginana*), bats (*Tadarida brasiliensis*), moles (*Condylura cristata, Parascalops breweri, Scalopus aquaticus*), shrews (*Blarina brevicauda, B. carolinensis, B. hylophaga; Cryptotis parva; Sorex cinereus, S. longirostris*), mice (*Clethrionomys gapperi; Microtus chrotorrhinus, M. ochrogaster, M. pennsylvanicus, M. pinetorum; Mus musculus; Napaeozapus insignis; Peromyscus gossypinus, P. leucopus, P. maniculatus; Reithrodontomys humulis, R. megalotis; Synaptomys cooperi; Zapus hudsonius*), small rats (*Neotoma floridana, Sigmodon hispidus*), chipmunks (*Tamias striatus*), young gray squirrels (*Sciurus carolinensis*), and young cottontails (*Sylvilagus floridanus*) (Anderson 1965; Atkinson 1901; Barbour 1950, 1962; Barton 1949; Beaupre and Roberts 2001; Brown 1979; Bush 1959; Campbell and Lamar 2004; Carpenter 1958; Chenowith 1948; Clark 1949; Collins 1980; Conant 1951; De Rageot 1957; Dundee and Rossman 1989; C. Ernst, personal observation; Ernst et al. 1997; Ettling 1986; Fitch 1960, 1982, 1999; Fitch and Clarke 2002; Garton and Dimmick 1969; Gibson 2001; Gloyd and Conant 1990; Godard et al. 2006; Graves 2002; Greding 1964; Greenbaum 2004; Greenbaum and Jorgensen 2004; Hamilton and Pollack 1955; Harris and Simmons 1977; Herreid 1961; Hook 1936; T. R. Johnson 1987; Klauber 1972; Klemens 1993; Lagesse and Ford 1996; Lee et al. 2008; Malnate 1944; McCrystal and Green 1986; Mitchell 1977, 1986, 1994; Moon et al. 2004b; Murphy 1964; Orth 1939; Palmer and Braswell 1995; Reid and Nichols 1970; Reiserer 2002; Savage 1967; Smith 1997; Strecker and Williams 1928; Surface 1906; Tennant 1984; Trauth et al. 2004; Trauth and McAllister 1995; Uhler et al. 1939; Walters et al. 1996; Wright and Wright 1957). Smaller individuals prey on various invertebrates (especially insects), salamanders, lizards, and small snakes, while adults prey more heavily on mammals. Both live prey and carrion are taken.

Agkistrodon contortrix mokasen, ambush position (Mason Neck National Wildlife Refuge, Fairfax County, Virginia; Steve W. Gotte)

Agkistrodon contortrix mokasen, juvenile (Rowan County, Kentucky; Roger W. Barbour)

A copperhead may consume more than twice its BM in prey each year; Schoener (1977) reported an active individual ate 8 meals, which totaled 1.25 times its BM during its annual activity period. Normally, gravid females do not eat, but, when they do, they consume smaller volumes of food than either males or nongravid females (Garton and Dimmick 1969). Carr (1926) reported that one captive female did not eat from 7 July to 17 June the next year, but did spend much time soaking in water, drank frequently, and shed six times during that period.

Differences in feeding patterns may occur within populations of copperheads. Garton and Dimmick (1969) found in a Tennessee population of *A. c. mokasen* that males fed mainly on voles (*Microtus*) and caterpillars; nongravid females ate white-footed mice (*Peromyscus*), voles, and birds (unidentified); while gravid females consumed lizards (*Sceloporus undulatus*) and shrews (*Blarina, Cryptotis*). The diet of the gravid females best represented the available species in their selected microhabitat.

Prey items also vary between populations according to availability. Bulk prey percentages recorded from the stomachs of Kentucky *A. c. mokasen* included *Peromyscus leucopus*, 58.3%; *Microtus pinetorum*, 16.6%; *Scincella lateralis*, 16.5%; and lepidopteran caterpillars, 8.3% (Bush 1959). The most frequently taken prey by *A. c. phaeogaster* in Kansas were voles (*Microtus ochrogaster* and *M. pinetorum*, 24%), cicada (*Tibicen pruinosa*, 15%), and white-footed mice (*Peromyscus* sp., 13%) (Fitch 1982). Voles and white-footed mice composed the greatest estimated biomass (1,051 g/ha), and so were the most readily available prey. In northeastern Texas *A. c. contortrix*, arthropods were found in 59.2% (insects, 50.7%; cicadas 33.8%), and vertebrates in 40.8% (Squamata, 22.5%; Scincidae, 14.1%; mammals, 2.8%) of the copperhead stomachs examined by Lagesse and Ford (1996).

Adult copperheads are mostly ambushers, although because they are primarily nocturnal during most of their annual cycle active hunting behavior may have been missed. Juveniles actively stalk much of their prey, but Neill (1960) has reported that the yellow tail of the newborn may be used to lure small frogs. Neonates use their

Agkistrodon contortrix phaeogaster (Douglas County, Kansas, John H. Tashjian)

yellow tails to lure small animals, and the type of prey seen or smelled probably stimulates the luring response (Ditmars 1931a; Neill 1960; Pycraft 1925; Reiserer 2002). The small snakes exhibited a greater response to moth caterpillars (100% ingestions) than to anurans (76.5%) and lizards (70.1%) in tests conducted by Reiserer (2002). Large prey are bitten, released, and tracked later after the venom has done its damage (Stiles et al. 2000, 2002). Small prey and birds are often retained in the mouth until dead. The average durations of fang contact during a predatory strike on an adult mouse, and defensive bites on an adult rat and on a model of a human limb, were 0.17, 0.07, and 0.16 seconds, respectively (Hayes et al. 2002).

Odor, sight, and heat reception all play roles in prey detection (Neill 1960; Reiserer 2002). Once the prey has been found and envenomated, its taste or odor becomes the dominant basis (Greenbaum 2004; Greenbaum and Jorgensen 2004; Stiles et al. 2000, 2002) for tracking the wounded prey; the vomeronasal organ may play a key role in recognizing and tracking bitten prey (Miller and Gutzke 1999).

Lutterschmidt et al. (2007) identified 16 bacteria belonging to the genera *Acinetobacter, Aeromonas* (3), *Citrobacter* (2), *Enterobacter, Escherichia, Klebsiella* (2), *Leclercia, Proteus, Providencia, Pseudomonas, Salmonella*, and *Xanthomonas* in the esophagus of 7 wild and 12 captive *A. contortrix*, but the possible role of these bacteria in digestion was not explained.

VENOM DELIVERY SYSTEM: A series of 214 copperheads, mostly *A. c. mokasen*, 17–110 cm in TBL and 1.2–2.7 cm in HL had 1.1–7.2 mm fangs (Ernst 1962, 1965, 1982b); FL increased linearly with growth in both body and head length; the means of TBL/FL and HL/FL were 136.6% and 6.9%, respectively, and no sexual dimorphism was evident (Ernst 1964). The 7.2 mm FL_{max} of *A. contortrix* placed it behind both *A. bilineatus* (12 mm) and *A. piscivorus* (11 mm). The calculated straight-line regression coefficients for the copperhead were FL = 0.005TBL + 0.979 (r = 0.95) and FL = 0.150HL + 0.285 (r = 0.98). The curvature of the fang increased gradually from juvenile to adult; the range in the estimated degree of curvature angle was 60–68°, but, due to the short FL of juveniles, it was extremely difficult to measure their curvature, so a greater error in the estimates exists for juvenile fangs. The maximum degree of curvature is less than that for *A. bilineatus* (73) and *A. piscivorus* (74). A large copperhead may produce puncture marks 22–28 mm apart (A. Savitsky, personal communication). The venom delivery morphology and embryonic development of the maxillary and prefrontal bones of *A. contortrix* are described and figured in A. H. Savitsky (1992).

Functional fangs are present in neonates, and they are capable of injecting venom (Boyer 1933; Stadelman 1929b). These fangs, however, do not remain with the snake throughout its life. Instead, they are shed and replaced periodically, an adaptation for replacing broken or loose fangs. A series of 5–7 (in less than 3% of snakes examined) replacement fangs occur in the gums behind and above the functional fang in alternating sockets on the maxillary bone (Ernst 1982b). The replacement fangs lie close together and those distal to the functional fang may be only 0.1 to 0.3 mm apart. In graduated lengths, they may range from a first reserve fang slightly longer than the functional fang (but never more than 0.2 mm longer) to only a short spike about 0.2 mm long in the last of the series. As with rattlesnakes (Klauber 1939), the graduated reserve fang series shifts forward to occupy an alternate series of sockets on the maxilla. These sockets are divided by a wall of tissue that separates the developing fangs that will enter the outer socket from those that will enter the inner socket. The socket of the first reserve fang is also separated from that of the functional fang by a membrane. Beside the functional fang is a vacant socket into which the first reserve fang migrates just prior to the shedding of the functional fang.

Such a replacement series is even found in the newborn. The fangs do not develop as a complete unit, but rather from the tip (represented by the most distal replacement fang) upward, and the hollow tube-shape is in evidence from the earliest period of development in which shape can be ascertained. Approximately 19% of the time, one or both fangs were in the process of replacement (Ernst 1982b). The mean replacement rate for the genus *Agkistrodon* is 21.1% (Ernst 1982b). The replacement rate for all subspecies of *A. contortrix* in Ernst's (1982b) sample was 20.1%. Fitch (1960, 1999) found 19.7% of the *A. c. phaeogaster* fangs he examined were being replaced, and reported a 33-day fang shedding cycle. Of 1,285 copperheads newly captured by Fitch (1999), 804 (62.6%) had only 1 fang on each side and 481 (37.4%) were in the process of fang replacement; 439 (34.2%) were replacing a fang on one side of the jaw, and 41 (3.2%) were replacing the fangs on both jaws simultaneously, but replacement was not necessarily at the same rate on both jaws. We have found individuals of *A. c. contortrix* and *A. c. mokasen* replacing both fangs simultaneously on several occasions.

VENOM AND BITES: The normal range of dry venom yield for *A. contortrix* is 40–75 mg (Altimari 1998; Ernst and Zug 1996; Minton and Minton 1969; Russell 1983; Russell and Puffer 1970), but Rehling (2002) reported a range of 3–61 mg and Brown (1973) one of 24–72 mg. Total venom yield increases with body length (Greenbaum et al. 2003; Jones and Burchfield 1971); McCue (2006) reported mean liquid extraction from 7 183–367 g copperheads was 193.7 (range, 111–376) mg; total volumes were allometrically correlated with BM, and the mean moisture content was approximately 70.9%. Minton (1953) reported liquid yields of 0.21–0.29 (mean, 0.238) mL from

4 healthy milked adult copperheads. Twenty-five to 75% of the contents of the venom glands may be discharged in one bite (Fitch 1960), and juveniles are as toxic as their parents (Minton 1967). Venoms of the five subspecies do not differ significantly in their biological activities, and interspecific differences are more pronounced than individual variation in enzymatic actions (Greenbaum et al. 2003; Tan and Ponnudurai 1990), but differences do exist in the subspecific venom electrophoretic patterns.

The venom of *A. contortrix* is highly hemolytic, and, in contrast to the reports that it rarely causes significant hemostatic abnormalities or hemorrhage (Burch et al. 1988; Van Mierop 1976b), mice or newborn rats we dissected 1–2 hours after having been bitten often exhibited massive hemorrhaging at the bite site and of internal organs. The venom is low in alkaline, and includes 31–37% solids and 65% protein; total baseline protein energy ranges from 0.166 to 0.561 (mean, 0.289; n = 7) kJ (McCue 2006). As with other snake venoms, the venom of *A. contortrix* has a high concentration of protein enzymes (both purine and pyrimidine), particularly fibrinogenases and metalloproteinases. Identified constituents include alfimeprase (a fibrinolytic zinc metalloproteinase), desoxyribonuclease, hyaluronidase, L-amino acid oxidase, various lipids, myotoxin α and Lys49 PLA_2 myotoxin, NAD-nucleosidase, 5′-nucleotidase, phosphodiesterase (but see below), phopholipase A_2, phosphomonoesterase, protein C activator (a serine proteinase), ribonuclease 1, venzyme (causes release of fibrinopetide B [and possibly A]). The venom produces high arginine ester hydrolase and hyaluronidase activity, myonecrosis, moderate disintegrin and hemorrhagin activity, and high fibrinolytic activity, but no phosphodiesterase activity.

Post-venom extraction, McCue's (2006) snakes demonstrated a mean 11% increase in their metabolic resting rate during the first 72 hours of venom replacement; this metabolic increase apparently resulted from metabolic costs involved in venom production, and was an order of magnitude greater than that predicted for producing an identical mass of mixed body growth.

A. contortrix has been one of the most popular species used in venom research, and many papers discussing its chemistry have been published; some of the most important of these are Ahmed et al. (1990a, 1990b), Aird (2005), Bertke et al. (1966), Bonilla and Horner (1969), Brown (1973), Datta et al. (1995), Dyr et al. (1989a, 1989b, 1990), Egen et al. (1987), Fletcher et al. (1996), Fletcher and Jiang (1998), García et al. (2004), Greenbaum et al. (2003), Guan et al. (1991), Herzig et al. (1970), Johnson and Ownby (1993a, 1993b), Li et al. (1993), Lomonte et al. (2003a, 2003b), Manning (1995), Markland (1988, 1991), Markland et al. (1988), McCue (2006), McMullen et al. (1989), Mebs and Samejima (1986), Meier and Stocker (1995), Minton and Minton (1969), Moran and Geren (1979), Murakami and Arni (2005), Orthner et al. (1988), Ramírez et al. (1999), Selistre de Araujo et al. (1996), Stocker et al. (1987), Stürzbecher et al. (1991), Swenson and Markland (2005), Takagi et al. (1988), Takahashi and Mashiko (1998), Tan and Ponnudurai (1990), Toombs (2001), Toombs and Deitcher (2006), Tu (1977), Wagner and Prescott (1966), and White (2005).

Minton and Minton (1969) reported LD_{50} values of 218 (intravenous), 102 (intraperitoneal), and 516 (subcutaneous) mg for 20 g mice; Friederich and Tu (1971) and Russell and Puffer (1970) gave the LD_{50} values (in mg/g) as 10.5 (intraperitoneal), 10.9 (intravenous), and 25.6 (subcutaneous), respectively; and Schmidt et al. (1972) reported an intramuscular LD_{50} of 20 mg/g for *A. c. laticinctus*. Arce et al. (2003) reported mean μg intraperitoneal and intravenous LD_{50} values of 9.17 (range, 6.97–12.07) and 10.17

(range, 8.16–12.67), respectively. Russell and Emery (1959) listed the mean mouse LD_{50} in mg/kg as 10.92 (range, 9.58–12.45) intravenous and 10.5 (range, 9.5–11.55) intraperitoneal. Greenbaum et al. (2003) reported that 15–20 g mice were immobilized in an average of 26.25, 39.50, and 62.83 minutes when injected with venom from Kansas [*A. c. phaeogaster*], Texas [subspecies?], and Louisiana [*A. c. contortrix*] copperheads, respectively; there is a regional or subspecific variation in the toxicity of the venom. The venom is neutralized in mice by injection of serum from the nonvenomous Japanese snake *Elaphe quadrivirgata* (Philpot 1954).

Brown rats (*Rattus norvegicus*) injected either intramuscularly or intraperitoneally with varying doses of *A. contortrix* venom show the largest amount of venom in the kidneys and liver at 24 hours, with a slight distribution in other organs, and the smallest amounts in the brain and spinal cord; however, those injected in their muscles show the highest level of venom at the injection site (Wingert et al. 1980). The mean LD_{50} for chicks (*Gallus domesticus*) is 1.79 (range, 1.40–2.20) mg/g intravenous and 8.12 (range, 6.49–10.15) mg/g intraperitoneal (Russell and Emery 1959).

An increase in resistance to venom from *A. c. contortrix* occurs with development in the bullfrog (*Rana catesbeiana*); tadpoles have an LD_{50} of about 2 mg/kg, at the limb bud stage it has risen to about 5 mg/kg, decreases to 4 mg/kg at premetamorphosis, then increases again to about 158–160 mg/kg at 1–2 weeks postmetamorphosis and more than 180 mg/kg at 4 weeks postmetamorphosis, but then decreases to about 120 mg/kg in the adult stage (Heatwole et al. 1999). The venom did not cause any in vivo effect on the liver of adult frogs, but a loss of epithelial integrity of Bowman's capsules and glomeruli occurred in the kidneys (Green et al. 2004).

When various doses of venom have been injected into snakes, *Agkistrodon contortrix, Nerodia sipedon, N. taxispilota, Opheodrys vernalis*, and *Thamnophis sirtalis* survived, while others—*A. contortrix, A. piscivorus, Charina bottae, Coluber constrictor, Contia tenuis, Crotalus horridus, Diadophis punctatus, Lampropeltis calligaster, L. triangulum*, some *Nerodia sipedon, Pantherophis obsoletus, Pituophis catenifer, Sistrurus catenatus*, and *Storeria dekayi*—died (Keegan and Andrews 1942; Swanson 1946).

Some prey and potential predators have a degree of immunity to copperhead venom. Kingsnakes (*Lampropeltis* sp.) are well known to be immune to copperhead venom (Ernst and Ernst 2003). Serum of Kansas prairie voles (*Microtus ochrogaster*) and eastern woodrats (*Neotoma floridana*) contains antihemorrhagic components to resist the venom of the copperhead (de Wit 1982), and the opossum (*Didelphis virginiana*) has innate immunity to the venom (Werner and Faith 1978).

Probably 100 mg or more of venom are needed to kill an adult human; Brown (1973) estimated the subcutaneous LD_{50} for a 70 kg human to be 1,470–1,792 mg. These dosages are more than are normally contained in the venom glands (see above).

Annually, *A. contortrix* is probably the cause of most human envenomations by snakes within its range. Parrish (1966) analyzed 2,836 cases of hospitalized patients for snakebite treatment for which detailed records were available; 811 (28.6%) were caused by copperheads. Of 596 venomous snakebites reviewed by Scalzo et al. (2007), 87.2% were by this species. Whitlow et al. (2007) reported the copperhead is responsible for approximately 13% of snake envenomations in the United States each year, but 25–29% is more realistic (Bush et al. 1997; Parrish 1966). In North Carolina, the species is responsible for 85% of snake envenomations (Lavonas et al. 2004).

Because the copperhead is so cryptic and remains still when approached, it is often unnoticed. It usually gives no warning of its presence. Most natural human envenomations come from persons placing their hands near or stepping beside an unseen one. C. Ernst has been almost bitten on several occasions under those circumstances, even though he was actively searching for the snake. Illegitimate (artificial) bites while handling the snake, particularly captives, are not uncommon; of 48 bites by 21 species that occurred during academic research in the United States and reported by Ivanyi and Altimari (2004), 9 (18.8%) involved *A. contortrix*.

Copperhead bites usually cause much discomfort, and about 33% produce only clinically significant local effects (Scharman and Noffsinger 2001). Besides pain (sometimes severe), symptoms of human bites include abscesses, breathing difficulty, decoloration of the skin, ecchymosis, either an increased or a weakened pulse, edema (often with swelling), fever, gangrene (if not treated), giddiness, headache, hemorrhage (sometimes severe), hypoglycemia, intestinal discomfort, dizziness, lethargy, nausea and vomiting, occasional hypotension and shock, scarring, stupor or unconsciousness, chills and sweating, and weakness (Burch et al. 1988; Campbell and Lamar 1989, 2004; Dart et al. 1992b; Hutchison 1929; Norris 2004; Scharman and Noffsinger 2001; Whitlow et al. 2007). Case histories of bites are given by Boyer (1933), Diener (1961), Fitch (1960), Keyler (1986d), McCauley (1945), Palmer (1992), Repp (1998b), White et al. (1988), and Whitlow et al. (2007).

While this snake is responsible for many bites each year, only approximately 11% are serious (Campbell and Lamar 1989). Mortality has occurred, but is extremely low and not normally expected from a bite (Scharman and Noffsinger 2001; Stickel 1952, White et al. 1988). The fatality rate is estimated to be only 0.01–0.03% (Brown 1973; Campbell and Lamar 1989; Minton and Minton 1969). Langley and Morrow (1997) estimated that about five deaths occur each year in the United States. We only found the following records of death producing bites: (1) a 14-year-old boy bitten on the finger while hunting rabbits in a burrow died after being treated too late with antivenom (do Amaral 1927); (2) a bite to a 6-year-old that caused death in several days; no further details given (Dart et al. 1992b); (3 and 4) a possible fatal copperhead bite to the finger of a 50-year-old; a 51-year-old who died from a possible copperhead bite (Dart 2006; presentation in the Snakebites in the New Millennium: A State-of-the-art Symposium, University of Nebraska Medical Center, Omaha); (5) two deaths in North Carolina, one of which involved a one-year-old child who was bitten on the hand and expired the next day (Kunzé 1883; Palmer and Braswell 1995); and (6) the death of an infant delivered by a woman who had been bitten six weeks before delivery (Entman and Moise 1984). Probably only the very young and old, or persons in poor health, need worry about this snake.

PREDATORS AND DEFENSE: Copperheads have several predators, especially when young. Bullfrogs (*Rana catesbeiana*), box turtles (*Terrapene carolina* [carrion]), alligators (*Alligator mississippiensis*), ophiophagous snakes (*Coluber constrictor; Drymarchon corais; Lampropeltis calligaster, L. getula, L. triangulum; Micrurus fulvius; Thamnophis sirtalis*), crows (*Corvus brachyrhynchos*), horned owls (*Bubo virginianus*), large hawks (*Buteo jamaicensis, B. platypterus*), opossums (*Didelphis virginiana*), moles (*Scalopus aquaticus* [captivity]), coyotes (*Canis latrans*), and feral cats (*Felis catus*) are natural enemies (Branson 1904; Campbell and Lamar 2004; Clark 1949; Cope 1900; Delavan 1939; C. Ernst, personal observation; Fitch 1960; Gloyd and Conant 1990; Greene 1984; Hurter 1911;

Jensen 1999; Keegan 1944; Kennedy 1964; Klauber 1972; Megonigal 1985; Minton 1944; Mitchell 1994; Montgomery et al. 2004; Palmer and Braswell 1995; Ross 1989; Sutton et al. 2009).

A young captive *A. contortrix* ate a dead litter mate of similar size (Mitchell 1977, 1986), but it is not known if this species is naturally cannibalistic. We have seen crows and ravens (*Corvus brachyrhynchos, C. ossifragus, C. corax*), starlings (*Sturnus vulgaris*), and vultures (*Cathartes aura, Coragyps atratus*) feed on copperhead carcasses, but do not know if these birds naturally prey on the snake or are merely opportunistic carrion eaters.

The copperhead is nonaggressive unless disturbed. Of the more than 300 copperheads (including several subspecies) we have encountered, none has ever advanced toward or attacked us. The snake remains motionless, usually in a coil, and increases its rate of tongue-flicking when an intruder is first detected, and this habit, along with its color and pattern camouflage, makes it very dangerous. Many bites have resulted from persons unwittingly stepping on, sitting on, or touching unseen snakes. Some vibrate their tails when agitated, and while doing so the rates of O_2 consumption and cytochrome oxidase activity and the amount of succinic dehydrogenase all increase in the tail (Moon 2001). If moving, the snake may pause or increase its speed to escape. When touched, it often quickly strikes, but at other times it just remains quiet or tries to crawl away. If prevented from escaping, it will go into an offensive mode, and is not restricted to striking from a coiled position. The warmer the snake is, the greater the defensive reaction. When handled, musk is usually sprayed upon the restrainer. Contrary to the folktale, this musk has its own odor and does not smell like cucumbers.

The snake reacts defensively with increased tongue-flicking to the odor of the predatory kingsnake, *Lampropeltis getula*. Much of this odor detection probably involves the snake's vomeronasal organ (Miller and Gutzke 1999). When confronted with the odor of the kingsnake, or the snake itself, the copperhead performs a body

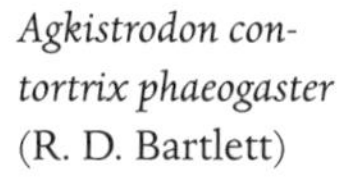
Agkistrodon contortrix phaeogaster (R. D. Bartlett)

Agkistrodon contortrix phaeogaster (R. D. Bartlett)

bridge to prevent the other snake from seizing it (C. Ernst, personal observation; Weldon and Burghardt 1979; Weldon et al. 1992).

PARASITES AND PATHOGENS: Ernst and Ernst (2006) list three trematodes (*Dasymetra natricis, Ochetosoma ancistrodontis, O. kansense*) and five nematodes (*Eustrongylides* sp., *Kalicephalus inermis, Physaloptera obtussima, Strongyloides gulae, S. serpentis*) as endoparasites of this species, and Fitch (1960) recovered a *Trombicula sylvilagi* from a copperhead in Kansas (Byrd and Denton 1938; Cable and Sanborn 1970; Crow 1913; Dyer and McNair 1974; Harwood 1932; Holl and Allison 1935; Hughes et al. 1941; Little 1966; MacCallum 1921; Schad 1962; Winsor 1948). Loomis (1956) reported parasitism by chiggers attached to the head of a Kentucky copperhead, and we have seen a tick (*Dermocenter* sp.). Fitch (1960) reported that individuals trapped in the fall after a wet summer had necrotic patches on their venters and occasional blister-like swellings on the dorsal scales.

POPULATIONS: The longest research study that has yielded significant population data was that of *A. c. phaeogaster* by Fitch over a period of more than 50 years at the University of Kansas Fitch Natural History Reservation and other nearby sites.

The total population of *A. c. phaeogaster*, based on a 10-year census, in Fitch's (1960) study area was approximately 1,664 individuals, a density for the 88 hectares of 18.8/ha. During the first 30 years, 2,681 copperheads were captured at the site (Fitch 1982). In 1982, Fitch estimated the total biomass of copperheads to be 0.80 kg/ha on the reservation. The adult to juvenile ratio and the adult male to female sex ratio were 4.26:1, and 1.22:1, respectively. Data collected to 1960 by Fitch, and later statistically analyzed by Vial et al. (1977), showed individuals of the smaller and younger size and age classes were more numerous than adults (However, this may have been influenced by the dates of field study; neonates enter the population in numbers in late summer and early fall, but many later perish.) Snakes older than 8 years represented only 5% of the population, while those no older than 2 years comprised 55%. Vial et al. (1977) constructed life tables for *A. c. phaeogaster*, and estimated a skewed sex ratio of 74% males to 26% females at birth, but calculated a greater mortality rate in the males, and predicted the sex ratio would reach unity by the eighth year. This is interesting because the sex ratios of litters are usually much closer to 1:1; Gloyd (1934) reported that in a total of 69 young in 16 litters, 29 (42%) were females and 40 (58%) were males.

Fitch and Echelle (2006) reported that in 1977 the density of Osage Copperheads at the Kansas reservation was 18.6 snakes/ha and the biomass 2,790 g/ha; these represented 0.89% and 13.4% of the total biomass of the total number of the 12 snake species present. Yearly densities in this population, which included more than 3,000 captured individuals by 1997, varied from 3.9 to 34.6 copperheads per hectare (Fitch 1999). The number of snakes varied by microhabitat; for example, a mean of 155.5 individuals occurred at one 10 ha lowland field, but 97.95 at a second 3.45 ha hilltop field near a quarry (which contained hibernacula; Fitch 1960). Density at the lowland field varied from 8.1/ha in 1977 to 1.1/ha in 1986, and at the smaller quarry plot from 21.6/ha in 1977 to 1.4/ha in 1988. Overall, the second, smaller, field maintained a higher mean density from 1977 to 1988. The numbers of snakes at the 2 microhabitat sites also varied by the date of field study, but the density was always estimated to be higher at the quarry site. In 2000, Fitch reported the maximum density per hectare was 4.9, and a biomass per hectare of 504.7 g, representing 5.6% of the total snake biomass on the reservation.

On 5 August at 1530 hours, at Clinton State Park, Kansas, Clarke found 7 *A. c. phaeogaster* lying in resting coils among rocks of a limestone outcrop along a trail. The snakes were closely associated in a group, in contact or separated by less than 5 cm. Another group of 4 snakes was about 1 m away, and a twelfth copperhead was about 1.5 m from its nearest neighbor (Fitch and Clarke 2002). The next day at 1500 hours, Fitch and Clarke returned to the site and captured 10 snakes there and 4 others from a position about 100 m farther west along the trail. On 9 August at 1140 hours Clarke returned and caught 2 more snakes at the sight of the large aggregation, and at 1500 hours caught another about 2 m away. The next day, she returned again and caught 3 more individuals at the same site. At least 17 copperheads were present at the original site of the large aggregation, including 15 gravid females, 1 male, and another snake that escaped. Four gravid females were found in the smaller group.

Elsewhere, *A. contortrix* also occurs in large aggregations, especially around hibernacula or sites with a high prey density. In July 1960, Barbour (1962) captured 7 adult *A. c. mokasen* in less than 15 minutes in an area no larger than 3 × 6 m in Breathitt County, Kentucky, and C. Ernst once caught 10 in as many minutes at another Rowan County, Kentucky, site. In northern Virginia, we have observed as many as 3 individuals of this subspecies basking together within a rotted stump of about 1 × 1 m in dimension, and as many as 15–25 individuals hibernating together annually at another location within that site. Martin (1976) found 243 DOR *A. c. mokasen* (44.6% of all snakes recorded) on the Blue Ridge Parkway and Skyline Drive in Virginia. Near Cades Cove in the Great Smokies National Park, Huheey and Stupka (1967) reported that a maintenance road crew killed approximately 150 northern copperheads within a few weeks, and that the snakes were so common that the mowers using hand scythes had to wear metal guards strapped to their legs to prevent being bitten.

The method of observation influences the total numbers of snakes detected. In contrast to the above population numbers, Wright (1987) recorded only 90 DOR

Agkistrodon contortrix pictigaster (Pecos River, Texas; R. D. Bartlett)

Agkistrodon contortrix pictigaster (Western Texas, Carl H. Ernst)

A. c. mokasen on the Blue Ridge Parkway in 2 years; elsewhere in Virginia, Clifford (1976) recorded only 17 DOR copperheads (6.1% of total snakes) in Amelia County, and Gibson and Merkle (2004) only 6 DOR (5.5% of total snakes) in Powhatan County. *A. c. mokasen* was reported as endangered in Iowa (Christiansen 1981). Ford and Lancaster (2007) collected only 41 *A. c. contortrix* (6.6% of a total of 621 snakes captured) by multiple methods in 3 years in a 2,300 ha bottomland deciduous forest in northeastern Texas. During a study from March to June of 1948, Fitch (1949) found only one *A. c. contortrix* on a road through wooded bottomland in Louisiana, and Meade (1940) noted that the subspecies was only occasionally encountered in the section of Louisiana along the Mississippi River between Baton Rouge and New Orleans.

The copperhead is still holding its own in most areas, is doing well in some areas, and even has some resilient populations close to large cities (Conant 1992). Over much of the range, though, habitat destruction, insecticide poisoning, and the automobile have severely reduced some populations, but collection for the pet trade has probably been negligible in reducing population numbers. Unfortunately, this species is often killed at once when found. Populations of *A. contortrix mokasen* are considered endangered in Iowa and Massachusetts.

REMARKS: In the past, the specific name *mokeson* has been used for the copperhead, but Gloyd and Conant (1990) and Smith and Gloyd (1963) concluded that *Cenchris mokeson* Daudin, 1803, is a junior synonym of *Agkistrodon contortrix mokasen* Palisot de Beuvois, 1799. Its intrageneric relationships are discussed in the genus account. *A. contortrix* has been reviewed by Campbell and Lamar (2004), Ernst and Ernst (2003), and Gloyd and Conant (1990).

Agkistrodon piscivorus (Lacépède, 1789)
Cottonmouth

RECOGNITION: This heavy-bodied, amphibious pitviper (TBL_{max}, 189.2 cm) is olive, dark brown, or black, and lacks both a loreal scale and broad dumbbell-shaped transverse dorsal bands. The 20–27 (mean, 13) straight to slightly wavy dorsal body bands of juveniles fade with age until absent in old, large adults. Juveniles are lighter olive or brown, but become progressively darker with age until, as adults, they are almost or totally black. Some populations are darker than others (Smith and List 1955; Strecker 1908), and individual variations in the color and pattern do occur (Lee 1996; Payne 1977; Strecker and Johnson 1935). The venter is tan to gray and heavily patterned with dark blotches. A light-bordered, dark cheek stripe and dark rostral bars may be present.

Dorsal body scales are keeled with two apical pits (scale surface ornamentation is illustrated in Chiasson et al. 1989), and occur in 25 or 27 (24–27) anterior rows, 25 (21–27) midbody rows, and 21(17–23) posterior rows. On the underside are 125–145 ventrals, 30–56 subcaudals forming an undivided row (Burkett [1966] reported a female with 17 undivided subcaudals, and Hampton [2005] reported a tail anomaly that reduced the number of subcaudal scutes to 23), and an undivided anal plate. Dorsal head scales consist of 9 large plates, as in all *Agkistrodon*; laterally are 2 nasals, no loreal, 2–3 preoculars, 3 (2–4) postoculars, 2–4 suboculars, 3–5 temporals, 7–8 (6–9) supralabials, and 10–11 (8–12) infralabials. Despite the absence of a loreal scale, a heat-sensitive pit is situated between the nostril and the eye. The skin is shed an average of 2.6 (range, 1–3) times per year in captivity (Stabler 1951), but the rate of ecdysis in the wild probably depends on the amount of food intake and growth during a year. The rate of ecdysis is not increased if the skin is injured (Neill 1949). Collins and Carpenter (1970) reported the ventral scale positions of internal organs in males and both nongravid and gravid females (see for details), Lillywhite and Smits (1992) described aspects of the cardiovascular system, and Walker (2003) described and illustrated the snake's trunk vertebrae.

The bilobed hemipenis has a nude base, a divided sulcus spermaticus that extends to the tip of each lobe, 12–40 large recurved spines near the base of each lobe, and calyces covering the distal half of each lobe and continuing down the sulcus to about the point of division (Campbell and Lamar 2004; Malnate 1990).

Kardong (1990) reported the following tooth counts: palatine, 5 (4–6); pterygoid, 15–16 (14–18); and dentary, 17–18 (15–20).

The larger males (TBL_{max}, 189.2 cm) have 125–145 (mean, 136) ventrals, 30–54 (mean, 46) subcaudals, and TLs 12–24 (mean, 17) % of TBL; the smaller females (TBL_{max}, 175 cm) have 128–144 (mean, 134) ventrals, 36–56 (mean, 43) subcaudals, and TLs 10–19 (mean, 15) % of TBL. Female *A. p. leucostoma* are sexually dichromatic, typically dark with little evidence of light cross bands, while males possess higher bands that are evident at least halfway up their sides (Zaidan 2001). In addition, males have longer quadrate bones and greater lateral head surfaces than females (Vincent et al. 2004a).

GEOGRAPHIC VARIATION: Three subspecies are recognized. *Agkistrodon piscivorus piscivorus* (Lacépède, 1789), the Eastern Cottonmouth, lacks a pattern on its light snout, and has its dorsal bands strongly contrasting with the relatively lighter ground color, 39–51 subcaudals in males and 40–50 in females. It is found along the Atlantic coastal plain from south-central and southeastern Virginia to east-central Alabama. *Agkistrodon p. conanti* Gloyd, 1969, the Florida Cottonmouth, has two conspicuous

Agkistrodon piscivorus piscivorus (Ted Borg, courtesy of J. Whitfield Gibbons, Savannah River Ecology Laboratory)

Agkistrodon piscivorus conanti, juvenile (R. D. Bartlett)

dark vertical rostral bars, transverse body bands lost in the relatively dark ground color, and 43–54 subcaudals in males and 41–49 in females. It occurs from southern Georgia and Mobile Bay, Alabama, south through peninsular Florida to the northern Keys. *Agkistrodon p. leucostoma* (Troost, 1836), the Western Cottonmouth, has no pattern on its dark brown to black snout, dorsal bands lost in the relatively dark body coloration, and 30–54 subcaudals in males and 36-56 in females (it is also the shortest subspecies; maximum TBL <160 cm while the other 2 may have TBLs >180 cm). It ranges in the Mississippi Valley from southern Indiana, southwestern Illinois, western Kentucky, southern Missouri, and southeastern Kansas south to the Gulf Coast from Mobile Bay to eastern Oklahoma and central Texas. *A. p. piscivorus* and *A. p. leucostoma* intergrade in the Florida panhandle, Alabama, and east-central Mississippi.

CONFUSING SPECIES: Use the above key to differentiate the copperhead (*A. contortrix*). Rattlesnakes (*Crotalus, Sistrurus*) have a tail rattle and loreal scales. Watersnakes (*Nerodia*) have round pupils, loreal scales, a single anal plate, two rows of subcaudals, but no facial pits. Also, when swimming, *A. piscivorus* inflates its lung, which results in much of its body floating on the surface; *Nerodia* elevates only the head and neck to the surface.

KARYOTYPE: The karyotype has 36 chromosomes: 16 macrochromosomes (4 metacentric, 8 submetacentric, 2 subtelocentric) and 20 microchromosomes; sex determination is ZW in females (the W chromosome is submetacentric) and ZZ in males (Baker et al. 1972; Bull et al. 1988; Cole 1990; Fischman et al. 1972; Zimmerman and Kilpatrick 1973). The karyotype of *A. piscivorus* differs from that of *A. contortrix* in having chromosome number 8 submetacentric rather than telocenteric as in the latter species; the cottonmouth's Z and W chromosomes are metacenteric or submetacentric, while those of the copperhead are submetacentric (Z) and telocentric or subtelocentric (W) (Cole 1990). *A. piscivorus* also possesses a cloned ZFY gene (a supposed testis-determining factor in mammals) that was hybridized to the DNA of reptiles with sex chromosomes and is expressed in embryos of both sexes (Bull et al. 1988). A triploid *A. p. leucostoma* was described by Tiersch and Figiel (1991).

FOSSIL RECORD: Fossil *A. piscivorus* are known from the Pleistocene Irvingtonian (Meylan 1995) and the Rancholabrean of Florida (Auffenberg 1963; Brattstrom 1953; Gilmore 1938; Holman 1958, 1959a, 1978, 1995, 1996; Tihen 1962; Webb and Simons 2006), Georgia (Holman 1967), and Texas (Hill 1971; Holman 1963, 1968, 1980, 1995).

DISTRIBUTION: The species ranges from the Piedmont of south-central (Chesterfield and Dinwiddie counties; Mitchell 1994; Watson 2006) and the coastal plain of southeastern Virginia south along the Atlantic coastal plain and the Piedmont to the northern Florida Keys, and west along the Gulf coastal plain, and south through the Mississippi Valley from southern Indiana (Forsyth et al. 1985; Wilson and Minton 1983), southern Illinois, western Kentucky, Missouri, and eastern Kansas (Cherokee County; Rundquist and Triplett 1993) to eastern Oklahoma and central Texas (Irion and Sterling counties; Price 1998). It also occurs on some barrier islands off the Atlantic and Gulf coasts. The collection of an *A. piscivorus* on a barge in Winona, Minnesota, shows that the snake may occasionally be passively carried up the Mississippi River on commercial boats (Cochran 2008).

HABITAT: Cottonmouths occupy almost any type of aquatic habitat, from brackish coastal marshes and saltwater islands to mangrove thickets (*Avicennia, Laguncularia, Rhizophora*), sand- or mud-bottomed freshwater cypress swamps (*Fraxinus* sp.; *Nyssa aquatica*; *Taxodium ascendens, T. distichum*), bayous, marshes (*Acer rubrum, Eichornia crassipes, Elocharis* sp., *Ilex opaca, Juncus roemerianus, Myricia* sp., *Pontederis cordata, Quercus* sp., *Sagittaria* sp., *Smilax* sp., *Typha* sp., *Zizaniopsis milliacea*), ponds, lakes, the banks of large rivers and streams (at both pools and riffles), periodically flooded pinewoods and palmetto thickets (*Pinus clausa, P. serotina; Sabal minor, S. palmetto*), open deciduous woodlands cut by small waterways (*Acer negundo, A. rubrum; Betula nigra; Liquidamber styraciflua; Quercus* sp.; *Ulmus americana*), and drainage ditches within

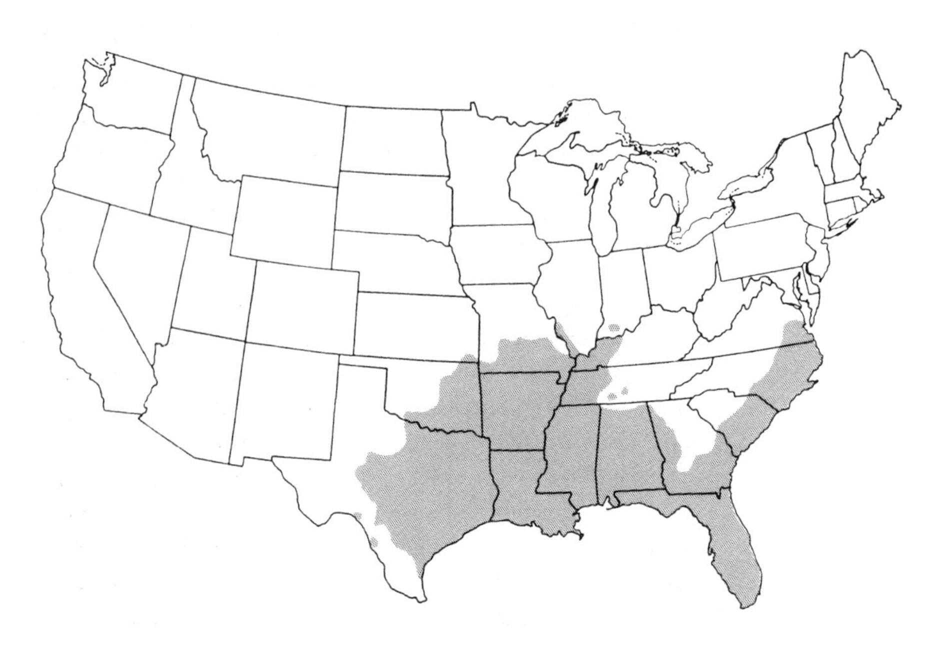

Distribution of *Agkistrodon piscivorus*.

some southern cities. We have even seen or collected them in such unlikely habitats as clear, gravelly, or rocky piedmont or mountain streams in Alabama, Arkansas, and Missouri. Occupied waterways usually have abundant mud, sand, or gravel substrates (Cook 1983), mud or sand banks, logs or brush piles for basking, and nearby upland hibernacula. Cottonmouths may be common in the rookeries of wading aquatic birds, like anhingas (*Anhinga anhinga*), egrets (*Bulbulcus ibis, Casmerodius albus, Dichromanassa rufescens, Egretta thula*), herons (*Ardea herodias, Butorides striatus, Florida caerulea, Hydranassa tricolor, Nyctanassa violacea, Nycticorax nycticorax*), ibis (*Eudocimus albus; Plegadis chihi, P. falcinellus*), whooping cranes (*Grus americanus*), and wood storks (*Mycteria americana*) (R. Barbour, personal communication; C. Ernst, personal observation; Wharton 1969), and may also occupy lodges of Florida round-tailed muskrats (*Neofiber alleni*) (Lee 1968).

The relative importance of a buffer zone to the riparian/stream habitat of Texas *A. p. leucostoma* was researched with radiotelemetry. Although 83% of all snake encounters were within 10 m of the water, the snakes were observed 0–94 m from it. Gravid and nongravid females had different patterns of spatial use. Gravid females inhabited the most distant sites in terrestrial habitats up to 94 m from the stream, and their distribution differed significantly ($p = <0.05$). Disturbance of the surrounding terrestrial habitat would likely impact gravid females the most, indicating a need for conservation buffer zones surrounding aquatic habitats (Roth 2005).

During a habitat modeling study using several statistical methods, Cross and Petersen (2001) found that *A. piscivorus* occupied areas that had significantly less vegetation cover, were closer to water, were nearer to understory trees, and had greater leaf cover and canopy closure and larger trees than random areas sampled. The most intensely used habitat was best determined by four variables: distance from water, leaf litter cover, vegetation cover, and distance to understory trees. In eastern Texas, cottonmouths are restricted to lowland floodplains, and have only a niche overlap value of 0.68 with the copperhead (Ford et al. 1991).

BEHAVIOR AND ECOLOGY: Over most of its range, at least some *A. piscivorus* are active from March into November. The farther south the population is located, the longer the annual activity cycle. Active *A. p. piscivorus* have been found in every month in the Carolinas (Gibbons and Semlitsch 1991; Palmer and Braswell 1995), but most activity in North Carolina is from May to October (Palmer and Braswell 1995). In Virginia, the main activity period is from April to September (Mitchell 1994). Southern Florida *A. p. conanti* may be active in every month, but less so in January to April (Dalrymple et al. 1991a). In eastern Texas, *A. p. leucostoma* seems to have its peak active period in September and October (Ford et al. 1991), and Tinkle (1959) reported that this same subspecies in Louisiana was often out early in the year at ATs of 10.0–12.5°C.

The annual cycle usually includes several stages. With the advent of warm weather in March–April, cottonmouths emerge from their hibernacula and return to their water bodies for the annual feeding and reproductive periods. In late August–early September they begin a rather leisurely exodus from the water to drier, usually more upland, hibernacula, and by October–November most have disappeared from their waterways. However, some emerge on warm winter days (Castleberry 2007).

The cottonmouth is more tolerant of cold than most snakes, and is one of the last to enter hibernation (Neill 1947, 1948). The winter is either spent alone in the South

(Neill 1948) or, particularly in the North, in groups (sometimes with *Crotalus horridus, Pantherophis obsoletus,* and other snakes; Sexton et al. 1992; Smith 1961), on shore at some upland sites, like a rock crevice on a hillside with a southern exposure, in a hollow log or stump, under the roots of overturned trees, in palmetto patches, buried in piles of leaves, or within crayfish, rodent, or gopher tortoise (*Gopherus polyphemus*) burrows.

During the spring and fall, activity is mostly diurnal, but in summer it is predominantly nocturnal; Virginia cottonmouths have a bipolar diel activity cycle, 0600–1000 hours and 1800–2100 hours (Blem and Blem 1995). During the day the snake basks (especially in the morning), remains undercover, or lies quietly in ambush beside logs or other objects.

BTs of active *A. piscivorus* average about 24–25 (range, 17–35) °C (Bothner 1973; Brattstrom 1965; Gibbons and Dorcas 2002; Wood 1954). Blem and Blem (1990) reported that *A. p. piscivorus* had a mean BT of 26.1°C when tested in a thermal gradient. The species frequently hunts at cool ATs of 11–15°C (Andreadis 2004), and Castleberry (2007) has observed them foraging in South Carolina in 5.9°C and 6.2°C waters. Wharton (1969) reported that the BTs of several *A. p. conanti* in an underground Florida hibernaculum were 4.2–16.5°C, and within a few degrees of the AT. He thought the STs kept the snake's BTs intermediate between the AT and ST. The mean BT of hibernating Virginia *A. p. piscivorus* was >6°C (Blem 1997), and Beaupre (*in* Zaidan 2003) found the BT of overwintering Arkansas *A. p. leucostoma* was approximately 7°C, but can drop to just above 0°C.

Crane and Greene (2008) studied thermoregulation by 13 female *A. p. leucostoma* during midsummer in southwestern Missouri. Their preferred BT range was 20.0–26.5 (mean, 23.4) °C measured on a laboratory thermal gradient, lower than those reported for most snake species. Gravid females used the thermal gradient more efficiently than nongravid ones by preferentially occupying the most thermally favorable habitats. They consistently maintained higher BTs than nongravid females, demonstrating the functional tie between habitat use and thermoregulation.

The snake's metabolic rate varies with the acclimation temperature, and seems circadian (Blem and Killeen 1993). *A. p. piscivorus* were acclimated by Blem and Blem (1990) to 22°C and 32°C. Those acclimated to the lower temperature were then measured at an ET of 22–32°C, and those acclimated at the higher temperature at 4–36°C. Statistically significant differences in metabolic rate between the two acclimation groups were observed, even though the metabolic rate was adjusted for the effects of BM and ET and all snakes were from the same area (22°C, kJ/hour = 0.0276ET + 0.0009BM − 0.4434; 32°C, kJ/hour = 0.0356ET + 0.0004BM − 0.4208). The rate of acclimation was a significant function of BM, with lighter individuals acclimating more rapidly than heavier ones.

In contrast, McCue and Lillywhite (2002) found no differences in the metabolic rate of island (Seahorse Key) versus mainland Florida *A. p. conanti* at four experimental ETs (15–30°C). Allometric BM exponents of the metabolic rate averaged 0.76 and were not affected by temperature. The volume of O_2 absorbed increased with temperature (Q_{10} = 2.4–2.8), and was elevated 29% during dark periods versus light periods. Neonates had elevated O_2 consumption compared to older juveniles of similar length, apparently due to assimilation of yolk that is present in the neonatal gut. Following feeding, adult consumption of O_2 increased 4 to 8 times, and the total energy devoted to specific dynamic action increased with the meal size and averaged 32.8% of total ingested energy. McCue and Lillywhite estimated that a 500 g adult Seahorse Key *A. p. conanti* uses 3,656 kJ of assimilated energy annually for maintenance and activity, which requires ingestion of about 1 kg of fish.

Zaiden (2003) studied the variation in the resting metabolic rates by measuring CO_2 production of southern populations and populations near the northern range limit of *A. p. leucostoma* along a latitudinal gradient. CO_2 production varied with photophase, with higher production occurring during the light periods versus the dark ones. When adjusted for BM, male and nongravid females did not differ in CO_2 production. Metabolic cold adaptation was indicated in northern populations that experienced hibernation typically longer than five months; but, while maintenance metabolism is more costly in northern populations during hibernation, it is not likely a limiting factor in the snake's geographic distribution, and may be used to fuel important processes other than activity metabolism. Arkansas cottonmouths allocated almost twice as much energy to resting metabolism during nonfeeding periods than did those from Louisiana.

The skin of *A. piscivorus* has low water permeability (20–50 μmoles/cm^2 per hour, abnormally low for an aquatic species), and it can apparently withstand much drying (Dunson and Freda 1985). *A. p. conanti* on Seahorse Key, Florida, stop foraging and retreat underground when they lose water amounting to somewhere between several percent and 12% of BM (Lillywhite and Zaidan 2004). During tests run by Moen et al. (2005), cottonmouths exhibited higher total evaporative water loss rates than did the copperhead, *A. contortrix*, and they thought this concordant with the evolutionary shift from an ancestral terrestriality to a more aquatic habitat. Cottonmouths in wet treatment had significantly higher rates of water loss, least squares mean rate, 153.0 (range, 128.2–178.4) mg/hour, than those in dry treatment, mean rate, 108.4 (range, 80.3–137.3) mg/hour. Similarly, the mean water influx rates of the neonates studied by Dunson and Freda (1985) were 35 (range, 10–90) and 94 (range, 8–658) μmol/cm^2 per hour for wet and dry treatments, respectively, and the water efflux rates under the same treatments were, respectively, 13 (range, 9–108) and 111 (range, 12–418) μmol/cm^2 per hour. We believe these results show the snake to be more amphibious than aquatic, and adapted to spending time in both wet habitats and drier terrestrial ones. Its annual cycle of occupying wet areas during the warm feeding period but hibernating in upland dry areas supports this hypothesis. Recently, studies of the effects of drought on *A. p. conanti* and other aquatic snakes in Florida by Willson et al. (2006) also support this; the cottonmouths that regularly migrate annually to and from wetland fare well as opposed to several species of *Nerodia* and *Seminatrix pygaea*, and reproduce during drought conditions while the populations of *Nerodia* decline and may not reproduce during drought conditions.

When crawling, cottonmouths characteristically keep the head and neck elevated above the ground. Snakes that crawl this way have a vertebral plexus, a network of spinal veins within the vertebral column interconnected with caval and portal veins lying beneath the column. When the snake is lying or crawling flat on the ground, the caval and portal veins are the major blood pathways; but during crawling with the head raised or climbing blood pressure drops in these veins, so the plexus ensures adequate blood circulation and pressure to the brain (Lillywhite 1993a, 1993b; Zippel et al. 2001).

In South Carolina, a population of *A. p. piscivorus* living in a Carolina bay used the wetland during the active season, left it in the fall to overwinter in more upland habitats, and then returned the following spring. Juveniles left the bay earlier than adults, but no difference in spring arrival time was evident. Spatially, captures of leaving and returning adults were nonrandom, with capture peaks corresponding to the directions to the nearest permanent water bodies. Juveniles left nondirectionally in the fall, but returned in the spring by the same routes as adults; apparently, they do not rely on adult scent trails to locate hibernacula, and in a region with moderate winter ETs, suitable overwintering sites may not be a limiting factor (Glaudas et al. 2007).

Home ranges of *A. p. conanti* on Sea Horse Key, Florida, were 0.04-1.22 ha; males had slightly larger home ranges (mean, 0.17 ha) than females (mean, 0.14 ha). Some made long movements (a 132 cm male crawled 320 m in 27 months; females moved 380 m, 450 m, and 498 m, and another left its home range to establish a new one on the other side of a mangrove inlet); but movements of 60–70 m between captures were most common (Wharton 1969).

Movements of 6 *A. p. leucostoma* (2 males, 4 females; SVL, 55–85 cm) in Montgomery County, Arkansas, were monitored by Metcalf et al. (2007). Total distance traveled by each snake averaged 604 (range, 468–826) m. Three of the females had noncontiguous summer and fall home ranges. The distance moved per day, mean 8 (range, 0.5–14.0) m, differed significantly between months, with the greatest movements occurring in October. Convex polygon home ranges averaged 0.45 (range, 0.13–1.14) ha. The snakes were found 7% of the time in a creek, 17% within 1 m of the creek, 52% within 5 m, and 75% within 10 m. Mean distance from water was 10.7 m. One female was found 85 m from the water. Proximity to water did not differ between the sexes, but did differ by month. Most movement was done in the fall during migration to hibernacula, but only averaged 8 m/day. Two males and a female moved to a hibernaculum beyond their normal home range. The greatest migration to a hibernaculum was 436 m, but all hibernacula were located within 10 m of the creek or a tributary in rocky outcroppings high on streamside banks or under embedded boulders. Four of the snakes periodically came to the surface during the winter.

Several other movement studies of *A. p. leucostoma* have been reported. A Louisiana Western Cottonmouth had a home range of approximately 0.16 ha (Tinkle 1959). At another Louisiana locality, the average home range of females was 0.93 ha and that of males 0.88 ha, and the snakes normally moved 10 m per night (Martin 1982). In a Texas stream habitat, males had a significantly larger linear home range area (mean, 1.86 ha) than either nongravid females (mean, 0.372 ha) or gravid females (mean, 0.963 ha), and gravid females occupied larger areas than nongravid females (Roth 2005). Males examined by Roth et al. (2006) had larger relative medial and dorsal cortex and telencephalon volumes than did females; but, although significant sexual differences in the dorsal cortex volume were not observed, males had significantly larger medial cortex relative to telencephalon volume. The significance of this to the snake's spatial ecology is unclear.

A. piscivorus is a good climber, but seldom ascends high out of the water, preferring instead to merely lie on the bank. However, we have seen an *A. p. conanti* climb about 1.6 m up a small tree in Florida. *A. p. leucostoma* may anchor its tail to aid in descending from basking perches (Burkett 1966).

Male cottonmouths participate in combat bouts, similar to those described under *A. contortrix*, either in shallow water or on land (Blem 1987; Burkett 1966; Carr and Carr 1942; Fogleman et al. 1986; Martin 1984; Perry 1978; Ramsey 1948; Wagner 1962). Apparently, combat behavior only manifests itself in males at least three years old, and is probably coincident with sexual maturity (Aldridge and Duvall 2002; Blem 1987). Blem (1987) observed the development of combat in two captive litters of *A. p. piscivorus*. The older litter contained 2 males born on 5 September 1981, and the younger litter had 3 males born 7 September 1982. The older 2 males were 51.4 cm and 59.6 cm when combat was first observed. At the same time the second set of males averaged only 40.6 (range, 38.7–47.4) cm SVL. Combat was first observed in the first litter males after the introduction of food (fish) and was observed at no other time. Minor bouts lasted 3 to 6 minutes, while prolonged ones took up to 45 minutes (average, 17 minutes). The second litter males showed less intense combat displays during this period, but short bouts of topping behavior occurred between them when food was present. Although females from both litters were present, no reproductive behavior was observed.

Bothner (1974) observed an apparent dominance of two *A. p. piscivorus* over a third that arrived late at a Georgia feeding pool.

Do cottonmouths exhibit "handedness" while coiling? To determine this, Roth (2003) examined the coiling posture of Texas *A. p. leucostoma*, classifying the behavior as clockwise, anticlockwise, outstretched, or random, and found that both juvenile and adult females and adult males preferred a clockwise coil, but that juvenile males all coiled anticlockwise. Unfortunately, he was unable to define the significance and mechanisms influencing this behavior.

REPRODUCTION: Maturity is attained by both sexes during their second or third year. A female raised in captivity produced young at an age of 2 years, 10 months (Conant 1933). Males of *A. p. piscivorus* and *A. p. leucostoma* are mature at SVLs of 60 cm and 52 cm, respectively (C. Ernst, personal observation); minimum SVLs of mature female *A. p. piscivorus, A. p. conanti,* and *A. p. leucostoma* are, respectively, 61.9 cm, 65.0 cm, and 45.5 and 49.0 cm (Blem 1997; Burkett 1966; Hampton and Ford 2005; Wharton, 1966).

The female sexual cycle is as described by Aldridge and Duvall (2002) for North American pitvipers. It begins with vitellogenesis in August–September; yolking ceases over winter, but resumes in the spring and continues until the follicles are of mature size; the eggs are ovulated in May–June (Blem and Blem 1995; Kofron 1979; Wharton 1966). Females are usually carrying young during the summer, particularly in August–September.

The anterior infundibulum and vagina undergo no seasonal changes in ultrastructure, but the posterior infundibulum, glandular uterus, and nonglandular uterus do. The oviduct becomes highly secretory at the beginning of vitellogenesis, including the luminal border of the uterus, the tubular glands of the glandular uterus, and the luminal border and sperm storage tubules of the posterior infundibulum. The secretory products vary among the regions in the oviduct and also among time periods in the same region of the oviduct. This is especially true in the sperm storage tubules, where secretory activity ceases after ovulation. However, the tubular glands of the glandular uterus remain secretory until parturition, at which time secretory activity in the various sections of the oviduct decreases noticeably. After parturition, the oviduct remains dormant until the next reproductive season (Seigel and Sever 2008b).

Egg dry mass averages 4.7 g and is composed of about 24% nonpolar lipids. Decrease in dry mass remains relatively constant during early development but accelerates during the third trimester; a mean 57% of the original nonpolar lipids remains in neonates as parental investment, and at an ET of 30°C stored lipids could serve the neonate's metabolism for about 22 days (Fischer et al. 1994).

Female reproduction is either annual (Blem 1981, 1982; Hill and Beaupre 2008; Kofron 1979) or biennial (Burkett 1966; Wharton 1966), and is probably determined by resource availability. Larger (older?) females are more likely to be gravid than smaller females in a population. Total lipids decline in males and nongravid females during the first year of a biennium, but gradually increase during the second year; gravid females have their highest lipid levels prior to ovulation, after which both total lipids and fat body lipids decline through the GP until they are at their lowest point in the summer (Blem 1997; Scott et al. 1995).

In Alabama males (probably intergrade *A. p. piscivorus* × *A. p. leucostoma*) testicular recrudescence occurs in April, spermiogenesis begins in June and peaks in July–August, and spermatogenesis ceases in October; sperm is stored over winter in the vas efferens, epididymides, and vas deferens. Hypertrophy of the renal sexual segment coincides with elevations in the plasma testosterone level. Tubules of the renal sexual segment are of greatest diameter in March and September; but the epithelium of the segment is highest in August, the lumina of the segment is of maximum diameter in March, and granular secretion by the epithelial cells of the tubule occurs in April and in the lumina in March–October (Johnson et al. 1982).

Renal sexual segments from male *A. piscivorus* collected in February–May and August–November are similar in appearance. The cells are eosinophilic and react with reagents for neutral carbohydrates and proteins. Ultrastructurally, the cells contain large (2 μm diameter), electron-dense secretory granules and smaller vesicles with diffuse material. These structures lie against the luminal border and upon clear vacuoles continuous with intercellular canaliculi. Both cycocrine and merocrine product release processes are occurring in the cells. These periods in which the renal sexual segment is hypertrophied correspond to the spring and fall mating seasons. In June–July, the renal sexual segment is significantly smaller in diameter, is largely basophilic, and has only scattered granules that react for neutral carbohydrates and proteins. Absent are electron-dense secretory granules and smaller vesicles with diffuse materials, but numerous condensing vacuoles and rough endoplasmic reticulum are present, indicating that active product synthesis is taking place (Sever et al. 2008).

In northwestern Arkansas, the plasma testosterone level of male *A. p. leucostoma* peaked in August (wild-caught, 34.4 ng/mL; laboratory controls, 14.1 ng/mL), coinciding with observed mating activity. An experimental group under photoperiod/ET conditions simulating those of a region north of the current range limit also exhibited a single testosterone peak (11.7 mg/mL), but a month later; possibly cottonmouths are limited in their northern distribution due to their reproductive physiology (Zaidan et al. 2003). Johnson et al. (1982) found circulating testosterone levels to be lowest in the spring (>1 ng/mL) and highest in August (~2.8 ng/mL).

A complete description of the courtship and mating behaviors has not been published, although chin-rubbing by the male has been observed (Hill and Beaupre 2008). Mating possibly occurs throughout the year in Florida (Wharton 1966), but elsewhere two breeding periods are evident, March–June and August–October (Aldridge and Duval [2002] classified *A. piscivorus* as a spring and summer breeder). In northwestern Arkansas *A. p. leucostoma*, bisexual pairing in late summer was observed in July (30), August (3, 12–13, 15, 17, 22), and September (3, 11) and in the spring in April (11, 20, 24) and May (3) (Hill and Beaupre 2008). We have observed bisexual pairing in western Kentucky *A. p. leucostoma* in August and September. Actual dates on which copulation has been observed in the wild or in captivity include 21 January, 10–11 March, 31 August, and 10 and 19 October (Allen and Swindell 1948; Anderson 1965; Dundee and Rossman 1989; Gloyd and Conant 1990; Wright and Wright 1957).

Sperm migrate up the oviduct to the sperm storage tubules, which have ciliated and secretory cells in their epithelium. The sperm align themselves in parallel rows with their nuclei facing an area of secretory cells at the base of the tubules. Sperm also embed inter- and intracellularly into the walls of the tubules. The secretions contain neutral carbohydrates. Sperm aggregated in the nonglandular section of the posterior uterus degrade before ovulation and take no part in fertilization (Seigel and Sever 2008a).

Uterine twisting occurs in females (Almeida-Santos and Salomão 2002), and they are capable of storing viable sperm for relatively long periods (Schuett 1992). A captive female separated from males for five years gave birth to young (Baker *in* Schuett 1992).

Female cottonmouths are ovoviviparous, and possibly provide nutrients and O_2 to their embryos. Birchard et al. (1984) found that hematocrit, hemoglobin concentration, O_2 capacity, Bohr effect, and Hill coefficients were not significantly different in fetal and maternal blood. A significant difference in fetal-maternal blood P_{50} was present (fetal, 19.5 Torr; maternal, 48.8 Torr). Nucleoside triphophate (NTP) levels were lower in the embryos than in maternal and juvenile snakes. The differences in the P_{50} and NTP levels disappeared soon after birth (juvenile P_{50}, 45.5 Torr). Apparently, the fetal-maternal shift in blood O_2 affinity is modulated in some way by NTP levels.

The GP is about 150 (range, 120–170) days. Parturition occurs from mid-August to early October. A positive correlation exists between the number and size of neonates and female length and BM (Blem 1981). Litters average 7.2 (range, 1–20; n = 74) young, but 5–8 young are probably most common; *A. p. leucostoma*, the shortest subspecies, produces the smallest litters (mean, 4.1–4.8 young; Ford et al. 2004; Hill and Beaupre 2008; Penn 1943). Females may congregate before giving birth, and remain with their brood after birth (Heinrich and Studenroth 1996; Walters and Card 1996; Wharton 1966). When born, the neonates are enclosed in fetal membranes from which they may take 30 minutes to more than an hour to emerge, although some may rupture the membranes in 5 to 8 minutes (C. Ernst, personal observation; Funk 1964c). RCMs of *A. p. leucostoma* litters of 4–6 young were 22.5%, 25.2%, 27.8%, and 35.9% (Ford et al. 1990, 2004).

Overall, neonates have 21.0–29.9 (mean, 26.2) cm TBLs, and weigh 9.6–20.2 (mean, 14.7) g. Newborn *A. p. leucostoma* are about 18.1–29.9 (mean, 21.2) cm long; those of *A. p. piscivorus* are 22.2–29.3 (mean, 26.0) cm; and neonates of *A. p. conanti* have 28.5–35.0 (mean, 33.5) cm TBLs. The young are light brown with darker brown transverse bands and yellow tails.

Hybridization with *A. contortrix* has occurred in captivity (Mount and Cecil 1982).

GROWTH AND LONGEVITY: In a northwestern Arkansas population of *A. p. leucostoma* studied by Hill and Beaupre (2008), the growth rates ranged from 0.13 to 1.04 cm/month, and were negatively correlated with initial SVL in both sexes. Mean growth per month was 0.151 cm and 0.178 cm for males and females, respectively. Both sexes <55 cm SVL had similar (not significantly different) growth rates (males, 0.109 cm/month; females, 0.051 cm/month). Females with >55 cm SVLs grew very little (mean, 0.313 cm/month), while males in that size range continued to grow (mean, 0.381 cm/month); females reached a maximum SVL of about 64 (mean, 54.9) cm, males about 78 (mean, 60.8) cm (interpreted from a figure in Hill and Beaupre 2008). These growth rates are probably the slowest recorded for a population of cottonmouths. Ford (2002) noted that in northeastern Texas neonates of this subspecies grow rapidly in their first fall and spring, but in later years growth is slower and mediated by prey availability and weather conditions. Both sexes grow at approximately the same rate for the first 3 years, but females begin to slow their growth rate at 24 months and their SVL becomes asymptotic between 40 and 55 cm.

Western Kentucky *A. p. leucostoma* have 26.0–29.8 cm TBLs at 7–8 months (having grown about 2.5 cm since birth) and 31.2–33.7 cm at 19–20 months (having grown about 4.5 cm), while those 31–32 months old average 42.5 cm (an increase of 9.5 cm) (Barbour 1956). Burkett (1966) reported an average growth rate of 0.72 cm/month for 3 *A. p. leucostoma*.

Yearling Virginia *A. p. piscivorus* have mean SVLs of 35 cm, second-year juveniles average 60 cm, and third-year individuals are all >60 cm (Blem and Blem 1995). At Cedar Key, Florida, first-year *A. p. conanti* grow an average of 1.46 cm/month, and those in their second year average 1.3 cm/month (Wharton 1966).

Vincent et al. (2004b) studied changes in intersexual head shape of *A. piscivorus*. All head measurements in adult males and females scale with negatively significant allometry, while those of juveniles typically scale isometrically, except for head volume (positive) and head length (negative). Juveniles have relatively broad, high, but short heads; large adults of both sexes have relatively small head dimensions overall. Juveniles experience a rapid change in head volume, which then slows considerably as sexual maturity is achieved, but multivariate analyses of size-adjusted head dimensions differ only slightly in head shape when juveniles and adults are compared.

The maximum known life span for *A. p. leucostoma* is 21 years, 4 months, and 10 days documented in a wild-caught adult male at the Dallas Zoo. A wild-caught male *A. p. piscivorus* survived an additional 13 years, 1 month, and 24 days at the North Carolina State Museum (Snider and Bowler 1992). We have had both subspecies live for more than 11 years in our laboratory.

DIET AND FEEDING HABITS: The cottonmouth is an opportunist that eats a wide variety of ecto- and endothermic prey, either live or as carrion, which are most available and easiest to catch at the time it is hungry in the wild or when offered to it in captivity.

Fish and amphibians are the most frequent wild prey. At least some reported marine fish prey was probably consumed as carrion dropped by birds. Himes (2003) recorded food items from the digestive tracts of 54 small wild *A. p. leucostoma,* which contained 22 lizards (31.9% of prey items), 18 fish (26.1%), 10 frogs (14.5%), 6 snakes (8.7%), 6 mammals (8.7%), 3 salamanders (4.3%), 2 insects (2.9%), a mollusk (1.4%), and a crustacean (1.4%); those of 55 wild adults contained 59 fish (55.1%), 15 frogs (14%), 8 snakes (7.5%), 6 birds (5.6%), 6 mammals (5.6%), 4 lizards (3.7%), 4 turtles (3.7%), 3 salamanders (2.8%), and 2 insects (1.9%). The fishes identified belonged to 7 families, with centrarchids comprising 53% of identified fish. In a second diet study, Himes (2004) examined the digestive tracts of preserved museum specimens of *A. p. leucostoma* from several Mississippi drainages. Those of cottonmouths <50 cm SVL contained 22 lizards (47% of recorded prey), 10 frogs (21%), 6 snakes (13%), 6 mammals (13%), and 3 salamanders (6%); those of >50 cm SVL cottonmouths contained 15 frogs (32%), 8 snakes (17%), 6 mammals (13%), 6 birds (13%), 4 lizards (9%), 4 turtles (9%), and 3 salamanders (7%). Klimstra (1959) reported the following prey by volume from southern Illinois *A. p. leucostoma*: fish (31.9%), amphibians (26%), reptiles (18.2%), mammals (17.9%), snails (1%), and miscellaneous items (5%); poikilothermic prey made up 76.1% of the total food volume. In contrast to the behavior of gravid females of other pitvipers, gravid *A. piscivorus* may continue to feed during gestation (Burkett 1966).

The following list includes reported prey consumed (either alive or as carrion) both in the wild and in captivity: algae (probably secondarily ingested with other prey); snails (*Euglandina rosacea*); freshwater clams and mussels (*Corbicula* sp., Viviparidae); conch; insects—beetles (Dytiscidae), cicadas (Cicadiae), damselflies (Odonata), grasshoppers (Orthoptera), hymenopterans (Hymenoptera), and moth larvae (*Actius luna*); spiders (Arachnida); crayfish (Cambaridae); fish—bowfin (*Amia calva*), gizzard shad (*Dorosoma cepedianium*), mudminnows (*Umbra limi*), eels (*Anguilla rostrata*), catfish (*Ameiurus melas, A. natalis, A. punctatus; Bagre marinus; Ictalurus furcatus, I. punctatus*), pickerel (*Esox americanus*), trout (*Onycorhynchus mykiss, Salvelinus fontinalis*), smelt (Osmeridae), live bearers (*Gambusia affinis, Heterandria formosa, Poecilia latipinna*), cyprinids (*Carassius auratus, Cyprinodon variegatus, Notemigonus crysoleucas, Notropis* sp., *Pimephales promelas*), killifish (*Fundulus dispar, F. grandis; Lucania* sp.), pirate perch (*Aphedoderus sayanus*), silversides (*Menedia* sp.), toadfish (*Opsanus* sp.), bass (*Acantharchus pomotis, Micropterus dolomieu*), black crappie (*Pomoxis nigromaculatus*), flier (*Centrachus micropterus*), pygmy sunfish (*Elassoma zonatum*), sunfish (*Lepomis cyanellus, L. gulosus, L. humilis, L. macrochirus, L. microlophus, L. symmetricus*), sea bass (*Morone americanus*), drum (*Aplodinotus grunniens, Cynoscion nebulosus*), perch (*Perca flavescens*), goby (*Gobionellus boleosoma*), and mullet (*Mugil* sp.); amphibians—salamanders (*Ambystoma opacum, A. talpoideum; Amphiuma means, A. tridactylum; Desmognathus brimleyorum; Eurycea longicaudata, E. lucifuga; Notophthalmus viridescens; Pseudotriton montanus; Siren intermedia, S. lacertina*) and anurans (*Acris crepitans, A. gryllus; Bufo terrestris; Gastrophryne carolinensis; Hyla avivoca, H. cinerea, H. gratiosa, H. versicolor; Osteopilus septentrionalis; Rana berlandieri, R. blairi, R. catesbeiana, R. clamitans, R. palustris, R. sphenocephala; Scaphiopus holbrookii, S. hurterii*); reptiles—juvenile alligators (*Alligator mississippiensis*), turtles (*Apalone ferox; Chelydra serpentina; Kinosternon subrubrum; Pseudemys concinna, P. peninsularis; Sternotherus odoratus; Terrapene carolina; Trachemys scripta*), lizards (*Anolis carolinensis; Eumeces inexpectatus, E. laticeps; Ophisaurus ventralis; Sceloporus olivaceus, S. undulatus; Scincella lateralis*), and snakes (*Agkistrodon piscivorus; Coluber constrictor; Crotalus horridus; Diadophis punctatus; Farancia abacura; Heterodon platirhinos; Lampropeltis getula, L. triangulum; Masticophis flagellum; Nerodia clarkii, N. cyclopion, N. erythrogaster, N. fasciata, N. rhombifer, N. sipedon, N. taxispilota; Opheodrys aestivus; Pantherophis obsoletus; Regina alleni, R. rigida; Sistrurus miliarius; Storeria dekayi; Thamnophis proximus,*

T. sauritus, T. sirtalis; Virginia striatula); birds, including eggs, nestlings, and fledglings (*Aix sponsa, Ammodramus maritimus, Amnospiza* sp., *Anas discors, Anhinga anhinga, Ardea alba, Baelophus bicolor, Cardinalis cardinalis, Charadrius* sp., *Columba livia* [egg], *Casmerodius albus, Corvus ossifragus, Egretta tricolor, Gallus gallus, Hydranassa tricolor, Hylocichla mustelina, Pahalacrocorax auritus, Parus carolinensis, Passer domesticus, Pipilo erythrophthalmus, Plegadis falcinellus, Podilymbus podiceps, Porzana carolina, Podilymbus podiceps, Sturnus vulgaris, Zenaida macroura*); and mammals—moles (*Scalopus aquaticus*), shrews (*Blarina brevicauda, B. carolinensis; Cryptotis parva; Sorex longirostris*), bats (species not named), muskrats (*Neofiber alleni, Ondatra zibethicus*), voles (*Microtus ochrogaster, M. pinetorum, M. pennsylvanicus*), murid rats and mice (*Mus musculus; Oryzomys palustris; Peromyscus leucopus, P. maniculatus; Rattus norvegicus, R. rattus; Sigmodon hispidus*), kangaroo mice and pocket mice (*Dipodomys* sp., *Perognathus hispidus*), squirrels (*Sciurus carolinensis, Sciurus* sp.; *Tamiasciurus hudsonicus*), cottontail rabbits (*Sylvilagus aquaticus, S. floridanus, S. palustris*), and pig carrion (*Sus scrofa*) (Allen and Swindell 1948; Barbour 1956; Berna and Gibbons 1991; Blem and Blem 1995; Bothner 1974; Burkett 1966; Campbell and Lamar 2004; Carpenter 1958; Carr 1936; Clark 1949; Collins 1980; Collins and Carpenter 1970; Cook 1983; Cross 2002; Davis 2002; Dundee and Rossman 1989; C. Ernst, personal observation; Ford 2002; Franklin 2006; Gibbons and Semlitsch 1991; Glaudas and Wagner 2004; Gloyd and Conant 1990; Goodman 1958; Guidry 1953; Hall 2008; Hamel 1996; Hamilton and Pollack 1955; Heinrich and Studenroth 1996; Himes 2003, 2004; Kauffeld 1957; Keiser 1993; Klimstra 1959; Kofron 1978; Langford and Borden 2007; Laughlin 1959; Leavitt 1957; Lillywhite et al. 2002; Lillywhite and Zaidan 2004; Lutterschmidt et al. 1996a; Mitchell 1986, 1994; Neill 1947, 1968; Palis 1993; Palmer and Braswell 1995; Penn 1943; Petzold and Engelmann 1975; Platt and Rainwater 2000; Price and LaDuc 2009; Rainwater et al. 2006; Roth et al. 2003; B. A. C. Savitsky 1992; Smith and List 1955; Stabler 1951; Trauth and McAllister 1995; Trowbridge 1937; Vincent et al. 2005; Wharton 1969; Williams et al. 2004; Yerger 1953). Other cottonmouths were consumed in both the wild and captivity, and both alive and as carrion.

From the many types of prey listed above, one forms the opinion that this snake does not care what it eats as long as it moves! This is not entirely correct, as it consumes carrion (Hall 2008) and has been known to refuse some types of prey in captivity. Neill (1947) noted that a captive *A. p. piscivorus*, a good feeder, ignored toads (*Bufo* ?) and an *Eumeces laticeps* presented to it.

Sometimes cottonmouths eat too much at one time. In one instance, a captive ate a *Nerodia taxispilota* about 15 cm longer and 230 g heavier than itself, but could not crawl and died the next day (Allen and Swindell 1948). Another swallowed 9 fish in 85 minutes (Bothner 1974), and we observed a young Florida *A. p. conanti* swallow a full-grown cotton rat (*Sigmodon hispidus*).

A foraging cottonmouth usually swims with its head elevated above the water. It occasionally explores pools with its head submerged (Bothner 1974; C. Ernst, personal observation), but, according to B. A. C. Savitsky (1992), such behavior is uncommon and the snake is mostly unsuccessful in capturing live or dead fish underwater. However, she reported *A. piscivorus* to be reasonably successful in capturing and handling live fish and that no live fish were lost after initial capture (presumably near the surface). Many times the fish's head is not found and it is swallowed tail first or eventually abandoned. Much of its prey capture occurs at the land-water interface, and cottonmouths that have captured prey in the water tend to carry it toward the shore before ingesting it. Several may congregate under wading bird rookeries and eat an occasional young bird or fish remains that fall from the nests (Lillywhite et al. 2002; Wharton 1969). The snakes are apparently drawn to the rookeries by odors, and crawl slowly with much tongue-flicking among the debris under the nests. They sometimes investigate and grasp objects like plant materials, bones, feathers, or soil that have had contact with avian excreta or fish fluids. The ingested wood reported by Goodman (1958) probably was either taken because it smelled of fish or was secondarily swallowed while taking other prey.

An odd feeding strategy was observed by Studenroth (1991). An *A. p. leucostoma* hooked and anchored its tail around the top of a pitfall trapline bucket and lowered itself into the bucket. It then swirled its head around in the debris at the bottom of the bucket several times as if searching for prey. More than 200 amphibians had been caught in the buckets of this trapline so the snake may have been attracted by amphibian odors.

Odor, sight, and heat radiation (from birds and mammals) are used to detect prey. The snake apparently does not use groundborne vibrations, like from fish remains falling from island bird rookeries, to detect carrion prey (Young et al. 2008). If in the water when prey is identified, the snake quickly swims to it, bites it, and may retain it if large (O'Connell et al. 1981). Small fish are bitten and swallowed immediately. On land, it either ambushes prey, or actively forages and pursues prey when detected. Small individuals have been observed to tail-lure potential prey (Carpenter and Gillingham 1990; Pycraft 1925). Kardong (1982b) reported that if more than one mouse is presented to a cottonmouth in close succession, the first is released after the bite, but mice subsequently bitten may be held.

Ambush site selection is nonrandom and strongly influenced by prey location as well as significant habitat features like cover and water depth. Cover may aid in camouflage from potential predators (Barber et al. 2005). Cottonmouths often occupy sites with a physical configuration that allows for more successful ambush predation. They use a coiled ambush posture when on a solid substrate, but a more open one with sigmoid flexures when ambushing near open or deeper water; they may chemically mark ambush sites by rubbing their faces against emergent objects (Andreadis 2004).

The following sequence of events occurs during a cottonmouth's predatory strike: (1) Search. (2) Approach—Tongue-flicking and the breathing rate increase; then the snake moves in loose curves after the prey, finally forming tight coils when near it. (3) Glide—A slight opening of the body curves allows the head to move toward the prey; elevated tongue-flicking and breathing continue. (4) Strike—A rapid straightening of the curves in the neck and trunk rapidly accelerates the head toward the prey; the jaws are extended to maximum gape just before prey contact. The upper jaw and cranium are rotated upward at least to a 45° angle, while the lower jaw is depressed about 20° in relation to the lateral trunk axis. The maxillary bone rotates the fangs to almost a 90° angle with the raised cranium and upper jaw. (5) Bite—The fangs contact the prey, penetrate, and are withdrawn. The tip of the upper jaw with fangs extended is rotated downward; the cranium and upper jaw move forward and downward on the prey as the rear of the cranium moves upward to drive the fangs deep into the prey. (6) Release—The fang and teeth disengage from the prey by a rapid opening of both jaws, and the head is withdrawn (Kardong 1975).

The predatory strike may be very fast; Walsh et al. (2005) witnessed an *A. p. leucostoma* strike and catch a jumping *Rana sphenocephala* in midair. Predatory strike acceleration and velocity do not differ whether on land (mice) or in water (fish), but instead are positively correlated with initial prey distance. However, the kinematics of terrestrial and aquatic strikes differ significantly in several aspects: maximum gape angle during the retraction phase, angular velocity of mouth closing during the strike, and the initial head angle before the strike. Strike success differs significantly between the two strike types; terrestrial strikes are much more successful than aquatic ones, possibly due to the relatively slower mouth-closing velocity in water (Vincent et al. 2005).

If the prey is large and struggles, it may be held in the mouth until the venom immobilizes it. This is dangerous for the snake, which may be chewed by its prey; we had an adult *A. p. leucostoma* have the side of its face and tongue destroyed while holding a small rat. Striking and tasting prey triggers a chemosensory search pattern, causing the snake to trail released prey with much tongue-flicking (Chiszar et al. 1979b, 1986b, 1991; O'Connell et al. 1981). Juveniles may use their yellow tails to lure frogs and other small prey (Wharton 1960).

Some potential prey avoid the cottonmouth when possible. Himes (2003) studied avoidance of the odor of *A. p. leucostoma* by *Nerodia sipedon* using a *Y* maze. A significant number of the watersnakes recognized and selected the odorless arm of the maze over that containing the odor of the cottonmouth.

The following bacteria have been isolated from the anterior digestive tract (mouth and esophageal tract) of two wild Texas *A. p. leucostoma*: *Acinetobacter baumannii; Citobacter freundii; Enterobacter cloacae; Klebsiella oxytoca, K. pneumonia; Kluyvera* sp.; *Providencia alcalifaciens; Pseudomonas putida*; and *Salmonella* sp. Four captives yielded *Citrobacter* sp., *Enterobacter cloacae, Klebsiella pneumonia*, and *Providencia rettgeri* (Lutterschmidt et al. 2007). Parrish et al. (1956) isolated the following bacterial flora from the mouth and venom glands of 25 North American pitvipers, including 10 *A. piscivorous*: *Aerobacter aerogenes, Clostridium* sp., *Corynebacterium* sp., *Escherichia coli, Micrococcus pyogenes* (variety, *aureus*), *Paracolon* sp., *Proteus vulgaris, Pseudomonas aeruginosa,* and *Streptococcus* sp. In addition to these bacteria, Hill et al. (2008) identified *Bacteroides fragilis* (and other *Bacteroides* sp.); *Fusobacterium* sp.; *Lactobacillus fermentum, L. hammerii, L. lactis;* and *Para bacteroides* sp. from intestines of the snake. The role these bacteria play in digestion is unknown. Jaksztien and Petzold (1959) isolated the following bacteria (identified only to genera) from the digestive tract of *A. piscivorus*: *Arizona, Charrau, Oranienburg*, and *Salmonella*.

VENOM DELIVERY SYSTEM: The cottonmouth's fang-strike mechanism is solenoglyphous. The maxillary bone's role in fang development and the strike is discussed by Kardong (1974) and A. H. Savitsky (1992). Mean FL for 210 *A. piscivorus* examined by Ernst (1962, 1965, 1982b) was 5.13 (range, 1.7–11.0) mm. FL is positively correlated with both TBL and HL ($FL = 0.008TBL + 0.634$, $r = 0.97$; $FL = 0.176HL + 0.045$, $r = 0.98$; mean TBL/FL = 141.4%; mean HL/FL = 6.86%); see the comments regarding allometry of head shape presented under "Growth and Longevity." No sexual dimorphism in FL exists (Ernst 1964). Adults (<60 cm TBL) have 6.03–11.0 mm fangs. Within the *Agkistrodon* complex, only *Deinagkistrodon acutus* (17.0 mm) and *Calloselasma rhodostoma* (15.1 mm) have achieved longer adult fangs. Neonates have mean FLs of 2.7 mm and fully developed venom glands (Ernst 1982b). The estimated range of change in fang curvature between juvenile and adult snakes is 63–74° (Ernst 1982b). Zamudio et al. (2000) reported a mean retracted interfang distance of 12.9 (range, 4.9–19.6) mm for 20 *A. piscivorus* with TBLs of 26.1–113.8 cm. The normal durations of fang contact during predatory strikes on a mouse and a rat by an adult cottonmouth are 0.21 seconds and 0.09 seconds, respectively (Hayes et al. 2002) and on humans by 50–92 cm SVL snakes 0.20 seconds (Rehling 2002).

At birth, the snake contains an entire series of 5–7 (in only 3% of individuals) replacement fangs (greater than the 4–5 reported by Allen and Swindell 1948), situated in sockets on each maxillary bone behind the functional fang. The functional fangs are shed periodically, and usually are replaced on one side at a time; 19% of the cottonmouths examined were replacing fangs (both sides, 1.9%; right side only, 9.0%; left side only, 8.1%) (Ernst 1982b). Replacement takes about five days, and during this time venom may be ejected through both the old and the new fangs, depending on the stage of development. We have found several cottonmouths with four functional fangs.

VENOM AND BITES: Several different amounts of the venom yield of *A. piscivorus* have been reported. It may contain 80–237 mg of venom (Brown 1973; Ernst and Zug 1996; Minton and Minton 1969; Russell and Brodie 1974; Russell and Puffer 1970); the record is 1,094 mg (4 mL) in a single extraction from a 152 cm TBL snake by Wolff and Githens (1939a). They (1939a, 1939b) reported the average yield per extraction of 0.55 mL with a dry weight of 158 mg; Altimari (1998) thought the average yield more likely to be 75 mg. The total amount of venom expelled during a single predatory bite on an adult mouse by 15.2–17.5 cm SVL cottonmouths averages 14 (range, 0–58) mg (Gennaro et al. 1961; Hayes et al. 2002); during a defensive bite by 50–92 SVL individuals it is 23 (range, 8–39) mg (Herbert 1998; Rehling 2002).

Cottonmouth venom contains 23–35% solids (Wolff and Githens 1939b), and 68% of its protein is contained within these solids (Brown 1973). It is moderately toxic; very hemolytic, destroying red blood cells and exhibiting strong overall anticoagulant activity; and highly proteolytic, with both fibrinogenolytic and fibrinolytic activity. Although clotting at the bite site has occurred, the venom's amino acid esterase has thrombin-like activity, but only cleaves fibrinopeptide B from fibrinogen, and so lacks significant procoagulant capability. Bradykinin is released, causing high phospholipase A activity.

Several components have been isolated from cottonmouth venom (most are enzymes): applaggin (platelet aggregation inhibitor), arginine ester hydrolase, ß-fibrinogenase, collagenase, disintegrins, hemorrhagins, hyaluronidase, kallikrin-like enzyme, L-amino acid oxidase, NAD-nucleotidase, 5′-nucleotidase, phospholipase A_2 (Lys49), phosphomonoesterase, phosphodiesterase, plasminogen activators, thrombin-like enzyme, and vascular endothelial growth factor receptor-binding protein. Data on the chemistry and biochemical actions of the venom are found in Aird et al. (1988), Arni and Ward (1996), Bonilla and Horner (1969), Brown (1973), Chao et al. (1989), Dhillon et al. (1987), Faria Ferreira et al. (1999), Flowers (1966), Gold et al. (2002), Hadidian (1956), Holland et al. (1990), Jia et al. (2008), Komori and Nikai (1998), Lomonte et al. (1999, 2003a, 2003b), Maraganore and Heinrikson (1986), Mebs and Samejima (1986), Moran and Geren (1979), Nikai et al. (1988a, 1988b), Norris (2004), Núñez et al. (2001), Ownby (1998), Ownby et al. (1999), Pichon-Prum et al. (1990), Ramírez et al. (1999), Richards et al. (1965), Russell and Brodie (1974), Savage et al.

(1990), Scott et al. (1986, 1992), Takahashi and Mashiko (1998), Tan and Ponnudurai (1990), Tu (1977), van den Bergh et al. (1988), Van Mierop (1976a, 1976b), Wagner and Prescott (1966), Welches et al. (1993), Yamazaki et al. (2005), and Yamazaki and Morita (2004).

The virulence of cottonmouth venom has been tested on several experimental animals. The mean LD_{50} of a mouse was reported as 4.00 (range, 3.42–4.68) mg/kg intravenous and 5.11 (range, 4.49–5.82) mg/kg intraperitoneal by Russell and Brodie (1974), Russell and Emery (1959), and Russell and Puffer (1970); 80 mg intravenous, 102 mg intraperitoneal, and 516 mg subcutaneous for a 20 g mouse by Minton and Minton (1969); and 2.04 mg/kg by Ernst and Zug (1996). Arce et al. (2003) reported mean intraperitoneal and intravenous LD_{50} values of 6.24 (range, 4.72–8.26) and 5.13 (range, 4.12–6.40), respectively, for 16–18 g mice. Pollard et al. (1952) reported it as 28–45 mg/kg for rats, and that dried venom remained relatively stable and may even become more toxic over time when stored at 15–20°C. The venom's phospholipase A_2 produces myonecrosis (Lomonte et al. 1999; Mebs and Samejima 1986). At envenomation, mammals exhibit an immediate but transitory constriction of the pupil (Gennaro and Casey 2005). The opossum (*Didelphis virginiana*) is apparently immune to the venom of *A. piscivorus* (Pérez and Sánchez 1999).

Russell and Emery (1959) reported the mean LD_{50} of a chick to be 0.91 (range, 0.76–1.12) mg/kg intravenous and 1.26 (range, 1.03–1.54) mg/kg intraperitoneal. The minimum lethal dose for a 350 g pigeon is 0.06–0.09 g; the normal amount of venom in an extraction could probably kill about 15,000 pigeons (Wolff and Githens 1939a).

Swanson (1946) injected the venom into 6 other cottonmouths and 3 juvenile *A. contortrix*. Two cottonmouths, 56 cm and 58 cm TBL, survived; but 4 juveniles, 24.4–26.9 cm TBL, died, as did the 3 copperheads, 22.5–35.5 cm TBL. When injected with the venom of *A. contortrix mokasen* and *A. c. laticinctus*, all 3 23.5–25.9 cm TBL cottonmouths died.

The LD_{50} of cottonmouth venom declines in adult bullfrogs (*Rana catesbeiana*); during metamorphosis the LD_{50} increases, but adults still have higher sensitivity (8.0 mg/kg) than the original level of tadpoles (2.0 mg/kg). Survival is longer in postmetamorphic juveniles than in adults after envenomation (Heatwole et al. 1999). When envenomated, a *Rana* frog experiences immediate and total constriction of the lung sacs, and loss of integrity of Bowman's capsules and of the capillaries of the glomeruli in the kidneys (Gennaro and Casey 2005; Green et al. 2004).

Langford and Borden (2007) reported that a wild 100 cm TBL three-toed amphiuma (*Amphiuma tridactylum*) in a funnel trap survived several cottonmouth bites on its head and midbody (during which the observers thought that the snake had expelled most of its venom), and in return bit the snake hard and shook it side to side, scraping dorsal scales off and breaking several of the snake's ribs. When released from the trap, the cottonmouth swam slowly away. Within seconds the amphiuma's head swelled to twice its normal size and the bites on the body swelled to golf ball size. It was transported to the laboratory and placed in a large tank. The bites secreted copious amounts of mucus for several hours after the incident, and remained swollen for two days, while the amphiuma remained motionless on the bottom of the tank, except to surface to breathe every 10 to 15 minutes. It fully recovered and was released on day 5. It is not known if the cottonmouth survived the amphiuma's retaliatory attack.

Burkett (1966) estimated that approximately 30% of the annual envenomations by snakes in the United States are by *A. piscivorus*; however, this is probably too high an estimate. Of 2,836 persons hospitalized for snakebite treatment for which detailed records were available, only 208 (7.3%) were bitten by cottonmouths (Parrish 1966), and only 1 of 48 bites during academic research was by a cottonmouth (Ivanyi and Altimari 2004).

Humans have experienced the following symptoms during cottonmouth envenomation: bruising and necrosis at the bite site, chills and sweating, difficulty in breathing, a drop in BT, general weakness, giddiness, heart failure, hemolysis, hemorrhage, lowered blood pressure and weakened pulse, muscle twitching, nausea and vomiting, nervousness, occasional paralysis, swelling, pupil contraction, tissue necrosis, pain (minimal to severe) at the site of the puncture wounds, and unconsciousness or stupor (Burch et al. 1988; Burkett 1966; Essex 1932; Hutchison 1929). Nasty secondary bacterial infections may also occur, like tetanus or gas gangrene (sometimes resulting in amputation of a limb, toes, or fingers) (Allen and Swindell 1948; Banner 1988; Burch et al. 1988; Dart et al. 1992b; Hulme 1952; McCollough and Gennaro 1968; Norris 2004; Roberts et al. 1985; Russell 1983; Watt 1985). Case histories of bites were published by Burkett (1966), Hulme (1952), and Shelton (1996).

The estimated lethal dose for a human is 100–150 mg (Minton and Minton 1969). Brown (1973) estimated that the subcutaneous lethal dose for a 70 kg human is 1,806 mg/kg × 70. Fatalities have occurred (Allen and Swindell 1948; Anderson 1965; Burkett 1966, Hutchison 1929). The mortality rate is about 17% (Willson 1908).

PREDATORS AND DEFENSE: Juveniles experience the most predation, but adults are not immune from attacks. Known predators (mostly of juveniles) are ghost crabs (*Ocypode quadrata*); fire ants (*Solenopus invicta*); fish (*Ictalurus* sp., *Lepisosteus occeus, Micropterus salmoides*); bullfrogs (*Rana catesbeiana*); snapping turtles (*Chelydra serpentina*); snakes (*Agkistrodon piscivorus, Drymarchon corais, Lampropeltis getula*); alligators (*Alligator mississippiensis*); birds—cranes (*Grus* sp.), egrets (*Casmeroides albus, Egretta thula*), herons (*Ardea herodias*), storks (*Mycteria americana*), crows (*Corvus* sp.), loggerhead shrikes (*Lanius ludovicianus*), owls (*Bubo virginianus*), turkey vultures (*Cathartes aura*), hawks (*Buteo* sp.), and eagles (*Haliaeetus leucocephalus*); and predatory mammals (*Canis familiaris, Felis catus, Lontra canadensis, Lynx rufus, Procyon lotor*) (Allen and Swindell 1948; Burkett 1966; Campbell and Lamar 2004; Camper 2005; Cross and Marshall 1998; Klauber 1972; Lönnberg 1894; Mitchell 1994; Penn 1943; Rainwater et al. 2007; Tompkins *in* Bent 1937). B. A. C. Savitsky (1992) thought large sunfish (*Lepomis* sp.) capable of preying on small cottonmouths.

Cottonmouths are large, sometimes aggressive, dangerous snakes. Individual cottonmouths vary in disposition from very timid to extremely aggressive; defense behaviors do not vary between the sexes. All bite if handled (Gibbons and Dorcas 1998), and they can and will bite underwater (Shelton 1996). Duration of fang contact averages 0.2 (range, 0.06–0.26) seconds (Herbert 1998). When first disturbed they either freeze in place or try to escape. If escaping is not possible, the snake will coil, flatten its body, flip over, vibrate the tail, and strike repeatedly (Carpenter and Gillingham 1975; Marchisin 1980; Neill 1947). In field tests conducted on South Carolina *A. p. piscivorus* by Gibbons and Dorcas (2002), 51% tried to escape, 78% used threat displays and other defensive tactics, 33% vibrated their tails, and 24% emitted detectable musk; 36% bit an artificial hand offered them. From these results, Gibbons and Dorcas thought the snake's aggressive reputation overrated. Striking behavior is independent of tail vibrations (Glaudas and Winne 2007), and is often the last act of defense (is venom too precious to waste?). However, if handled, cottonmouths thrash about violently, strike, and spray musk; they may even bite themselves. Although many are bluffers, because they are common and some do have violent tempers, cottonmouths should be approached with care and not handled.

Another frequently reported threat display is its habit of mouth-gaping to show the inner pinkish-white lining, from which it gets its common name cottonmouth. Schuett (*in* Gloyd and Conant 1990) has reported that gaping is probably innate, as neonates do it, and that it is more frequently displayed by cool (BT <18°C) snakes (no relationship between BT and the tendency to bite was noted by Gibbons and Dorcas 2002). The display may be given by snakes that are either coiled or lying spread out on the ground (C. Ernst, personal observation). Gibbons and Dorcas (2002) recorded mouth-gaping in only 64% of the *A. p. piscivorus* that did not flee. We have not found this threat display to be common in *A. p. leucostoma*; certainly less than 50% of the several hundred wild western cottonmouths that we encountered have mouth-gaped. However, cottonmouths that mouth-gape are more likely to strike than those that do not (Glaudas and Winne 2007).

Odor probably also plays a major role in predator detection, and is mediated by the snake's vomeronasal system (Miller and Gutzke 1999), but visual cues are the most important stimuli for initiating a strike. Thermal cues do not influence the chances that the snake will strike, the number of times it will strike, or the latency of the strike (Glaudas and Gibbons 2005).

Approach by humans and confinement elicit stress responses involving significant releases of corticosterone; such responses do not occur to low-level disturbances (Bailey et al. 2009). Odors or sight of ophiophagus snakes, like *Lampropeltis getula*, result in defensive behaviors (Carpenter and Gillingham 1975; Gutzke 2001), and often result in body-bridging, where the cottonmouth turns its body sideways to the potential attacker and raises a single body loop off the ground. Such defensive displays occur both in the wild and in the laboratory, and in all age classes and both sexes. The response is greater in juveniles (66.7% response) and subadults and adults (81.3%) than in neonates (46.6%), but, surprisingly, only 7.7% of large adults responded (Gutzke 2001). Possibly some experience is involved. During tests run by Glaudas et al. (2006) neonate and adult *A. p. piscivorus* were repeatedly challenged with potential threats. The two groups differed in their tendencies to habituate; neonates did not decrease their defensiveness, but adults did. Adults exhibited habituation of strike components but not warning displays. In agreement, Roth and Johnson (2004) found that defensive responses by *A. p. leucostoma* declined significantly with increasing SVL, but, after controlling for body size, no differences between the sexes were detected. These results support the hypothesis that there may be ontogenetic differences in predator perception.

Glaudas (2004) studied the behavior of *A. p. piscivorus* during and after handling by humans. His snakes changed significantly in defensive behavior between day 1 and day 5 of the experiment, noticeably calming and making fewer displays or bites, and they did not significantly revert to their original behavior 16 days later. Nevertheless, those who have kept individuals of all three subspecies in their laboratories can attest that this does not always occur, and that individual cottonmouths have different temperaments, with some never calming down. A >75 cm SVL one that we had for several years could not even be removed while its cage was being cleaned without it repeatedly striking and spraying venom on us and all over the floor and nearby objects; needless to say, it was handled with extreme care.

PARASITES AND PATHOGENS: Cottonmouths are the hosts of several endoparasitic helminths: Trematoda—*Leptophyllum tamiamiensis; Ochetosoma ancistrodontis, O. aniarum, O. kansense, O. laterotrema, O. monostruosum, O. septicum; Styphlodora agkistrodontis, S. aspina, S. bascaniensis,* and *S. simplex;* Cestoda—*Ophiotaenia* sp.; *Proteocephalus agkistrodontis, P. grandis, P. marenzelleri,* and *P. perspicua;* Acanthocephala—*Macracanthorhynchus ingens;* and Nematoda—*Capillaria heterodontis; Cosmocercoides dukae; Gnathostoma procyonis; Hexametra boddaërtii; Kalicephalus inermis, K. rectiphilus; Physaloptera obtussima; Rhabdias eustreptos; Strongyloides gulae, S. serpentis;* and *Terranova caballeroi;* and we have found a leech (*Placobdella* sp.) in the mouth of one. In addition, the linguatulid tongue worm, *Porocephalus crotali,* is known to parasitize *A. piscivorus* (Ash 1962; Barrios 1898; Bennett 1938; Bowman 1984; Brooks 1978; Burkett 1966; Byrd and Denton 1938; Byrd et al. 1940a, 1940b; Byrd and Roudabush 1939; Collins 1969; Detterline et al. 1984; Dyer and McNair 1974; Elkins and Nickol 1983; Ernst and Ernst 2006; Esslinger 1962; Fontenot and Font 1996; Foster et al. 2000; Freze 1965; Harwood 1932; Hughes et al. 1941; Kuzmin et al. 2003; Leiper and Atkinson 1914; Little 1966; MacCallum 1921; McIntosh 1939; Rabalias 1969; Railliet 1899; Riley and Self 1979; Roberts 1956; Schad 1962; Schmidt and Hubbard 1940; Sogandares-Bernal and Grenier 1971; Sprent 1979; Zehnder and Mariaux 1999).

Karstad (1961) found the eastern equine encephalomyelitis arbovirus in the serum of a cottonmouth, and Jaksztien and Petzold (1959) reported one infected with *Salmonella* bacteria; so the snake is a potential reservoir host for these diseases. Neoplasms and other tumors have been reported from cottonmouths by Cowan (1968), Effron et al. (1977), Ippen (1972), and Wadsworth (1960).

POPULATIONS: *A. piscivorus* may be the most common reptile in some habitats. At certain Illinois hibernacula, *A. p. leucostoma* formerly occurred in the hundreds (Perkins *in* Sexton et al. 1992), and at Murphy's Pond, Hickman County, Kentucky, it formerly occurred in densities of more than 700/ha (Barbour, 1956). We have seen up to 20 individuals of this subspecies basking on the same log or driftwood cluster along the Mississippi River at a site in Union County, Kentucky. Hill and Beaupre (2008) captured 142 *A. p. leucostoma* 283 times from August 1996 through September 2003 at a riffle-pool creek system in northwestern Arkansas. Viosca once collected 114 western cottonmouths in one day on Delacroix Island, St. Bernard Parish, Louisiana (*in* Dundee and Rossman 1989), and Penn (1943) collected 25 in a few hours on 6 consecutive days at a site in Cameron Parish. *A. p. leucostoma* composed 8.6% of the snakes collected by Clark (1949) in the Louisiana uplands, and Fitch (1949) reported the subspecies abundant along some small woodland streams in western Louisiana. It made up 15.1% of the snakes captured by Ford et al. (1991) at a site in northeastern Texas. The number of snakes and their density at the latter locality varied from 130–170 individuals and 4.04–5.28/ha, respectively, from 1984 to 1997 (Ford 2002). At still another 2,300 ha bottomland deciduous woodland site in northeastern Texas, 128 cottonmouths were collected in 1998–2000, making the Western Cottonmouth the third most commonly captured snake; it comprised 20.6% of the total recorded snake population of 14 species (Ford and Lancaster 2007).

A. p. conanti comprised 4.3% and 12.6%, respectively, of the snake assemblages at 2 sites in Florida's Everglades (Dalrymple et al. 1991b), and we have seen as many as 8 *A. c. conanti* basking on a fallen cyprus tree in Collier County, Florida.

Immature individuals made up 32.5% of *A. p. leucostoma* examined by Burkett (1966) and about 45% of the Murphy's Pond population (Barbour, 1956). Females comprised 53% of adults and also of 48 embryos examined by Burkett (1966); but the subspecies had essentially even adult sex ratios in populations in northeastern Texas (Ford 2002) and northwestern Arkansas (Hill and Beaupre 2008). Blem (1981) reported a 1.8:1.0 male to female ratio in Virginia *A. p. piscivorus*.

Ford (2002) calculated the survival rates for a small cohort of neonates that were pit-tagged and released in 1989 and for another in 1990 at his northeastern Texas study site. Of 12 snakes released in 1989, 3 were captured in 1990 and 1 in 1992, yielding a minimum survival rate of 30% for the first year. Of 6 released in 1990, none were collected in 1991, but 3 were eventually found again, bringing the annual survival rate of the small snakes to 50%. Individuals with SVLs of 35–50 cm had annual survival rates of 33–89%, and adults (SVL >50 cm) had annual survival rates of 75–100%. Some snakes were not recaptured until several years had passed, so actual annual survival rates could be higher. No sexual differences in survival rates were evident.

Over most of its range, populations of *A. piscivorus* are sustaining, and some are doing well. It is probably the most common venomous snake in most areas of the South. The species does, however, face challenges in the future (Conant 1992). Habitat destruction by humans (Palmer 1971), particularly in Florida, is the most serious threat; but natural phenomena like drought (Cypert 1961; Ford 2002), hurricanes and other strong flooding storms, and the continuing invasion by seawater into the freshwater areas of the Mississippi Delta also destroy habitat. Fires, natural or manmade, contribute to the death of cottonmouths caught on land. The continued chemical pollution of our water bodies is also a serious problem. Residues of polychlorinated biphenyls and other pesticides (aldrin, DDD, DDE, DDT, dieldrin, endrin, heptochlor, methoxychlor, toxaphane) and heavy metals and other chemicals (arsenic, copper, chromium, mercury, nickel, selenium, zinc) have been isolated from *A. piscivorus* (Clark

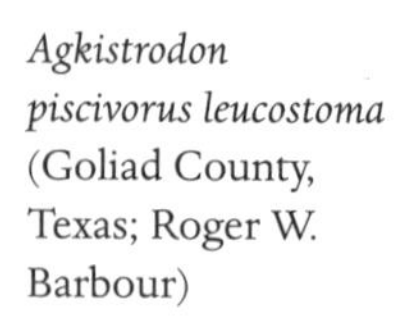

Agkistrodon piscivorus leucostoma (Goliad County, Texas; Roger W. Barbour)

Agkistrodon piscivorus leucostoma (R. D. Bartlett)

et al. 2000; Ford and Hill 1991; Hall 1980; Munro 1949a; Rainwater et al. 2005; Tiersch et al. 1990). One *A. p. leucostoma* examined by Rainwater et al. (2005) contained the highest mercury concentration (8.61 ng/g) yet reported for a snake.

Construction of roads fragments habitat, and the resulting increase in motorized traffic causes more cottonmouths to be killed while crossing the roads (Andrews and Gibbons 2005). Random killings and collections for the pet and venom trades are probably of negligible impact at this time. Since the alligator has been protected and automobile traffic has increased, there has been a noticeable decrease in cottonmouth populations at certain areas of Florida; but periods of severe drought have occurred in the last two decades and many wetlands dried completely, and this may have played a role in the decline in cottonmouth numbers (C. Ernst, personal observation). The species is protected as threatened in Indiana.

Laughlin and Wilks (1962) used sodium pentobarbital to anesthetize *A. p. leucostoma*, and recommend its use in population and dietary studies when it is necessary to handle the snake.

REMARKS: *A. piscivorus* was reviewed by Campbell and Lamar (2004) and Gloyd and Conant (1990).

Agkistrodon taylori Burger and Robertson, 1951
Taylor's Cantil
Metapil

RECOGNITION: This Mexican east coast representative of *Agkistrodon* is very similar to *A. bilineatus*, of which it was once considered a subspecies. The following description is a compendium from Campbell and Lamar (2004), Gloyd and Conant (1990), and our own experience.

Agkistrodon taylori (R. D. Bartlett)

Agkistrodon taylori, juvenile (Tamaulipas; Houston Zoo, John H. Tashjian)

It is a thick-bodied snake with a TBL_{max} of 96 cm, but most individuals are shorter than 85 cm (an estimate of TBL of 137 cm is considered unreliable; Logan *in* Gloyd and Conant 1990). The tail is relatively long (see below), but shorter than that of *A. bilineatus*. Body color changes with age. Adult ground color varies from grayish (sometimes with a lavender hue), brown, olive-brown, reddish-brown, orangish-red, maroon, to black; juveniles are more brightly colored. A series of 11–16 (mean, 13) gray, brown, to olive-brown or black bands cross the dorsum. The bands are bordered with irregular white, yellow, or orangish flecks. The tip of the long, rattleless tail, at least in juveniles, is light-colored, usually greenish-gray. The venter is dark gray to black with white, yellow, or orange markings. A broad, medial, light bar runs downward from the rostrum to the upper lip. Laterally, the face is patterned with two light stripes. The first, a dorsal one, extends backward from the rostrum over the orbit to the level of the corner of the mouth and then turns obliquely downward onto the neck. The lower, second one, is prominently broad and runs from either the tip of the rostrum or from below the naris backward over the supralabials to the corner of the mouth, where it turns downward before continuing posteriorly on the neck to often contact the downturned upper light stripe. The keeled body scales are as described for *A. bilineatus*; the lowermost two rows may be smooth. Dorsal scale rows are as follows: 25 (23) anterior, 23 (21) midbody, and 21 (19–20) posterior. The underside contains 127–138 ventrals, 45–61 subcaudals, and an undivided anal plate. Dorsal head scalation is composed of the same 9 enlarged plates as on the other 3 species of *Agkistrodon*. Laterally, there are 2 nasals, 1 loreal, 2 (3) preoculars, 3–4 (2–5) combined suboculars and postoculars, several rows of temporals, 8 (7–9) supralabials, and 10–11 (9–12) infralabials. On the chin lie a mental scale and 2 elongated chin shields. The pupils are vertical ellipses. The tip of the tail is downturned.

The bilobed hemipenis of *A. taylori* is similar to that of *A. bilineatus*, but is longer (13 mm to 8–10 mm), and has longer lobes (9 mm to 6–7 mm). The calyculate area of the lobes covers 50% in *A. taylori*, but 75% in *A. bilineatus*. The length of the retractor muscle is 22 mm versus 22–27 mm, and the length of the retractor fork 3–6 mm versus 4–6 mm in *A. taylori* compared to *A. bilineatus* (Malnate 1990).

Kardong (1990) describes the dentition, but does not differentiate between *A. bilineatus* and *A. taylori*. Tooth counts consist of, excluding the maxillary fang, 3–4 (5) palatines, 15 (14–17) pterygoids, and 15 (11–17) dentaries. The 2 *A. taylori* we examined had dentition falling within these ranges.

Sexual dichromatism is pronounced (Shine 1993). Females are more boldly patterned, with light cross bands containing some yellow, orangish, or reddish-brown pigment, alternating with very dark brown or black (normally) ones that are boldly patterned with yellow to orangish-red. The dorsum of the head and the cheek stripe are dark gray or reddish-brown. Males are overall more melanistic, usually black, with any bright pigmentation present restricted to the lower sides. The top of their head and the cheek stripe are blackish. In addition, females (TBL_{max}, 92.5 cm) have 130–138 ventrals, 40–47 subcaudals (19–35 undivided), and tails 13–18 (mean, 15.2) % of TBL. Males (TBL_{max}, 96 cm) have 127–137 ventrals, 45–61 subcaudals (27–46 undivided), and tails 15.3–19.0 (mean, 17–18) % of TBL.

GEOGRAPHIC VARIATION: No subspecies are currently recognized. Smith and Chiszar (2001) described a long-preserved specimen, from central Veracruz, Mexico, which they named *Agkistrodon bilineatus lemoespinali*. Bryson and Mendoza-Quijano (2007), however, showed that the majority of the scale differences listed for this taxon by Smith and Chiszar fell within the ranges of *A. taylori*, and rejected its validity.

CONFUSING SPECIES: The rattlesnakes (*Crotalus, Sistrurus*) normally have a tail rattle. *Agkistrodon b. bilineatus* can be distinguished by using the key above. *A. b. russeolus* has a much narrower lower facial stripe that does not extend downward to include the margin of the upper lip, and its narrow medial anterior stripe does not extend far onto the chin. Watersnakes (*Nerodia*) lack a facial pit and stripes, and their tail tip is not light-colored.

Distribution of *Agkistrodon taylori*.

KARYOTYPE: The karyotype has not been described, but is probably like that of *A. bilineatus*.

FOSSIL RECORD: No fossils have been reported.

DISTRIBUTION: *A. taylori* occurs along the northeastern Atlantic slope of Mexico in east-central Nuevo León, central and southern Tamaulipas, east-central San Luis Potosí, northwestern and central Veracruz, and northeastern Hidalgo (Bryson and Mendoza-Quijano 2007; Burchfield 1982; Burger and Robertson 1951; Campbell and Lamar 2004; Gloyd and Conant 1990; Martin 1958; Smith and Chiszar 2001; Tovar-Tovar and Mendoza-Quijano 2001).

HABITAT: It seems to prefer open canopy wooded, limestone areas with outcroppings, but according to Burchfield (1982), in southwestern Tamaulipas this snake inhabits ecotones between arid tropical scrub and tropical semideciduous forests. Local woodcutters told Burchfield that rock-strewn hillsides are also used. The more prevalent plants of the area include various acacias (including *Acacia berlandierii*, *A. rigidula*, and *A. wrightii*), chavaja (*Elaphrium simaruoa*), ebony (*Pithecellobium flexicaule*), mala mujer (*Jatropha multiloba*), mesquite (*Prosopus juliflora*), palms (*Sabal texana*), pineapple (*Bromelia balansae*), prickly pear cactus (*Opuntia* sp.), strangler figs (*Ficus* sp.), and tillandsia (*Tillandsia* sp.). Campbell and Lamar (2004) mention mesquite grasslands and hardwood forests as habitats. Most rainfall in the region occurs during the winter months. Habitat elevations range from near sea level to about 2,500 m (Gloyd and Conant 1990).

BEHAVIOR AND ECOLOGY: The only detailed study of life history components of *A. taylori* was published by Burchfield (1982), and we draw heavily on it below. Too many aspects of this snake's behavior, ecology, and reproduction are unknown, and we could benefit from a comprehensive study.

Seasonally, *A. taylori* is most active during the cool wet season from late October to March. It has not been reported if the animal estivates during the hot, dry summer. Rainfall brings the snake out, and it is often observed on overcast or rainy days (Campbell and Lamar 2004). It is probably more crepuscular or nocturnal during the drier parts of the year. Martin (1958) remarked that three were removed from a "den" by laborers clearing palm forest, but does not give the date or additional details.

Temperature data are few and associated with breeding in captivity. Jungnickel (2002) successfully kept and bred Taylor's cantil under an illumination regime at 900–2100 hours and ATs of 22–24°C in winter and 28–35°C in summer, with a localized hot spot of 27–38°C. Likewise, Burchfield (1982) reported the species mated and delivered young when kept at an AT of 30–34°C and 65% relative humidity.

Male-male combat, usually associated with breeding, has been observed in *A. taylori*. The males face each other, while twitching their bodies and vibrating their tails. Much tongue-flicking occurs at this time and later. They then approach each other with heads and tails raised, and intertwine their bodies, keeping the anterior portion of their bodies free and facing one another. Pushing then occurs, as one male attempts to attain a superior position over the other and pin his opponent to the ground and then lie on top of it. The loser usually quickly retreats after being pinned (C. Ernst, personal observation; Hubbard *in* Burchfield 1982).

REPRODUCTION: The exact size and age of maturity by the 2 sexes of *A. taylori* are unknown, but females with 63.8–88.6 cm TBLs have successfully bred (Burchfield 1982); Burchfield (*in* Gloyd and Conant 1990) reported minimum TBLs of 63.8 cm for captive adults of each sex at the Brownsville Zoo. Likewise, data on the reproductive cycles of both sexes are lacking.

Courtship and copulation have been observed in captives (Burchfield 1982; C. Ernst, personal observation; Jungnickel 2002). These matings occurred from 14 September to 29 February, with most occurring between November and February. Burchfield (1982) thought that this latter period may also be the mating season in the wild. The male actively pursues the female with much tongue-flicking; if receptive, she may raise her tail to expose her vent. When the female is reached, the male brings his body beside hers and begins rubbing his chin along her anterior back while crawling onto her back and aligning their heads. This is done spasmodically, with the goal of positioning his head directly over that of the female. Head-jerking consists of bobbing up and down, moving side to side, or anterior-posterior jerking. Tongue-flicking is almost constant during this stage. If the female tries to escape, the male flattens his anterior body against her to hold her down (we have not observed the convulsive thrashing described by Burchfield). The male's posterior body and tail are then moved alongside those of the female, and the male entwines their tails. This is followed by the male performing a series of caudocephalic waves and wriggling his body. Intromission soon occurs. Copulations have lasted 1.5–3.0 hours. It is not known if *A. taylori* practices long-term sperm storage.

Captive neonates have been born during the dates 6–12 May and 21 June to 21 July (Burchfield 1982; Jungnickel 2002; Peterson 1982b [1981]). Litters contain 3–11 (mean, 8) young. The young are bright cream, yellow, or salmon-colored, and have light gray to yellowish-green tail tips. Twenty-five neonates mentioned in the literature had 17.2–27.0 (mean, 23.7) cm TBLs, 11.0–23.4 (mean, 19.5) cm SVLs, 3.4–4.4 (mean, 4.2) cm TLs, and BMs of 3.1–16.0 (mean, 12.1) g. Juveniles have TLs of 16–20% of TBL (males—mean, 18.6%; females—mean, 17.2%) (Gloyd and Conant 1990).

GROWTH AND LONGEVITY: The only growth data available are those of an underdeveloped neonate born 6 May that died the next day at 18.0 cm TBL, 14.8 cm SVL, and 7.2 g (Peterson 1982b [1981]). A male was still alive on 31 May 1999 after 15 years, 7 months, and 13 days in captivity at the Houston Zoo (Snider and Bowler 1992). Other longevity data are reported in Gloyd and Conant (1990).

DIET AND FEEDING BEHAVIOR: Feces of Tamaulipan *A. taylori* contained grasshopper (Orthoptera) remains and mammal hairs; two regurgitated a Mexican spiny pocket mouse (*Liomys irroratus*) and a white-footed mouse (*Peromyscus leucopus*), respectively (Burchfield 1982). Captives have fed on hamsters (*Mesocricetus auratus*), house mice (*Mus musculus*), and brown rats (*Rattus norvegicus*) (C. Ernst, personal observation; Jungnickel 2002; Peterson 1982b [1981]). Captive neonates have fed on fish, small anurans, and newborn house mice (Peterson 1982b [1981]), and cannibalism has been observed in captivity (Burchfield *in* Gloyd and Conant 1990).

Adult hunting is probably accomplished by using the same methods and stimuli as mentioned for *A. bilineatus*; the young employ their light-colored tails to lure small animals (Strimple 1995b).

VENOM DELIVERY SYSTEM: Two male specimens with TBLs of 55.8 and 68.4 cm that we examined had FLs of 6.05 and 7.45 mm, respectively.

VENOM AND BITES: The only specific venom data for *A. taylori* were published by Possani et al. (1980), but most likely at least some of the venom studies mentioned under *A. bilineatus* included data taken from specimens of *A. taylori,* especially as at the time most were conducted, *A. taylori* was considered only a subspecies of *A. bilineatus*. Probably the general chemical composition and symptomatic effects of its venom are similar to those discussed under *A. bilineatus* (see above).

Possani et al. (1980) reported a mouse intraperitoneal LD_{50} of 2.3 µg/g, and that the venom of *A. taylori* has hemolytic, hydrolase (esterase), phospholipase, and protein activities. However, it did not have measurable hyaluronidase or a direct lytic action on human erythrocytes. They noted that their composition data were complementary to those for *A. bilineatus,* and that the venom of *A. taylori* is similar to that of *A. bilineatus*. Venom from *A. taylori* shows at least 12 electrophoretic bands at alkaline pH, but only 2 bands at acidic pH. Burchfield (1982) reported an intraperitoneal LD_{50} of 47 per 20 g mouse.

A captive male *A. taylori* was apparently killed by the bite of another male during a combat bout that caused severe hemorrhaging (Hubbard *in* Burchfield 1982). Case reports of human envenomations by this species have not been published, but the hemolytic symptoms are probably like those caused by the venom of *A. bilineatus*. Although no fatalities have been reported, *A. taylori* should be respected.

PREDATORS AND DEFENSE: No data are available concerning predation on this snake. It is aggressive and hot-tempered, and behaves defensively like *A. bilineatus*.

PARASITES AND PATHOGENS: None have been reported for *A. taylori*.

Agkistrodon taylori, juvenile (R. D. Bartlett)

POPULATIONS: Again, no detailed information is available. Depending on the season, *A. taylori* may be common or scarce at the same locality. Peterson (*in* Gloyd and Conant 1990) found 2 dead within 3 m of each other within 45 minutes near where a small stream crossed under a road in Tamaulipas, and workers found 3 together at another site in that Mexican state (Martin 1958), but the vast majority of reports involve individuals.

Like *A. bilineatus*, this species faces a bleak future due to habitat destruction for agriculture. Killing by humans and collecting for the pet trade seem to be having negligible effects on the various populations.

REMARKS: Using both mtDNA and tRNA, Parkinson et al. (2000) found enough points of separation from *A. bilineatus*, along with its allopatry from that snake, to elevate *A. taylori* from a subspecies of it to a full, separate, species.

Sistrurus Garman, 1883
Pygmy Rattlesnakes
Viboritas de Cascabeles

Molecular, morphological, and serological studies have all shown the genus *Sistrurus* is most closely related to the larger rattlesnakes of the genus *Crotalus* (Beaman and Hayes 2008; Bushar et al. 2001; Cadle 1992; Dowling et al. 1996; Gutberlet and Harvey 2002; Minton 1992; Murphy et al. 2002; Parkinson et al. 2002). Klauber (1972) thought *Sistrurus* ancestral to *Crotalus*. Within the *Agkistrodon* complex, *Sistrurus* seems closest to *Deinagkistrodon* (Knight et al. 1992).

The earliest fossil remains of the genus *Sistrurus* are trunk vertebrae of an unidentified species from late Miocene (Clarendonian NALMA) in Brown County, Nebraska (Parmley and Holman 2007); so these snakes have been present on the central Great Plains for at least five million years. The genus probably evolved in Mexico and later migrated north to the United States (Klauber 1972). Parmley and Holman (2007) reviewed the vertebral characteristics.

Gibbs and Rossiter (2008) described the rapid evolution of the PLA_2 venom genes isolated from genomic DNA of *Sistrurus*. Their development was possibly related to divergent diets among the species; please see their paper for details.

KEY TO THE SPECIES OF *SISTRURUS*

1a. Prefrontal scale not in contact with loreal scale, preocular scale in contact with postnasal scale. . . . . *catenatus*

1b. Prefrontal scale in broad contact with loreal scale, preocular scale not in contact with postnasal scale . . . . *miliarius*

Sistrurus catenatus (Rafinesque, 1818)
Massasauga
Cascabel de Massasauga

RECOGNITION: The massasauga has a TBL_{max} of 100.3 cm, but most individuals are shorter than 55 cm. It is gray to light brown with a dorsal series of 21-50 (mean, 40) dark brown to black blotches, 3 rows of small brown to black spots on each side of the body, and the tail with alternate dark and light bands. Some individuals may be melanistic (Harris 2006), and rarely some are striped (Lardie and Lardie 1976). Individuals from loamy plains are usually grayish-brown, while those from Arizona have a more reddish appearance. Occasionally melanism or albinism occurs. The black venter is mottled with yellow, cream, or white marks, or it may be nearly all black. A dark, light-bordered stripe runs backward from the eye, and another dark mid-dorsal, light-bordered stripe extends posteriorly on the back of the head. Dorsal body scales con-

Sistrurus catenatus catenatus (Cincinnati, Ohio; Dallas Zoo; John H. Tashjian)

Sistrurus catenatus catenatus (Case County, Michigan; James H. Harding)

tain apical pits and are strongly keeled (Gray 2006; Harris 2005), and lie in 23 (24–27) anterior rows, 23 or 25 (21–27) midbody rows, and 17–19 rows anterior to the anal plate (the skin is shed approximately every 6 months; Stabler 1939). Ventral scales total 129–160, subcaudals 19–36, and the anal plate is complete. Cephalic dorsal scales include a higher than wide rostral scale, followed by 9 enlarged plates—2 internasals, 2 prefrontals, 1 large frontal, 2 supraoculars, and 2 parietals. Laterally are 2 nasals, 1 loreal, 2 preoculars (the upper touches the postnasal), 3–4 (2–5) postoculars, 1–2 suboculars, 11–12 (9–14) supralabials, and 11–13 (10–16) infralabials.

The hemipenis is bifurcate with a divided sulcus spermaticus, about 33 recurved spines and 23 fringes per lobe, and some spines between the lobes (Gloyd 1940); it is illustrated in Campbell and Lamar (2004), Murphy and Crabtree (1988), and Wright (1941).

The dental formula consists of 1 maxillary fang, 1–3 (mean, 2.6) palatines, 5–9 (mean, 5–6) pterygoids, and 9–11 (mean, 10) dentaries on each side (Brattstrom 1964; Campbell and Lamar 2004; Klauber 1972).

Males have 129–155 (mean, 144) ventrals, 24–36 (mean, 32) subcaudals, 5–11 (mean, 8) dark tail bands, and TLs 9.0–12.5 (mean, 10.8) % of TBL; females have 132–160 (mean, 148) ventrals, 19–29 (mean, 27) subcaudals, 3–8 (mean, 6) dark tail bands, and TLs 7.5–9.5 (mean, 9) % of TBL.

GEOGRAPHIC VARIATION: Three subspecies are recognized (Minton 1983). *Sistrurus c. catenatus* (Rafinesque, 1818), the Eastern Massasauga, ranges from southern Ontario and central New York west to Iowa and eastern Missouri. It normally has 25 midbody scale rows, 129–157 ventrals, 24–33 subcaudals in males and 19–29 in females, 20–40 dorsal body blotches, the venter mostly dark gray or black, and 55–80 cm adult TBLs. *Sistrurus c. edwardsi* (Baird and Girard, 1853), the Desert Massasauga or Massasauga del Desierto, lives in southeastern Colorado, the Texas Panhandle, extreme southwestern Texas, and eastern and southern New Mexico in the United States, and in the Cuatro Cienegas Basin of Coahuila and at Aramberri, Nuevo León, in Mexico. It averages 23 midbody scale rows, and has 137–152 ventrals, 28–36 subcaudals in males and 24–29 in females, 27–41 dorsal body blotches, a whitish to cream-colored venter with only a few small dark spots, and a TBL_{max} of 58.8 cm (Holycross 2002). *Sistrurus c. tergeminus* (Say, 1823), the Western Massasauga, is found from southwestern Iowa and northwestern Missouri southwest through extreme southeastern Nebraska, east and central Kansas, and western Oklahoma to western Texas, and south from Oklahoma through east-central Texas to the Gulf Coast. It normally has 24 midbody scale rows, 138–160 ventrals, 27–34 subcaudals in males and 21–28 in females, 28–50 dorsal body blotches, a whitish or cream-colored venter with dark lateral blotches,

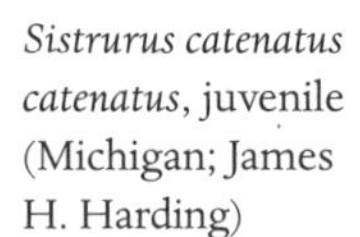

Sistrurus catenatus catenatus, juvenile (Michigan; James H. Harding)

and 45–65 cm adult TBLs. The subspecies of the massasauga have been reviewed by Campbell and Lamar (2004), Ernst (1992), Ernst and Ernst (2003), and Gloyd (1955).

Morphological intergradation zones occur between *S. c. catenatus* and *S. c. tergeminus* in south-central Iowa and adjacent Missouri, and between *S. c. tergeminus* and *S. c. edwardsi* from southwestern Colorado through extreme western Oklahoma and the central Panhandle of Texas.

Several studies have been conducted of the molecular variation between populations of *S. catenatus*. Gibbs et al. (1994) reported that diversity between four southern Canadian (Ontario) populations was small and not significant when tested using randomly amplified polymorphic DNA (RAPD) markers and randomized procedures. Further study of six microsatellite DNA markers within a single Ontario population revealed significant heterozygote deficiency at all loci in that population (Gibbs et al. 1998). However, comparison of six microsatellite DNA markers from populations in Ontario, New York, and Ohio by Gibbs et al. (1997) showed these populations to differ significantly in allele frequencies (even though some populations were <50 km apart), and may contain distinct subpopulations <2 km apart, have an average of 23% of alleles that are population specific, and have significant F_{IS} values (overall $F_{IS} = 0.194$) probably due to a combination of the presence of null alleles. Lougheed et al. (2000) examined these same scattered regional populations (>100 km) using both microsatellite and RAPD markers, and also found differences among them. The above data indicate that populations may be structured on a very fine scale due to limited dispersal of individuals and consequentially little gene exchange.

CONFUSING SPECIES: The large rattlesnakes (*Crotalus*) have numerous small scales on the crown between the two supraocular scales. *S. miliarius* has its prefrontal in broad contact with the loreal scale, 8–13 supralabials, 9–14 infralabials, 122–148 ventrals, 25–39 subcaudals, 19–25 body blotches, and 7–9 pterygoid teeth (Campbell and Lamar 2004). Hognose snakes (*Heterodon*) and the glossy snake (*Arizona elegans*) lack a rattle and loreal pits, and have rounded pupils.

KARYOTYPE: According to Zimmerman and Kilpatrick (1973), the karyotype of *S. catenatus* consists of 36 chromosomes: 16 macrochromosomes (4 large to medium metacentrics, 6 large to medium-sized submetacentrics, 4 medium to small subtelocentrics) and 20 microchromosomes. Sex determination is ZZ in males and ZW in females. The metacentric Z chromosome is the fourth largest chromosome. The acrocentric W chromosome is comparable to the medium-sized submetacentrics.

FOSSIL RECORD: Both Pliocene and Pleistocene remains of *S. catenatus* have been found. Pliocene (Blancan) fossils have been reported from Kansas, Nebraska, and Texas (Brattstrom 1967; Holman and Schloeder 1991; Rogers 1976, 1984). Pleistocene Irvingtonian fossils are known from Nebraska and West Virginia (Holman 1982, 1995), and Rancholabrean remains have been found in Kansas (Holman 1972; Preston 1979). The massasauga migrated northeastward into its present range after the end of Pleistocene glaciation (Conant 1978).

DISTRIBUTION: The massasauga ranges from southern Ontario, central New York, and northwestern Pennsylvania, west to eastern Iowa, and southwest to western Texas, southern New Mexico, and southeastern Arizona. Disjunct Mexican populations occur in Coahuila (McCoy and Minckley 1969) and Nuevo León. Its distribution has been

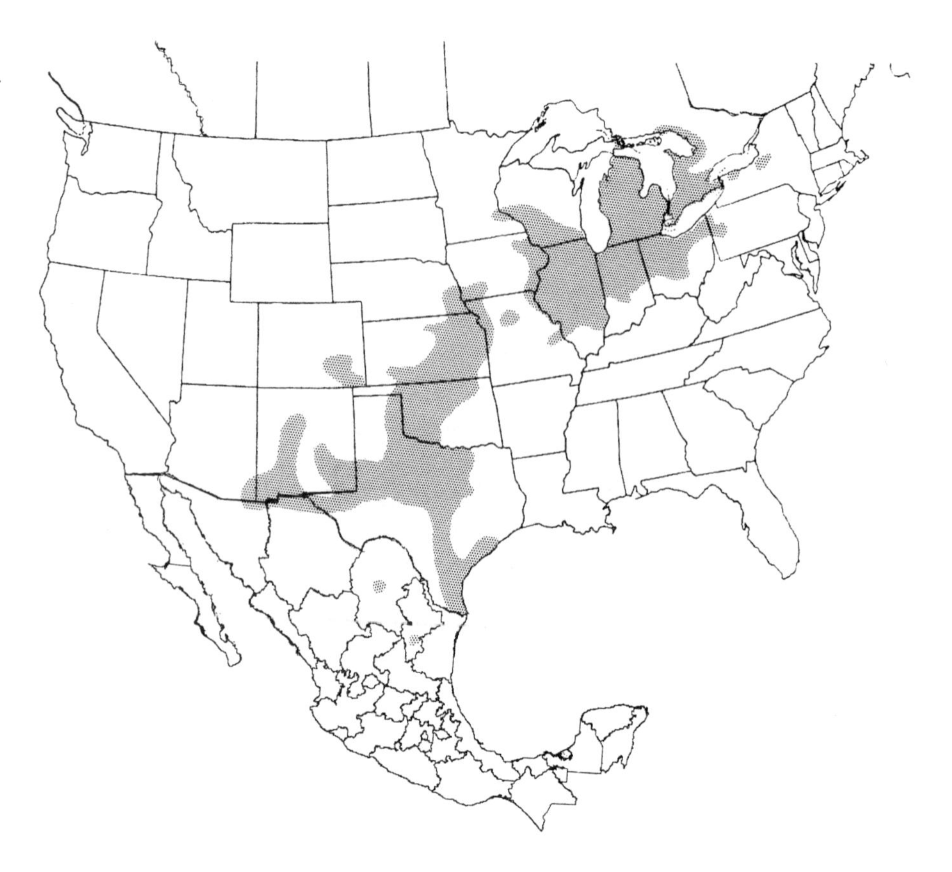

Distribution of *Sistrurus catenatus*.

reviewed by Beltz (1993) and Campbell and Lamar (2004). Northern populations may be limited in range because of a lack of suitable hibernation sites (Harvey and Weatherhead 2006a).

HABITAT: The massasauga occurs from sea level to an elevation of about 2,100 m. Over most of the eastern range it is found in moist habitats: (1) swamps and marshes dominated by sedges (*Carex* sp.), cinquefoil (*Potentilla fructicosa*), goldenrod (*Solidago* sp.), lady's slipper (*Cypripedium* sp.), thoroughwort (*Eupatorium* sp.), rushes (*Juncus* sp.), cattails (*Typha latifolia*), and stands of dogwood (*Cornus* sp.), tamarack (*Larix larcina*), poplar (*Populus* sp.), oaks (*Quercus* sp.), and *Rhus* sp.; (2) bogs and fens containing sedges, cattails, swamp milkweed (*Asclepias incarnata*), manna grass (*Glyceria* sp.), blue grass (*Poa compressa*), and alder thickets (*Alnus* sp.); (3) minerotrophic shrubby peatlands, with bryophytes (*Polytrichum* sp., *Sphagnum* sp.), leatherleaf (*Chamaedaphne calyculata*), blueberry (*Vaccinium corymbosum*), chokeberry (*Aronia melanocarpa*), mountain-holly (*Nemopanthus mucronata*), birch (*Betula alleghaniensis, B. pendula*), elm (*Ulmus americana*), white ash (*Fraxinus americana*), red maple (*Acer rubrum*), black gum (*Nyssa sylvatica*), hemlock (*Tsuga canadensis*), spruce (*Picea mariana*), pine (*Pinus strobus*), and tamarack; (4) wet meadows, with *Carex* tussocks and cordgrass (*Spartina* sp.); (5) seasonally moist grasslands (*Phalaris arundinacea*); and (6) wet, often floodplain, woods (*Cornus* sp., *Cratergus mollis, Poa compressa*). These listed plants are only a few of the more important ones in a given habitat.

In such microhabitats massasaugas use low, poorly drained areas near the hibernaculum, often a crayfish burrow, in the spring and fall, but in the summer gravid females use more dry habitats with low or sparse vegetation (Reinert and Kodrich 1982). In Missouri, they move from prairie habitats in spring to upland old fields and deciduous woods in summer, and then return to the prairies in the fall (Seigel 1986). During the active season in northeastern Indiana, massasaugas prefer emergent wetland vegetation, with wooded areas and meadows used to a lesser extent. Gravid females use less cattail microhabitat compared to nongravid females, which use more *Eupatorium*/*Solidago* habitat than gravid females (Marshall et al. 2006).

Ontario *S. catenatus* were found in sparse forest 24.2% of the time, open pasture or abandoned fields 22.5% of the time, and dense deciduous forests 16.5% of the time; no other microhabitat was used more than 9.1% of the time. Wetlands involved only <1–3.8% of observations. *S. catenatus* select habitats primarily by microhabitat preferences, and microhabitats close to retreat sites or shrubs are more heavily used. Forested habitats were used during winter hibernation, and the snakes steadily increased their use of open, wetland, and edge habitats to a peak in midsummer. Gravid females are most selective, using sites with more rock cover and less canopy closure than sites used by males and nongravid females (Harvey and Weatherhead 2006b).

Many of the eastern minerotrophic peatlands are becoming unsuitable for the *S. catenatus* due to succession. Johnson and Leopold (1998) established study plots to evaluate the effects of cutting, burning, herbivore exclosure following cutting and burning, and herbicide application on the snake's habitat. After the first growing season, shrub density, basal area, and height were greater in cut-only areas than in those that were cut and burned. After three years, only shrub height was greater at the cut-only sites. The shrub mountain-holly became less important in the plots over time, but still maintained the highest importance over the three-year study period. At mammal-browsed sites, shrub height increased in all treatment types, and at the end of one year was significantly greater in cut-over sites. At the end of year 2, this relationship held only for fenced-in compared to unfenced plots, indicating that browsing had a significant effect on growth in shrub height. Abundance and diversity of small mammals increased in herbicide (glyphosate)-treated areas in shrub peatlands. Data were inconclusive as to the massasauga's use of these cleared areas, but 9 of 89 (10.1%) surface radiolocations from 9 individual snakes tracked one year occurred in or around treatment areas that constituted only 2.5% of the total peatland area.

In its drier range in the Southwest and Mexico, the snake is restricted to habitats like river bottoms; dry desert grasslands, with salt grass (*Distichlis* sp.), clumped tobosa (*Hilaria mutica*), creosote (*Larrea* sp.), mesquite (*Prosopsis* sp.), and desert willow (*Chilopsus linearis*); short grass prairies, with sage (*Artemesia filafolia*) and grasses (*Bouteloua gracilis, Buchloe dactyloides*); and plains of mesquite, various grasses, yucca (*Yucca* sp.), creosote, and various cacti (Cactaceae); pinyon-juniper woods (*Pinus edulis, Juniperus* sp.); and scrub oak (*Quercus*) woods. It is able to survive in such arid regions because it uses rodent burrows, which provide a humid microclimate that retards moisture loss, as daytime retreats and hibernacula.

BEHAVIOR AND ECOLOGY: The total annual duration of activity by *S. catenatus* depends on latitude, elevation, and yearly weather patterns. In the United States and Canada, the snake is surface-active from mid-March or April to October or mid-November (Conant 1938; Greene and Oliver 1965; Hobert et al. 2004; Lowe et al. 1986;

Sistrurus catenatus edwardsi (Cochise County, Arizona; Steve W. Gotte)

Mackessy 2005; Seigel 1986). It is annually above ground for about 192 days in Pennsylvania (Reinert 1978) and 193–202 days in Missouri (Seigel 1986). Possibly the disjunct Mexican populations are active all year, but less so in the summer and winter. Although Conant (1938) reported that in northwestern Ohio most surface activity occurs in May–August, annual activity is usually bimodal with peaks in late April–June and late September–October. Activity increases after summer rain events.

In the East, the winter is usually spent in crayfish burrows, in old stumps or rotten logs within peat bogs and other wetlands, or in old root systems, rodent burrows, or rock crevices in moist, often coniferous, woodlands; in the dry Southwest, where wetlands are seasonal and crayfish scarce or absent, hibernation takes place in rock crevices and rodent burrows. In northeastern Ohio and Wisconsin, massasaugas are capable of maintaining BTs above ET for 45 minutes, and can withstand freezing BTs for short periods without harm, but usually hibernate in wet crayfish holes at depths below the frost line (Maple and Orr 1968; Vogt 1981). Few Ontario massasaugas use the same hibernaculum in consecutive years, but most (>70%) hibernate within 100 m of their previous-year site (Harvey and Weatherhead 2006a). The snakes use deep hibernacula in Ontario; consequently, the overwinter mortality rate (23%) is similar to that of the active season (21%) (Harvey and Weatherhead 2006a).

Prior and Weatherhead (1994) recorded positive disturbance responses at 15–31°C BTs from active Ontario massasaugas to 15°C (2%), 16–20°C (25%), 21–25°C (31%), 25–30°C (45%), and 31+°C (71%). Maple and Orr (1968) reported that Ohio massasaugas hibernating at depths of 30–90 cm had a mean BT of 2.6 (range, 0.5–7.0) °C.

During the spring and fall, Texas *S. catenatus* actively forages or basks (often on grass tussocks) during the day; but when daytime summer ATs reach 34°C it becomes more crepuscular or nocturnal (Tennant 1985). It first appears on roads shortly before dusk (0800 hours) and may remain resting on warm surfaces for several hours (Knopf and Tinkle 1961). Colorado massasaugas shift the time of activity from primarily diurnal in the cooler periods of April and late September–October to mostly evening and early night during the warmer period of May to mid-September; there they are active at ATs of 14–30°C, and most activity occurs between 20 and 26 (mean, 22.1) °C ATs (Hobert 1997; Hobert et al. 2004; Mackessy 2005). Arizona snakes are predominantly nocturnal in summer and are most often found moving in the early night hours (Holycross 2003; Lowe et al. 1986).

In northwestern Missouri, most spring activity is from 1200 to 1600 hours (55%), 1600 to 2000 hours (24%), and 0800 to 1200 hours (17%); little activity (4%) occurs from 2000 to 2400 hours; summer activity is from 1600 to 2000 hours (42%), 1200 to 1600 hours (33%), and 2000 to 2400 hours (18%), and in the fall the most activity (70%) is from 1200 to 1600 hours (Seigel 1986). Pennsylvania *S. catenatus* are mainly summer-active from 0900 to 1500 hours (Reinert 1978).

Illinois massasaugas gradually stabilize their BT between a photoperiod of 10–14 hours, and achieve the warmest daily BTs during mid-day with >12.5 hours of light; contrary to some previous reports, gravid females there do not achieve warmer BTs than nongravid females and males. The snakes rarely exceed a BT of 35°C, and the largest peaks in BT were between 26 and 28°C, 29 and 33°C, and 30 and 34°C for nongravid females, gravid females, and males, respectively (Dreslik 2004).

Movement data for *S. catenatus* are plentiful. It normally moves some distance between the hibernaculum and summer activity range, and, although males move greater overall distances and have larger home ranges, often it is a gravid female who migrates the greatest distance (King 1999). The snakes move freely during the annual activity period, but their daily distances moved and home range areas are dependent on the amount of available habitat.

During a Colorado study by Mackessy (2005), radio-tracked massasaugas moved 1–350 m per day, and had 45–413 ha home ranges. Over an entire season the snakes moved as much as 2–4 km. Movements of up to 2 km away from or to a hibernaculum were essentially made in a straight line.

In Illinois, the mean home range in a limited habitat was only 2.38 ha (Wilson and Mauger 1999). Only 3 (1 female, 2 males; 1 male was killed crossing a road) of 40 radio-tracked adult Illinois *S. catenatus* crossed roads, significantly less often than predicted by chance, indicating that roads are apparently barriers to dispersal between fragmented habitats (Shepard et al. 2008a, 2008b). Dreslik (2004) found no significant differences in Illinois activity ranges between the sexes or size classes, but early in the active season, gravid females preferred more open habitats compared with males and nongravid females. Gravid females made about 50% of their constrained movements between June and July, and 50% of their random movements before this period.

Male Wisconsin massasaugas had a mean home range of 161.5 ha, nongravid females 6.7 ha, and gravid females 2.8 ha; as expected, neonates had the shortest home ranges and made shorter daily or seasonal movements (King 1999). Michigan snakes had average home ranges of 1.3 (males, 1.09–2.37; females, 0.25–4.52) ha, and average daily movement rates of 6.9 (males, 0.84–19.27; females, 2.68–11.20) m per day (Moore and Gillingham 2006). At an Indiana fen, males had longer mean home ranges (417.5 m) than nongravid females (317.6 m), which had longer home ranges than gravid females (276.2 m); activity center estimations followed the same trend. Males (78.1%, 15.1 m per day, 1,653 m) and nongravid females (72.8%, 10.1 m per day, 1,183.5 m) also differed from gravid females (48.4%, 6.2 m per day, 636.9 m) in mean frequency of daily movement, distance moved per day, and total seasonal distance moved, but

Sistrurus catenatus edwardsi (Bernadino, Arizona; Charles Hanson, Tucson, Arizona; John H. Tashjian)

males only differed from nongravid females in distance moved per day (Marshall et al. 2006).

Mean home range area and length in western Pennsylvania were 9,794 m² and 89 m, respectively, and the mean distance moved per day was 9.1 m, with no significant differences between the sexes—but gravid females had significantly shorter home range lengths than nongravid females (Reinert and Kodrich 1982). Mean home range size and maximum range length, respectively, of western Pennsylvania neonates were 0.36 (range, 0.11–0.54) ha and 116.9 (range, 62.1–182.3) m, and the mean daily distance moved was 5.27 (range, 3.15–7.74) m; the small snakes returned to their general parturition area to overwinter, and one remained within 3.8 m of its mother's path for 18 consecutive days (Jellen and Kowalski 2007). For 16 days after their initial ecdysis, they were found within 11.1 m of the mother's gestation site.

In New York these 3 classes had mean home ranges of 27.8 ha, 41.4 ha, and 2 ha, respectively (Johnson 2000). Ontario home ranges averaged 0.25 km² (Weatherhead and Prior 1992). Mean daily distances for New York and Ontario snakes that had moved were 7–167 m, and per season, 752–3,712 m, with gravid females traveling less distance than either nongravid females or males (Johnson 2000; Weatherhead and Prior 1992). Massasaugas are good swimmers and readily enter water.

Males engage in ritualized dominance bouts associated with nearby sexually receptive females. The males approach each other with head-bobbing and tongue-flicking (much tongue activity occurs throughout the combat sequence). Then, facing each other, the males elevate their heads and anterior portions of their bodies into an *S*-shape with the snout pointing upward. Next, the venters are pressed together, and about 33% of their bodies are intertwined. The snakes sway their bodies back and forth laterally. One male eventually wraps his body over that of the other in an attempt to get on top of the other, and begins to push the inferior male downward to pin it to the ground. This sequence may be repeated several times until one male is finally pinned. Aggression then ceases, and the vanquished foe usually retreats; the larger male normally wins the bout. Bouts may last from 30 minutes to more than an hour (Collins and Collins 1993; C. Ernst, personal observation; Shepard et al. 2003; Vandewalle 2005).

REPRODUCTION: The sexes are mature when TBLs are 40–54 cm (Stebbins 2003; Wright 1941); such lengths are reached at 3–4 years. The smallest mature male and female examined by Goldberg and Holycross (1999) had 28.0 cm and 32.9 cm SVLs, respectively. Mackessy (2005) reported a minimal mature female SVL of 32.9 cm in his Colorado study population. Reproductive males in Illinois had 49.7–72.0 cm SVLs (Jellen et al. 2007b).

The annual reproductive cycles of both sexes are essentially like those reported for other North American pitvipers by Aldridge and Duvall (2002) and Aldridge et al. (2008). Males initiate spermiogenesis in June, with it peaking in August–September. The diameter of the seminiferous tubules and overall size of the testes are small in May but increase to their maximum in late July or early September. Size of the sexual segment of the kidney, which passes a plug into the female, parallels the diameter of the testes; sexual segment tubules are lowest in the early part of the activity season (spring) and peak in diameter and secretory activity in August–September, when they are enlarged to the ureter. Both male combat bouts and mating occur when the sexual segment of the kidney is hypertrophied (Aldridge et al. 2008; Goldberg and Holycross 1999; Mackessy 2005). Some sperm is present in the vas deferens throughout the year, so males are capable of fertilizing a female at any time.

Females initiate vitellogenesis in the summer and fall, similar to other temperate zone pitvipers. Vitellogenic follicles reach 20 mm in length by late September, remain this size over winter, and resume growth in the spring. Guthrie (1927) reported that an Iowa female passed 13 "solid yellow" eggs in two installments of 7 and 6 on 21 and 22 October. Ovulation and fertilization occur later in the spring, and embryos are present from May through August. Parturition begins in August (Aldridge et al. 2008; Goldberg and Holycross 1999; Mackessy 2005; Reinert 1981; Roberts and Quarters 1947).

First-summer Wisconsin females show no follicular growth; those in their second summer have follicles about 7 mm in diameter; 50% of third-summer females are gravid and 25% are postpartum (one collected 31 August had yellow follicles about 25 cm long); and only about 3% of fourth-summer females are not gravid or have not already given birth (Keenlyne 1978). Only 7% of third-summer or older Wisconsin females, and only 3% of fourth-summer or older females, are nonreproductive. This suggests an annual female reproductive cycle. However, Goldberg and Holycross (1999), Reinert (1981), and Seigel (1986) reported a biennial reproductive cycle in females from Arizona, Colorado, Missouri, and Pennsylvania. The annual percentage of reproductive females in Arizona and Colorado is only about 15%; in Missouri, 33–71%; and in Pennsylvania, only 52–58%. Seigel et al. (1998) found fewer females (23%) gravid in the Missouri population after a severe flood event. Obviously, the number of reproducing females in any population varies from year to year. Significant size differences exist between females from Arizona, Colorado, and Pennsylvania and those from Wisconsin, the Wisconsin females being larger—perhaps SVL influences the breeding cycle. Ectopic pregnancies can occur; Jellen et al. (2007a) reported a case in which the monitored female, which had given normal birth to three offspring, showed complete degeneration of a remaining embryo in the peritoneal cavity.

Published observations of mating activity, including those of captives, seem to indicate that the breeding period extends from March to November (see Ernst 1992), but in the wild some mating occurs in the spring (Guthrie 1927 [on 7 May, within two days of capture]; Jellen et al. 2007b; T. R. Johnson 1987; Wright 1941), but more often in July–September (Aldridge et al. 2008; Jellen et al. 2001, 2007b; Mauger and Wilson 2005; Shepard et al. 2003).

During the latter period males become quite active; Illinois males moved a mean of 21.8 (range, 7.4–57.7) m per day as compared to 13.3 (range, 4.2–32.7) m per day in the nonbreeding season, and their movement frequency increased from 63 (range, 38–100) % to 77 (range, 43–100) % (Jellen et al. 2007b). Both movement rate and body size were positively related to the number of females acquired; heavier males were seen accompanying females as the mating season progressed. Females were accompanied by up to seven males per season, but only 18% of the males located more than one female during a single mating season.

S. catenatus can recognize odors of its own kind (Wellborn et al. 1982), so species odor and probably also sexual pheromones play an important role in bringing the sexes together for mating. During courtship, the male lies on the female's back with his tail coiled around her tail, and frequently rubs her head and neck with his chin and writhes his body. He then massages the posterior portion of the female's tail with his tail loop by tightening it and stroking it posteriorly until her rattle is touched, and then reversing the stroke until he reaches the place of his original grip. This entire cycle is repeated several times, and copulation may last several hours. The chin-rubbing rate is relatively constant, and continues for the entire period of time between successive tail-stroke cycles (Chiszar et al. 1976b; Mackessy 2005). Some males spasmodically twitch, mainly the posterior body, but sometimes the entire body (Guthrie 1927). If mating occurs in late summer or early fall and the young are not born until the following summer, the sperm may be stored in the female's oviducts and not used until the following spring (Schuett 1992). A Coahuilan female containing three fully developed embryos (TBLs, 10.2 cm, 11.2 cm, and 12.4 cm) collected on 18 June by McCoy and Minckley (1969) may have been fertilized from such stored sperm; otherwise, mating may occur earlier in the spring at the southern end of the species' range.

The GP is about 100 (range, 71–115) days; the ovoviviparous young are born from late July to mid-October (Hobert et al. 2004), but most births occur in August to mid-September. From 1 to 10 minutes elapse between individual births, and the young rupture the fetal membrane within the first few minutes. Their first act after emergence from the membrane is to stretch their jaws as if yawning (Anderson 1965).

Litters average 7.8 young (n = 77), and vary from 2 (Stebbins 1954) to 20 (Anton 2000; Vogt 1981). The smaller females from southwestern populations produce fewer young per litter, usually 5–6 (Goldberg and Holycross 1999), and Seigel (1986) found a significant positive relationship between female length and litter size in Missouri. Seigel et al. (1986) and Seigel and Fitch (1984) reported an RCM of 24.7%, and Hobert et al. (2004) noted that a female had lost 26% of her BM during parturition. An RCM of 40.3% was reported by Anton (2000). The neonates may remain near their mother for a short period of time.

Neonates (n = 86) have 13.5–33.8 (mean, 21.6) cm TBLs, 7.7–25.7 (mean, 18.3) cm SVLs, and BMs of 2.7–15.9 (mean, 10.2) g. The tail and rattle button are pink or yellow.

A natural hybrid *S. catenatus* × *Crotalus horridus* was reported by Bailey (1942).

Sistrurus catenatus edwardsi (R. D. Bartlett)

GROWTH AND LONGEVITY: At birth, neonate *S. catenatus* normally have SVLs of 18.8–24.4 cm and BMs of 3.4–13.3 g. During a study in western Pennsylvania, neonate BM did not differ at parturition, but free-moving males gained significantly more weight (mean BM, 18.3 g) than free-moving females (mean BM, 12.7 g) over the first 50 days, but no difference in SVL was apparent (Jellen and Kowalski 2007). Yearlings from Illinois have 39–43 cm TBLs, an increase of about 65% from birth (Wright 1941), and Missouri yearlings have SVLs of 30–40 cm (Seigel 1986). Massasaugas with 50–54 cm SVLs are probably 3–4 years old (Keenlyne 1978; Seigel 1986; Wright 1941). An adult (SVL, 32 cm; BM, 24.2 g) Colorado male grew 5.6 cm (117%) and 27 g (112%) in 149 days (Hobert et al. 2004). This indicates that the growth rate is probably closely tied to resource (food) availability and the ET (see Beck 1996). Growth is apparently not affected by human disturbance in the habitat (Parent and Weatherhead 2000).

The longevity record for this species is for a male who survived 20 years and 5 days at the Staten Island Zoo (Snider and Bowler 1992). Mackessy (2005) estimated a greatest wild longevity in Colorado of 15 years.

DIET AND FEEDING BEHAVIOR: The massasauga has a broad diet consisting mainly of the small vertebrates found within its habitat. Reported natural prey are crayfish; centipedes (*Scolopendra sp.*) and insects; fish; amphibians—anurans (*Bufo* sp.; *Pseudacris crucifer; Rana berlandieri, R. clamitans, R. pipiens, R. sylvatica; Spea bombifrons*); reptiles—lizards (*Cnemidophorus gularis, C. sexlineatus, C. uniparens, Crotaphytus collaris; Eumeces obsoletus, Gambelia wislizenii; Holbrookia maculata; Phrynosoma cornutum; Sceloporus olivaceus, S. undulatus; Scincella lateralis; Urosaurus ornatus; Uta stansburiana*) and snakes (*Heterodon nasicus; Opheodrys vernalis; Sistrurus catenatus; Storeria dekayi, S. occipitomaculata; Sonora semiannulata; Thamnophis sirtalis; Tantilla nigriceps; Tropidoclonion lineatum*); birds (including eggs and nestlings)—quail (*Colinus virginianus*), sparrows (*Chondestes grammacus, Melospiza melodia, Spizella pusilla*), blackbirds (*Agelaius phoeniceus*), and a "warbler"; and mammals—shrews (*Blarina brevicauda, B. carolinensis;*

Cryptotis parva; Notiosorex crawfordi; Sorex cinereus), voles (*Clethrionomys gapperi; Microtus ochrogaster, M. pennsylvanicus*), murid mice (*Baiomys taylori; Mus musculus; Onychomys leucogaster; Peromyscus leucopus, P. maniculatus; Perognathus flavescens; Reithrodontomys megalotis, R. montanus*), jumping mice (*Napaeozapus insignis, Zapus hudsonius*), pocket mice (*Perognathus hispidus, P. merriami*), and young snowshoe hares (*Lepus americana*) (Applegate 1995; Atkinson and Netting 1927; Best 1978; Brush and Ferguson 1986; Campbell and Lamar 2004; Conant 1951; Degenhardt et al. 1996; Greene and Oliver 1965; Holycross 2003; Holycross and Mackessy 2002; Keenlyne and Beer 1973; Klauber 1972; Lardie 1976; Lowe et al. 1986; Mauger and Wilson 1999; McKinney and Ballinger 1966; Minton 1972; Reinert 1978; Seigel 1986; Shepard et al. 2004; Tennant 1985; Tinkle 1967; Vandewalle and Vandewalle 2008; Vogt 1981; Weatherhead and Prior 1992; Webb 1970; Wright 1941; Wright and Wright 1957).

Most data concerning the massasauga's prey are based on single observations; however, several more comprehensive studies have been reported. Nearly 95% of the prey of Wisconsin massasaugas was warm-blooded, and 85.7% of the entire diet consisted of voles; other prey were deer mice, 4.4%; garter snakes, 4.4%; jumping mice, 2.2%; red-winged blackbirds, 1.1%; shrews, 1.1%; and an unidentified snake, 1.1%. Food items by sex and percentages of snakes containing prey were males, 83.6%; nongravid females, 55.6%; and gravid females, 10.4% (Keenlyne and Beer 1973). Two major categories of food were consumed by Missouri *S. catenatus*: small rodents and snakes (Seigel 1986). Illinois free-ranging neonates feed primarily on short-tailed shrews (*Blarina brevicauda*) (Shepard et al. 2004).

In two feeding trials conducted by Shepard et al. (2004), Illinois neonates preferred snakes (*Thamnophis sirtalis*) and mice (*Mus musculus*), showed limited interest in lizards (*Eumeces fasciatus*), and were not interested in anurans (*Acris crepitans*) or insects (short-horned grasshoppers, Acrididae). Although they attacked the mice, because of gape limitations the neonates had difficulty swallowing them. Snakes are comparably easier to ingest and are the most common neonate prey item throughout the species range. Prey recovered from free-ranging Illinois neonates consisted primarily of the shrew *Blarina carolinensis*.

Fowlie (1965) reported that in Arizona *S. c. edwardsi* is primarily an amphibian predator, but Greene (1990), after examining museum specimens (mostly roadkills), concluded that they eat small lizards and pocket mice, and Lowe et al. (1986) reported that mice and lizards (especially whiptails and earless) are taken.

Carrion is not passed up. Ruthven (*in* Klauber 1972) thought snakes devoured were probably carrion. Greene and Oliver (1965) found a Texas massasauga attempting to swallow a recently road-killed snake; Shepard et al. (2004) reported that a neonate contained a shrew with maggots on it, indicating that it was taken when dead; and Schwammer (1983) observed one eating carrion in Colorado. *S. catenatus* has consumed its own species both as carrion and when alive (Klauber 1972).

Adults usually capture prey by ambush, but some active foraging is probably also important. Some individuals may hold on to the prey after biting it (Curran and Kauffeld 1937). Envenomated rodent prey are preferred over those not previously bitten (Duvall et al. 1978). Gravid females rarely feed during gestation.

Young *S. catenatus* tail-lure by waving their yellowish tails back and forth over their heads to attract small frogs (Schuett et al. 1984). Warm-blooded prey are probably detected by the heat sensory facial pit, but sight (movement is the primary cue in elic-

Sistrurus catenatus tergeminus (Barry Mansell)

iting exploratory behavior; Scudder and Chiszar 1977) and olfaction are also important feeding cues (Chiszar et al. 1976a, 1979a, 1981). Most prey are struck and then eaten only after they die, but anurans may be swallowed alive.

VENOM DELIVERY SYSTEM: Massasauga fangs are short; Klauber (1939) reported FLs for 5, 62.5–78.7 cm TBL *S. c. catenatus* were 5.0–5.9 (mean, 5.4) mm, the mean TBL/FL was 131%, the mean HL/FL was 5.78%, and the mean angle of fang curvature was 61°. Similarly, for 5 47.7–62.5 cm *S. c. tergeminus*, the FLs were 47.7–62.5 (mean, 5.0) mm, the mean TBL/FL was 109%, the mean HL/FL was 5.51%, and the mean angle of fang curvature was 69°. The fangs are developed, and neonates are capable of envenomating mice immediately after birth.

VENOM AND BITES: The total dry venom yield for an adult massasauga is 14–45 mg (Brown 1973; Glenn and Straight 1982; Klauber 1972; Russell 1983); however, the yield per bite is probably closer to 5–6 mg.

The percentages of protein recovered by fractions after partial purification of the venom by Bonilla et al. (1971) were fraction I, 20.5%; fraction II, 28.2%; and fraction III, 13.3%. The venom contains proteolytic enzymes (including hemolysin and peptide inhibitors), 5′-nucleotidase, phosphodiesterase, phosphomonoesterase, phopholipase A_2, L-amino oxidase (Aird 2005; Glenn and Straight 1985; Graham et al. 2008; McCue 2005; Minton 1956; Munekiyo and Mackessy 2005; Russell 1983). Bober et al. (1988) detected myotoxin α-like proteins in venoms of two *S. c. catenatus*, but not in the venom from a *S. c. tergeminus*; but the presynaptic neurotoxin, sistruxin, has now been identified in the venom of both subspecies (Chen et al. 2004; Sanz et al. 2006). Clotting factors are practically nonexistent; hemagglutinin occurs at a titer of only <1.4 (Minton 1956).

In spite of the adult snake's relatively small size, it can deliver a very painful and serious bite. The venom is 5 to 10 times more toxic than that of *Crotalus atrox* and almost equal to that of *C. scutulatus* (Le Ray 1930; Lowe et al. 1986). It is largely hemolytic and causes much ecchymosis as capillary walls are destroyed, but a neurotoxic Mojave toxin (probably sistruxin) is also present (Chen et al. 2004; Sanz et al. 2006; Tan and Ponnudurai 1991; Tubbs et al. 2000).

Symptoms recorded in human envenomations include an initial burning sensation, pain, skin sensitivity, discoloration and swelling spreading from the bite site, swollen and painful lymph glands, bleeding from the puncture wounds, numbness at the bite site, a cold sweat, fever, faintness, metallic taste in the mouth, nausea and loss of appetite, tremors (sometimes severe) and chills, headache, depression, and nervousness (Allen 1956; Anon 1910; Atkinson and Netting 1927; Baldwin 1999; Burgess and Dart 1991; Campbell and Lamar 2004; Dodge and Folk 1960; Henderson and Dujan 1973; Hutchison 1929; Jaffe 1957; Klauber 1972; La Pointe 1953; Menne 1959; Wright 1941).

A house mouse (*Mus musculus*) bitten by *S. catenatus* usually dies in 20–25 minutes (C. Ernst, personal observation). The LD_{50} for a mouse has been estimated as 2.85–2.9 mg/kg by Ernst and Zug (1996) and Russell (1983); but Minton (1956) reported an intraperitoneal LD_{50} of only 0.22 mg/kg (subcutaneous, 5.25 mg/kg). Intraperitoneal LD_{50} values of 0.76 mg/20 g by Bonilla et al. (1971) and 0.90 mg/kg by Githens and Wolff (1939) have been reported. Schoettler (1951) reported a subcutaneous LD_{50} of 6.8 mg/kg. Young massasaugas only a few days old have venom toxic enough to cause bitten mice to die within several hours (Conant 1951).

Humans have died from massasauga bites (Anon 1910; Branson 1904; Froom 1964; Lyon and Bishop 1936; Menne 1959; Minton 1972; Stebbins 1954). The estimated human lethal dose is 30–40 mg (Minton and Minton 1969). Brown (1973) estimated the subcutaneous LD_{50} to be 350 mg for a 70 kg human, and that the number of lethal doses per massasauga bite is 0.04–0.09.

The massasauga seems immune to the venom of its own species; but, when its venom was injected into other snakes, some died (*Agkistrodon contortrix, Diadophis punctatus, Nerodia sipedon, Storeria dekayi*) and others either experienced no effects or recovered (*Coluber coluber; Lampropeltis triangulum; Nerodia sipedon; Thamnophis radis, T. sirtalis*) (Keegan and Andrews 1942; Swanson 1946).

PREDATORS AND DEFENSE: Predation data concerning *S. catenatus* are scarce. Reported or potential predators are large fish [?], bullfrogs (*Rana catesbeiana*), turtles [?], snakes (*Coluber constrictor; Lampropeltis getula, L. triangulum; Masticophis flagellum*), loggerhead shrikes (*Lanius ludovicianus*), large hawks (*Buteo* sp., *Circus cyaneus*), golden eagles (*Aquila chrysaetos*), owls (*Bubo virginianus*), roadrunners (*Geococcyx californianus*), turkeys (*Meleagris gallopavo*), large wading birds (*Ardea herodius, Botaurus lentiginosus*, etc.), raccoons (*Procyon lotor*), canids (*Canis domesticus, C. latrans; Urocyon cinereoargenteus; Vulpes macrotus, V. vulpes*), bobcats (*Felis catus, Lynx rufus*), badgers (*Taxidea taxus*), skunks (*Mephitis* sp., *Spilogale* sp.), weasels (*Mustela frenata, M. vison*), and hogs (*Sus scrofa*) (Campbell and Lamar 2004; Chapman and Casto 1972; Collins and Collins 1993; Hammerson 1999; Johnson 1992; Mackessy 2005; Minton 1972; Moore and Gillingham 2006; Ross 1989; Vogt 1981; Wright 1941). Cattle (*Bos taurus*) and deer (*Odocoileus virginianus*) probably occasionally trample them.

S. catenatus reacts strongly to the odors of potentially dangerous kingsnakes (*Lampropeltis* sp.) and, when confronted with one, either flees, freezes in place, or, if attacked, bridges its body (Inger *in* Weldon et al. 1992).

S. catenatus is usually rather sluggish and mild mannered, and normally becomes aggressive only if provoked. Its color and pattern blend in well with some backgrounds, and it does not always rattle when approached and so is often unnoticed. Warm snakes are more likely to rattle; Ontario massasaugas rattled 2% of the time when their BT was at or below 15°C, 25–31% at BTs of 16–25°C, 45% at BTs of 26–30°C, and 71% at BTs at or below 31°C (Prior and Weatherhead 1994). Most positive responses involved snakes rattling while fleeing (66%) as opposed to simply rattling when approached (34%). At another Ontario site, as human disturbance occurred, gravid females became less visible to observation, but the chances of seeing a male or nongravid female remained about the same (Parent and Weatherhead 2000). Mean distance moved per day decreased and mean time between moves greater than 10 m increased in all 3 sex classes with increasing exposure to human disturbance, but no differences in the condition, growth rate, or litter size of gravid females were evident.

However, differences in disposition between individuals do occur—a large female that we found in Missouri was very alert and irritable, rattling her tail and striking whenever approached; C. Ernst will never forget the sight of this striking snake's open mouth and exposed fangs streaking at him through the lens of his camera as he tried to photograph it. Lee et al. (2004) have also experienced aggression; on 18 December, a warm Missouri *S. catenatus* that had been basking advanced on Lee and struck his boot twice.

The SVL-adjusted loudness of the warning rattling is about 70 dB, and the sound frequencies are 6.4–19.4 kHz with a dominant frequency of 13.5 kHz (Cook et al. 1994; Fenton and Licht 1990; Young and Brown 1993). Disturbed populations become less visible (Parent and Weatherhead 2000).

PARASITES AND PATHOGENS: The snake is parasitized by the nematodes *Hexametra boddaërtii*, *Physaloptera* sp., and *Physocephalus* sp. (Ernst and Ernst 2006; Goldberg et al. 2001).

Diseases also occur in wild populations. Allender et al. (2006) found that 100% of the 20 Illinois massasaugas they tested carried antibodies for ophidian paramyxovirus (see also Allender et al. 2008), but none were seropositive for West Nile virus. A captive *S. c. catenatus* developed a large benign smooth muscle tumor (leiomyoma) paralleling its atrophied intestinal tract and apparently died as a result (Christiansen et al. 2008), and another had an ovarian tumor (Effron et al. 1977).

POPULATIONS: Populations are normally rather small but dense (Conant 1951; Fitch 1993; Hurter 1911; Kingsbury 1996; Mauger and Wilson 1999; Seigel 1986; Seigel et al. 1998). Overall, the populations in Colorado are the largest known (Hobert et al. 2004; Mackessy 1998a).

Some data on massasauga population dynamics have been published. Knopf and Tinkle (1961) collected 15 one mid-May evening on 25.6 km of a Texas road (a density of 0.58/km), and Campbell and Lamar (2004) recorded 43 in one early May evening while road-cruising west of Fort Worth, Texas. A Colorado population contained 254 snakes (199 adults, 55 juveniles) (Hobert et al. 2004). The average SVL of adult males was 35.5 cm, that of adult females 36.4 cm. In Missouri, a population studied by Seigel (1986), mostly adults with SVLs >50 cm, were caught during April–May, but the numbers of small juveniles (SVL, 20–30 cm) increased dramatically in August–October. Mackessy (2005) reported that the female subset of a Colorado population contained 54.9% neonates, 13.4% prereproductive juveniles (age 2), 6.7% first reproductive snakes

Sistrurus catenatus tergeminus (Fort Worth, Texas; Houston Zoo; John H. Tashjian)

(age 3), and the following reproductive age classes: 4 years, 5.4%; 5 years, 4.3%; 10 years, 1.8%; and 15 years, 0.05%. Nonreproductive females made up about 4% of the total population.

Seigel (1986) reported an adult sex ratio of 1.16 (32.8% of total sample) females to 1.00 (28.1%) males in Missouri, but a contrasting male to female ratio of 1.50:1.00 was recorded in Colorado by Hobert et al. (2004). However, recorded data on sex distributions in litters and natural populations show the male to female ratio to be essentially 1:1; the sex ratio may be equal at birth (Klauber 1936). The juvenile to adult ratio may comprise about 30–40%, especially from August to hibernation.

Populations crowded into small areas make *S. catenatus* more prone to extirpation (Breisch 1984; Campbell and Lamar 2004; Greene and Campbell 1992; Seigel and Sheil 1999). Small populations suffer from inbreeding, and may develop a genetic bottleneck if gene flow with adjacent populations is interrupted. In addition, the species is very sensitive to changes in adult and juvenile survival rates. Individual survivorship seems low, in some cases not exceeding four years total age (Mackessy 2005). Mortality in the neonate and juvenile classes is particularly high.

In the past, fairly dense populations of these snakes probably occurred at suitable sites (Hurter 1911). However, human (Bushey 1985; Campbell and Lamar 2004; Greene and Campbell 1992; Hoy 1883; Hurter 1911; Kingsbury 1996; Le Ray 1930; Loomis 1948; Seigel 1986) and natural (Hurter 1911; Seigel et al. 1998) habitat destruction and roadkills have severely damaged or destroyed many colonies (Kingsbury 1996; Menne 1959; Seigel et al. 1998; Shepard et al. 2008a). Past drainage of prairie wetlands has been particularly destructive to colonies in Missouri (Hurter 1911). Mowing and prescribed summer burnings of woody plants often result in direct or indirect mortality of the snake, and mowing before burning renders grassland habitats unsuitable for it (Durbian 2006; Durbian and Lenhoff 2004). In the Southwest, loss of prairie grasslands to overgrazing has eliminated much of the snake's original habitat. Populations have severely declined where in the past bounties have been paid for dead massasaugas.

Humans probably eliminate more massasaugas each year than all natural predators combined.

During a Wisconsin experimental repatriation study to negate population declines, *S. catenatus* released during late July had lower mortality rates, had larger home ranges, and gained more BM than snakes released in September. July-released females successfully reproduced; those released in September did not. So, such an early release regime may be a viable method of restoring massasauga populations (King et al. 2004).

Sistrurus catenatus is now considered threatened or endangered over most of its eastern and mid-western range. It is considered endangered in Illinois, Iowa, New York, Pennsylvania, and Wisconsin; is considered threatened in Indiana; and is protected in Arizona and Colorado; *S. c. catenatus* is listed as endangered in Missouri and of special concern in Michigan. Unfortunately, it has apparently been extirpated in Minnesota, and is currently losing ground rapidly in Illinois, Indiana, and Iowa.

Suitable habitat must be preserved, populations protected, and roadkills eliminated or alleviated, if it is to remain a viable species. A vital part of the conservation effort is to educate the public about its survival status and the problems that have brought on its decline. Good model education programs are in place in Michigan and Ontario, and should be copied by other states where the snake is in trouble.

REMARKS: Anderson (2006) tested six microsatellite DNA primer pairs developed for *S. catenatus* on a population of *Crotalus horridus*, and found that the primer set successfully amplified at each locus for all loci at an annealing temperature of 57°C. In spite of some limitations, this primer set may be a useful complement to those developed for the *Crotalus* rattlesnakes, and seems to provide justification for further studies of cross-species amplification of microsatellite loci in the two rattlesnake genera.

The type locality of *Sistrurus catenatus* (Rafinesque, 1818) was re-evaluated, and restricted to "the prairies between the Cannonball and Heart rivers, within 40 km of the Missouri River in North Dakota" by Holycross et al. (2008). This causes several taxonomic problems, as the new type locality lies within the range of the Western Massasauga, *S. c. tergeminus* (Say, 1823), and invalidates that name and replaces it with *S. c. catenatus* (Rafinesque, 1818). This leaves the Eastern Massasauga (formerly *S. c. catenatus*) without a valid name. The oldest available name for that taxon is the unfamiliar one, *Crotalus messasaugus* Kirtland, 1838, or *Sistrurus messasaugus* (Kirtland, 1838). To conserve the long-term current usage, Holycross et al. are preparing an appeal to the International Commission of Zoological Nomenclature to allow a neotype of *S. catenatus* to be designated from the type locality ("Ohio") of *Crotalus messasaugus*.

Synopses of the taxonomy and literature of *S. catenatus* were published by Beaman and Hayes (2008), Campbell and Lamar (2004), and Minton (1983). Campbell and Lamar (2004), Dreslik (2004), and Ernst and Ernst (2003) reviewed the snake's life history.

Sistrurus miliarius (Linnaeus, 1766)
Pygmy Rattlesnake

RECOGNITION: *S. miliarius* is small (maximum TBL, 83.2 cm, but typically 40–55 cm), with a short tail rattle about 2.7 mm long consisting of little more than a button;

38–39% of individuals are rattleless (about 52% of *S. m. miliarius*) (Rowe et al. 2002). Nine enlarged scales are present on the dorsal surface of the head (see *S. catenatus*); the snake has a series of 22–45 (mean, 32–33) dark dorsal body blotches, 1–2 (3) rows of dark spots on its sides, and a red to orange vertebral stripe. Some pattern variation occurs (Shupe 1977). The body is normally gray or tan, but reddish-orange to brick-red individuals occur along the northeastern edge of the range, chiefly in Beaufort and Hyde counties, North Carolina, and probably both albinistic and melanisistic individuals occur. The venter is whitish to cream-colored with a moderate to heavy pattern of dark blotches. A black or reddish-brown bar extends backward from the eye to beyond the corner of the mouth, and two dark longitudinal, often wavy, stripes are present on the back of the head. Dorsal body scales have keels, and occur in 25 (23–24) anterior rows, 23 (19–25) midbody rows, and 17–18 (15–19) rows near the vent. The venter has 122–148 ventral scutes, 24–40 subcaudals, and an undivided anal plate. On the side of the head are 2 nasals, 1 loreal (lying between the postnasal and upper preocular), 2 preoculars, 3–4 (5–6) postoculars, 4–5 rows of temporals, 10–11 (8–13) supralabials, and 11 (9–14) infralabials.

The hemipenis is illustrated in Campbell and Lamar (2004), Dowling (1975), and Murphy and Crabtree (1988). It is similar to that described for *S. catenatus*, except the sulcus grooves extend to the tips of the two short lobes, the region of large spines covers an area from the naked base to the lower portion of each lobe, and the basal spines are smaller.

The dental formula consists of 1–3 (mean, 2–3) palatine, 7–9 (mean, 7.8) pterygoid, 9–11 (mean, 10.4) dentary teeth, and 1 maxillary fang on each side (Brattstrom 1964; Klauber 1972).

Males are usually smaller (TBL_{max}, 80.3 cm; Snellings and Collins 1996), and have longer tails that are thicker at the base (Bishop et al. 1996). They also have 122–148 (mean, 130) ventrals, 28–39 (mean, 34) subcaudals, 7–14 (mean, 11–12) dark tail bands, and TLs 10–15 (mean, 12.5)% of TBL; the larger females (TBL_{max}, 83.2 cm) have 123–

Sistrurus miliarius miliarius, red morph (Moore County, North Carolina; R. D. Bartlett)

Sistrurus miliarius miliarius, red morph (Barry Mansell)

148 (mean, 135) ventrals, 24–36 (mean, 30) subcaudals, 6-13 (mean, 9) dark tail bands, and TLs 7–12 (mean, 10.5)% of TBL.

GEOGRAPHIC VARIATION: Three subspecies have been described (Beaman and Hayes 2008; Gloyd 1935; Palmer 1978). *Sistrurus miliarius miliarius* (Linneaus, 1766), the Carolina Pygmy Rattlesnake, is gray to reddish-brown with a well-marked head pattern, 1–2 rows of lateral spots, rounded or somewhat quadrangular dorsal spots, usually 25 anterior and 23 midbody scale rows, and a cream-colored venter with the dark ventral spots at least 2 scutes wide. It ranges from Hyde and Beaufort counties, North Carolina, southwestward to central Alabama. *Sistrurus m. barbouri* Gloyd, 1935, the Dusky Pygmy Rattlesnake, is dark gray with an obscured head pattern, 3 rows of lateral spots, rounded dorsal spots, usually 25 anterior and 23 midbody scale rows, and a heavily dark-spotted cream to gray venter. Christman (1980) found that in Florida individuals ventral and subcaudal counts decrease clinally toward the north, and that coastal populations have higher dorsal scale row and blotch counts, and larger, more rounded, dorsal blotches than inland populations. The subspecies' range is from extreme southwestern South Carolina south through peninsular Florida and west through southern Georgia, the Florida Panhandle, and southern Alabama to southeastern Mississippi. *Sistrurus m. streckeri* Gloyd, 1935, the Western Pygmy Rattlesnake, is gray to tan with a well-marked head pattern, the dorsal spots often reduced to irregular cross bars, 1–2 rows of lateral spots, usually 23 anterior and 21 midbody scale rows, and diffuse ventral blotches about 1 scale wide on the white to light gray venter. It occurs from the "Land between the Lakes" in western Kentucky and Tennessee, southern Missouri, and south-central Oklahoma south to the Gulf Coast of Louisiana and central Texas.

CONFUSING SPECIES: *Sistrurus miliarius* can be distinguished from *S. catenatus* by the key presented above, and larger rattlesnakes (*Crotalus*) have small scales between their supraoculars. Hognose snakes (*Heterodon*) have upturned rostral scales and lack a rattle and facial pits. The rattleless yellow-bellied kingsnake (*Lampropeltis calligaster*) and short-tailed snake (*Stilosoma extenuatum*) have smooth scales, round pupils, and divided subcaudals.

KARYOTYPE: Zimmerman and Kilpatrick (1973) described the diploid complement as 16 macrochromosomes (the autosomes include 2 pairs of large to medium-sized metacentrics, 3 pairs of large to medium-sized submetacentrics, and 2 pairs of medium-sized to small subtelocentrics), and 20 microchromosomes; sex determination is ZZ in males and ZW in females.

FOSSIL RECORD: Pleistocene fossils have been recovered from Irvingtonian deposits in Citrus and Hillsbourgh counties, Florida (Auffenberg 1963; Holman 1995, 2000; Hulbert and Morgan 1989; Meylan 1982, 1995), and Rancholabrean sites in Alachua, Marion, and Levy counties, Florida (Auffenberg 1963; Gut and Ray 1963; Holman 1959a, 1996, 2000; R. A. Martin 1974; Tihen 1962).

DISTRIBUTION: *S. miliarius* ranges from the Albemarle Sound in northeastern North Carolina south to the Florida Keys, west along the Gulf Coast to Louisiana and southeastern Texas, north in the Mississippi Valley to the "Land between the Lakes" in southwestern Kentucky, and west to southern Missouri, south-central Oklahoma, and central Texas.

HABITAT: *S. miliarius* lives in a variety of sandy-soiled habitats at elevations from sea level near the coast to about 500 m on the Piedmont. All of its eastern habitats contain open water or wet zones. Included are cedar glades (*Juniperus virginiana*); mixed turkey oak (*Quercus laevis*)–longleaf/loblolly pine (*Pinus palustris, P. taeda*) forest (with *Andropogon* sp.; *Acer rubrum; Baccharis halimifolia; Callicarpa americana; Crategus* sp.; *Diospyros* sp.; *Liquidamber styraciflua; Lonicera* sp.; *Myrica cerifera; Quercus alba, Q. falcata, Q. michauxii, Q. stellata, Smilax* sp.; and *Toxicodendron* sp.); scrub pine (*Pinus virginiana*) woods; pine flatwoods, like *Pinus palustris–P. taeda-Astida stricta* communities

Sistrurus miliarius miliarius, red morph (Hyde County, North Carolina; R. D. Bartlett)

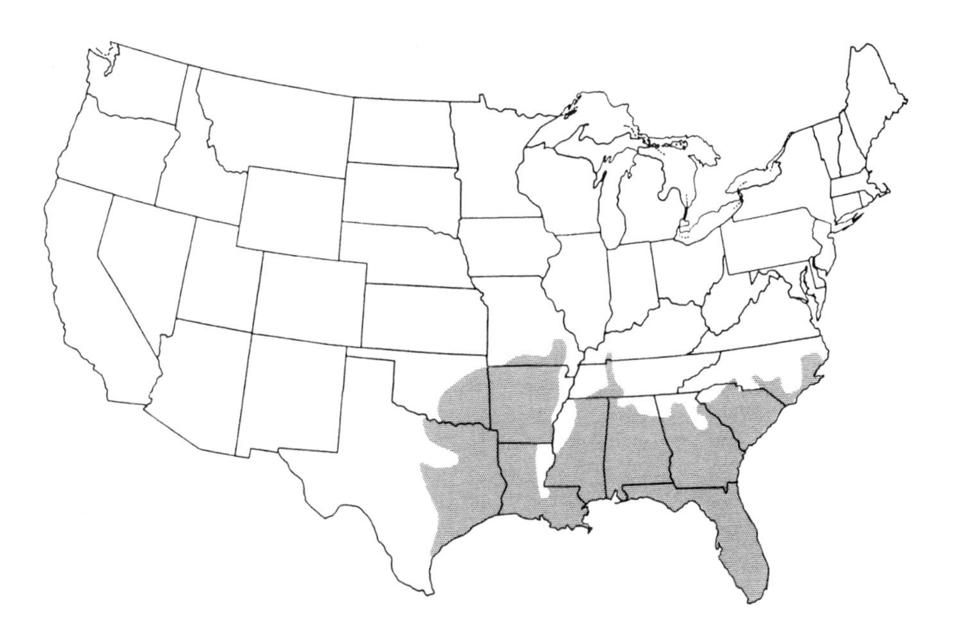

Distribution of *Sistrurus miliarius*.

(with *Andropogon* sp., *Acer rubrum, Liquidamber styraciflua, Magnolia virginiana, Myrica cerifera, Persea borbonia,* and *Oxydendrum arboreum*); sandhills (with, in the Carolinas, *Andropogon* sp.; *Astrida stricta; Cladonia* sp.; *Gaylussacia* sp.; *Heterotheca gossypinus; Opuntia compressa; Petalostemum pinnatum; Pinus palustris; Quercus laevis, Q. merilandica, Q. pumila, Q. virginianus*; and *Vaccinium* sp.), sawgrass (*Cladium jamaicensis*), wiregrass (*Aristida stricta*), and palmetto (*Sabal* sp.) flatwoods. In the western, drier, parts of its range, the snake is restricted to mesic grasslands.

In north-central Florida, the pygmy rattler is one of the most common snakes in upland xeric habitats (Dodd and Franz 1995). Populations may be dense in isolated patches of semievergreen broad-leaved hammocks surrounded by freshwater floodplain marsh (May et al. 1997). Farther south in the Everglades, it is seldom encountered in pinewoods or other dry habitats, but flooding may force it to higher ground like canal banks and roads (Duellman and Schwartz 1958). There it is often found hiding beneath cabbage palm (*Sabal palmetto*) leaves.

The burrows of the gopher tortoise (*Gopherus polyphemus*) and small mice and rats (*Peromyscus polionotus, Podomys floridanus, Sigmodon hispidus*) and gophers (*Geomys* sp.), and the lodges of *Neofiber alleni* are used as retreats (C. Ernst, personal observation; Seyle and Williamson 1982; Witz et al. 1991). Shelters are very important habitat features. They are often found by olfaction, especially if anuran (a choice prey group) odors are initially detected. This was true in laboratory tests by Bevelander et al. (2006); however, after 23 hours only the shelters themselves seemed to be important (perhaps the anuran, *Acris* sp., odors had dissipated). Mouse (*Mus musculus*) and lizard (*Anolis* sp.) odors were not of much interest.

BEHAVIOR AND ECOLOGY: Except for studies by the May-Farrell group in Florida and Palmer and Williamson (1971) in North Carolina, *S. miliarius*, although often common, has been largely ignored, so many aspects of its life history are only generally known.

From North Carolina to Florida, the species may remain active all year, but at lower numbers during the winter. Over the entire range, most surface activity occurs in June–September, but in North Carolina, 82% of activity is in July–October (Palmer and Braswell 1995). The earliest emergence there was on 8 March (Palmer and Williamson 1971). Similarly, Chamberlain (1935) reported that 82% of the *S. miliarius* he collected in South Carolina were taken from July through September. Florida snakes are active in all months, but have a bimodal seasonal movement pattern—peaking in March–April, moving least in December–February and July–August (May et al. 1996). There, the highest incidence of surface activity occurs during periods of high water table. Missouri pygmy rattlers are active from mid-April to mid-October (T. R. Johnson 1987).

Little is known of the snake's overwintering behavior, but hibernating individuals have been found in saw dust piles and old logs (Klauber 1972; Neill 1948; Palmer and Williamson 1971), and they probably also wait out the cold weather in small mammal and gopher tortoise burrows. Those in Florida do not go deep (May et al. 1997), often just under leaf litter or other debris (C. Ernst, personal observation).

Missouri *S. miliarius* are diurnal, especially in the afternoon, during the cooler days of spring; but, as the daytime temperatures rise during the summer, they become more crepuscular or nocturnal (T. R. Johnson 1987). Eastern *S. miliarius* are active from the late afternoon into the night, often basking in the morning. In the fall, winter, and early spring, activity is primarily in the afternoon, when it usually basks. More than 75% of individuals found are coiled either in an ambush mode or are basking, and both juveniles and adults heat themselves in this manner. May et al. (1996, 1997) reported it surface-active in Florida at ATs of 13–32°C and BTs of 15–37°C (BT averaged 1.9°C above AT); microsites used averaged 0.4°C above surrounding STs. Clark (1949) found them basking in Louisiana at ATs as high as 38–45°C. Captives have bred when maintained in a temperature range of 22°C at night to 28°C during the daytime (Smetsers 1990).

The home range is small. Individuals monitored by Hudnall (1979) moved a maximum distance of 9–242 m from their first capture point. A male averaged 179.6 m between recaptures, but 2 others only averaged 81.0–89.5 m. Gravid females are rather sessile; during 34 days in Texas, one was never found more than 2 m from its original capture point (Fleet and Kroll 1978). Individuals with pit tags and those without moved averages of 98.9 m and 92.5 m, respectively, between captures, and 1.76 m per day and 1.54 m per day, respectively (Jemison et al. 1995).

Although usually found on the ground, *S. miliarius* will climb into trees and bushes; Klauber (1972) reported that one had ascended 8 m up a tree, and May et al. (1997) have found them as high as 3.6 m in palm trees during flood events. The snake is, however, a very good swimmer.

Males participate in dominance combat dances similar to those reported for other rattlesnakes (C. Ernst, personal observation; Lindsey 1979; Palmer and Williamson 1971). Such contests are most prevalent in late August and September during the mating season. Two will rise up, face each other, sometimes sway back and forth, lunge toward each other, entwine the posterior 40–50% of their bodies, and push with the raised anterior body until one of the two is pinned firmly to the ground. The vanquished snake then crawls away unharmed. Such bouts may last for a few minutes to more than two hours, and may be continuous or sporadic.

Sistrurus miliarius barbouri (Collier County, Florida; Roger W. Barbour)

REPRODUCTION: Females are mature at TBLs of 33.2–38.1 cm and masses of 50–150 g (Clark 1949; Conant and Collins 1998; May et al. 1997; Sabath and Worthington 1959). Males are mature at a minimum TBL of 38 cm (Conant and Collins 1998). Such lengths are probably reached in 2–3 years in nature, but 1- to 2-year-old captive females have given birth (Smetsers 1990).

The male gametic cycle has not been described, but it probably follows the stages outlined for other pitvipers by Aldridge and Duvall (2002). Some data on the female cycle exist. Females apparently store sperm obtained during fall matings in infundibular tubules (Seigel and Sever 2006), and fertilization occurs the following spring (Montgomery and Schuett 1989; Schuett 1992). May et al. (1997) found Florida *S. m. barbouri* gravid in early April. As their embryos grew, these snakes sought exposed places and basked to raise their BT to enhance embryonic growth. In more than 100 Florida litters, all but a few young were born in August. Iverson (1978) reported that Florida females have had enlarged ovarian follicles on 7 January, partly developed embryos on 2 July, nearly full-term fetuses on 16 July, and had given birth on 15 July, 2 August, and 4 October. Florida females experience both an annual and a biennial reproduction (Farrell et al. 1995), and in any given year 30–70% of the females are gravid (Farrell et al. 2008). North Carolina *S. m. miliarius* have contained oviductal young on 27 May, 4 July, and during August (Palmer and Braswell 1995). A wild Louisiana *S. m. streckeri* gave birth in August (Smith and List 1955), and captive *S. m. streckeri* from Louisiana have reproduced annually (Raymond 2004).

All observations of mating in the wild were made from mid-August to early November, primarily September–October in Florida. Although no spring dates have been reported, it is possible that some matings occur after emergence from hibernation. Captives have mated in December and January (Drent 1989; Verkerk 1986, 1987).

All observations of associated bisexual pairs of Florida *S. miliarius* by May et al. (1997) were between September and January. These interactions consist of two phases.

First, there is an extended mate-guarding sequence, in which the male remains in close physical proximity to a reproductive female; the pair may remain in contact, often with one snake coiled on top of the other, for several days at a time without moving. The second phase is the actual mating. During courtship, a male waves his tail as he crawls after a female. When she is reached, he examines her with tongue-flicks, crawls over her, and eventually inserts one hemipenis; copulation may last more than six hours (Montgomery and Schuett 1989; May et al. 1997; Smetsers 1990).

After a long GP of 107–294 (mean, 172; n = 7) days, the ovoviviparous young are born from late July through September, with August being the most important month. In Florida, females giving birth to clutches with large masses undergo parturition earlier than those producing clutches of lighter mass (Farrell et al. 1995). Captive births have occurred in January and from April to late August (Bronsgeest 1992b; Drent 1989; Raymond 2004; Verkerk 1986, 1987). At birth, the young are enclosed in a sheath-like membrane, and appear at intervals of 30–80 (mean, 46) minutes after a series of posteriorly progressing peristaltic contractions (Fleet and Kroll 1978). The young gape their jaws and appear to yawn (their first breath?) after breaking through the fetal membranes. Messenger (2008) described a morphological anomaly consisting of a short head and underdeveloped eyes that bulged out laterally.

Litters contain 2–32 young (Carpenter 1960; Farrell et al. 1995; Verkerk 1987), and average 5.9 (n = 68) young. Northern *S. miliarius* produce smaller litters (Fitch 1985). RCMs of several litters averaged 32.7 (range, 19.0–50.2) % (Farrell et al. 1995; Ford et al. 1990; Seigel et al. 1986; Seigel and Fitch 1984; Smetsers 1990). Both litter size and clutch mass are positively correlated with female body size (Bishop et al. 1996; Farrell et al. 1995). The proportion of viable young per litter is about 88–97%. Neonates have 7.8–23.0 (mean 12.3; n = 118) cm TBLs, BMs of 0.8–6.2 (mean, 2.4; n = 45) g, and cream, yellow, or orangish tails.

After 10 years of studying wild *S. m. barbouri,* Farrell et al. (2009) determined that females have repeatability of reproductive traits from one litter to the next: neonate SVL, mean mass, litter size and total mass, birth dates in days after 15 July, and female postpartum condition.

An apparent defense of young by a captive female was witnessed by Verkerk (1987). When disturbed, the young hid behind her back while she rattled and tried to bite the observer. During laboratory and field experiments conducted by Greene et al. (2002), 59 captive female *S. miliarius*, when separated by size-selected barriers under 3 experimental conditions, returned to their litters and their neonates returned to them after the barriers were removed. Also, the attending females were more aggressive toward a natural predator than before birth or after neonatal ecdysis. During field studies, 25 free-living females remained with their young during the postpartum ecdysis cycle.

GROWTH AND LONGEVITY: In Florida, the mean annual growth rate of juvenile male *S. m. barbouri* is 3.5 cm, and that of juvenile females is 2.6 cm; but the growth rate is not significantly different between the sexes (Bishop et al. 1996). Florida juveniles grow about 10.5 cm and increase their BM about 75 g between birth and their third year. Most growth in Florida occurs within the first 2–3 years of life; the body length of adults remains relatively stable (May et al. 1997). A male *S. m. barbouri,* 23.5 cm in TBL when collected, grew to 60.9 cm in 2 years in our laboratory. Another male of that subspecies, 65 cm long when captured by Snellings and Collins (1996) in March 1981,

Sistrurus miliarius barbouri (Florida, R. D. Bartlett)

grew to 80.3 cm by 1987 and was thought to be about 20 years old when it died in April 1993. A Mississippi neonate *S. m. streckeri* grew to 17.3 cm in 6 months (Smith and List 1955).

A lack of growth in adult *S. m. barbouri* and the tendency of some females to not reproduce each year appears to foster the repeatability of reproductive traits previously discussed (Farrell et al. 2009).

Insertion of a pit tag apparently does not influence the rate of growth (Jemison et al. 1995).

A wild-caught female *S. m. streckeri* lived an additional 16 years, 1 month, and 4 days at the Houston Zoo (Snider and Bowler 1992).

DIET AND FEEDING BEHAVIOR: Wild *S. miliarius* eat a variety of small animals. Reported prey include beetles (*Copris minutus*), crickets, moth and butterfly caterpillars, spiders, centipedes (*Scolopendra heros*), minnows [?] (Cyprinidae), salamanders (*Hemidactylium scutatum*), anurans (*Acris* sp.; *Bufo* sp.; *Gastrophryne carolinensis; Hyla* sp.; *Rana clamitans, R. sphenocephala*), lizards (*Anolis* sp.; *Cnemidophorus sexlineatus; Eumeces egregius, E. inexpectatus; Scincella lateralis*), small snakes (*Carphophis amoenus; Coluber constrictor; Diadophis punctatus; Nerodia* sp.; *Sistrurus miliarius; Storeria dekayi; Thamnophis sauritus. T. sirtalis; Virginia valeriae*), nestling birds, shrews (*Blarina brevicauda, B. carolinensis; Sorex dispar*), and small rodents (*Microtus pinetorum, Oryzomys palustris, Peromyscus maniculatus, Reithrodontomys humulis*) (Anderson 1965; Chamberlain 1935; Clark 1949; Cook 1954; Dundee and Rossman 1989; C. Ernst, personal observation; Fann 1997; Farrell et al. 1995; Hamilton and Pollack 1955; Jacob 1981; Klauber 1972; May et al. 1997; Mitchell 1903; Palmer and Braswell 1995; Palmer and Williamson 1971; Trauth and Cochran 1991; Trauth et al. 2004; Trauth and McAllister 1995; Travers et al. 2008; Werler and Dixon 2000; Wright and Bishop 1915; Wright and Wright 1957). Neill and Allen (1956) thought the invertebrate prey may have been secondarily ingested during the swallowing of lizards.

Captives readily take crickets, anurans (*Pseudacris crucifer*), lizards (*Anolis carolinensis; Eumeces fasciatus, E. inexpectatus; Hemidactylus turcicus; Sceloporus undulates; Scincella lateralis*), snakes (*Carphophis amoenus, Diadophis punctatus, Nerodia sipedon, Sonora semiannulata, Storeria dekayi, Thamnophis sirtalis, Virginia striatula*), small mice (*Mus musculus, Peromyscus maniculatus*), and small rats (*Rattus norvegicus*) (Anderson 1965; C. Ernst, personal observation; T. R. Johnson 1987; Kennedy 1964; May et al. 1997; Montgomery and Schuett 1989; Palmer and Williamson 1971; Verkerk 1987).

Twelve of 16 Georgia *S. miliarius* examined by Hamilton and Pollack (1955) contained prey; reptiles were found in 50% of the stomachs, centipedes in 33%, and mammals in 17%. In Florida, the proportion of snakes with prey is highest in March, May, and June, is lower in the summer, but increases again in September–October; snakes make up less than 10% of the winter food, but 20–25% in late spring or fall (May et al. 1996, 1997).

The isotopic composition of young *S. miliarius barbouri* is highly variable (<95%) between litters from different Florida localities; within-litter differences only spanned 2.8%. Pilgrim (2007) interpreted the high variation between sites as indicating largely the retention of a maternal signal rather than differences in prey utilization among the young snakes. Adults mostly ambush their prey, lying quietly along a prey trail by a grass clump or at the side of a log with their heads oriented upward. Such sites are probably chosen because of their strong prey odor, particularly that left by anurans (Roth et al. 1999).

Juveniles lure both frogs and lizards by waving their yellowish tails back and forth (C. Ernst, personal observation; Jackson and Martin 1980; Neill 1960; Rabatsky and Waterman 2005), although Reiserer (2002) reported that they only lure frogs. The prey is attracted within striking distance and then bitten. For the tail-luring to be effective, light levels must conceal the body while illuminating the bright tail (Rabatsky and Farrell 1996). During tests using both adults and juveniles, only juveniles tail-lured, and those that lured the longest time were the most successful in attracting prey (Rabatsky and Waterman 2005). Both sexes lured at equal frequencies, but females took longer to achieve a similar level of foraging success as males. TL is sexually dimorphic in *S. miliarius* (see "Recognition"), and Rabatsky and Waterman (2005) suggested that the longer tail of males may be a more effective luring adaptation. It apparently has an optical advantage over the shorter tail of the female (Schuett et al. 1984). Rabatsky (2008) noted that length of the rattle segment does not affect luring success; so probably tail-luring was not a precursor in the development of the rattle of *Sistrurus* and *Crotalus*. Instead, use as a warning to predators may have been the evolutionary stimulus.

Some prey, however, are actively sought. Many pygmy rattlesnakes we have found were moving and seemed to be searching for prey. C. Ernst observed one in Florida stalk (with much tongue-flicking), strike, and swallow an *Anolis carolinensis*. Again, prey odor is most likely the leading cue, but vision cannot be discounted, especially when the prey is moving.

However, Bevelander et al. (2006) reported that the *S. m. barbouri* they tested, using anuran (*Acris* sp.), lizard (*Anolis* sp.), and mouse (*Mus musculus*) odors, and water as a control, initially preferred anuran odor over the others, but that prey odor preference soon declined. The snakes indicated no prey odor preference during the first 2 hours of a total 23 hours of testing. Their pattern of circadian foraging activity had two peaks, one shortly before the beginning of the nocturnal phase and a second just before 2400 hours. After this, activity gradually declined, and most of the snakes had entered shelters by morning, when the diurnal phase began. Those in shelters coiled into their stereotypic ambush postures facing the shelter entrance, where a prey was most likely to pass by.

Unlike the larger North American *Crotalus*, gravid female *S. miliarius* normally feed until near the time of parturition.

VENOM DELIVERY SYSTEM: *S. miliarius* has short fangs. Five *S. m. barbouri* with 66.6–70.2 cm TBLs had 5.2–6.3 (mean, 5.6) mm FLs, a mean TBL/FL of 121%, and a mean HL/FL of 5.7% (Klauber 1939); and the mean angles of fang curvature of *S. m. barbouri* and *S. m. streckeri* were 65° and 61°, respectively (Klauber 1939). FL is directly proportional to the snake's SVL (C. Ernst, personal observation). The fang is about 1.0 to 1.3 times longer than the height of the maxilla (Campbell and Lamar 2004).

VENOM AND BITES: The venom is extremely virulent, but the snake is small. Although Klauber (1972) reported a total dried venom yield of 125 mg, the average total yield is as low as 12–35 mg (Ernst and Zug 1996; Klauber 1972; Minton and Minton 1969), and a typical bite probably involves only about 18.0–34.6 mg (Allen and Maier 1941; do Amaral 1928b; Klauber 1972).

The venom does not cause strong neurotoxic symptoms, but does contain strong hemorrhagins and some fibrinolytic activity occurs (Bonilla et al. 1971; Kitchens 1992; Scarborough et al. 1991; Tan and Ponnudarai 1991; Van Mierop 1976a, 1976b). It reacts electrophoretically at a pH of 2.9, and Bonilla et al. (1971) recovered the following percentages of protein fractions from partially purified venom: fraction I, 24.7%; fraction II, 46.5%; and fraction III, 15.8%. Included are enzymes that degrade all classes of biomolecules: ATPase, barbourin (a protease similar to that of some true vipers that inhibits blood-clotting), 5′-nucleotidase, kallikrein-like proteases, L-amino acid oxidase, metalloproteases, myotoxin α-like proteins, phosphodiesterase, phospholipase A_2, phosphomonoesterase, serotonin, and trypsin-like (tryptamine) compounds are present (Aird 2005; Bober et al. 1988; Bonilla et al. 1971; Jackson 2005; Scarborough et al. 1991; Welsh 1966).

It is very toxic to mammals. Various reported mouse (*Mus musculus*) LD_{50} values are intravenous—2.80–4.47 mg/kg (Friederich and Tu 1971; Kocholaty et al. 1971; Vick 1971), 6.00–6.84 mg/kg (Githens and Wolff 1939; Kocholaty et al. 1971), and 0.28 mg/20 g (Bonilla et al. 1971), and subcutaneous—24.25 mg/kg (Minton 1956). Small mice bitten by our adult *S. miliarius* have died in 30–90 seconds. A *Crotalus horridus* bitten by a captive adult female *S. miliarius* was not affected by the venom and only suffered from the resulting wound punctures (Munro 1947).

Human envenomation by the pygmy rattlesnake is not uncommon, particularly in Florida (44% of envenomations). Most bites are on the hand or fingers, and occur as the snake is either handled (in captivity, and often in the wild when mistaken for a juvenile hognose snake, *Heterodon* sp.) or the hand is accidentally brought into close proximity to an undetected snake in the wild. Bites have also been afflicted to bare feet (C. Ernst, personal observation; Kauffman 1928); going barefoot is a bad practice where this small snake is common.

Symptoms experienced have included serum seeping from the bite punctures, swelling, pain (often immediately like the stinging sensation from a bee sting), weakness,

Sistrurus miliarius barbouri, juvenile (Florida, R. D. Bartlett)

giddiness, rarely defibrination, respiratory difficulty, a drop in the blood platelet count (thrombocytopenia), hemorrhage, nausea, ecchymosis, the passage of bloody urine (hematuria) with rare renal failure, and prolonged numbness at the bite site (Banner et al. 2007; Chamberlain 1935; Harris 1965; Hutchison 1929; Keyler 1988; Kitchens 1992; Klauber 1972; Norris 2004; Russell 1983; Schmidt and Inger 1957; Van Mierop 1976b; Vick 1971). The bite is more serious in children than adults, and small children may require a period of hospitalization when bitten (Guidry 1953). Antivenom administration is not very effective. Case histories of bites to humans are given by Chamberlain (1935), Harris (1965), Klauber (1972), and Schmidt and Inger (1957).

The lethal dose for an adult human has not been calculated, but is probably more than the total venom capacity of the snake.

PREDATORS AND DEFENSE: Although few cases of predation have been reported, it is certain that *S. miliarius* has many natural enemies, especially when neonate or subadult. Known or potential predators include snakes (*Coluber constrictor, Drymarchon corais, Lampropeltis getula, Masticophis flagellum, Micrurus fulvius, Sistrurus miliarius*), hawks (*Buteo* sp.), shrikes (*Lanius ludovicianus*), owls (*Bubo virginianus*), opossums (*Didelphis virginianus*), raccoons (*Procyon lotor*), bears (*Ursus americanus*), skunks (*Mephitis mephitis, Spilogale putorius*), canids (*Canis familiarius, C. latrans; Urocyon cinereoargenteus; Vulpes vulpes*), and felids (*Felis catus, Lynx rufus*) (Allen and Neill 1950b; Babis 1949; Campbell and Lamar 2004; Klauber 1972; Printiss 1994; Tennant 1997).

The dark spotted pattern and grayish coloration of this species helps camouflage it. This is especially true of Florida *S. m. barbouri* from areas where pine trees have been partially burned and ashes are abundant. There, its color closely matches the sandy soil, and its black markings resemble pieces of charred wood.

In our experience, *S. miliarius* is a nasty little snake, with a fiery temper! When disturbed, the small snake tries to flee, quickly coils, and bobs its head (Marchisin 1978), or flattens its body and strikes sideways (our experience with all three subspecies). It rarely rattles, but does move its tail about, particularly individuals with fewer than two rattle segments, possibly to distract the disturber (Rabatsky and Waterman 2005). Those not coiled when first found are more active in their defense than those coiled, which usually remain calm until further disturbed. When approached too closely, the snake bobs its head and strikes with little warning.

In contrast to our observations, Glaudas et al. (2005) found Florida *S. m. barbouri* to be rather nonaggressive when encountered. Upon detection, they tapped the snout of 336 of the snakes with a gloved hand, and recorded if the snakes struck or fled. Only 27 (8%) bit the glove. Initial posture influenced the striking behavior; uncoiled snakes struck significantly more than those that were coiled. Fleeing behavior was affected by sex (females fled more often than males), ecdysis state (those about to shed tried to flee more often than those between sheddings), and initial posture (uncoiled individuals fled more often than coiled ones). Perhaps, as suggested by Glaudas et al. (2005), most of the pygmy rattlers that we encountered were not coiled but were moving at the time; although we remember several that were coiled, and those kept in our laboratory repeatedly struck from a coiled position.

The species has a very small rattle that can barely be heard. The SVL adjusted loudness is barely over 40 dB, and the frequency is 2.85–24.38 kHz (most common frequency, 9.95–14.3 kHz) (Cook et al. 1994; Young and Brown 1993). While rattling, the body and tail of *S. miliarius* consume averages of 2.4 and 10.9 μmL/mg BM of O_2 per hour, respectively; similarly, mean changes in the optical densities of 53.8 and 138.1 $\times 10^{-6}$/mg BM of succinic dehydrogenase, and 67.3 and 299.3 $\times 10^{-4}$/minute/mg BM of cytochrome oxidase occur, respectively, in the two body regions (Moon 2001).

With the aid of its vomeronasal organ (Miller and Gutzke 1999), *S. miliarius* recognizes the odor of a predatory snake, like *Drymarchon corais* and various species of *Lampropeltis* (Weldon et al. 1992), and when introduced to it will either hide its head, thrash about, or body-bridge (Carpenter and Gillingham 1975; Gutzke et al. 1993).

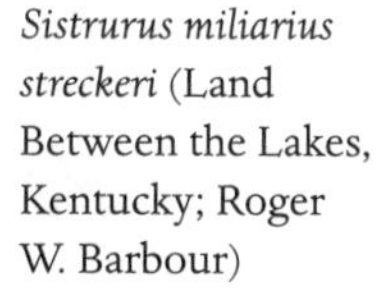

Sistrurus miliarius streckeri (Land Between the Lakes, Kentucky; Roger W. Barbour)

PARASITES AND PATHOGENS: Pygmy rattlers are parasitized by the trematode *Ochetosoma kansense*, and it and the nematode *Kallicephalus inermis* have been found in snakes only identified to *Sistrurus* sp. (Ernst and Ernst 2006). Cheatwood et al. (2003) reported an outbreak of fungal dermatitis (*Galaetomyces geotrichum, Geotrichum candidum, Pestalotia pezizoides, Sporothrix schenckii*) in a wild population in Florida. Lee (1969) found the maggots of the fruit fly, *Drosophila hydei*, in skin wounds of captive *S. m. barbouri*, but such parasitism must be very rare at best in the wild.

POPULATIONS: Some data are available. In the proper habitat, this species can be quite numerous; the Florida population of *S. m. barbouri* studied by May et al. (1996, 1997) had 400–500 within about 8 ha, a density of 50–62.5/ha; this is a closed population with recruitment of <50–200 per year through birth (Farrell et al. 2008). In mesic upland habitats at the Katherine Ordway Preserve-Swisher Memorial Sanctuary in Florida, *S. m. barbouri* comprised 26 (11.3%) of a total of 230 snakes trapped in 1989 and 1990 by Dodd and Franz (1995). It ranked fourth behind *Coluber constrictor, Heterodon platirhinos*, and *Masticophis flagellum*, respectively; however, the snake may have been more common than this figure shows, as it is possibly less prone to being trapped than the other species. At another xeric upland habitat in Florida, Enge and Wood (2002) caught 406 snakes along a 6 km stretch of road in 2.8 years; of these only 3 (0.7%) were this subspecies, which comprised 16.7% of the 18 snake species recorded. Viosca (*in* Dundee and Rossman 1989) collected 103 *S. m. streckeri* in 12 days on Delacroix Island, Louisiana, when the snakes were forced onto levees by severe flooding.

However, at other places, particularly on the Piedmont (Platt et al. 1999), it may be uncommon to rare. Clark (1949) recorded only 11 *S. m. streckeri* in a sample of more than 2,000 snakes from the upland parishes of Louisiana, and Fann (1997) found only 8 *S. m. streckeri* in 4 years (3 DOR, 2 females) in western Tennessee.

Populations are dominated by adults, but in August and September during the birthing season the numbers of subadults increase dramatically. Recruitment of neo-

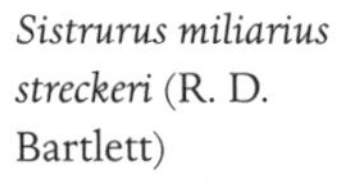

Sistrurus miliarius streckeri (R. D. Bartlett)

nates into the population varies annually, depending on the number of potentially reproductive females in the population, the percentage of these females that are gravid, the mean number of viable offspring per litter, and the proportion of neonates that survive parturition. Mean number of neonates per litter and mean percentage of viable young vary little between years, but the number of mature females in the population and the percentage of those gravid are highly variable (Farrell et al. 2008).

The sex ratio does not differ significantly from 1:1 (Bishop et al. 1996; Carpenter 1960; May et al. 1996).

Although the species is still plentiful over much of its range, especially in Florida, it is in trouble in some areas, including northeastern North Carolina, southwestern Kentucky, and Tennessee (*S. m. streckeri* is considered threatened and is now legally protected in Tennessee). The three major causes of its decline in these regions are habitat destruction (particularly clear-cut lumbering), motorized vehicles, and the pet trade.

REMARKS: Dorsal scale microdermoglyphics indicate *S. miliarius* is more closely related to *Crotalus pricei, C. cerastes,* and *C. mitchellii* than to *S. catenatus* and *C. ravus* (Stille 1987). Aspects of the snake's cardiovascular system were reported by Lillywhite and Smits (1992). The species was reviewed by Palmer (1978).

Glossary of Scientific Names

Agkistrodon Hooked tooth, referring to the fang and other curved teeth

alvarezi A patronym honoring the Mexican herpetologist Dr. Miguel Alvarez del Toro, who collected the type specimen

australis Southern, referring to the snake's distribution

barbouri A patronym honoring the Harvard University herpetologist Thomas Barbour

bilineatus Two lines, referring to the two facial stripes

catenatus Chain-like, referring to the body pattern

charlesbogerti A patronym honoring the American Museum of Natural History herpetologist Dr. Charles M. Bogert

cinctum Banded, girdled; referring to the banded body pattern

conanti A patronym honoring the noted herpetologist Dr. Roger Conant

contortrix To twist, full of motion; referring to the snake's agility

distans Separated, apart; referring to the snake's widely separated red and yellow body bands

edwardsii A patronym honoring U.S. Army surgeon L. A. Edwards, who collected the type specimen

euryxanthus Broad yellow, referring to the yellow bands in the body pattern

exasperatum Irritating, referring to the lizard's scalation, which is difficult to distinguish from other subspecies of *Heloderma horridum*

fitzingeri A patronym honoring Dr. Leopold Fitzinger of the Naturhistorisches Museum Wien, Vienna, Austria

fulvius Reddish-yellow, orange; referring to the banded body pattern

Heloderma Nail skin, referring to the bead-like body scales

horridum Horrid, dreadful; referring to the lizard's venomous condition

howardgloydi A patronym honoring rattlesnake expert Dr. Howard K. Gloyd

laticinctus Broad band, referring to the wide body bands

leucostoma White mouth, referring to the white or cream-colored oral lining displayed when the mouth is gaped

maculatus Spotted, variegated; referring to the black-tipped scales or spots on the red body bands

michoacanensis Belonging to the state of Michoacán, Mexico

microgalbineus Small, yellow bands; referring to the body pattern

Micruroides Like or similar to *Micrurus*; referring to the similarity in banded body patterns

Micrurus Small tail

miliarius Millet-like, referring to the body pattern

mokasen Moccasin, a Native American name for a snake of the genus *Agkistrodon*

neglectus Neglected, poorly known

Pelamis Fish-like, referring to the marine habitat

phaeogaster Dark belly; referring to the gray, dark-mottled venter

pictigaster Painted belly, referring to the reddish-brown to black mottling on the cream-colored venter

piscivorus Fish eating, referring to dietary preference

platura Flat tail, referring to the laterally flattened, oar-like tail

russeolus Somewhat red, referring to the reddish-brown body color

Sistrurus Rattle tail

streckeri A patronym honoring Baylor University herpetologist John K. Strecker, Jr.

suspectum Suspected, distrusted; referring to the venomous nature
tener Having bands, referring to the body pattern
tergeminus Triple, threefold; referring to the spotted body pattern
zweifeli A patronym honoring American Museum of Natural History herpetologist Dr. Richard G. Zweifel

Bibliography

Entries are in alphabetical order by names of senior authors. Under a given senior author, single-author works are arranged by ascending year; works by two authors are ordered alphabetically by last name of second author; works by three or more authors are ordered by year only (with a, b, c extensions used to distinguish et al. citations within the same year).

Abbuhl, B. 2008. *Agkistrodon contortrix mokeson* (northern copperhead): Mid-winter thermoregulation. Herpetol. Rev. 39: 465.

Adler, K. K. 1960. On a brood of *Sistrurus* from northern Indiana. Herpetologica 16: 38.

Ahmed, N. K., K. D. Tennant, F. S. Markland, and J. P. Lacz. 1990a. Biochemical characteristics of fibrolase, a fibrinolytic protease from snake venom. Haemostasis 20: 147–154.

Ahmed, N. K., R. R. Gaddis, K. D. Tennant, and J. P. Lacz. 1990b. Biological and thrombolytic properties of fibrolase—a new fibrinolytic protease from snake venom. Haemostasis 20: 334–340.

Aird, S. D. 1985. A quantitative assessment of variation in venom constituents within and between three nominal rattlesnake subspecies. Toxicon 23: 1000–1004.

———. 2002. Ophidian envenomation strategies and the role of purines. Toxicon 40: 335–393.

———. 2005. Taxonomic distribution and quantitative analysis of purine and pyrimidine nucleosides in snake venoms. Comp. Biochem. Physiol. 140B: 109–126.

———. 2008a. Nucleoside composition of *Heloderma* venoms. Comp. Biochem. Physiol. 150B: 183–186.

———. 2008b. Snake venom dipeptidyl peptidase IV: Taxonomic distribution and quantitative variation. Comp. Biochem. Physiol. 150B: 222–228.

Aird, S. D., and N. J. da Silva, Jr. 1991. Comparative enzymatic composition of Brazilian coral snake (*Micrurus*) venoms. Comp. Biochem. Physiol. 99B: 287–294.

Aird, S. D., C. S. Seebart, and I. I. Kaiser. 1988. Preliminary fractionation and characterization of the venom of the Great Basin rattlesnake (*Crotalus viridis lutosus*). Herpetologica 44: 71–85.

Alagón, A. C., M. E. A. Maldonado, J. Z. Juliá, C. R. Sánchez, and L. D. Possani. 1982. Venom from two sub-species of *Heloderma horridum* (Mexican beaded lizard): General characterization and purification of *N*-benzoyl-L-arginine ethyl ester hydrolase. Toxicon 20: 463–475.

Alagón, A. C., L. D. Possani, J. Smart, and W. D. Schleuning. 1986. Helodermatinae, a kallikrein-like, hypotensive enzyme from the venom of *Heloderma horridum horridum* (Mexican beaded lizard). J. Exp. Med. 164: 1835–1845.

Alape-Girón, A., B. Lomonte, B. Gustafsson, N. J. da Silva, and M. Thelestam. 1994. Electrophoretic and immunochemical studies of *Micrurus* snake venoms. Toxicon 32: 713–723.

Albritton, D. C., H. M. Parish, and E. R. Allen. 1970. Venenation by the Mexican beaded lizard (*Heloderma horridum*): Report of a case. South Dakota J. Med. 23: 9–11.

Alcala, A. C. 1986. Guide to Philippine flora and fauna: vol. X. Amphibians and reptiles. Natural Resources Management Center, Ministry of Natural Resources and Univ. Philippines, Manila.

Aldridge, R. D. 1975. Environmental control of spermatogenesis in the rattlesnake, *Crotalus viridis*. Copeia 1975: 493–496.

———. 1979a. Seasonal spermatogenesis in sympatric *Crotalus viridis* and *Arizona elegans* (Reptilia, Serpentes) in New Mexico. J. Herpetol. 13: 187–192.

———. 1979b. Female reproductive cycles of the snakes *Arizona elegans* and *Crotalus viridis*. Herpetologica 35: 256–261.

———. 2002. The link between mating season and male reproductive anatomy in the rattlesnakes *Crotalus viridis oreganus* and *Crotalus viridis helleri*. J. Herpetol. 36: 295–300.

Aldridge, R. D., and W. S. Brown. 2002. Male reproductive cycle, age at maturity, and cost of reproduction in the timber rattlesnake *(Crotalus horridus)*. J. Herpetol. 29: 399–407.

Aldridge, R. D., and D. Duvall. 2002. Evolution of the mating season in the pitvipers of North America. Herpetol. Monogr. 16: 1–25.

Aldridge, R. D., B. C. Jellen, M. C. Allender, M. J. Dreslik, D. B. Shepard, J. M. Cox, and C. A. Phillips. 2008. Reproductive biology of the massasauga (*Sistrurus catenatus*) from south-central Illinois, pp. 403–412. *In* W. K. Hayes, K. R. Beaman, M. D. Cardwell, and S. P. Bush (eds.), The biology of rattlesnakes. Loma Linda Univ. Press, Loma Linda, California.

Al-Joufi, A. M. H., and G. S. Bailey. 1994. A survey of kininase, tyrosine esterase, kininogenase and arginine esterase activities in some snake venoms. Comp. Biochem. Physiol. 108B: 221–224.

Allen, E. R. 1949. Observations of the feeding habits of the juvenile cantil. Copeia 1949: 225–226.

Allen, E. R., and E. Maier. 1941. The extraction and processing of snake venom. Copeia 1941: 248–252.

Allen, E. R., and W. T. Neill. 1950a. The vertical position of the pupil in crocodilians and snakes. Herpetologica 6: 95–96.

———. 1950b. The pigmy rattlesnake. Florida Wildl. 5(4): 10–11.

———. 1950c. Coral snake. Florida Wildl. 5(5): 14–15, 22.

Allen, E. R., and D. Swindell. 1948. Cottonmouth moccasin in Florida. Herpetologica 4(suppl. 1): 1–16.

Allen, W. B. 1955. Some notes on reptiles. Herpetologica 11: 228.

———. 1956. The effects of a massasauga bite. Herpetologica 12: 151.

Allender, M. C., M. A. Mitchell, C. A. Phillips, K. Gruszynski, and V. R. Beasley. 2006. Hematology, plasma biochemistry, and antibodies to select viruses in wild-caught eastern massasauga rattlesnakes (*Sistrurus catenatus catenatus*) from Illinois. J. Wildl. Dis. 42: 107–114.

Allender, M. C., M. A. Mitchell, M. J. Dreslik, C. A. Phillips, and V. R. Beasley. 2008. Measuring agreement and discord among hemagglutination inhibition assays against different ophidian paramyxovirus strains in the eastern massasauga (*Sistrurus catenatus catenatus*). J. Zoo Wildl. Med. 39: 358–361.

Almeida-Santos, S. M., and M. G. Salomão. 2002. Reproduction in neotropical pitvipers, with emphasis on species of the genus *Bothrops*, pp. 445–462. *In* G. W. Schuett, M. Höggren, M. E. Douglas, and H. W. Greene (eds.), Biology of the vipers. Eagle Mountain Publ., Eagle Mountain, Utah.

Altimari, W. 1998. Venomous snakes: A safety guide for reptile keepers. Soc. Stud. Amphib. Rept. Herpetol. Circ. 26: i–iv, 1–24.

Alvarez, F., and A. Celis. 2004. On the occurrence of *Conchoderma virgatum* and *Dosima fascicularis* (Cirripedia, Thoracica) on the sea snake, *Pelamis platurus* (Reptilia, Serpentes) in Jalisco, Mexico. Crustaceana (Leiden) 77: 761–764.

Álvarez del Toro, M. 1960. Los reptiles de Chiapas. Inst. Zool. Estudo, Tuxtla Gutierrez, Chiapas, Mexico.

———. 1972 [1973]. Los reptiles de Chiapas. 2nd ed. Inst. Hist. Nat. Estado, Tuxtla Gutiérrez, México.

———. 1982. Los reptiles de Chiapas. 3rd ed. Publicación del Instituto de Historia Natural, Tuxtla Gutiérrez, Chiapas, Mexico.

Alvarez-León, R., and J. I. Hernández-Camacho. 1998. Notas sobre la occurrencia de *Pelamis platurus* (Reptilia: Serpentes: Hydrophiidae) en el Pacífico Colombiano. Caldasia 20: 93–102.

Ambrosio, A. L. B., M. C. Nonato, H. S. Selistre de Auaújo, R. Arni, R. J. Ward, C. L. Ownby, D. H. F. de Souza, and R. C. Garratt. 2005. A molecular mechanism for Lys^{49}-phospholipase A_2 activity based on ligand-induced conformational change. J. Biol. Chem. 280: 7326–7335.

Anderson, C. D. 2006. Utility of a set of microsatellite primers developed for the massasauga rattlesnake (*Sistrurus catenatus*) for population genetic studies of the timber rattlesnake (*Crotalus horridus*). Mol. Ecol. Notes 6: 514–517.

Anderson, P. 1942. Amphibians and reptiles of Jackson County, Missouri. Bull. Chicago Acad. Sci. 6: 203–220.

Anderson, P. C. 1998. Bites by copperhead snakes in mid-Missouri. Missouri Med. 95: 629–632.

Anderson, P. K. 1965. The reptiles of Missouri. Univ. Missouri Press, Columbia.

Andreadis, P. 2004. Ay, there's the rub: Hunting by an impatient ambusher, pp. 7–8. *In* Proceedings of the Snake Ecology Group 2004 Conference, Jackson County, Illinois.

Andrews, E. H., and C. B. Pollard. 1953. Report of snake bites in Florida and treatment: Venoms and antivenoms. J. Florida Med. Assn. 40: 388–437.

Andrews, K. M., and J. W. Gibbons. 2005. How do highways influence snake movement? Behavioral responses to roads and vehicles. Copeia 2005: 772–782.

Anon. 1910. Massasauga fatally bites little girl. Marshall News, Marshall, Michigan 14(19): 1.

———. 1929a. The venom of *Sistrurus catenatus*. Bull. Antiv. Inst. Am. 2: 108–109.

———. 1929b. The effect of snake venoms upon Protozoa. Bull. Antiv. Inst. Am. 3:89.

Anton, A. H., and J. F. Gennaro, Jr. 1965. Norepinephrine and serotonin in the tissues and venoms of two pit vipers. Nature (London) 208: 1174–1175.

Anton, T. G. 1994. Observation of predatory behavior in the regal ringneck snake (*Diadophis punctatus regalis*) under captive conditions. Bull. Chicago Herpetol. Soc. 29: 95.

———. 2000. *Sistrurus catenatus* (massasauga): Litter size. Herpetol. Rev. 31: 248.

———. 2004. Management of urban massasauga (*Sistrurus catenatus*) populations: An incitement to discussion, p. 8. *In* Proceedings of the Snake Ecology Group 2004 Conference, Jackson County, Illinois.

Antonio, F., and E. C. Greiner. 2003. *Agkistrodon contortrix contortrix* (southern copperhead): Endoparasites. Herpetol. Rev. 34: 59–60.

Appleby, E. C., and W. G. Siller. 1960. Tails of Gila monsters and beaded lizards, pp. 39–44. *In* Proceedings of the 1991 Northern California Herpetological Society Conference on Captive Propagation and Husbandry of Reptiles and Amphibians.

Applegate, R. D. 1995. *Sistrurus catenatus catenatus* (eastern massasauga). Food habits. Herpetol. Rev. 26: 206.

Applegate, R. W. 1999. *Heloderma suspectum-horridum*: The horrible, suspicious, venomous lizards of the new world. Reptilia 7: 28–33.

Arce, V., E. Rojas, C. L. Ownby, G. Rojas, and J. M. Gutiérrez. 2003. Preclinical assessment of the ability of polyvalent (Crotalinae) and anticoral (Elapidae) antivenoms produced in Costa Rica to neutralize the venoms of North American snakes. Toxicon 41: 851–860.

Ariano Sánchez, D. 2003. Distribución e historia natural del Escorpión, *Heloderma horridum charlesbogerti* Campbell y Vannini (Sauria: Helodermatidae), en Zacapa, Guatemala y caracterización de su veneno. Tesis Licenciado en Biología, Universidad del Valle de Guatemala, Guatemala, Guatemala.

Arnberger, L. P. 1948. Gila monster swallows quail eggs whole. Herpetologica 4: 209–210.

Arnett, J. R. 1976. Mexican beaded lizards in captivity, pp. 107–107d. *In* American Association of Zoological Parks and Aquariums, Regional Conference Proceedings, American Zoo and Aquarium Association, Silver Spring, Maryland.

Arni, R. K., and R. J. Ward. 1996. Phospholipase A_2—a structural review. Toxicon 34: 827–841.

Arnold, R. E. 1982. Treatment of rattlesnake bites, pp. 315–338. *In* A. T. Tu (ed.), Rattlesnake venoms: Their actions and treatment. Marcel Dekker, New York.

Arrington, O. N. 1930. Notes on the two poisonous lizards with special reference to *Heloderma suspectum*. Bull. Antiv. Inst. Am. 4: 29–35.

Arthur, W. J., III, and D. H. Janke. 1986. Radionuclide concentrations in wildlife occurring at a solid radioactive waste disposal area. Northwest Sci. 60: 154–159.

Ash, L. R. 1962. Development of *Gnathostoma procyonis* Chandler, 1942, in the first and second intermediate hosts. J. Parasitol. 48: 298–305.

Ashton, R. E., Jr., and P. S. Ashton. 1981. Handbook of reptiles and amphibians of Florida. Part 1. The snakes. Windward Publ., Miami, Florida.

Atkinson, D. A. 1901. The reptiles of Allegheny County, Pennsylvania. Ann. Carnegie Mus. 1: 145–157.

Atkinson, D. A., and M. G. Netting. 1927. The distribution and habits of the Massasauga. Bull. Antiv. Inst. Am. 1: 40–44.

Auffenberg, W. 1963. The fossil snakes of Florida. Tulane Stud. Zool. 10: 131–216.

Austin, C. R. 1965. Fine structure of the snake sperm tail. J. Ultrastructure Res. 12: 452–462.

Avery, R. A. 1982. Field studies of body temperatures and thermoregulation, pp. 93–166. *In* C. Gans and F. H. Pough (eds.), Biology of the Reptilia, vol. 12: Physiological ecology. Academic Press, New York.

Babb, R. D. 1989. *Micruroides euryxanthus* (Arizona coral snake). Size. Herpetol. Rev. 2: 53.

Babb, R. D., and E. A. Dugan. 2008. *Agkistrodon bilineatus* (cantil). México: Sonora. Herpetol. Rev. 39: 110.

Babcock, H. L. 1928. Notes on the treatment of rattlesnake bites during colonial times in Massachusetts. Bull. Antiv. Inst. Am. 2: 77–78.

Babis, W. A. 1949. Notes on the food of the indigo snake. Copeia 1949: 147.

Backshall, S. 2008. Venomous animals of the world. Johns Hopkins Univ. Press, Baltimore, Maryland.

Bailey, F. C., V. A. Cobb, T. R. Rainwater, T. Worrall, and M. Klukowski. 2009. Adrenocortical effects of human encounters on free-ranging cottonmouths (*Agkistrodon piscivorus*). J. Herpetol. 43: 260–266.

Bailey, R. M. 1942. An intergeneric hybrid rattlesnake. Am. Nat. 76: 376–385.

Baird, E. 1946. They shall take up serpents. Woman 17(2): 36–39.

Baird, S. F., and C. Girard. 1853. Catalogue of North American reptiles in the Museum of the Smithsonian Institution. Part 1. Serpents. Smithsonian Misc. Coll. 2(5): 1–172.

Bajwa, S. S., H. Kirakossian, K. N. N. Reddy, and F. S. Markland. 1982. Thrombin-like and fibrinolytic enzymes in the venoms from the Gaboon viper (*Bitis gabonica*), eastern cottonmouth moccasin (*Agkistrodon p. piscivorus*) and southern copperhead (*Agkistrodon c. contortrix*) snakes. Toxicon 20: 427–432.

Baker, M. R. 1987. Synopsis of the Nematoda parasitic in amphibians and reptiles. Memorial Univ. Newfoundland Occ. Pap. Biol. No. 11.

Baker, R. J., G. A. Mengden, and J. J. Bull. 1972. Karyotypic studies of thirty-eight species of North American snakes. Copeia 1972: 257–265.

Bakken, G. S., and A. R. Krochmal. 2007. The imaging properties and sensitivity of the facial pits of pitvipers as determined by optical and heat-transfer analysis. J. Exp. Biol. 210: 2801–2810.

Balderas-Valdivia, C. J., and A. Ramírez-Bautista. 2005. Aversive behavior of beaded lizard, *Heloderma horridum*, to sympatric and allopatric predator snakes. Southwest. Nat. 50: 24–31.

Baldwin, A. S. 1999. Case report of an untreated human envenomation by the western massasauga rattlesnake *Sistrurus catenatus tergeminus*. Bull. Maryland Herpetol. Soc. 35: 14–20.

Banner, W. 1988. Bites and stings in the pediatric patient. Current Probl. Pediatrics 18: 9–69.

Banner, W., L. Baxter, and J. Adkins. 2007. Envenomations due to pygmy rattlesnakes in Oklahoma: Treatment options. *In* Venom week 2007 (program abstracts). Univ. Arizona Colleges of Medicine, Arizona Health Sci. Center, Off. Continuing Medical Educ., Tucson, Arizona.

Barber, T., P. Andreadis, and J. C. Gillingham. 2005. Effectiveness of two types of artificial cover objects for sampling terrestrial salamander populations. Michigan Academ. 36: 121.

Barbour, R. W. 1950. The reptiles of Big Black Mountain, Harlan County, Kentucky. Copeia 1950: 100–107.

———. 1956. A study of the cottonmouth, *Ancistrodon piscivorus leucostoma* in Kentucky. Trans. Kentucky Acad. Sci. 17: 33–41.

———. 1962. An aggregation of copperheads, *Agkistrodon contortrix*. Copeia 1962: 640.

Barbour, T. 1926. Reptiles and amphibians: Their habits and adaptations. Houghton Mifflin, Boston.

Barme, M. 1968. Venomous sea snakes (Hydrophiidae), pp. 285–308. *In* W. Burcherl, E. E. Buckley, and V. Deulofeu (eds.), Venomous animals and their venoms, vol. 1: Venomous vertebrates. Academic Press, New York.

Barrett, S. L., and J. A. Humphrey. 1986. Agonistic interactions between *Gopherus agassizii* (Testudinidae) and *Heloderma suspectum* (Helodermatidae). Southwest. Nat. 31: 261–263.

Barrios, T. C. 1898. Sur quelques ichthyotenias parasites des serpentes. Mem. Soc. Sci. Agric. Arts, Lille 2: 1–4.

Barron, J. N. 1997. Condition-adjusted estimator of reproductive output in snakes. Copeia 1997: 306–318.

Bartlett, R. D. 1988. In search of reptiles & amphibians. E. J. Brill, Leiden.

Bartlett, R. D., and P. Bartlett. 2003. Florida's snakes: A guide to their identification and habits. Univ. Press Florida, Gainesville.

Barton, A. J. 1949. Ophiophagy by a juvenile copperhead. Copeia 1949: 232.

———. 1950. Replacement fangs in newborn timber rattlesnakes. Copeia 1950: 235–236.

Beaman, K. R., and W. K. Hayes. 2008. Rattlesnakes: Research trends and annotated checklist, pp. 5–16. *In* W. K. Hayes, K. R. Beaman, M. D. Cardwell, and S. P. Bush (eds.), The biology of rattlesnakes. Loma Linda Univ. Press, Loma Linda, California.

Beaman, K. R., D. D. Beck, and B. M. McGurty. 2006. The beaded lizard (*Heloderma horridum*) and Gila monster (*Heloderma suspectum*): A bibliography of the family Helodermatidae. Smithson. Herpetol. Inform. Serv. 136: 1–66.

Beasley, R. J., E. L. Ross, P. M. Nave, D. H. Sifford, and B. D. Johnson. 1993. Phosphodiesterase activities of selected crotalid venoms. SAAS Bull. Biochem. Biotech. 6: 48–53.

Beaupre, S. J. 1993. An ecological study of oxygen consumption in the mottled rock rattlesnake, *Crotalus lepidus lepidus* and the black-tailed rattlesnake, *Crotalus molossus molossus*, from two populations. Physiol. Zool. 66: 437–454.

———. 1995. Effects of geographically variable thermal environment on bioenergetics of mottled rock rattlesnakes. Ecology 76: 1655–1665.

———. 1996. Field metabolic rate, water flux, and energy budgets of mottled rock rattlesnakes, *Crotalus lepidus*, from two populations. Copeia 1996: 319–329.

———. 2008. Annual variation in time-energy allocation by timber rattlesnakes (*Crotalus horridus*) in relation to food acquisition, pp. 111–122. *In* W. K. Hayes, K. R. Beaman, M. D. Cardwell, and S. P. Bush (eds.), The biology of rattlesnakes. Loma Linda Univ. Press, Loma Linda, California.

Beaupre, S. J., and D. Duvall. 1998a. Variation in oxygen consumption of the western diamondback rattlesnake (*Crotalus atrox*): Implications for sexual size dimorphism. J. Comp. Physiol. 168B: 497–506.

———. 1998b. Integrative biology of rattlesnakes: Contributions to biology and evolution. Bioscience 48: 531–538.

Beaupre, S. J., and F. Zaidan. 2001. Scaling of CO_2 production in the timber rattlesnake (*Crotalus horridus*) with comments on the cost of growth in neonates and comparative patterns. Physiol. Biochem. Zool. 74: 757–768.

Beaupre, S. J., and K. G. Roberts. 2001. *Agkistrodon contortrix contortrix* (southern copperhead): Chemotaxis, arboreality and diet. Herpetol. Rev. 32: 44–45.

Bebarta, V. S., and R. C. Dart. 2005. Effectiveness of delayed use of Crotalidae polyvalent immune Fab (ovine) antivenom. *In* Snakebites in the new millennium: A state-of-the-art symposium (program abstracts). Univ. Nebraska Medical Center, Center for Continuing Educ., Omaha.

Beck, D. D. 1985a. *Heloderma suspectum cinctum* (banded Gila monster): Pattern/coloration. Herpetol. Rev. 16: 53.

———. 1985b. The natural history, distribution, and present status of the Gila monster in Utah. Report to the Utah Division of Wildlife Research, Nongame Section, Salt Lake City, Utah.

———. 1986. The Gila monster in Utah: Bioenergetics and natural history considerations. Master's thesis, Utah State Univ., Logan.

———. 1990. Ecology and behavior of the Gila monster in southwestern Utah. J. Herpetol. 24: 54–68.

———. 1993. A retrospective of "the Gila monster and its allies." *In* C. M. Bogert and R. Martín del Campo, The Gila monster and its allies: The relationships, habits, and behavior of the lizards of the family Helodermatidae. Originally published in 1956. Reprinted by the Soc. Stud. Amphib. Rept.

———. 1994. A field study of the Gila monster in southwestern New Mexico. Progress Rep. Prof. Serv. Contract No. 93-516.6-3, New Mexico Dept. Game Fish, Santa Fe.

———. 1996. Effects of feeding on body temperatures of rattlesnakes: A field experiment. Physiol. Zool. 69: 1442–1445.

———. 2004a. Overview of the family Helodermatidae for varanophiles, pp. 517–520. *In* E. R. Pianka, D. R. King, and R. A. King (eds.), Varanoid lizards of the world. Indiana Univ. Press, Bloomington.

———. 2004b. *Heloderma horridum* (Wiegmann 1829), pp. 521–527. *In* E. R. Pianka, D. R. King, and R. A. King (eds.), Varanoid lizards of the world. Indiana Univ. Press, Bloomington.

———. 2004c. *Heloderma suspectum* (Cope 1869), pp. 528–534. *In* E. R. Pianka, D. R. King, and R. A. King (eds.), Varanoid lizards of the world. Indiana Univ. Press, Bloomington.

———. 2004d. Venomous lizards of the desert. Nat. Hist. 113(6): 32–37.

———. 2005. Biology of the Gila monsters and beaded lizards. Univ. California Press, Berkeley.

———. 2009. Family Helodermatidae: Gila monsters and beaded lizards, pp. 496–502. *In* L. L. C. Jones and R. E. Lovich (eds.), Lizards of the American Southwest: A photographic field guide. Rio Nuevo Publ., Tucson, Arizona.

Beck, D. D., and R. D. Jennings. 2003. Habitat use by Gila monsters: The importance of shelters. Herpetol. Mongr. 17: 112–130.

Beck, D. D., and C. H. Lowe. 1991. Ecology of the beaded lizard, *Heloderma horridum*, in a tropical dry forest in Jalisco, México. J. Herpetol. 25: 395–406.

———. 1994. Resting metabolism of helodermatid lizards: Allometric and ecological relationships. J. Comp. Physiol. 164B: 124–129.

Beck, D. D., and A. Ramírez-Bautista. 1991. Combat behavior of the beaded lizard, *Heloderma h. horridum*, in Jalisco, México. J. Herpetol. 25: 481–484.

Beck, D. D., M. R. Dohm, T. Garland, Jr., A. Ramírez-Bautista, and C. H. Lowe. 1995. Locomotor performance and activity energetics of helodermatid lizards. Copeia 1995: 577–585.

Becke, L. 1909. Neath Austral skies. John Milne, London.

Beckers, G. J. L., T. A. A. M. Leenders, and H. Strijbosch. 1996. Coral snake mimicry: Live snakes not avoided by a mammalian predator. Oecologia (Berlin) 106: 461–463.

Behler, J. L., and F. W. King. 1979. The Audubon Society field guide to North American reptiles and amphibians. Alfred A. Knopf, New York.

Bellairs, A. d'A., and G. Underwood. 1951. The origin of snakes. Biol. Rev. 26: 193–237.

Bellman, L., B. Hoffman, N. R. Levick, and K. D. Winkel. 2007. U.S. snakebite mortality, 1979–2005. *In* Venom week 2007 (program abstracts). Univ. Arizona Colleges of Medicine, Arizona Health Sci. Center, Off. Continuing Medical Educ., Tucson, Arizona.

Belson, M. S. 2000. *Drymarchon corais couperi* (eastern indigo snake) and *Micrurus fulvius fulvius* (eastern coral snake). Predator-prey. Herpetol. Rev. 31: 105.

Belt, P. J., A. Malhotra, R. S. Thorpe, D. A. Warrell, and W. Wüster. 1997. Russell's viper in Indonesia: Snakebite and systematics, pp. 219–234. *In* R. S. Thorpe, W. Wüster, and A. Malhotra (eds.), Venomous snakes: Ecology, evolution and snakebite. Clarendon Press, Oxford.

Beltz, Ellin. 1993. Distribution and status of the eastern massasauga rattlesnake, *Sistrurus catenatus catenatus* (Rafinesque, 1818), in the United States and Canada, pp. 26–31. *In* B. Johnson and V. Menzies (eds.), International symposium and workshop on the conservation of the eastern massasauga rattlesnake. Toronto Zoo, Toronto, Ontario.

Bennett, H. J. 1938. A partial check list of the trematodes of Louisiana vertebrates. Proc. Louisiana Acad. Sci. 4: 178–181.

Bent, A. C. 1937. Life histories of North American birds of prey. Part 1. U.S. Natl. Mus. Bull. 167.

Berna, H. J., and J. W. Gibbons. 1991. *Agkistrodon piscivorus piscivorus* (eastern cottonmouth). Diet. Herpetol. Rev. 22: 130–131.

Bernstein, P. 1999. Morphology of the nasal capsule of *Heloderma suspectum* with comments on the systematic position of helodermatids (Squamata: Helodermatidae). Acta Zool. 80: 219–230.

Bertke, E. M., D. D. Watt, and T. Tu. 1966. Electrophoretic patterns of venoms from species of Crotalidae and Elapidae snakes. Toxicon 4: 73–76.

Bertram, N., and K. W. Larsen. 2004. Putting the squeeze on venomous snakes: Accuracy and precision of length measurements taken with the "squeeze box." Herpetol. Rev. 35: 235–238.

Best, I. B. 1978. Field sparrow reproductive success and nesting ecology. Auk 95: 9–22.

Bevelander, G., T. L. Smith, and K. V. Kardong. 2006. Microhabitat and prey odor selection in the foraging pigmy rattlesnake. Herpetologica 62: 47–55.

Bhullar, B.-A. S., and K. T. Smith. 2008. Helodermatid lizard from the Miocene of Florida, the evolution of the dentary in Helodermatidae, and comments on dentary morphology in Varanoidea. J. Herpetol. 42: 286–302.

Biardi, J. E. 2000. Adaptive variation and coevolution in California ground squirrel (*Spermophilus beecheyi*) and rock squirrel (*Spermophilus variegatus*) resistance to rattlesnake venom. Ph.D. diss., Univ. California, Davis.

———. 2006. Small mammals as a natural source of snake venom metalloprotease inhibitors, p. 42. *In* S. A. Seifert (ed.), From Snakebites in the new millennium: A state-of-the-art symposium, University of Nebraska Medical Center, 21–23 October 2005, Omaha, Nebraska. J. Med. Toxicol. 2: 29–45.

———. 2008. The ecological and evolutionary context of mammalian resistance to rattlesnake venoms, pp. 557–568. *In* W. K. Hayes, K. R. Beaman, M. D. Cardwell, and S. P. Bush (eds.), The biology of rattlesnakes. Loma Linda Univ. Press, Loma Linda, California.

Biardi, J. E., D. C. Chien, and R. G. Coss. 2005. California ground squirrel (*Spermophilus beecheyi*) defenses against rattlesnake venom digestive and hemostatic toxins. J. Chem. Ecol. 31: 2501–2518.

Bieber, A. L. 1979. Metal and nonprotein constituents in snake venoms, pp. 295–306. *In* C. Y. Lee (ed.), Snake venoms. Springer-Verlag, Berlin.

Bieber, A. L., R. H. McParland, and R. R. Becker. 1987. Amino acid sequences of myotoxins from *Crotalus viridis concolor* venom. Toxicon 25: 677–680.

Birchard, G. F., C. P. Black, G. W. Schuett, and V. Black. 1984. Foetal-maternal blood respiratory properties of an ovoviviparous snake the cottonmouth, *Agkistrodon piscivorus*. J. Exp. Biol. 108: 247–255.

Bird, W., and P. Peak. 2007. A snake hunting guide: Methods, tools, and techniques for finding snakes. ECO Publications, Lansing, Michigan.

Bishop, L. A., T. M. Farrell, and P. G. May. 1996. Sexual dimorphism in a Florida population of the rattlesnake *Sistrurus miliarius*. Herpetologica 52: 360–364.

Bissinger, B. E., and C. A. Simon. 1979. Comparison of tongue extrusions in representatives of six families of lizards. J. Herpetol. 13: 133–139.

Blanchard, F. N., and E. B. Finster. 1933. A method of marking living snakes for future recognition with discussion of some problems and results. Ecology 14: 334–347.

Blaney, R. M. 1971. An annotated check list and biogeographic analysis of the insular herpetofauna of the Apalachicola region, Florida. Herpetologica 27: 406–430.

Blem, C. R. 1981. Reproduction of the eastern cottonmouth *Agkistrodon piscivorus piscivorus* (Serpentes: Viperidae) at the northern edge of its range. Brimleyana 5: 117–128.

———. 1982. Biennial reproduction in snakes: An alternative hypothesis. Copeia 1982: 961–963.

———. 1987. Development of combat rituals in captive cottonmouths. J. Herpetol. 21: 64–65.

———. 1997. Lipid reserves of the eastern cottonmouth (*Agkistrodon piscivorus*) at the northern edge of its range. Copeia 1997: 53–59.

Blem, C. R., and K. L. Blem. 1990. Metabolic acclimation in three species of sympatric, semiaquatic snakes. Comp. Biochem. Physiol. 97A: 259–264.

———. 1995. The eastern cottonmouth (*Agkistrodon piscivorus*) at the northern edge of its range. J. Herpetol. 29: 391–398.

Blem, C. R., and K. B. Killeen. 1993. Circadian metabolic cycles in eastern cottonmouths and brown water snakes. J. Herpetol. 27: 341–344.

Bober, M. A., J. L. Glenn, R. C. Straight, and C. L. Ownby. 1988. Detection of myotoxin *a*-like proteins in various snake venoms. Toxicon 26: 665–673.

Boerema, H. 1989. The keeping and breeding of *Agkistrodon bilineatus bilineatus*. Litt. Serpent. Engl. Ed. 9: 267–268.

———. 1990a. *Agkistrodon contortrix contortrix*. Litt. Serpent. Engl. Ed. 10: 54.

———. 1990b. *Agkistrodon contortrix mukeson* [sic]. Litt. Serpent. Engl. Ed. 10: 148.

———. 1991a. *Agkistrodon contortrix contortrix*: Copperhead. Litt. Serpent. Engl. Ed. 11: 114.

———. 1991b. *Agkistrodon contortrix mokeson*: Copperhead. Litt. Serpent. Engl. Ed. 11: 114.

Bogdan, G. M., R. C. Dart, S. C. Falbo, J. McNally, and D. Spaite. 2000. Recurrent coagulopathy after antivenom treatment of crotalid snakebite. South. Med. J. 93: 562–566.

Bogert, C. M. 1943. Dentitional phenomena in cobras and other elapids with notes on adaptive modifications of fangs. Bull. Am. Mus. Nat. Hist. 81: 285–360.

———. 1949. Thermoregulation in reptiles: A factor in evolution. Evolution 3: 195–211.

———. 1960. The influence of sound on the behavior of amphibians and reptiles, pp. 137–320. *In* W. E. Lanyon and W. N. Tavolga (eds.), Animal sounds and communication. Am. Inst. Biol. Sci. Publ. 7.

Bogert, C. M., and R. Martin del Campo. 1956. The Gila monster and its allies. The relationships, habits, and behavior of the lizards of the family Helodermatidae. Bull. Am. Mus. Nat. Hist. 109: 1–238.

Bogert, C. M., and J. A. Oliver. 1945. A preliminary analysis of the herpetofauna of Sonora. Bull. Am. Mus. Nat. Hist. 83: 297–426.

Bolaños, R. 1972. Toxicity of Costa Rican snake venoms for the white mouse. Am. J. Trop. Med. Hyg. 21: 360–363.

———. 1983 [1982]. Serpientes venenosos de Centro América: Distribución, caraterísticas, y patrones cardiológicos. Mem. Inst. Butantan 46: 275–291.

Bolaños, R., and J. R. Montero. 1968. *Agkistrodon bilineatus* Günther from Costa Rica. Rev. Biol. Trop. 16: 277–279.

Bolaños, R., A. Flores, R. T. Taylor, and L. Cerdas. 1974. Color patterns and venom characteristics in *Pelamis platurus*. Copeia 1974: 909–912.

Bolaños, R., L. Cerdas, and J. W. Abalos. 1978. Venoms of coral snakes (*Micrurus* ssp.): Report on a multivalent antivenin for the Americas. Bull. Pan Am. Health Org. 12: 23–27.

Bon, C. 1991. Venins de serpents et serums antivenimeux. Bull. Soc. Herp. Fr. 57: 1–18.

———. 2007. Comparative action of specific Fab and Fab2 antibodies to pharmacokinetics of viper and scorpion venoms during envenomations. *In* Venom week 2007 (program abstracts). Univ. Arizona Colleges of Medicine, Arizona Health Sci. Center, Off. Continuing Medical Educ., Tucson, Arizona.

Bonilla, C. A. 1975. Defibrinating enzyme from timber rattlesnake (*Crotalus h. horridus*) venom: A potential agent for therapeutic defibrination. I. Purification and properties. Thromb. Res. 6: 151–169.

Bonilla, C. A., and M. K. Fiero. 1971. Comparative biochemistry and pharmacology of salivary gland secretions: II. Chromatographic separation of the basic proteins from some North American rattlesnake venoms. J. Chromatog. 56: 253–263.

Bonilla, C. A., and N. V. Horner. 1969. Comparative electrophoresis of *Crotalus* and *Agkistrodon* venoms from North American snakes. Toxicon 7: 327–329.

Bonilla, C. A., W. Seifert, and N. Horner. 1971. Comparative biochemistry of *Sistrurus miliarius barbouri* and *Sistrurus catenatus tergeminus* venoms, pp. 203–209. *In* W. Bücherl and E. E. Buckley (eds.), Venomous animals and their venoms, vol. II: Venomous vertebrates. Academic Press, New York.

Bonnaterre, J. P. 1789–1790. Encyclopédie méthodique. Tableau encyclopédique et méthodique des trois règnes de la nature. 2 vols. Paris: Panckoucke.

Boquet, P. 1948. Venins de serpents et antivenins. Coll. Inst. Pasteur [Paris].

Bothner, R. C. 1973. Temperatures of *Agkistrodon p. piscivorus* and *Lampropeltis g. getulus* in Georgia. HISS News-J. 1: 24–25.

———. 1974. Some observations on the feeding habits of the cottonmouth in southeastern Georgia. J. Herpetol. 8: 257–258.

Bou-Abboud, C. F., and D. G. Kardassakis. 1988. Acute myocardial infarction following a Gila monster (*Heloderma suspectum cinctum*) bite. West. J. Med. 148: 577–579.

Boulenger, G. A. 1885. Catalogue of lizards in the British Museum (Natural History), vol. II. 2nd ed. Trustees (British Museum), London.

———. 1893–1896. Catalogue of the snakes in the British Museum (Natural History). 3 vols. Trustees (British Museum), London.

Boundy, J. 1997. Snakes of Louisiana. Louisiana Dept. Wildl. Fish., Baton Rouge.

Bowler, J. K. 1977. Longevity of reptiles and amphibians in North American collections. Soc. Stud. Amphib. Rept. Misc. Publ. Herpetol. Circ. 6: 1–32.

Bowman, D. D. 1984. *Hexametra leidyi* sp. n. (Nematoda: Ascarididae) from North American pit vipers (Reptilia: Viperidae). Proc. Helminthol. Soc. Washington 51: 54–61.

Boyer, D. A. 1933. A case report on the potency of the bite of a young copperhead. Copeia 1933: 97.

Boyer, L. V., S. A. Seifert, R. F. Clark, J. T. McNally, S. R. Williams, S. P. Nordt, F. G. Walter, and R. C. Dart. 1999. Recurrent and persistent coagulopathy following pit viper envenomation. Arch. Intern. Med. 159: 706–710.

Boyer, L. V., S. A. Seifert, and J. S. Cain. 2001. Recurrence phenomena after immunoglobulin therapy for snake envenomations: Part 2. Guidelines for clinical management with crotaline Fab antivenom. Ann. Emerg. Med. 37: 196–201.

Bragança, B. M., and J. H. Quastel. 1952. Amino acid oxidations by snake venoms. Arch. Biochem. Biophys. 40: 130–134.

Bragança, B. M., Y. M. Sambray, and R. Y. Sambray. 1970a. Isolation of polypeptide inhibitor of phospholipase A from cobra venom. Eur. J. Biochem. 13: 410–145.

Bragança, B. M., N. Patel, and T. Patel. 1970b. Cobra venom protein preferentially cytotoxic to certain tumors and its effects on the surface constituents on susceptible cells, p. 403. *In* Proceedings of the 10th International Cancer Congress.

Branch, B. 1998. Field guide to snakes and other reptiles of southern Africa. Ralph Curtis Books, Sanibel Island, Florida.

Branch, W. R. 1979. The venomous snakes of southern Africa. Part 2. Elapidae and Hydrophidae. The Snake 11: 199–225.

———. 1981. The venomous snakes of southern Africa. Part 2. Elapidae and Hydrophidae. Bull. Maryland Herpetol. Soc. 17: 1–47.

———. 1982. Hemipeneal morphology of platynotan lizards. J. Herpetol. 16: 16–38.

Branson, E. B. 1904. Snakes of Kansas. Univ. Kansas Sci. Bull. 2: 353–430.

Brattstrom, B. H. 1953. Records of Pleistocene reptiles and amphibians from Florida. Quart. J. Florida Acad. Sci. 16: 243–248.

———. 1954. Amphibians and reptiles from Gypsum Cave, Nevada. Bull. So. California Acad. Sci. 53: 8–12.

———. 1964. Evolution of the pit vipers. Trans. San Diego Soc. Nat. Hist. 13: 185–268.

———. 1965. Body temperatures of reptiles. Am. Midl. Nat. 73: 376–422.

———. 1967. A succession of Pliocene and Pleistocene snake faunas from the High Plains of the United States. Copeia 1967: 188–202.

Breisch, A. R. 1984. Just hanging in there: The eastern massasauga in danger of extinction. Conservationist (New York) 39(3): 35.

Breithaupt, H., and E. Habermann. 1973. Biochemistry and pharmacology of phospholipase A from *Crotalus terrificus* venom as influenced by crotapotin, pp. 83–88. *In* E. Kaiser (ed.), Animal and plant toxins. Goldmann, München.

Brennan, G. A. 1924. A case of death from *Heloderma* bite. Copeia 1924(129): 45.

Brennan, T. C., and A. T. Holycross. 2006. A field guide to amphibians and reptiles in Arizona. Arizona Game Fish Dept., Phoenix.

Brimley, C. S. 1923a. North Carolina herpetology. Copeia 1923(114): 3–4.

———. 1923b. The copperhead moccasin at Raleigh, N.C. Copeia 1923(123): 113–116.

———. 1941–1942. The amphibians and reptiles of North Carolina: The snakes. Carolina Tips 4–5(19–26).

Brinkhous, K. M., and S. V. Smith. 1988. Platelet-aggregating noncoagulant snake venom fractions, pp. 363–376. *In* H. Pirkle and F. S. Markland (eds.). Hematology, vol. 7: Hemostasis and animal venoms. Marcel Dekker, New York.

Broadley, D. G. 1957. Fatalities from the bites of *Dispholidus* and *Thelotornis* and personal case history. J. Herp. Assoc. Rhodesia 1: 5.

Brodie, E. D., III. 1993. Differential avoidance of coral snake banded patterns by free-ranging avian predators in Costa Rica. Evolution 47: 227–235.

Brodie, E. D., III., and F. J. Janzen. 1995. Experimental studies of coral snake mimicry: Generalized avoidance of ringed snake patterns by free-ranging avian predators. Funct. Ecol. 1995(9): 186–190.

Bronsgeest, H. 1992a. *Sistrurus catenatus tergeminus*—western massasauga. Litt. Serpent. Engl. Ed. 12: 23.

———. 1992b. *Sistrurus miliarius barbouri*—dusky pygmy rattlesnake. Litt. Serpent. 12: 24.

Brooks, D. R. 1978. Systematic status of protocephalid cestodes from reptiles and amphibians in North America with descriptions of three new species. Proc. Helminthol. Soc. Washington 45: 1–28.

Browder, R. 2005. *Agkistrodon contortrix mokasen* (northern copperhead). Catesbeiana 25(1): 31–32.

Brown, B. C. 1950. An annotated check list of the reptiles and amphibians of Texas. Baylor Univ. Press, Waco, Texas.

Brown, B. C., and H. M. Smith. 1942. A new subspecies of Mexican coral snake. Proc. Biol. Soc. Washington 55: 63–66.

Brown, D. E. (ed.). 1994. Biotic communities: Southwestern United States and northwestern Mexico. Univ. Utah Press, Salt Lake City.

Brown, D. E., and N. B. Carmony. 1999. Gila monster: Facts and folklore of America's Aztec lizard. Univ. Utah Press, Salt Lake City.

Brown, E. E. 1979. Some snake food records from the Carolinas. Brimleyana 1: 113–124.

Brown, J. H. 1973. Toxicology and pharmacology of venoms from poisonous snakes. Charles C. Thomas, Springfield, Illinois.

Brown, T. K., J. M. Lemm, J.-P. Montagne, J. A. Tracey, and A. C. Alberts. 2008. Spatial ecology, habitat use, and survivorship of resident and translocated red diamond rattlesnakes (*Crotalus ruber*), pp. 377–394. *In* W. K. Hayes, K. R. Beaman, M. D. Cardwell, and S. P. Bush (eds.), The biology of rattlesnakes. Loma Linda Univ. Press, Loma Linda, California.

Brown, W. H., and C. H. Lowe, Jr. 1954. Physiological effects of labial gland secretion from *Heloderma* (Gila monster) upon *Heloderma*. Am. J. Physiol. 177: 539–540.

———. 1955. Technique for obtaining maximum yields of fresh labial gland secretions from the lizard *Heloderma suspectum*. Copeia 1955: 63.

Brown, W. S. 1993. Biology, status, and management of the timber rattlesnake *(Crotalus horridus):* A guide for conservation. Soc. Stud. Amphib. Rept. Herpetol. Circ. 22: *i–vi,* 1–78.

Brugger, K. E. 1989. Red-tailed hawk dies with coral snake in talons. Copeia 1989: 508–510.

Brunson, E. M., B. D. Johnson, and D. H. Sifford. 1978. NAD nucleosidase of *Agkistrodon bilineatus* venom. Am. J. Trop. Med. Hyg. 27: 365–368.

Brush, S. W., and G. W. Ferguson. 1986. Predation on lark sparrow eggs by a massasauga rattlesnake. Southwest. Nat. 31: 260–261.

Bryson, R. W., Jr., and F. Mendoza-Quijano. 2007. Cantils of Hidalgo and Veracruz, Mexico, with comments on the validity of *Agkistrodon bilineatus lemosespinali*. J. Herpetol. 41: 536–539.

Bull, J. J., D. M. Hillis, and S. O'Steen. 1988. Mammalian ZFY sequences exist in reptiles regardless of sex-determining mechanism. Science (New York) 242: 567–568.

Bullock, T. H., and R. B. Cowles. 1952. Physiology of an infrared receptor: The facial pit of pit vipers. Science (New York) 115: 541–543.

Bullock, T. H., and F. P. J. Diecke. 1956. Properties of an infra-red sense organ in the facial pit of pit vipers. J. Physiol. 134: 47–87.

Bullock, T. H., and W. Fox. 1957. The anatomy of the infra-red sense organ in the facial pit vipers. Quart. J. Micro. Sci. 98: 219–234.

Burch, J. M., R. Agarwal, K. L. Mattox, D. V. Feliciano, and G. L. Jordan. 1988. The treatment of crotalid envenomation without antivenin. J. Trauma Inj. Infect. Crit. Care 28: 35–43.

Burchfield, P. M. 1982. Additions to the natural history of the crotaline snake *Agkistrodon bilineatus taylori*. J. Herpetol. 16: 376–382.

Burger, W. L., and W. B. Robertson. 1951. A new subspecies of the Mexican moccasin, *Agkistrodon bilineatus*. Univ. Kansas Sci. Bull. 34: 213–219.

Burgess, J. L., and R. C. Dart. 1991. Snake venom coagulopathy: Use and abuse of blood products in the treatment of pit viper envenomation. Ann. Emerg. Med. 20: 795–801.

Burgess, J. L., R. C. Dart, N. B. Egen, and M. Mayersohn. 1992. Effects of constriction bands on rattlesnake venom absorption: A pharmacokinetic study. Ann. Emerg. Med. 21: 1086–1093.

Burkett, R. D. 1966. Natural history of the cottonmouth moccasin, *Agkistrodon piscivorus* (Reptilia). Univ. Kansas Publ. Mus. Nat. Hist. 17: 435–491.

Burnett, J. W., G. J. Calton, and R. J. Morgan. 1985. Gila monster bites. Cutis 35(4): 323.

Burns, B. 1969. Oral sensory papillae in sea snakes. Copeia 1969: 617–619.

Burns, B., and G. V. Pickwell. 1972. Cephalic glands in sea snakes (*Pelamis, Hydrophis*, and *Laticauda*). Copeia 1972: 547–559.

Bush, F. M. 1959. Foods of some Kentucky herptiles. Herpetologica 15: 73–77.

Bush, S. P. 2005. Pre-hospital, first aid and field considerations in rattlesnake bite, pp. 21–22. *In* Biology of the Rattlesnakes Symposium (program abstracts). Loma Linda Univ., Loma Linda, California.

Bush, S. P., and P. W. Jansen. 1995. Severe rattlesnake envenomation with anaphylaxis and rhabdomyolysis. Ann. Emerg. Med. 25: 845–848.

Bush, S. P., T. L. Thomas, and E. S. Chin. 1997. Envenomations in children. Ped. Emerg. Med. Rep. 2: 1–12.

Bush, S. P., V. H. Wu, and S. W. Corbett. 1999. Rattlesnake venom-induced thrombocytopenia response to Antivenin (Crotalidae) Polyvalent. Academic Emerg. Med. 6: 393.

Bush, S. P., K. G. Hegewald, S. M. Green, M. D. Cardwell, and W. K. Hayes. 2000. Effects of a negative pressure venom extraction device (Extractor) on local tissue injury after artificial rattlesnake envenomation in a porcine model. Wild. Environ. Med. 11: 180–188.

Bush, S. P., S. M. Green, T. A. Laack, W. K. Hayes, M. D. Cardwell, and D. A. Tanen. 2005. Pressure-immobilization delays mortality and increases intra-compartmental pressure after artificial intramuscular rattlesnake envenomation in a porcine model, p. 22. *In* Biology of the Rattlesnakes Symposium (program abstracts). Loma Linda Univ., Loma Linda, California.

Bushar, L. M., M. Maliga, and H. K. Reinert. 2001. Cross-species amplification of *Crotalus horridus* microsatellites and their application in phylogenetic analysis. J. Herpetol. 35: 532–537.

Bushar, L. M., H. K. Reinert, and A. H. Savitzky. 2005. Isolation and reduced genetic variation in the timber rattlesnake, *Crotalus horridus,* of the New Jersey Pine Barrens, pp. 22–23. *In* Biology of the Rattlesnakes Symposium (program abstracts). Loma Linda Univ., Loma Linda, California.

Bushey, C. L. 1985. Man's effect upon a colony of *Sistrurus c. catenatus* (Raf.) in northeastern Illinois (1834–1975). Bull. Chicago Herpetol. Soc. 20: 1–12.

Byrd, E. E., and J. F. Denton. 1938. New trematodes of the subfamily Reniferinae, with a discussion of the systematics of the genera and species assigned to the subfamily group. J. Parasitol. 24: 379–401.

Byrd, E. E., and R. L. Roudabush. 1939. *Leptophyllum ovalis* n. sp., a trematode from the brown water-snake. J. Parasitol. 25: 471–473.

Byrd, E. E., M. V. Parker, and R. J. Reiber. 1940a. A new genus and two new species of digenetic trematodes, with a discussion of the systematics of these and certain related forms. J. Parasitol. 26: 111–122.

———. 1940b. Taxonomic studies on the genus *Styphlodora* Looss, 1899 (Trematoda: Styphlodorinae) with descriptions of four new species. Trans. Am. Micro. Soc. 59: 294–326.

Caballero, C. E. 1954. Nemátodos de los reptiles de México. XI. Nuevo especie de filaria de íguanidos. Riv. Parasitol. (Rome) 15: 305–313.

Caballero y Caballero, E. 1954. Estudios helmintológicas de la región oncocercosa de México y de la República de Guatemala. Nematod. 8ª Parte. Anal. Inst. Biol. Univ. Nac. Aut. Mexico 25: 259–274.

Cable, R. M., and C. R. Sanborn. 1970. Two oviduct flukes from reptiles in Indiana: *Telorchis compactus* sp. n. and a previously described species. Proc. Helminthol. Soc. Washington 37: 211–215.

Cadle, J. E. 1992. Phylogenetic relationships among vipers: Immunological evidence, pp. 41–48. *In* J. A. Campbell, and E. D. Brodie, Jr. (eds.), Biology of the pitvipers. Selva, Tyler, Texas.

Cadle, J. E., and V. M. Sarich. 1981. An immunological assessment of the phylogenetic position of New World coral snakes. J. Zool. (London) 195: 157–167.

Cagle, F. R. 1968. Reptiles, pp. 213–268. *In* W. F. Blair, A. P. Blair, P. Brodkorb, F. R. Cagle and G. A. Moore (eds.),Vertebrates of the United States. 2nd ed. McGraw-Hill, New York.

Caldwell, G. S., and R. W. Rubinoff. 1983. Avoidance of venomous sea snakes by naive herons and egrets. Auk 100: 195–198.

Callahan, N. 1952. Smoky Mountain country. Duell, Sloan & Pearce, New York.

Calmette, A., A. Saenz, and L. Costi. 1933. Effects du venin de cobra sur les greffes cancereuses et sur le cancer spontane (adeno-carcinoma) de la souris. Comp. Rend. Acad. Sci. 197: 205–210.

Calvete, J. J., C. Marcinkiewicz, D. Monleon, V. Esteve, B. Celda, P. Juarez, and L. Sanz. 2005. Snake venom disintegrins: Evolution of structure and function. Toxicon 45: 1063–1074.

Camilleri, C., S. Offerman, R. Gosselin, and T. Albertson. 2005. Conservative management of delayed, multicomponent coagulopathy following rattlesnake envenomation. Clin. Toxicol. 43: 201–206.

Campbell, H. 1953a. Probable strychnine poisoning in a rattlesnake. Herpetologica 8: 184.

———. 1977. The coral snake: A New Mexico treasure. New Mexico Wildl. 22(4): 2–5, 26–29.

Campbell, J. A. 1973. A captive hatching of *Micrurus fulvius tenere* (Serpentes, Elapidae). J. Herpetol. 7: 312–315.

Campbell, J. A., and E. D. Brodie, Jr. (eds.). 1992. Biology of the pitvipers. Selva, Tyler, Texas.

Campbell, J. A., and W. W. Lamar. 1989. The venomous reptiles of Latin America. Comstock Publ. Assoc., Cornell Univ. Press, Ithaca, New York.

———. 2004. The venomous reptiles of the western hemisphere in two volumes. Comstock Publ. Assoc., Cornell Univ. Press, Ithaca, New York.

Campbell, J. A., and J. P. Vannini. 1988. A new subspecies of beaded lizard, *Heloderma horridum*, from the Motagua Valley of Guatemala. J. Herpetol. 22: 457–468.

Campbell, J. A., and D. H. Whitmore, Jr. 1989. A comparison of the skin keratin biochemistry in vipers with comments on its systematic value. Herpetologica 45: 242–249.

Campbell, J. A., D. R. Formanowicz, Jr., and E. D. Brode, Jr. 1989. Potential impact of rattlesnake roundups on natural populations. Texas J. Sci. 41: 301–317.

Camper, J. D. 2005. Observations on problems with using funnel traps to sample semi-aquatic snakes. Herpetol. Rev. 36: 288–290.

Camper, J. D., and B. G. Hanks. 1995. Variation in the nucleolus organizer region among New World snakes. J. Herpetol. 29: 468–471.

Canseco-Márquez, L., and A. L. Nolasco-Vélez. 2008. *Agkistrodon bilineatus* (cantil). Herpetol. Rev. 39: 369.

Cantrell, F. L. 2003. Envenomation by the Mexican beaded lizard: A case report. J. Toxicol. Clin. Toxicol. 41: 241–244.

Caras, R. 1974. Venomous animals of the world. Prentice-Hall, Englewood Cliffs, New Jersey.

Caravati, E. M. 2004. Copperhead bites and Crotalidae Polyvalent Immune Fab (Ovine): Routine use requires evidence of improved outcomes. Ann. Emerg. Med. 43: 207–208.

Caravati, E. M., and S. C. Hartsell. 1994. Gila monster envenomation. Ann. Emerg. Med. 24: 731–735.

Card, W., and D. Mehaffey. 1994. A radiographic sexing technique for *Heloderma suspectum*. Herpetol. Rev. 25: 17–19.

Cardwell, M. D., S. P. Bush, and R. T. Clark. 2005. Males biting males: Does testosterone shape both sides of the snakebite equation?, p. 23. *In* Biology of the Rattlesnakes Symposium (program abstracts). Loma Linda Univ., Loma Linda, California.

Cardwell, M. E., and D. L. Ryerson. 1940. A new species of the genus *Pseudomonas* pathogenic for certain reptiles. J. Bacteriol. 39: 323–336.

Carmichael, R. L. 2008. The grass is rattling: A rattlesnake conservation education program and exhibit made possible by a private-public partnership, pp. 485–494. *In* W. K. Hayes, K. R. Beaman, M. D. Cardwell, and S. P. Bush (eds.), The biology of rattlesnakes. Loma Linda Univ. Press, Loma Linda, California.

Carpenter, C. C. 1958. Reproduction, young, eggs and food of Oklahoma snakes. Herpetologica 14: 113–115.

———. 1960. A large brood of western pigmy rattlesnakes. Herpetologica 16: 142–143.

Carpenter, C. C., and G. W. Ferguson. 1977. Variation and evolution of stereotyped behavior in reptiles, pp. 335–554. *In* C. Gans and D. W. Tinkle (eds.), Biology of the Reptilia, vol. 7. Academic Press, London.

Carpenter, C. C., and J. C. Gillingham. 1975. Postural responses to kingsnakes by crotaline snakes. Herpetologica 31: 293–302.

———. 1990. Ritualized behavior in *Agkistrodon* and allied genera, pp. 523–531. *In* H. K. Gloyd and R. Conant, Snakes of the *Agkistrodon* complex: A monographic review. Soc. Stud. Amphib. Rept. Contrib. Herpetol. 6.

Carr, A. F., Jr. 1936. The Gulf-Island cottonmouths. Proc. Florida Acad. Sci. 1: 86–90.

———. 1940. A contribution to the herpetology of Florida. Univ. Florida Biol. Sci. Ser. 3: 1–118.

Carr, A. F., Jr., and M. H. Carr. 1942. Notes on the courtship of the cottonmouth moccasin. Proc. New England Zool. Club 20: 1–6.

Carr, A. F., Jr., and C. J. Goin. 1955. Guide to the reptiles, amphibians, and freshwater fishes of Florida. Univ. Florida Press, Gainesville.

Carr, W. H. 1926. The fasting of a copperhead. Copeia 150: 104.

Carrigan, P., F. E. Russell, and J. Wainschel, Jr. 1978. Clinical reactions to antivenin, pp. 457–465. *In* P. Rosenberg (ed.), Toxins: Animal, plant and microbial. Pergamon Press, New York.

Carroll, R. R., E. L. Hall, and C. S. Kitchens. 1997. Canebrake rattlesnake envenomation. Ann. Emerg. Med. 30: 45–48.

Casas-Andreu, G. 1997. Distribución de la culebra de mar *Pelamis platurus* en el Pacífico Mexicano. Rev. Soc. Mexicana Hist. Nat. 47: 157–166.

Cashman, J. L., M. Peirce, and P. R. Krausman. 1992. Diets of mountain lions in southeastern Arizona. Southwest. Nat. 37: 324–326.

Castleberry, J. S. 2007. *Agkistrodon piscivorous piscivorous* (eastern cottonmouth): Brumation. Herpetol. Rev. 38: 202–203.

Castoe, T. A., and C. L. Parkinson. 2006. Bayesian mixed models and the phylogeny of pitvipers (Viperidae: Serpentes). Mol. Phylogenet. Evol. 39: 91–110.

Castro, H. C., and C. R. Rodrigues. 2006. Current status of snake venom thrombin-like enzymes. Toxin Rev. 25: 291–318.

Cecchini, A. L., S. Marcussi, L. B. Silveira, C. R. Borja-Oliveira, L. Rodrigues-Simioni, S. Amara, R. G. Stábeli, J. R. Giglio, E. C. Arantes, and A. M. Soares. 2005. Biological and enzymatic activities of *Micrurus* sp. (coral) snake venoms. Comp. Biochem. Physiol. 140A: 125–134.

Chabaud, A. G., and W. Frank. 1961a. Nouvelle filaire parasite des artères de *l'Heloderma suspectum* Cope: *Macdonaldius andersoni* n. sp. (Nematodes, Onchocercidae). Ann. Parasitol. Human Comp. 36: 127–133.

———. 1961b. Les filaires de l'heloderme. Ann. Parasitol. Human Comp. 36: 804–805.

Chamberlain, E. B. 1935. Notes on the pygmy rattlesnake, *Sistrurus miliarius* Linnaeus, in South Carolina. Copeia 1935: 146–147.

Chance, B. 1970. A note on the feeding habits of *Micrurus fulvius fulvius*. Bull. Maryland Herpetol. Soc. 6: 56.

Chang, C. C., and K. H. Tseng. 1978. Effect of crotamine, a toxin of South American rattlesnake venom, on the sodium channel of murine skeletal muscle. Br. J. Pharmacol. 63: 551–559.

Chang, L. S., H. B. Huang, and S. R. Lin. 2000. The multiplicity of cardiotoxins from *Naja naja atra* (Taiwan cobra) venom. Toxicon 38: 1065–1076.

Chao, B. H., J. A. Jakubowski, B. Savage, E. P. Chow, U. M. Marzec, L. A. Harker, and J. M. Maraganore. 1989. *Agkistrodon piscivorus piscivorus* platelet aggregation inhibitor: A potent inhibitor of platelet activation. Proc. Natl. Acad. Sci. USA 86: 8050–8054.

Chapman, B. R., and S. D. Casto. 1972. Additional vertebrate prey of the loggerhead shrike. Wilson Bull. 84: 496–497.

Cheatwood, J. L., E. R. Jacobson, P. G. May, T. M. Farrell, B. L. Homer, D. A. Samuelson, and J. W. Kimbrough. 2003. An outbreak of fungal dermatitis and stomatitis in a free-ranging population of pigmy rattlesnakes (*Sistrurus miliarius barbouri*) in Florida. J. Wildl. Dis. 39: 329–337.

Chen, C. J. 1996. New development in acute anticoagulant therapy: What improvements over traditional heparin are on the horizon? Postgrad. Med. 99: 129.

Chen, T., H. F. Kwok, C. Ivanyi, and C. Shaw. 2006. Isolation and cloning of exendin precursor cDNAs from single samples of venom from the Mexican beaded lizard (*Heloderma horridum*) and the Gila monster (*Heloderma suspectum*). Toxicon 47: 288–295.

Chen, Y. E., and D. J. Drucker. 1997. Tissue-specific expression of unique mRNAs that encode proglucagon-derived peptides or exendin 4 in the lizard. J. Biol. Chem. 272: 4108–4115.

Chen, Y. H., Y. M. Wang, M. J. Hseu, and I. H. Tsai. 2004. Molecular evolution and structure-function relationships of crotoxin-like and asparagine-6-containing phospholipases A_2 in pit viper venoms. Biochem. J. 381: 25–34.

Chen, Z. X., H. L. Zhang, Z. L. Gu, B. W. Chen, R. Han, P. F. Reid, L. N. Raymond, and Z. H. Qin. 2006. A long-form alpha-neurotoxin from cobra venom produces potent opioid-independent analgesia. Acta Pharmacol. Sin. 27: 402–408.

Chenowith, W. L. 1948. Birth and behavior of young copperheads. Herpetologica 4: 162.

Chermock, R. L. 1952. A key to the amphibians and reptiles of Alabama. Univ. Alabama Mus. Pap. 33: 1–88.

Chiang, H. S., R. S. Yang, and T. F. Huang. 1996. Thrombin enhances the adhesion and migration of human colon adenocarcinoma cells via increased beta-3 integrin expression on the tumor cell surface and their inhibition by the snake venom peptide, rhodostomin. Br. J. Cancer 73: 902–908.

Chiasson, R. B. 1982. The apical pits of *Agkistrodon* (Reptilia: Serpentes). J. Arizona-Nevada Acad. Sci. 16: 69–73.

Chiasson, R. B., D. L. Bentley, and C. H. Lowe. 1989. Scale morphology in *Agkistrodon* and closely related crotaline genera. Herpetologica 45: 430–438.

Chiodini, R. J., J. P. Sundberg, and J. A. Czikowsky. 1982. Gross anatomy of snakes. Veterin. Med. Sm. Anim. Clin. 77: 413–419.

Chippaux, J. P. 2006. Snake venoms and envenomations. Kreiger Publ. Co., Melbourne, Florida.

Chiszar, D., and C. W. Radcliffe. 1976. Rate of tongue flicking by rattlesnakes during successive stages of feeding on rodent prey. Bull. Psychon. Soc. 7: 485–486.

Chiszar, D., K. Scudder, and L. Knight. 1976a. Rate of tongue flicking by garter snakes (*Thamnophis radix haydeni*) and rattlesnakes (*Crotalus v. viridis, Sistrurus catenatus tergeminus,* and *Sistrurus catenatus edwardsii*) during prolonged exposure to food odors. Behav. Biol. 18: 273–283.

Chiszar, D., K. Scudder, H. M. Smith, and C. W. Radcliffe. 1976b. Observations of courtship behavior in the western massasauga (*Sistrurus catenatus tergeminus*). Herpetologica 32: 337–338.

Chiszar, D., K. Scudder, and H. M. Smith. 1979a. Chemosensory investigation of fish mucus odor by rattlesnakes. Bull. Maryland Herpetol. Soc. 15: 31–36.

Chiszar, D., L. Simonsen, C. Radcliffe, and H. M. Smith. 1979b. Rate of tongue flicking by cottonmouths (*Agkistrodon piscivorus*) during prolonged exposure to various odors, and strike-induced chemosensory searching by the cantil (*Agkistrodon bilineatus*). Trans. Kansas Acad. Sci. 82: 49–54.

Chiszar, D., S. W. Taylor, C. W. Radcliffe, H. M. Smith, and B. O'Connell. 1981. Effects of chemical and visual stimuli upon chemosensory searching by garter snakes and rattlesnakes. J. Herpetol. 15: 415–423.

Chiszar, D., C. W. Radcliffe, R. Overstreet, T. Poole, and T. Byers. 1985. Duration of strike-induced chemosensory searching in cottonmouths (*Agkistrodon piscivorus*) and a test of the hypothesis that striking prey creates a specific search image. Can. J. Zool. 63: 1057–1061.

Chiszar, D., D. Dickman, and J. Colton. 1986a. Sensitivity to thermal stimulation in prairie rattlesnakes (*Crotalus viridis*) after bilateral anesthetization of the facial pits. Behav. Neurol. Biol. 45: 143–149.

Chiszar, D., C. W. Radcliffe, R. Boyd, A. Radcliffe, H. Yun, H. M. Smith, T. Boyer, B. Atkins, and F. Feiler. 1986b. Trailing behavior in cottonmouths (*Agkistrodon piscivorus*). J. Herpetol. 20: 269–272.

Chiszar, D., H. M. Smith, J. L. Glenn, and R. C. Straight. 1991. Strike-induced chemosensory searching in venomoid pit vipers at Hogle Zoo. Zoo Biol. 10: 111–117.

Christel, C. M., and D. F. De Nardo. 2006. Release of exendin-4 is controlled by mechanical action in Gila monsters, *Heloderma suspectum*. Comp. Biochem. Physiol. 143A: 85–88.

———. 2007. Absence of exendin-4 effects on postprandial glucose and lipids in the Gila monster, *Heloderma suspectum*. J. Comp. Physiol. 177B: 129–134.

Christel, C. M., D. F. De Nardo, and S. M. Secor. 2007. Metabolic and digestive response to food ingestion in a binge-feeding lizard, the Gila monster (*Heloderma suspectum*). J. Exp. Biol. 210: 3430–3439.

Christiansen, J. L. 1981. Population trends among Iowa's amphibians and reptiles. Proc. Iowa Acad. Sci. 88: 24–27.

Christiansen, J. L., J. M. Grzybowski, and E. R. Jacobson. 2008. *Sistrurus catenatus catenatus* (eastern Massasauga): Leiomyoma. Herpetol. Rev. 39: 358–359.

Christie, T. 1994. Snake bite! Wyoming Wildl. 58(6): 22–29.

Christman, S. P. 1980. Patterns of geographic variation in Florida snakes. Bull. Florida St. Mus. Biol. Sci. 25: 157–256.

Christopher, D. G., and C. B. Rodning. 1986. Crotalidae envenomation. South. Med. J. 79: 159–162.

Cifelli, R. L., and R. L. Nydam. 1995. Primitive, helodermatid-like platynotan from the early Cretaceous of Utah. Herpetologica 51: 286–291.

Clark, D. R., Jr. 1963. Variation and sexual dimorphism in a brood of the western pigmy rattlesnake (*Sistrurus*). Copeia 1963: 157–159.

Clark, D. R., Jr., J. W. Bickham, D. L. Baker, and D. F. Cowman. 2000. Environmental contaminants in Texas, USA, wetland reptiles: Evaluation using blood samples. Environ. Toxicol. Chem. 19: 2259–2265.

Clark, R. F. 1949. Snakes of the hill parishes of Louisiana. J. Tennessee Acad. Sci. 24: 244–261.

Clark, R. F., S. R. Williams, S. P. Nordt, and L. V. Boyer-Hassen. 1997. Successful treatment of crotalid-induced neurotoxicity with a new polyspecific crotalid Fab antivenom. Ann. Emerg. Med. 30: 54–57.

Clarke, R. F. 1958. An ecological study of reptiles and amphibians in Osage County, Kansas. Emporia State Res. Stud. 7: 1–52.

Cliburn, J. W. 1979. Range revisions for some Mississippi reptiles. J. Mississippi Acad. Sci. 24: 31–37.

Clifford, M. J. 1976. Relative abundance and seasonal activity of snakes in Amelia County, Virginia. Virginia Herpetol. Soc. Bull. 79: 4–6.

Cloudsley-Thompson, J. 2006. The colouration and displays of venomous reptiles: A review. Herpetol. Bull. 95: 25–30.

Cochran, D. A. 1954. Our snake friends and foes. Natl. Geogr. 106: 333–364.

Cochran, D. M. 1943. Poisonous reptiles of the world: A wartime handbook. Smithsonian Inst. Press, Washington, D.C.

———. 1944. Dangerous reptiles. Smithsonian Inst. Press, Washington, D.C.

Cochran, D. M., and C. J. Goin. 1970. The new field book of reptiles and amphibians. Putnam, New York.

Cochran, P. A. 2008. A cottonmouth (*Agkistrodon piscivorus*) in Minnesota, and historical reports of other pit vipers unexpected in the Upper Midwest. Northeast. Nat. 15: 461–464.

Cogger, H. C. 1992. Reptiles & amphibians of Australia. Cornell Univ. Press, Ithaca, New York.

Cohen, P., and E. B. Seligmann, Jr. 1966. Immunologic studies of coral snake venom. Mem. Inst. Butantan 33: 339–347.

Cole, C. J. 1990. Chromosomes of *Agkistrodon* and other viperid snakes, pp. 533–538. *In* H. K. Gloyd and R. Conant, Snakes of the *Agkistrodon* complex: A monographic review. Soc. Stud. Amphib. Rept. Contrib. Herpetol. 6.

Coleman, G. E. 1928. Rattlesnake venom antidote of the Hopi Indians. Bull. Antiv. Inst. Am. 1: 97–99.

Collins, J. T. 1990. Standard common and current scientific names for North American amphibians and reptiles. 3rd ed. Soc. Stud. Amphib. Rept. Herpetol. Circ. 19: 1–41.

———. 1991. Viewpoint: A new taxonomic arrangement for some North American amphibians and reptiles. Herpetol. Rev. 22: 42–43.

———. 2006. A re-classification of snakes native to Canada and the United States. J. Kansas Herpetol. 19: 18–20.

Collins, J. T., and S. L. Collins. 1993. Reptiles and amphibians of Cheyenne Bottoms. Hirth Publishing, Hillsboro, Kansas.

Collins, R. F. 1969. The helminths of *Natrix* spp. and *Agkistrodon piscivorus piscivorus* (Reptilia: Ophidia) in eastern North Carolina. J. Elisha Mitchell Sci. Soc. 85: 141–144.

———. 1980. Stomach contents of some snakes from eastern and central North Carolina. Brimleyana 4: 157–159.

Collins, R. F., and C. C. Carpenter. 1970. Organ position-ventral scute relationship in the water moccasin (*Agkistrodon piscivorus leucostoma*), with notes on food habits and distribution. Proc. Oklahoma Acad. Sci. 49: 15–18.

Conant, R. 1929. Notes on a water moccasin in captivity (*Agkistrodon piscivorus*) (female). Bull. Antiv. Inst. Am. 3: 61–64.

———. 1933. Three generations of cottonmouths, *Agkistrodon piscivorus* (Lacépède). Copeia 193: 43.

———. 1938. On the seasonal occurrence of reptiles in Lucas County, Ohio. Herpetologica 1: 137–144.

———. 1951. The reptiles of Ohio. 2nd ed. Univ. Notre Dame Press, Notre Dame, Indiana.

———. 1975. A field guide to reptiles and amphibians of eastern and central North America. 2nd ed. Houghton Mifflin, Boston.

———. 1978. Distributional patterns of North American snakes: Some examples of the effects of Pleistocene glaciations and subsequent climatic changes. Bull. Maryland Herpetol. Soc. 14: 241–259.

———. 1982. The origin of the name "cantil" for *Agkistrodon bilineatus*. Herpetol. Rev. 13: 118.

———. 1984. A new subspecies of the pit viper, *Agkistrodon bilineatus* (Reptilia: Viperidae) from Central America. Proc. Biol. Soc. Washington 97: 135–141.

———. 1986. Phylogeny and zoogeography of the genus *Agkistrodon* in North America, pp. 89–92. *In* Z. Roček (ed.), Studies in herpetology. Proc. European Herpetol. Mtng. (3rd Ordinary General Meeting of the Societas Europaea Herpetologica), Charles Univ., Prague.

———. 1990. The fossil history of the genus *Agkistrodon* in North America, pp. 539–543. *In* H. K. Gloyd and R. Conant, Snakes of the *Agkistrodon* complex: A monographic review. Soc. Stud. Amphib. Rept. Contrib. Herpetol. 6.

———. 1992. Comments on the survival status of members of the *Agkistrodon* complex. Greater Cincinnati Herpetol. Soc. Contrib. Herpetol. 1992: 29–33.

Conant, R., and W. Bridges. 1939. What snake is that? Appleton Century, New York.

Conant, R., and J. T. Collins. 1998. A field guide to reptiles and amphibians: Eastern and central North America. 3rd ed., expanded. Houghton Mifflin, Boston.

Conceição, L. G., N. M. Argôlo Neto, A. P. Castro, L. B. A. Faria, and C. O. Fonterrada. 2007. Anaphylactic reaction after *Crotalus* envenomation treatment in a dog: Case report. J. Venom. Anim. Toxins incl. Trop. Dis. 13: 549–557.

Conlon, J. M., S. Patterson, and P. R. Flatt. 2006. Major contributions of comparative endocrinology to the development and exploitation of the incretin concept. J. Exp. Zool. 305A: 781–786.

Consroe, P., N. B. Egen, F. E. Russell, K. Gerrish, D. C. Smith, A. Sidki, and J. T. Landon. 1995. Comparison of a new ovine antigen binding fragment (Fab) for United States Crotalidae with the commercial antivenin for protection against venom-induced lethality in mice. Am. J. Trop. Med. Hyg. 53: 507–510.

Cook, D. G. 1983. Activity patterns of the cottonmouth water moccasin, *Agkistrodon piscivorus* Lacepede, on a northwest Florida headwater stream. Master's thesis, Univ. Florida, Gainesville.

Cook, F. A. 1954. Snakes of Mississippi: Survey bulletin. Mississippi Game Fish Comm., Wildl. Mus., Jackson.

Cook, P. M., M. P. Rowe, and R. W. Van Devender. 1994. Allometric scaling and interspecific differences in the rattling sounds of rattlesnakes. Herpetologica 50: 358–368.

Cooke, M. E., G. V. Odell, S. A. Hudiburg, and J. M. Gutierrez. 1986. Analysis of fatty acids in the lipid components of snake venom. Proceedings of the second American symposium on animal, plant and microbial toxins. Tempe, Arizona.

Coombs, E. M. 1977. Wildlife observations of the hot desert region, Washington Co., Utah, with emphasis on reptilian species and their habitat in relation to livestock grazing. Utah Div. Wildl. Res. for USDI/BLM, Utah State Off. Contract No. YA-512-CT 6-102.

Cooper, R. H. 1968. Melanoma in *Heloderma suspectum* Cope. Proc. Indiana Acad. Sci. 78: 466–467.

Cooper, R. H., and J. C. List. 1979. Further information on the health and longevity of the Gila monster (*Heloderma suspectum* Cope). Proc. Indiana Acad. Sci. 88: 434–435.

Cooper, W. E., Jr. 1989. Prey odor discrimination in the varanoid lizards *Heloderma suspectum* and *Varanus exanthematicus*. Ethology 81: 250–258.

Cooper, W. E., Jr., and J. Arnett. 1995. Strike-induced chemosensory searching in the Gila monster. Copeia 1995: 89–96.

———. 2001. Absence of discriminatory tongue-flicking responses to plant chemicals by helodermatid lizards. Southwest. Nat. 46: 405–409.

———. 2003. Correspondence between diet and chemosensory responsiveness by helodermatid lizards. Amphibia-Reptilia 24: 86–91.

Cooper, W. E., Jr., C. S. Deperno, and J. Arnett. 1994. Prolonged poststrike elevation in tongue-flicking rate with rapid onset in Gila monster, *Heloderma suspectum*: Relation to diet and foraging and implications for evolution of chemosensory searching. J. Chem. Ecol. 20: 2867–2881.

Coote, J. 1981. The California mountain kingsnake (*Lampropeltis zonata*) and the coral snake mimic problem. Herptile 6: 17–19.

Cope, E. D. 1869. Remarks on *Heloderma suspectum*. Proc. Acad. Nat. Sci. Philadelphia 21: 4–5.

———. 1892. A critical review of the characters and variations of snakes of North America. Proc. U.S. Natl. Mus. 14(882): 589–694.

———. 1900. The crocodilians, lizards, and snakes of North America. Ann. Rept. U.S. Natl. Mus. 1898: 153–1294.

Cope, F. O. 1979. Purification of antivenin (Crotalidae) polyvalent (IgG-Equine) using ultrafiltration: Isolation of a venom-neutralizing inhibitory factor. Proc. Pennsylvania Acad. Sci. 53: 43–46.

Corrigan, P., F. E. Russell, and J. Wainschel. 1978. Clinical reactions to antivenin. Toxicon 1 (suppl.): 457–465.

Coues, E. 1875. Synopsis of the reptiles and batrachians of Arizona, pp. 585–663. *In* Report upon United States geographical surveys west of the one hundredth meridian (Wheeler Report), vol. 5: Zoology. U.S. Gov't. Print. Off., Washington, D.C.

Cowan, D. F. 1968. Diseases of captive reptiles. J. Am. Vet. Med. Assoc. 153: 848–859.

Cozzi, C. A. 1980. The absence of sea snakes in the Atlantic Ocean. Bull. Maryland Herpetol. Soc. 16: 113–118.

Crane, A. L., and B. D. Greene. 2008. The effect of reproductive condition on thermoregulation in female *Agkistrodon piscivorus* near the northwestern range limit. Herpetologica 64: 156–167.

Crosman, A. M. 1956. A longevity record for Gila monster. Copeia 1956: 54.

Cross, C. L. 2002. *Agkistrodon piscivorus piscivorus* (eastern cottonmouth): Diet. Herpetol. Rev. 33: 55–56.

Cross, C. L., and C. Marshall. 1998. *Agkistrodon piscivorus piscivorus* (eastern cottonmouth): Predation. Herpetol. Rev. 29: 43.

Cross, C. L., and C. E. Petersen. 2001. Modeling snake microhabitat from radiotelemetry studies using polytomous logistic regression. J. Herpetol. 35: 590–597.

Cross, J. K., and M. S. Rand. 1979. Climbing activity in wild-ranging Gila monsters, *Heloderma suspectum* (Helodermatidae). Southwest. Nat. 24: 703–705.

Crother, B. I. (Chair.). 2000. Scientific and standard English names of amphibians and reptiles of North America north of Mexico, with comments regarding confidence in our understanding. Soc. Stud. Amphib. Rept. Herpetol. Circ.(29): 1–82.

Crow, H. E. 1913. Some trematodes of Kansas snakes. Kansas Univ. Sci. Bull. 7: 123–134.

Cruz, G. A., L. D. Wilson, and J. Espinosa. 1979. Two additions to the reptile fauna of Honduras, *Eumeces managuae* Dunn and *Agkistrodon bilineatus* (Gunther), with comments on *Pelamis platurus* (Linnaeus). Herpetol. Rev. 10: 26–27.

Cuesta Terrón, C. 1930. Los crotalianos Mexicanos. An. Inst. Biol., Univ. Nac. Autónoma México 1: 187–199.

Cuffey, R. J. 1971. Pacific sea snakes—a highly mobile newly recognized substrate for bryozoans. Proc. Sec. Int. Bryozoan Assoc. 1 p. (abstract).

Culotta, W. A., and G. V. Pickwell. 1991. The venomous sea snakes: A comprehensive bibliography. Krieger, Melbourne, Florida.

Cundall, D. 2002. Envenomation strategies, head form, and feeding ecology in vipers, pp. 149–162. *In* G. W. Schuett, M. Höggren, M. E. Douglas, and H. W. Greene (eds.), Biology of the vipers. Eagle Mountain Publ., Eagle Mountain, Utah.

Curran, C. H., and C. Kauffeld. 1937. Snakes and their ways. Harper Bros., New York.

Curry, S. C., D. Horning, P. Brady, R. Requa, D. B. Kunkel, and M. V. Vance. 1989. The legitimacy of rattlesnake bites in central Arizona. Ann. Emerg. Med. 18: 658–663.

Curtis, L. 1949a. The snakes of Dallas County, Texas. Field & Lab. (Southern Methodist Univ.) 17: 1–13.

———. 1949b. Notes on the eggs of *Heloderma horridum*. Herpetologica 5: 148.

———. 1952. Cannibalism in the Texas coral snake. Herpetologica 8: 27.

Cypert, E. 1961. The effect of fires in the Okefenokee Swamp in 1954 and 1955. Am. Midl. Nat. 66: 485–503.

DaLie, D. A. 1953. Poisonous snakes of America. J. Forest. 51: 243–248.

Dalrymple, G. H., T. M. Steiner, R. J. Nodell, and F. S. Bernardino, Jr. 1991a. Seasonal activity of the snakes of Long Pine Key, Everglades National Park. Copeia 1991: 294–302.

Dalrymple, G. H., F. S. Bernardino, Jr., T. M. Steiner, and R. J. Nodell. 1991b. Patterns of species diversity of snake community assemblages, with data on two Everglades snake assemblages. Copeia 1991: 517–521.

Daltry, J. C., W. Wüster, and R. S. Thorpe. 1997. The role of ecology in determining venom variation in the Malayan pitviper, *Calloselasma rhodostoma*, pp. 155–171. *In* R. S. Thorpe, W. Wüster, and A. Malhotra (eds.), Venomous snakes: Ecology, evolution and snakebite. Clarendon Press, Oxford.

Darlington, P. J., Jr. 1957. Zoogeography: The geographical distribution of animals. John Wiley and Sons, New York.

Dart, R. C. 2006. Clinical controversies in North American crotaline snakebite, pp. 31–32. *In* S. A. Seifert (ed.), Snakebites in the new millennium: A state-of-the-art symposium, University of Nebraska Medical Center, 21–23 October 2005, Omaha, Nebraska. J. Med. Toxicol. 2: 29–45.

Dart, R. C., and J. McNally. 2001. Efficacy, safety, and use of snake antivenoms in the United States. Ann. Emerg. Med. 37: 181–188.

Dart, R. C., P. C. O'Brien, B. S. Garcia, J. C. Jarchow, and J. McNally. 1992a. Neutralization of *Micrurus distans distans* venom by antivenom (*Micrurus fulvius*). J. Wild. Med. 3: 377–381.

Dart, R. C., J. T. McNally, D. W. Spaite, and R. Gustafson. 1992b. The sequelae of pitviper poisoning in the United States, pp. 395–404. *In* J. A. Campbell and E. D. Brodie, Jr. (eds.), Biology of the pitvipers. Selva, Tyler, Texas.

Dart, R. C., K. M. Hurlbut, R. Garcia, and J. Boren. 1996. Validation of a severity score for the assessment of crotalid snakebite. Ann. Emerg. Med. 27: 321–326.

Dart, R. C., S. A. Seifert, L. Carroll, R. F. Clark, E. Hall, L. V. Boyer-Hassen, S. C. Curry, C. S. Kitchens, and R. A. Garcia. 1997. Affinity-purified, mixed monospecific crotalid antivenom ovine Fab for the treatment of crotalid venom poisoning. Ann. Emerg. Med. 30: 33–39.

Dart, R. C., C. Stanford, V. S. Bebarta, C. P. Holstege, S. P. Bush, W. H. Richardson, and D. Olsen. 2007. The efficacy and safety of CroFab in the treatment of severe crotaline snake envenoming. *In* Venom week 2007 (program abstracts). Univ. Arizona Colleges of Medicine, Arizona Health Sci. Center, Off. Continuing Medical Educ., Tucson, Arizona.

Datta, G., and A. T. Tu. 1997. Structure and other chemical characterizations of Gila toxin, a lethal toxin from lizard venom. J. Peptide Res. 50: 443–450.

Datta, G., A. Dong, J .Witt, and A. T. Tu. 1995. Biochemical characterization of basilase, a fibrinolytic enzyme from *Crotalus basiliscus basiliscus*. Arch. Biochem. Biophys. 317: 365–373.

Daudin, F. M. 1801–1803. Historie naturelle, général et particulière des reptiles. 7 vols. F. Dufart, Paris.

Davenport, J. W. 1943. Field book of snakes of Bexar County, Texas and vicinity. White Mem. Mus., San Antonio, Texas.

David, P., and I. Ineich. 1999. Les serpents venimeux du monde: Systématique et répartition. Dumerilia 3: 3–499.

Davidson, T. M., and J. Eisner. 1996. United States coral snakes. Environ. Med. 1: 38–45.

Davis, J. B. 2002. *Agkistrodon piscivorus* (cottonmouth): Predation. Herpetol. Rev. 33: 136–137.

Davis, J. R., and D. F. De Nardo. 2007. The urinary bladder as a physiological reservoir that moderates dehydration in a large desert lizard, the Gila monster *Heloderma suspectum*. J. Exp. Biol. 210: 1472–1480.

Davis, R. A. 1980. Vipers among us. Cincinnati Mus. Nat. Hist. Quart. 17(2): 8–12.

Davis, W. B., and H. M. Smith. 1953. Snakes of the Mexican State of Morelos. Herpetologica 8: 133–143.

Dawson, W. R. 1967. Interspecific variation in physiological responses of lizards to temperature, pp. 230–257. *In* W. W. Milstead (ed.), Lizard ecology: A symposium. Univ. Missouri Press, Columbia.

Dean, B. 1938. Note on the sea-snakes, *Pelamis platurus* (Linnaeus). Science (New York) 88: 144–145.

Deckert, R. F. 1918. A list of reptiles from Jacksonville, Florida. Copeia 1918(54): 30–33.

de Cock Bunning, T. 1983. Thermal sensitivity as a specialization for prey capture and feeding in snakes. Am. Zool. 23: 363–375.

Degenhardt, W. G., C. W. Painter, and A. H. Price. 1996. Amphibians and reptiles of New Mexico. Univ. New Mexico Press, Albuquerque.

Delavan, W. 1939. *Corvus* and a copperhead. Field Ornithol. 1: 6–7.

Delezenne, C. 1919. Le zinc, constituant cellulaire de l'organisme animal. Sa présence et son rôle dans le venin des serpents. Annls Inst. Pasteur, Paris 33: 68–136.

Demeter, B. J. 1986. Combat behavior in the Gila monster (*Heloderma suspectum cinctum*). Herpetol. Rev. 17: 9–11.

De Nardo, D. 2005. Gila monsters: Surviving in the desert with air-conditioning, plumbing, and supermarkets. Sonoran Herpetol. 18: 97–101.

De Nardo, D. F., T. E. Zubal, and T. C. M. Hoffman. 2004. Cloacal evaporative cooling: A previously undescribed means of increasing evaporative water loss at higher temperatures in a desert ectotherm, the Gila monster *Heloderma suspectum*. J. Exp. Biol. 207: 945–953.

Denson, K. W. E., F. E. Russell, D. Almagro, and R. C. Bishop. 1972. Characterization of the coagulant activity of some snake venoms. Toxicon 72: 557–562.

De Rageot, R. 1957. Predation on small mammals in the Dismal Swamp, Virginia. J. Mammal. 38: 281.

Deraniyagala, A. W. C. T. 1955. A colored atlas of some vertebrates from Ceylon, vol. 3. Gov't. Press, Colombo.

de Roodt, A. R., S. Litwin, and S. O. Angel. 2003. Hydrolysis of DNA by 17 snake venoms. Comp. Biochem. Physiol. 135C: 469–479.

De Smet, W. H. O. 1978. The chromosomes of 23 species of snakes. Acta Zool. Pathol. (Antverp.) 70: 85–118.

Detterline, J. L., J. S. Jacob, and W. E. Wilhem. 1984. A comparison of helminth endoparasites in the cottonmouth (*Agkistrodon piscivorus*) and three species of water snakes (*Nerodia*). Trans. Am. Micro. Soc. 103: 137–143.

Deufel, A., and D. Cundall. 2003. Prey transport in "palatine-erecting" elapid snakes. J. Morphol. 258: 358–375.

Deutsch, H. F., and C. R. Diniz. 1955. Some proteolytic activities of snake venoms. J. Biol. Chem. 216: 17–26.

Devi, A. 1968. The protein and non-protein constituents of snake venoms, pp. 119–165. *In* W. Bücherl, E. E. Buckley, and V. Deulofeu (eds.), Venomous animals and their venoms. Academic Press, New York.

de Wit, C. A. 1982. Yield of venom from the Osage copperhead, *Agkistrodon contortrix phaeogaster*. Toxicon 20: 525–527.

Dhillon, D. S., E. Condrea, J. M. Maraganore, R. L. Heinrikson, S. Benjamin, and P. Rosenberg. 1987. Comparison of enzymatic and pharmacological activities of lysine-49 and aspartate-49 phospholipases A_2 from *Agkistrodon piscivorus piscivorus* snake venom. Biochem. Pharmacol. 36: 1723–1730.

Dickman, J. D., J. S. Colton, D. Chiszar, and C. A. Colton. 1987. Trigeminal responses to thermal stimulation of the oral cavity in rattlesnakes (*Crotalus viridis*) before and after bilateral anesthetization of the facial pit organs. Brain Res. 400: 365–370.

Diener, R. A. 1961. Notes on the bite of the broadbanded copperhead, *Ancistrodon contortrix laticinctus* Gloyd and Conant. Herpetologica 17: 143–144.

Dijkstra, B. W., J. Drenth, and K. H. Kalk. 1981. Active site and catalytic mechanism of phospholipase A_2. Nature (London) 289: 604–606.

Ditmars, R. L. 1923. Reptiles of the Southwest. Bull. New York Zool. Soc. 26(2): 22–30.

———. 1927. Occurrence and habits of our poisonous snakes. Bull. Antiv. Inst. Am. 1: 3–5.

———. 1931a. The reptile book. Doubleday Doran and Co., Garden City, New York.

———. 1931b. Snakes of the world. Macmillan, New York.

———. 1936. The reptiles of North America. Doubleday, Doran and Co., Garden City, New York.

———. 1939. A field book of North American snakes. Doubleday, Doran and Co., New York.

Divers, S. J., and P. J. Lennox. 1998. Veterinary investigation into radiographic sexing and blood parameters of Gila monsters (*Heloderma suspectum*). British Herpetol. Soc. Bull. 65: 36–37.

Dixon, J. R., M. Sabbath, and R. Worthington. 1962. Comments on snakes from central and western Mexico. Herpetologica 18: 91–100.

do Amaral, A. 1927. The anti-snake-bite campaign in Texas and in the sub-tropical United States. Bull. Antiv. Inst. Am. 1: 77–85.

———. 1928a. Improved process of venom extraction. Bull. Antiv. Inst. Am. 1: 100–102.

———. 1928b. Studies on snake venoms. I. Amounts of venom secreted by Nearctic pit vipers. Bull. Antiv. Inst. Am. 1: 103–104.

Dodd, C. K., Jr. 1987. Status, conservation, and management, pp. 478–513. *In* R. A. Seigel, J. T. Collins, and S. S. Novak (eds.), Snakes: Ecology and evolutionary biology. McGraw-Hill, New York.

———. 1993. Strategies for snake conservation, pp. 363–393. *In* R. A. Seigel and J. T. Collins (eds.), Snakes: Ecology and behavior. McGraw-Hill, New York.

Dodd, C. K., and G. G. Charest. 1988. The herpetofaunal community of temporary ponds in North Florida sandhills: Species composition, temporal use, and management implications, pp. 87–97. *In* R. C. Szaro, K. E. Severson, and D. R. Patton (eds.), Management of amphibians, reptiles, and small mammals in North America. U.S.D.A. For. Serv. Gen. Tech. Rep. RM-166.

Dodd, C. K., Jr., and R. Franz. 1995. Seasonal abundance and habitat use of selected snakes trapped in xeric and mesic communities of north-central Florida. Bull. Florida Mus. Nat. Hist. 38, Pt. I(2): 43–67.

Dodd, C. K., Jr., K. M. Enge, and J. N. Stuart. 1989. Reptiles on highways in north-central Alabama, U.S.A. J. Herpetol. 23: 197–200.

Dodge, C. H., and G. E. Folk, Jr. 1960. A case of rattlesnake poisoning in Iowa with a description of early symptoms. Proc. Iowa Acad. Sci. 67: 622–624.

Doery, H. M., and J. E. Pearson. 1961. Haemolysins in venoms of Australian snakes: Observations on the haemolysins of the venoms of some Australian snakes and the separation of phospholipase A from the venom of *Pseudechis porphyriacus*. Biochem. J. 78: 820–827.

Dolensek, E. P., and R. A. Cook. 1987. Clinical challenge [Radiographs of enteritis due to *Salmonella* infection in a Gila monster (*Heloderma suspectum*)]. J. Zoo Anim. Med. 18: 168–170.

Dolley, J. S. 1939. An anomalous pregnancy in the copperhead. Copeia 1939: 170.

Douglas, M. E., M. R. Douglas, G. W. Schuett, D. D. Beck, and B. K. Sullivan. 2003. Molecular biodiversity of Helodermatidae (Reptilia, Squamata). Abstract and presentation. Joint Meeting of Ichthyologists and Herpetologists, Manaus, Amazonas, Brazil.

Dowling, H. G. 1951a. A proposed method of expressing scale reductions in snakes. Copeia 1951: 131–134.

———. 1951b. A proposed standard system of counting ventrals in snakes. British J. Herpetol. 1: 97–99.

———. 1957. A review of the amphibians and reptiles of Arkansas. Occ. Pap. Univ. Arkansas Mus. 3: 1–51.

———. 1959. Classification of the Serpentes: A critical review. Copeia 1959: 38–52.

——— (ed.). 1975. Yearbook of herpetology. HISS, New York.

Dowling, H. G., and J. M. Savage. 1960. A guide to the snake hemipenis: A survey of basic structure and systematic characteristics. Zoologica (New York) 45: 17–28.

Dowling, H. G., C. A. Hass, S. B. Hedges, and R. Highton. 1996. Snake relationships revealed by slow-moving proteins: A preliminary survey. J. Zool. (London) 240: 1–28.

Downey, D. J., G. E. Omer, and M. S. Moneim. 1991. New Mexico rattlesnake bites: Demographic review and guidelines for treatment. J. Trauma 31: 1380–1386.

Draud, M. J. 1993. The dusky pigmy rattlesnake: A ruby in the sand. Rept. Amphib. Mag. May–June: 77–78, 80.

Drda, W. J. 1968. A study of snakes wintering in a small cave. J. Herpetol. 1: 64–70.

Drent, J. 1989. *Sistrurus miliarius barbouri*. Litt. Serpent. Engl. Ed. 9: 275–276.

———. 1991. *Agkistrodon bilineatus bilineatus*. Litt. Serpent. Engl. Ed. 11: 145.

Dreslik, M. J. 2004. Ecology of the eastern massasauga (*Sistrurus catenatus catenatus*) from Carlyle Lake, Clinton County, Illinois. Ph.D. diss., Univ. Illinois, Urbana-Champaign.

Dubnoff, J. W., and F. E. Russell. 1970. Isolation of lethal protein and peptide from *Crotalus viridis helleri* venom. Proc. West. Pharm. Soc. 13: 98.

Duellman, W. E. 1950. A case of *Heloderma* poisoning. Copeia 1950: 151.

———. 1954. The amphibians and reptiles of Jorullo Volcano, Michoacán, México. Occ. Pap. Mus. Zool. Univ. Michigan 560: 1–24.

———. 1961. The amphibians and reptiles of Michoacán, México. Univ. Kansas Publ. Mus. Nat. Hist. 15: 1–148.

———. 1965a. Amphibians and reptiles from the Yucatan Peninsula, México. Univ. Kansas Publ. Mus. Nat. Hist. 15: 577–614.

———. 1965b. A biogeographic account of the herpetofauna of Michoacán, México. Univ. Kansas Publ. Mus. Nat. Hist. 15: 627–709.

Duellman, W. E., and A. Schwartz. 1958. Amphibians and reptiles of southern Florida. Bull. Florida St. Mus. Biol. Sci. 3: 181–324.

Dugan, E., and S. A. Meyer. 2007. *Agkistrodon bilineatus* (Cantil): Reproduction. Herpetol. Rev. 38: 85–86.

Dugès, A. 1891. *Elaps diastema* var. *michaacanensis*. La Naturaleza (Mexico), ser. 2, 1: 487.

———. 1899. Vénin de l'*Heloderma horridum* (Wiegm.), pp. 134–137. *In* Cinquantenaire Soc. Biol., Paris.

Duke, J. A. 1986. Handbook of medicinal herbs. CRC Press, Boca Raton, Florida.

Dullemeijer, P. 1959. A comparative functional-anatomical study of the heads of some Viperidae. Morphol. Jarhb. 99: 881–985.

———. 1969. Growth and size of the eye in viperid snakes. Netherlands J. Zool. (London) 19: 249–276.

Dundee, H. A., and W. L. Burger, Jr. 1948. A denning aggregation of the western cottonmouth. Nat. Hist. Misc. (21): 1–2.

Dundee, H. A., and D. A. Rossman. 1989. The amphibians and reptiles of Louisiana. Louisiana State Univ. Press, Baton Rouge.

Dunson, M. K., and W. A. Dunson. 1975. The relation between plasma Na concentration and salt gland Na-K ATPase content in the diamondback terrapin and the yellow-bellied sea snake. J. Comp. Physiol. 101: 89–97.

Dunson, W. A. 1968. Salt gland secretion in the pelagic sea snake *Pelamis*. Am. J. Physiol. 215: 1512–1517.

———. 1971. The sea snakes are coming. Nat. Hist. 80: 52–60.

———. 1975. The biology of sea snakes. Univ. Park Press, Baltimore, Maryland.

Dunson, W. A., and G. W. Ehlert. 1971. Effects of temperature, salinity, and surface water flow on distribution of the sea snake *Pelamis*. Limnol. Oceanogr. 16: 845–853.

Dunson, W. A., and J. Freda. 1985. Water permeability of the skin of the amphibious snake, *Agkistrodon piscivorus*. J. Herpetol. 19: 93–98.

Dunson, W. A., and G. D. Robinson. 1976. Sea snake skin: Permeable to water but not to sodium. J. Comp. Physiol. 108: 303–311.

Dunson, W. A., and G. D. Stokes. 1983. Asymmetrical diffusion of sodium and water through the skin of sea snakes. Physiol. Zool. 56: 106–111.

Dunson, W. A., R. K. Packer, and M. K. Dunson. 1971. Sea snakes: An unusual salt gland under the tongue. Science (New York) 173: 437–441.

Duque-Osorio, J. F., A. Sanchez, L. Fierro, S. Garzon, and R. S. Castano. 2007. Snake venom and antivenom molecules. Revista de la Academia Colombiana de Ciencias Exactas Fisicas y Naturales 31(118): 109–137.

Duran-Reynals, F. 1939. A spreading factor in certain snake venoms and its relation to their mode of action. J. Exp. Med. 69: 69–81.

Durbian, F. E. 2006. Effects of mowing and summer burning on the massasauga (*Sistrurus catenatus*). Am. Midl. Nat. 155: 329–334.

Durbian, F. E., III, and L. Lenhoff. 2004. Potential effects of mowing prior to summer burning on the eastern massasauga (*Sistrurus c. catenatus*) at Squaw Creek National Wildlife Refuge, Holt County, Missouri, U.S.A. Trans. Missouri Acad. Sci. 38: 21–25.

Durkin, J. P., G. V. Pickwell, J. T. Trotter, and W. T. Shier. 1981. Phospholipase A_2 electrophoretic variants in reptile venoms. Toxicon 19: 535–546.

Duvall, D., D. Chiszar, and J. Trupiano. 1978. Preference for envenomated rodent prey by rattlesnakes. Bull. Psychon. Soc. 11: 7–8.

Dyer, W. G., and D. M. McNair. 1974. Ochetostomatid flukes of colubrid snakes from Illinois and Central America. Trans. Illinois Acad. Sci. 67: 463–464.

Dyr, J. E., B. Blombäck, B. Hessel, and F. Kornalík. 1989a. Conversion of fibrinogen to fibrin induced by preferential release of fibrinopeptide B. Biochem. Biophys. Acta 990: 18–24.

Dyr, J. E., B. Hessel, J. Suttnar, F. Kornalík, and B. Blombäck. 1989b. Fibrinopeptide-releasing enzymes in the venom from the southern copperhead snake (*Agkistrodon contortrix contortrix*). Toxicon 27: 359–373.

Dyr, J. E., J. Suttnar, J. Šimák, H. Fortová, and F. Kornalík. 1990. The action of a fibrin-promoting enzyme from the venom of *Agkistrodon contortrix contortrix* on rat fibrinogen and plasma. Toxicon 28: 1364–1367.

Edmund, A. G. 1960. Tooth replacement phenomena in the lower vertebrates. Royal Ontario Mus. Life Sci. Contrib. 52: 1–190.

Edstrom, A. 1992. Venomous and poisonous animals. Krieger, Malabar, Florida.

Effron, M., L. Griner, and K. Bernirschke. 1977. Nature and rate of neoplasia in captive wild mammals, birds, and reptiles at necropsy. J. Natl. Canc. Inst. 59: 185–198.

Egen, N. B., F. E. Russell, D. W. Sammons, R. C. Humphreys, A. L. Guan, and F. S. Markland, Jr. 1987. Isolation by preparative isoelectric focusing of a direct acting fibrinolytic enzyme from the venom of *Agkistrodon contortrix contortrix* (southern copperhead). Toxicon 25: 1189–1198.

Eidenmüller, B. 1993. Kurze Mitteilungen: Ergänzende Bemerkungen zu einem Nachzuchtbericht von *Heloderma suspectum*. Salamandra 29: 258–260.

Eidenmüller, B., and R. Wicker. 1992. Über eine Nachzucht von *Heloderma suspectum* (Cope, 1869). Salamandra 28: 106–111.

Elkins, C. A., and B. B. Nickol. 1983. The epizootiology of *Macracanthorhynchus ingens* in Louisiana. J. Parasitol. 69: 951–956.

Elliott, W. B. 1978. Chemistry and immunology of reptilian venoms, pp. 163–436. *In* C. Gans and K. A. Gans (eds.), Biology of the Reptilia, vol. 8: Physiology B. Academic Press, London.

El-Refael, M. F., and N. H. Sarkar. 2009. Snake venom inhibits the growth of mouse mammary tumor cells in vitro and in vivo. Toxicon 54: 33–41.

Endo, T., and N. Tamiya. 1991. Structure-function relationships of postsynaptic neurotoxins from snake venoms, pp. 165–222. *In* A. L. Harvey (ed.), Snake toxins. Pergamon Press, New York.

Eng, J., P. C. Andrews, W. A. Kleinman, L. Singh, and J.-P. Raufman. 1990. Purification and structure of exendin-3, a new pancreatic secretagogue isolated from *Heloderma horridum* venom. J. Biol. Chem. 265: 20259–20262.

Eng, J., W. A. Kleinman, L. Singh, G. Singh, and J.-P. Raufman. 1992. Isolation and characterization of exendin-4, an exendin-3 analogue, from *Heloderma suspectum* venom. J. Biol. Chem. 267: 7402–7405.

Enge, K. M., and K. N. Wood. 2002. A pedestrian road survey of an upland snake community in Florida. Southeast. Nat. 1: 365–380.

Engelhardt, G. P. 1914. Notes on the Gila monster. Copeia 1914(7): 1–2.

———. 1932. Notes on poisonous snakes in Texas. Copeia 1932: 37–38.

Engelman, W., and F. J. Obst. 1981. Snakes: Biology, behavior and relationships to man. Exeter Books, New York.

Entman, S. S., and K. J. Moise. 1984. Anaphylaxis in pregnancy. South. Med. J. 77: 402.

Ernst, C. H. 1962. The comparative fang lengths of Nearctic snakes of the genus *Agkistrodon*. Master's thesis, West Chester Univ., West Chester, Pennsylvania.

———. 1964. Sexual dimorphism in American *Agkistrodon* fang lengths. Herpetologica 20: 214.

———. 1965. Fang length comparisons of American *Agkistrodon*. Trans. Kentucky Acad. Sci. 26: 12–18.

———. 1982a. A study of the fangs of Russell's viper (*Vipera russellii*). J. Herpetol. 16: 67–71.

———. 1982b. A study of fangs of snakes belonging to the *Agkistrodon*-complex. J. Herpetol. 16: 72–80.

———. 1992. Venomous reptiles of North America. Smithsonian Inst. Press, Washington, D.C.

Ernst, C. H., and R. W. Barbour. 1989. Snakes of eastern North America. George Mason Univ. Press, Fairfax, Virginia.

Ernst, C. H., and E. M. Ernst. 2003. Snakes of the United States and Canada. Smithsonian Inst. Press, Washington, D.C.

———. 2006. Synopsis of helminths endoparasitic in snakes of the United States and Canada. Soc. Stud. Amphib. Rept. Herpetol. Circ. 34: 1–86.

Ernst, C. H., and J. E. Lovich. 2009. Turtles of the United States and Canada. Johns Hopkins Univ. Press, Baltimore, Maryland.

Ernst, C. H., and G. R. Zug. 1996. Snakes in question: The Smithsonian answer book. Smithsonian Inst. Press, Washington, D.C.

Ernst, C. H., S. C. Belfit, S. W. Sekscienski, and A. F. Laemmerzahl. 1997. The amphibians and reptiles of Ft. Belvoir and Northern Virginia. Bull. Maryland Herpetol. Soc. 33: 1–62.

Esnouf, M. P., and G. W. Tunnah. 1967. The isolation and properties of the thrombin-like activity from *Agkistrodon rhodostoma* venom. British J. Haematol. 13: 581–590.

Espinosa-Avilés, D., V. M. Salomón-Soto, and S. Morales-Martínez. 2008. Hematology, blood chemistry, and bacteriology of the free-ranging Mexican beaded lizard (*Heloderma horridum*). J. Zoo Wildl. Med. 39: 21–27.

Essex, H. E. 1932. The physiologic action of the venom of the water moccasin (*Agkistrodon piscivorus*). Bull. Antiv. Inst. Am. 5: 81.

Essghaier, M. F. A., and D. R. Johnson. 1975. Aspects of the bioenergetics of Great Basin lizards. J. Herpetol. 9: 191–195.

Esslinger, J. H. 1962. Morphology of the egg and larva of *Porocephalus crotali* (Pentastomida). J. Parasitol. 48: 457–462.

Estes, R. 1964. Fossil vertebrates from the Late Cretaceous Lance Formation, eastern Wyoming. Univ. California Publ. Geol. Sci. 49: 1–180.

Etheridge, R. 1967. Lizard caudal vertebrae. Copeia 1967: 699–721.

Ettling, J. A. 1986. Cicada feeding by captive copperheads (*Agkistrodon contortrix*). Bull. Chicago Herpetol. Soc. 21: 32.

Evans, G. M. 1987. The coral snake question. Herptile 12: 105–107, 110.

Evans, P. D., and H. K. Gloyd. 1948. The subspecies of the massasauga, *Sistrurus catenatus*, in Missouri. Bull. Chicago Acad. Sci. 8: 225–232.

Fairley, N. H. 1929. The present position of snake bite and the snake bitten in Australia. Bull. Antiv. Inst. Am. 3: 65–76.

Fann, E. C. 1997. Population density and present habitat range of the western pygmy rattlesnake (*Sistrurus miliarius streckeri*) in Tennessee. Am. Zoo Aquar. Assoc. Reg. Conf. Proc. 1997: 34–35.

Faria Ferreira, L. A., H. Auer, E. Haslinger, C. Fedele, and G. G. Habermehl. 1999. Spatial structures of the bradykinin potentiating peptide F from *Agkistrodon piscivorus piscivoris* venom. Toxicon 37: 661–676.

Farrell, T. M., P. G. May, and M. A. Pilgrim. 1995. Reproduction in the rattlesnake, *Sistrurus miliarius barbouri*, in central Florida. J. Herpetol. 29: 21–27.

Farrell, T. M., M. A. Pilgrim, and P. G. May. 2008. Annual variation in neonate recruitment in a Florida population of the pigmy rattlesnake, *Sistrurus miliarius*, pp. 257–264. *In* W. K. Hayes, K. R. Beaman, M. D. Cardwell, and S. P. Bush (eds.), The biology of rattlesnakes. Loma Linda Univ. Press, Loma Linda, California.

Farrell, T. M., P. G. May, and M. A. Pilgrim. 2009. Repeatability of female reproductive traits in pigmy rattlesnakes (*Sistrurus miliarius*). J. Herpetol. 43: 332–335.

Faure, G. 1999. Les phospholipases A_2 des venins de serpents. Bull. Soc. Zool. France 124: 149–168.

Fayrer, J. 1872. The Thantophidia of India. J. & A. Churchill, London.

Fazelat, J., S. Teperman, and M. Touger. 2007. Case report: Recurrent hemorrhage after western diamondback rattlesnake envenomation treated with Crotalidae polyvalent immune Fab (ovine). *In* Venom week 2007 (program abstracts). Univ. Arizona Colleges of Medicine, Arizona Health Sci. Center, Off. Continuing Medical Educ., Tucson, Arizona.

Feltoon, A. R., T. Guerra, C. Lehn, and M. R. J. Forstner. 2007. Isolation and characterization of six microsatellites in the Mexican beaded lizard *Heloderma horridum*. Mol. Ecol. Notes 7: 433–435.

Fenton, M. B., and L. E. Licht. 1990. Why rattle snake? J. Herpetol. 24: 274–279.

Fiero, M. K., M. W. Seifert, T. J. Weaver, and C. A. Bonilla. 1972. Comparative study of juvenile and adult prairie rattlesnake (*Crotalus viridis viridis*) venoms. Toxicon 10: 81–82.

Finn, R. 2001. Snake venom protein paralyzes cancer cells. J. Ntl. Cancer Inst. 93: 261–262.

Finneran, L. C. 1953. Aggregation behavior of the female copperhead, *Agkistrodon contortrix mokeson*, during gestation. Copeia 1953: 61–62.

Fischer, J. G. 1882. Anatomische Notizen über *Heloderma horridum*, Wiegm. Berh. Vergl. Naturw. Unterh. Hamburg 5: 2–16.

Fischer, R. U., D. E. Scott, J. D. Congdon, and S. A. Busa. 1994. Mass dynamics during embryonic development and parental investment in cottonmouth neonates. J. Herpetol. 28: 364–369.

Fischman, H. K., J. Mitra, and H. Dowling. 1972. Chromosome characteristics of 13 species in the order Serpentes. Mammal Chromosome Newsl. 13: 72–73.

Fischthal, J. H., and R. E. Kuntz. 1975. Some trematodes of amphibians and reptiles from Taiwan. Proc. Helminthol. Soc. Washington 42: 1–13.

Fisher, C. B. 1973. Status of the flat-headed snake, *Tantilla gracilis* Baird and Girard, in Louisiana. J. Herpetol. 7: 136–137.

Fitch, H. S. 1949. Road counts of snakes in western Louisiana. Herpetologica 5: 87–90.

———. 1956. Temperature responses in free-living amphibians and reptiles in northeastern Kansas. Univ. Kansas Publ. Mus. Nat. Hist. 8: 417–476.

———. 1958. Home ranges, territories, and seasonal movements of vertebrates of the Natural History Reservation. Univ. Kansas Publ. Mus. Nat. Hist. 11: 63–326.

———. 1959. A patternless phase of the copperhead. Herpetologica 15: 21–24.

———. 1960. Autecology of the copperhead. Univ. Kansas Publ. Mus. Nat. Hist. 13: 85–288.

———. 1970. Reproductive cycles in lizards and snakes. Univ. Kansas Mus. Nat. Hist. Misc. Publ. 52: 1–247.

———. 1981. Sexual size differences in reptiles. Univ. Kansas Mus. Nat. Hist. Misc. Publ. 70: 1–72.

———. 1982. Resources of a snake community in prairie-woodland habitat of northeastern Kansas, pp. 93–97. *In* N. J. Scott, Jr. (ed.), Herpetological communities. U.S. Fish Wildl. Serv. Wildl. Res. Rep. 13.

———. 1985. Variation in clutch and litter size in New World reptiles. Univ. Kansas Mus. Nat. Hist. Misc. Publ. 76: 1–76.

———. 1993. Relative abundance of snakes in Kansas. Trans. Kansas Acad. Sci. 96: 213–224.

———. 1998. The Sharon Springs Roundup and prairie rattlesnake demography. Trans. Kansas Acad. Sci. 101: 101–113.

———. 1999. A Kansas snake community: Composition and changes over 50 years. Krieger, Malabar, Florida.

———. 2002. A comparison of growth and rattle strings in three species of rattlesnakes. Univ. Kansas Mus. Nat. Hist. Sci. Pap. 24: 1–6.

———. 2003. Reproduction in snakes of the Fitch natural history reservation in northeastern Kansas. J. Kansas Herpetol. 6: 21–24.

———. 2005. Observations on wandering of juvenile snakes in northeastern Kansas. J. Kansas Herpetol. 13: 11–12.

Fitch, H. S., and A. L. Clarke. 2002. An exceptionally large natural assemblage of female copperheads (*Agkistrodon contortrix*). Herpetol. Rev. 33: 94–95.

Fitch, H. S., and J. T. Collins. 1985. Intergradation of Osage and broad-banded copperheads in Kansas. Trans. Kansas Acad. Sci. 38: 135–137.

Fitch, H. S., and A. F. Echelle. 2006. Abundance and biomass of twelve species of snakes native to northeastern Kansas. Herpetol. Rev. 37: 161–165.

Fitch, H. S., and G. R. Pisani. 1993. Life history traits of the western diamondback rattlesnake (*Crotalus atrox*) studied from roundup samples in Oklahoma. Occ. Pap. Mus. Nat. Hist. Univ. Kansas 156: 1–24.

Fitch, H. S., and H. W. Shirer. 1971. A radiotelemetric study of spatial relationships in some common snakes. Copeia 1971: 118–128.

Fitz Simons, D. C., and H. M. Smith. 1958. Another rear-fanged South African snake lethal to humans. Herpetologica 14: 198–202.

Fitz Simons, F. W. 1919. The snakes of South Africa: Their venom and the treatment of snake bite. Maskew Miller, Capetown.

Fitz Simons, V. F. W. 1978. A field guide to the snakes of southern Africa. Collins, St. James Place, London.

Fix, J. D. 1980. Venom yield of the North American coral snake and its clinical significance. South. Med. J. 73: 737–738.

Fix, J. D., and S. A. Minton, Jr. 1976. Venom extraction and yields from the North American coral snake, *Micrurus fulvius*. Toxicon 14: 143–145.

Fleckenstein, A., and H. Gerkhardt. 1952. Über die biologische Bedeutung des hohen Zinkgehaltes in Schlangengiften: Zink als Schlangengift-Ihnibitor. Naunyn-Schmiedebergs Arch. exp. Path. Pharmak. 214: 134–146.

Fleckenstein, A., and W. Jaeger. 1952. Weiter Ergebnisse über die Blockierung der Bienengift- und Schlangengiftwirkung durch Zinksalze. Naunyn-Schmiedebergs Arch. Exp. Path. Pharmak. 215: 163–176.

Fleet, R. R., and J. C. Kroll. 1978. Litter size and parturition behavior in *Sistrurus miliarius streckeri*. Herpetol. Rev. 9: 11.

Fletcher, J. E., and M.-S. Jiang. 1998. Lys^{49} phospholipase A_2 myotoxins lyse cell cultures by two distinct mechanisms. Toxicon 36: 1549–1555.

Fletcher, J. E., M. Hubert, S. J. Wieland, Q.-H. Gong, and M.-S. Jiang. 1996. Similarities and differences in mechanisms of cardiotoxins, melittin and other myotoxins. Toxicon 34: 1301–1311.

Flowers, H. H. 1966. Effects of X-irradiation on the antigenic character of *Agkistrodon piscivorus* (cottonmouth moccasin) venom. Toxicon 3: 301–304.

Fogell, D. D., T. J. Leonard, and J. D. Fawcett. 2002. *Agkistrodon contortrix phaeogaster* (Osage copperhead): Breeding. Herpetol. Rev. 33: 209–210.

Fogleman, B., W. Byrd, and E. Hanebrink. 1986. Observations of the male combat dance in the cottonmouth (*Agkistrodon piscivorus*). Bull. Chicago Herpetol. Soc. 21: 26–28.

Fontana, F. 1795. Treatise on the venom of the viper; on the American poisons; and the cherry laurel, and some other vegetable poisons. Vol. 1, 2nd ed., translated from the original French by Joseph Skinner and printed for John Cuthell, London. [Richerche Fisiche Sopra il Veleno

Della Vipera. Lucca, J. Giusti, 1767; Treatise on the Venom of the Viper. 2 vols., Paris, 1781; Traite sur le Venin de la Vipere. 2 vols, London, J. Murray, 1787.]

Fontenot, L. W., and W. F. Font. 1996. Helminth parasites of four species of aquatic snakes from two habitats in southeastern Louisiana. Proc. Helminthol. Soc. Washington 63: 66–75.

Ford, K. M., III. 1992. Herpetofauna of the Albert Athens local fauna (Pleistocene: Irvingtonian), Nebraska. Master's thesis, Michigan St. Univ., East Lansing.

Ford, N. B. 2002. Ecology of the western cottonmouth (*Agkistrodon piscivorus leucostoma*) in northeastern Texas, pp. 167–177. *In* G. W. Schuett, M. Höggren, M. E. Douglas, and H. W. Greene (eds.), Biology of the vipers. Eagle Mountain Publ., Eagle Mountain, Utah.

Ford, N. B., and G. M. Burghardt. 1993. Perceptual mechanisms and the behavioral ecology of snakes, pp. 117–164. *In* R. A. Seigel and J. T. Collins (eds.), Snakes: Ecology and evolutionary biology. McGraw-Hill, New York.

Ford, N. B., and D. L. Lancaster. 2007. The species-abundance distribution of snakes in a bottomland hardwood forest of the southern United States. J. Herpetol. 41: 385–393.

Ford, N. B., V. Cobb, and W. W. Lamar. 1990. Reproductive data on snakes from northeastern Texas. Texas J. Sci. 42: 355–368.

Ford, N. B., V. Cobb, and J. Stout. 1991. Species diversity and seasonal abundance of snakes in a mixed pine-hardwood forest in eastern Texas. Southwest. Nat. 36: 171–177.

Ford, N. B., F. Brischoux, and D. Lancaster. 2004. Reproduction in the western cottonmouth, *Agkistrodon piscivorus leucostoma*, in a floodplain forest. Southwest. Nat. 49: 465–471.

Ford, W. M., and E. P. Hill. 1991. Organochlorine pesticides in soil sediments and aquatic animals in the Upper Steele Bayou watershed of Mississippi. Arch. Environ. Contam. Toxicol. 20: 161–167.

Forsyth, B. J., C. D. Baker, T. Wiles, and C. Weilbaker. 1985. Cottonmouth, *Agkistrodon piscivorus*, records from the Blue River and Potato Run in Harrison County, Indiana (Ohio River drainage, U.S.A.). Proc. Indiana Acad. Sci. 94: 633–634.

Foster, G. W., P. E. Moler, J. M. Kinsella, S. P. Terrell, and D. J. Forrester. 2000. Parasites of eastern indigo snakes (*Drymarchon corais couperi*) from Florida, U.S.A. Comp. Parasitol. 67: 124–128.

Foster, S., and R. A. Caras. 1994. A field guide to venomous animals and poisonous plants: North America north of Mexico. Houghton Mifflin, Boston.

Foster, S., and J. A. Duke. 1990. A field guide to medicinal plants: Eastern and central North America. Houghton Mifflin, Boston.

Fowlie, J. A. 1965. The snakes of Arizona. Azul Quinta, Fallbrook, California.

Fox, H. 1913. Anatomy of the poison gland of *Heloderma*, pp. 17–28. *In* L. Loeb, C. L. Alsberg, E. Cooke, E. P. Corson-White, M. S. Fleisher, H. Fox, T. S. Githens, S. Leopold, M. K. Meyers, M. E. Rehfuss, D. Rivas, and L. Tuttle (eds.), The venom of *Heloderma*. Carnegie Inst. Washington Publ. 117.

Fox, J. W., and S. M. Serrano. 2005. Structural considerations of the snake venom metalloproteinases, key members of the M12 reprolysin family of metalloproteinases. Toxicon 45: 969–985.

Fox, J. W., M. Elzinga, and A. T. Tu. 1979. Amino acid sequence and disulfide bond assignment of myotoxin-a isolated from the venom of prairie rattlesnake (*Crotalus virdis viridis*). Biochemistry 18: 678–684.

Fox, W. 1956. Seminal receptacles of snakes. Anat. Rec. 124: 519–540.

Francis, B., C. Seebart, and I. I. Kaiser. 1992. Citrate is an endogenous inhibitor of snake venom enzymes by metal-ion chelation. Toxicon 30: 1239–1246.

Frank, N., and E. Ramus. 1994. State, federal, and C.I.T.E.S. regulations for herpetologists. Publ. Rept. Amphib. Mag., Pottsville, Pennsylvania.

Franklin, C. J. 2006. *Agkistrodon piscivorus leucostoma* (western cottonmouth): Diet. Herpetol. Rev. 37: 93.

Franz, R. 1974. Parasites of reptiles. Part II: digenetic trematodes inhabiting the respiratory and upper digestive tracts of snakes. Bull. Maryland Herpetol. Soc. 10: 7–15.

Freitas, M. A., P. W. Geno, L. W. Sumner, M. E. Cooke, S. A. Hudiburg, C. L. Ownby, I. I. Kaiser, and G. V. Odell. 1992. Citrate is a major component of snake venoms. Toxicon 30: 461–464.

Freze, V. I. 1965. Proteocephalata in fish, amphibians, and reptiles, pp. 1–597. *In* K. I. Skrjabin (ed.), Essentials of cestodology: vol. V. Acad. Sci. U.S.S.R. Helminthol. Lab.

Friederich, C., and A. T. Tu. 1971. Role of metals in snake venoms for hemorrhagic, esterase and proteolytic activities. Biochem. Pharmacol. 20: 1549–1556.

Froom, B. 1964. The massasauga rattlesnake. Can. Audubon 26: 78–80.

Frost, D. R., and J. T. Collins. 1988. Nomenclatural notes on reptiles of the United States. Herpetol. Rev. 19: 73–74.

Fry, B. 2009. A central role for venom in predation by *Varanus komodoensis* (Komodo dragon) and the extinct giant *Varanus (Megalania) prisca*. Presentation (Abstract 1046, Herp Reproduction & Behavior) at Joint Meeting of Ichthyologists and Herpetologists, Portland, Oregon.

Fry, B. G., K. D. Winkel, J. C. Wickramaratna, W. C. Hodgson, and W. Wüster. 2003a. Effectiveness of snake antivenom: Species and regional venom variation and its clinical impact. J. Toxicol. Toxin Rev. 22: 23–34.

Fry, B. G., W. Wüster, R. M. Kini, V. Brusic, A. Khan, D. Venkataraman, and A. P. Rooney. 2003b. Molecular evolution and phylogeny of elapid snake venom three-finger toxins. J. Mol. Evol. 57: 110–129.

Fry, B. G., N. G. Lumsden, W. Wüster, J. C. Wickramaratna, W. C. Hodgson, and R. Manjunatha Kini. 2003c. Isolation of a neurotoxin (α-colubritoxin) from a nonvenomous colubrid: Evidence for early origin of venom in snakes. J. Mol. Evol. 57: 446–452.

Funk, R. S. 1964a. On the food of *Crotalus m. molossus*. Herpetologica 20: 134.

———. 1964b. On the reproduction of *Micruroides euryxanthus* (Kennicott). Copeia 1964: 219.

———. 1964c. Birth of a brood of western cottonmouths, *Agkistrodon piscivorus leucostoma*. Trans. Kansas Acad. Sci. 67: 199.

———. 1966. Notes about *Heloderma suspectum* along the western extremity of its range. Herpetologica 22: 254–258.

Furiani, M. 1989. The reproduction of *Agkistrodon bilineatus* (Gunther, 1888) in captivity. Litt. Serpent. Engl. Ed. 9: 142–144.

Furman, J. 2007. Timber rattlesnakes in Vermont & New York: Biology, history, and the fate of an endangered species. Univ. Press New England, Lebanon, New Hampshire.

Furukawa, K., and S. Ishimaru. 1990. Use of thrombin-like snake venom enzymes in the treatment of vascular occlusive diseases, pp. 161–174. *In* K. F. Stocker (ed.), Medical use of snake venom proteins. CRC Press, Boca Raton, Florida.

Gabe, M., and H. Saint Girons. 1965. Contribution à la morphologie comparée du cloaque et des glandes épidermoïdes de la région cloacale chez les lépidosauriens. Mem. Mus. Natl. Hist. Nat. Ser. A Zool. 33: 151–292.

Gadow, H. 1911. Isotely and coralsnakes. Zool. Jarb., Abt. Syst. 31: 1–24.

Gaige, H. B. T. 1936. Some reptiles and amphibians from Yucatan and Campeche, Mexico. Carnegie Inst. Washington Publ. 457: 289–304.

Galán, J. A., E. E. Sánchez, A. Rodríguez-Acosta, and J. C. Pérez. 2004. Neutralization of venoms from two southern Pacific rattlesnakes (*Crotalus helleri*) with commercial antivenoms and endothermic animal sera. Toxicon 43: 791–799.

Gallardo, L. I., D. F. De Nardo, and E. N. Taylor. 2002. Habitat selection and movement patterns in Gila monsters, *Heloderma suspectum*. Poster Presentation, American Society of Ichthyologists and Herpetologists, Annual Meeting, Kansas City, Missouri.

Galligan, J. H., and W. A. Dunson. 1979. Biology and status of timber rattlesnake (*Crotalus horridus*) populations in Pennsylvania. Biol. Conserv. 15: 13–58.

Gans, C., and W. B. Elliott. 1968. Snake venoms: Production, injection, action. Adv. Oral Biol. 3: 45–81.

García, L. T., L. T. Parreiras e Silva, P. H. P. Ramos, A. K. Carmona, P. A. Bersanetti,, and H. S. Selistre-de-Araujo. 2004. The effect of post-translational modifications on the hemorrhagic activity of snake venom metalloproteinases. Comp. Biochem. Physiol. 138C: 23–32.

Gardner, S. C., and E. Oberdörster. 2006. Toxicology of reptiles. CRC Press, Boca Raton, Florida.

Garman, S. W. 1883 [1884]. North American Reptilia. Part I: Ophidia. Mem. Mus. Comp. Zool. 8(3): 1–185.

Garrett, J. M., and D. G. Barker. 1987. A field guide to reptiles and amphibians of Texas. Texas Monthly Press, Austin.

Garton, J. S., and R. W. Dimmick. 1969. Food habits of the copperhead in middle Tennessee. J. Tennessee Acad. Sci. 44: 113–117.

Gasmi, A., M. Karoui, Z. Benlastar, H. Karoui, M. El Ayeb, and K. Dellagi. 1991. Purification and characterization of a fibrinogenase from *Vipera lebetina* (desert adder) venom. Toxicon 29: 827–836.

Gasmi, A., A. Chabchoub, S. Guermazi, H. Karoui, M. El Ayeb, and K. Dellagi. 1997. Further characterization and thrombolytic activity in a rat model of a fibrinogenase from *Vipera lebetina* venom. Thromb. Res. 86: 233–242.

Gates, G. O. 1956a. A record length for the Arizona coral snake. Herpetologica 12: 155.

———. 1956b. Mating habits of the Gila monster. Herpetologica 12: 184.

———. 1957. A study of the herpetofauna in the vicinity of Wickenburg, Maricopa County, Arizona. Trans. Kansas Acad. Sci. 60: 403–418.

Gawade, S. P. 2004. Snake venom neurotoxins: Pharmacological classification. J. Toxicol. Toxin Rev. 23: 37–96.

Gehlbach, F. R. 1972. Coral snake mimicry reconsidered: The strategy of self-mimicry. Forma et Functio 5: 311–320.

Gennaro, J. F., Jr., and E. R. Casey, Jr. 2005. Two neutrotropic effects of pit viper venoms, p. 29. *In* Biology of the rattlesnakes symposium (program abstracts). Loma Linda Univ., Loma Linda, California.

Gennaro, J. F., R. S. Leopold, and T. W. Merriam. 1961. Observations on the actual quantity of venom introduced by several species of crotalid snakes in their bite. Anat. Rec. 139: 303.

Gennaro, J. F., Jr., J. W. Skimming, L. H. S. Van Mierop, and C. S. Kitchens. 2006. Treatment of envenomation by the southeastern coral snake (*Micrurus fulvius*), p. 30. *In* S. A. Seifert (ed.), Snakebites in the new millennium: A state-of-the-art symposium, University of Nebraska Medical Center, 21–23 October 2005, Omaha, Nebraska. J. Med. Toxicol. 2: 29–45.

Gensler, P. A. 2001. The first fossil *Heloderma* from the mid-Pleistocene (Late Irvingtonian) Coyote Badlands, Anza-Borrego Desert State Park, southern California, p. 71. *In* R. E. Reynolds (ed.), The changing face of the east Mojave Desert. Abstracts of the 2001 Desert Symposium, Desert Studies Center, California St. Univ.

George, I. D. 1930. Notes on the extraction of venom at the serpentarium of the Antivenin Institute at Tela, Honduras. Bull. Antiv. Inst. Am. 4: 57–59.

Gibbons, J. W. 1977. Snakes of the Savannah River Plant with information about snakebite prevention and treatment. ERDA's Savannah River Nat. Environ. Res. Park. SRO-NERP-1.

Gibbons, J. W., and M. Dorcas. 1998. Cowards, bluffers, and warriors. Nat. Hist. Mag. November 1998: 56–57.

———. 2002. Defensive behavior of cottonmouths (*Agkistrodon piscivorus*) toward humans. Copeia 2002: 195–198.

———. 2005. Snakes of the Southeast. Univ. Georgia Press, Athens.

Gibbons, J. W., and R. D. Semlitsch. 1991. Guide to the reptiles and amphibians of the Savannah River Site. Univ. Georgia Press, Athens.

Gibbons, J. W., R. R. Haynes, and J. L. Thomas. 1990. Poisonous plants and venomous animals of Alabama and adjoining states. Univ. Alabama Press, Tuscaloosa.

Gibbs, H. L., and W. Rossiter. 2008. Rapid evolution by positive selection and gene gain and loss: PLA_2 venom genes in closely related *Sistrurus* rattlesnakes with divergent diets. J. Mol. Evol. 66: 151–166.

Gibbs, H. L., K. A. Prior, and P. J. Weatherhead. 1994. Genetic analysis of populations of threatened snake species using RAPD markers. Mol. Ecol. 3: 329–337.

Gibbs, H. L., K. A. Prior, P. J. Weatherhead, and G. Johnson. 1997. Genetic structure of populations of the threatened eastern massasauga rattlesnake, *Sistrurus c. catenatus*: Evidence from microsatellite DNA markers. Mol. Ecol. 6: 1123–1132.

Gibbs, H. L., K. Prior, and C. Parent. 1998. Characterization of DNA microsatellite loci from a threatened snake: The eastern massasauga rattlesnake (*Sistrurus c. catenatus*) and their use in population studies. J. Heredity 89(2): 169–173.

Gibson, A. R. 1977. *Agkistrodon* venom variation: A criticism. Copeia 1977: 607–608.
Gibson, J. D. 2001. Amphibians and reptiles of Powhatan County, Virginia. Catesbeiana 21: 3–28.
Gibson, J. D., and D. A. Merkle. 2004. Road mortality of snakes in central Virginia. Banisteria 24: 8–13.
Gienger, C. M. 2003. Natural history of the Gila monster in Nevada. Master's thesis, Univ. Nevada, Reno.
Gienger, C. M., and D. D. Beck. 2007. Heads or tails? Sexual dimorphism in helodermatid lizards. Can. J. Zool. 85: 92–98.
Gienger, C. M., and C. R. Tracy. 2008a. *Heloderma suspectum* (Gila monster): Prey. Herpetol. Rev. 39: 224–226.
———. 2008b. Ecological interactions between Gila monsters (*Heloderma suspectum*) and desert tortoises (*Gopherus agassizii*). Southwest. Nat. 53: 265–268.
Gienger, C. M., G. W. Johnson, M. McMillan, S. Sheldon, and C. R. Tracy. 2005. Timing of hatching in beaded lizards (*Heloderma horridum*). Sonoran Herpetol. 18: 93–94.
Gilmore, C. W. 1928. Fossil lizards of North America. Mem. Natl. Acad. Sci. 22: 1–201.
———. 1938. Fossil snakes of North America. Geol. Soc. Am. Spec. Pap. 9: 1–93.
Giorgi, R., M. M. Bernardi, and Y. Cury. 1993. Analgesic effect evoked by low molecular weight substances extracted from *Crotalus durissus terrificus* venom. Toxicon 31: 1257–1266.
Githens, T. S. 1931. Antivenin: Its preparation and standardization. Bull. Antiv. Inst. Am. 4: 81–85.
Githens, T. S., and N. O. Wolff. 1939. The polyvalency of crotalidic antivenins. J. Immunol. 37: 33–51.
Glass, T. G., Jr. 1969. Cortisone and immediate fasciotomy in the treatment of severe rattlesnake bite. Texas Med. 69(7): 40–47.
———. 1976. Early debridement in pit viper bites. J. Am. Med. Assoc. 235: 2513–2516.
———. 1982. Management and treatment of the western diamondback rattlesnake bite, pp. 339–360. *In* A. T. Tu (ed.), Rattlesnake venoms: Their actions and treatment. Marcel Dekker, New York.
Glaudas, X. 2004. Do cottonmouths (*Agkistrodon piscivorus*) habituate to human confrontations? Southeast. Nat. 3: 129–138.
Glaudas, X., and J. W. Gibbons. 2005. Do thermal cues influence the defensive strike of cottonmouths (*Agkistrodon piscivorus*)? Amphibia-Reptilia 26: 264–267.
Glaudas, X., and D. L. Wagner. 2004. *Agkistrodon piscivorus piscivorus* (eastern cottonmouth): Diet. Herpetol. Rev. 35: 272.
Glaudas, X., and C. T. Winne. 2007. Do warning displays predict striking behavior in a viperid snake, the cottonmouth (*Agkistrodon piscivorus*)? Can. J. Zool. 85: 574–578.
Glaudas, X., T. M. Farrell, and P. C. May. 2005. Defensive behavior of free-ranging pygmy rattlesnakes (*Sistrurus miliarius*). Copeia 2005: 196–200.
Glaudas, X., C. T. Winne, and L. A. Fedewa. 2006. Ontogeny of anti-predator behavioral habituation in cottonmouths (*Agkistrodon piscivorus*). Ethology 112: 608–615.
Glaudas, X., K. M. Andrews, J. D. Wilson, and J. W. Gibbons. 2007. Migration patterns in a population of cottonmouths (*Agkistrodon piscivorus*) inhabiting an isolated wetland. J. Zool. (London) 271: 119–124.
Glenn, J. L., and R. C. Straight. 1982. The rattlesnakes and their venom yield and lethal toxicity, pp. 3–119. *In* A. T. Tu (ed.), Rattlesnake venoms: Their actions and treatment. Marcel Dekker, New York.
———. 1985. Distribution of proteins immunologically similar to Mojave toxin among species of *Crotalus* and *Sistrurus*. Toxicon 23: 28.
Glenn, J. L., R. C. Straight, and C. C. Snyder. 1972. Yield of venom obtained from *Crotalus atrox* by electrical stimulation. Toxicon 10: 575–579.
Gloyd, H. K. 1934. Studies on the breeding habits and young of the copperhead, *Agkistrodon mokasen* Beauvois. Pap. Michigan Acad. Sci. Arts Lett. 19: 587–604.
———. 1935. The subspecies of *Sistrurus miliarius*. Occ. Pap. Mus. Zool. Univ. Michigan 322: 1–7.
———. 1937. A herpetological consideration of faunal areas in southern Arizona. Bull. Chicago Acad. Sci. 5: 79–136.

———. 1938. A case of poisoning from the bite of a black coral snake. Herpetologica 1: 121–124.

———. 1940. The rattlesnakes, genera *Sistrurus* and *Crotalus*. Chicago Acad. Sci. Spec. Publ. 4: 1–266.

———. 1947. Notes on the courtship and mating behavior of certain snakes. Chicago Acad. Sci. Nat. Hist. Misc. 12: 1–4.

———. 1955. A review of the massasaugas, *Sistrurus catenatus*, of the southwestern United States (Serpentes: Crotalidae). Bull. Chicago Acad. Sci. 10: 84–98.

———. 1969. Two additional subspecies of North American snakes, genus *Agkistrodon*. Proc. Biol. Soc. Washington 82: 219–232.

———. 1972. A subspecies of *Agkistrodon bilineatus* (Serpentes: Crotalidae) on the Yucatán Peninsula, México. Proc. Biol. Soc. Washington 84: 327–334.

Gloyd, H. K., and R. Conant. 1934. The broad-banded copperhead: A new subspecies of *Agkistrodon mokasen*. Occ. Pap. Mus. Zool. Univ. Michigan 283: 1–5.

———. 1943. A synopsis of the American forms of *Agkistrodon* (copperheads and moccasins). Bull. Chicago Acad. Sci. 7: 147–170.

———. 1990. Snakes of the *Agkistrodon* complex: A monographic review. Soc. Stud. Amphib. Rept. Contrib. Herpetol. 6: i–vi, 1–614.

Godard, R. D., C. M. Wilson, C. M. Rock, and W. B. Cash. 2006. *Agkistrodon contortrix* (copperhead): Diet. Herpetol. Rev. 37: 476.

Goin, C. J., O. B. Goin, and G. R. Zug. 1978. Introduction to herpetology. 3rd ed. Freeman, San Francisco.

Golay, P. 1985. Checklist and keys to the terrestrial proteroglyphs of the world (Serpentes: Elapidae-Hydrophiidae). Elapsoidea, Gilbert Ray, Geneva, Switzerland.

Gold, B. S., R. C. Dart, and R. A. Barish. 2002. Bites of venomous snakes. N. Engl. J. Med. 347: 347–356.

Goldberg, S. R. 1997. Reproduction in the western coral snake, *Micruroides euryxanthus* (Elapidae), from Arizona and Sonora, México. Great Basin Nat. 57: 363–365.

———. 2004. Note on reproduction of the yellowbelly sea snake, *Pelamis platurus* (Serpentes: Elapidae) from Costa Rica. Bull. Maryland Herpetol. Soc. 40: 91–93.

Goldberg, S. R., and D. D. Beck. 2001. *Heloderma horridum* (beaded lizard): Reproduction. Herpetol. Rev. 32: 255–256.

Goldberg, S. R., and C. R. Bursey. 1990. Redescription of the microfilaria *Piratuba mitchelli* (Smith) (Onchocercidae) from the Gila monster, *Heloderma suspectum* Cope (Helodermatidae). Southwest. Nat. 35: 458–460.

———. 1991. Gastrointestinal helminths of the reticulate Gila monster, *Heloderma suspectum suspectum* (Sauria: Helodermatidae). J. Helminthol. Soc. Washington 58: 146–149.

———. 2000. *Micuroides euryxanthus* (western coral snake). Endoparasites. Herpetol. Rev. 31: 105–106.

Goldberg, S. R., and A. T. Holycross. 1999. Reproduction in the desert massasauga, *Sistrurus catenatus edwardsii*, in Arizona and Colorado. Southwest. Nat. 44: 531–535.

Goldberg, S. R., and C. H. Lowe. 1997. Reproductive cycle of the Gila monster, *Heloderma suspectum*, in southern Arizona. J. Herpetol. 31: 161–166.

Goldberg, S. R., C. R. Bursey, and A. T. Holycross. 2001. *Sistrurus catenatus edwardsi* (desert massasauga). Herpetol. Rev. 32: 365.

Gomez, H. F., and R. C. Dart. 1995. Clinical toxicology of snakebite in North America, pp. 619–644. *In* J. Meier and J. White (eds.), Handbook of clinical toxicology of animal venoms and poisons. CRC Press, Boca Raton, Florida.

Gomez, F., A. Vandermeers, M.-C. Vandermeers-Piret, R. Herzog, J. Rathe, M. Stievenart, J. Winand, and J. Christophe. 1989. Purification and characterization of five variants of phospholipase A_2 and complete primary structure of the main phospholipase A_2 variant in *Heloderma suspectum* (Gila monster) venom. Eur. J. Biochem. 186: 23–33.

Gonçalves, J. M. 1956. Purification and properties of crotamine, pp. 261–274. *In* E. E. Buckley and N. Proges (eds.), Venoms. Am. Assoc. Advan. Sci., Washington, D.C.

González-Ruiz, A., E. Godinez-Cano, and I. Rojas-González. 1996. Captive reproduction of the Mexican acaltetepon, *Heloderma horridum*. Herpetol. Rev. 27: 192.

Goode, M., and J. F. Smith. 2005. Calm before the storm: Tiger rattlesnakes and urban development, pp. 29–30. *In* Biology of the Rattlesnakes Symposium (program abstracts). Loma Linda Univ., Loma Linda, California.

Goodman, J. D. 1958. Material ingested by the cottonmouth, *Agkistrodon piscivorus*, at Reelfoot Lake, Tennessee. Copeia 1958: 149.

Grace, T. G., and G. E. Omer. 1980. The management of upper extremity pit viper wounds. J. Hand Surg. 5: 168–177.

Graham, G. L. 1977. The karyotype of the Texas coral snake, *Micrurus fulvius tenere*. Herpetologica 33: 345–348.

Graham, J. B. 1974a. Aquatic respiration in the sea snake *Pelamis platurus*. Resp. Physiol. 21: 1–7.

———. 1974b. Body temperature of the sea snake *Pelamis platurus*. Copeia 1974: 531–533.

Graham, J. B., I. Rubinoff, and M. K. Hecht. 1971. Temperature physiology of the sea snake *Pelamis platurus*: An index of its colonization potential in the Atlantic Ocean. Proc. Natl. Acad. Sci. USA 68: 1360–1363.

Graham, J. B., J. H. Gee, and F. S. Robison. 1975. Hydrostatic and gas exchange functions of the lung of the sea snake *Pelamis platurus*. Comp. Biochem. Physiol. 50A: 477–482.

Graham, J. B., J. H. Gee, J. Motta, and I. Rubinoff. 1987a. Subsurface buoyance regulation by the sea snake *Pelamis platurus*. Physiol. Zool. 60: 251–261.

Graham, J. B., W. R. Lowell, I. Rubinoff, and J. Motta. 1987b. Surface and subsurface swimming of the sea snake *Pelamis platurus*. J. Exp. Biol. 127: 27–44.

Graham, R. L. J., C. Graham, S. McClean, T. Chen, M. O'Rourke, D. Hirst, D. Theakston, and C. Shaw. 2005. Identification and functional analysis of a novel bradykinin inhibitory peptide in the venoms of New World *Crotalinae* pit vipers. Biochem. Biophys. Res. Commun. 338: 1587–1592.

Graham, R. L. J., C. Graham, D. Theakston, G. McMullan, and C. Shaw. 2008. Elucidation of trends within venom components from the snake families Elapidae and Viperidae using gel filtration chromatography. Toxicon 51: 121–129.

Grant, G. 1952. Probably the first legislation to protect a poisonous animal. Herpetologica 6: 64–65.

Grant, M. L., and L. J. Henderson. 1957. A case of Gila monster poisoning with a summary of some previous accounts. Proc. Iowa Acad. Sci. 64: 686–697.

Graves, G. R. 2002. Copperhead preys on star-nosed mole in the Great Dismal Swamp. Banisteria 20: 70.

Gray, B. S. 2006. The serpent's cast: A guide to identification of shed skins from snakes of the Northeast and Mid-Atlantic states. Cent. N. Am. Herptile Monogr. 1: 1–88.

Greding, E. J., Jr. 1964. Food of *Agkistrodon c. contortrix* in Houston and Trinity counties, Texas. Southwest. Nat. 9: 105.

Green, J. M., H. Heatwole, B. Black, and N. Poran. 2004. Effect of the venoms of two viperid snakes, the copperhead (*Agkistrodon contortrix*) and the cottonmouth (*Agkistrodon piscivorus*), on the liver and kidney of the bullfrog (*Rana catesbeiana*). Russian J. Herpetol. 11: 21–29.

Greenbaum, E. 2004. The influence of prey-scent stimuli on predatory behavior of the North American copperhead *Agkistrodon contortrix* (Serpentes: Viperidae). Behav. Ecol. 15: 345–350.

Greenbaum, E., and M. Jorgensen. 2004. Envenomated-invertebrate prey preference of the viperid *Agkistrodon contortrix* during strike-induced chemosensory searching. Amphibia-Reptilia 25: 165–172.

Greenbaum, E., N. Galeva, and M. Jorgensen. 2003. Venom variation and chemoreception of the viperid *Agkistrodon contortrix*: Evidence for adaptation? J. Chem. Ecol. 29: 1741–1755.

Greene, H. W. 1973a. The food habits and feeding behavior of New World coral snakes. Master's thesis, Univ. Texas, Arlington.

———. 1973b. Defensive tail display by snakes and amphisbaenians. J. Herpetol. 7: 143–161.

Greene, H. W. 1976. Scale overlap, a directional sign stimulus for prey ingestion by ophiophagous snakes. Z. Tierpsychol. 41: 113–120.

———. 1983. Dietary correlates of the origin and radiation of snakes. Am. Zool. 23: 431–441.

———. 1984. Feeding behavior and diet of the eastern coral snake, *Micrurus fulvius*, pp. 147–162. *In* R. A. Seigel, L. E. Hunt, J. L. Knight, L. Malert, and N. L. Zuschlag (eds.), Vertebrate ecology and systematics: A tribute to Henry S. Fitch. Univ. Kansas Mus. Nat. Hist. Spec. Publ. 10.

———. 1988. Antipredator mechanisms in reptiles, pp. 1–152. *In* C. Gans and R. B. Huey (eds.), Biology of the Reptilia, vol. 16: Ecology B. Defense and life history. Alan R. Liss, New York.

———. 1990. A sound defense of the rattlesnake. Pacific Discovery 43(4): 10–19.

———. 1997. Snakes: The evolution of mystery in nature. Univ. California Press, Berkeley.

Greene, H. W., and J. A. Campbell. 1992. The future of pitvipers, pp. 421–428. *In* J. A. Campbell and E. D. Brodie, Jr. (eds.), Biology of pitvipers. Selva, Tyler, Texas.

Greene, H. W., and R. W. McDiarmid. 1981. Coral snake mimicry: Does it occur? Science (New York) 213: 1207–1212.

Greene, H. W., and G. V. Oliver, Jr. 1965. Notes on the natural history of the western massasauga. Herpetologica 21: 225–228.

Greene, H. W., and W. F. Pyburn. 1973. Comments on aposematism and mimicry among coral snakes. Biologist 55: 144–148.

Greene, H. W., P. G. May, D. L. Hardy, Sr., J. M. Sciturro, and T. M. Farrell. 2002. Parental behavior by vipers, pp. 179– 205. *In* G. W. Schuett, M. Höggren, M. E. Douglas, and H. W. Greene (eds.), Biology of the vipers. Eagle Mountain Publ., Eagle Mountain, Utah.

Greer, G. 1998. Maintenance, care & observations of an eastern coral snake, *Micrurus fulvius*, in captivity. Rept. Amphib. Mag. 54: 28–31.

Gregory, P. T. 1983. Identification of sex of small snakes in the field. Herpetol. Rev. 14: 42–43.

Gregory, V. M., F. E. Russell, J. R. Brewer, and L. R. Zawadski. 1984. Seasonal variations in rattlesnake venom proteins. Proc. West. Pharmacol. Soc. 27: 233–236.

Gregory-Dwyer, V. M., N. B. Egen, A. B. Bosisio, P. G. Righetti, and F. E. Russell. 1986. An isoelectric focusing study of seasonal variation in rattlesnake venom proteins. Toxicon 24: 995–1000.

Griffin, P. R., and S. D. Aird. 1990. A new small myotoxin from the venom of the prairie rattlesnake (*Crotalus viridis viridis*). FEBS Lett. 274: 43–47.

Griner, L. A. 1983. Pathology of zoo animals. Zool. Soc. San Diego, San Diego, California.

Grismer, L. L. 1994a. The evolutionary and ecological biogeography of the herpetofauna of Baja California and the Sea of Cortez, Mexico. Ph.D. diss., Loma Linda Univ., Loma Linda, California.

———. 1994b. Geographic origins for the reptiles on islands in the Gulf of California, México. Herpetol. Nat. Hist. 2: 17–40.

———. 1999. An evolutionary classification of reptiles on islands in the Gulf of California, México. Herpetologica 55: 446–469.

———. 2002. Amphibians and reptiles of Baja California, including its Pacific islands and the islands in the Sea of Cortés, México: Natural history, distribution and identification. Univ. California Press, Berkeley.

Grobman, A. B. 1978. An alternative solution to the coral snake mimic problem (Reptilia, Serpentes, Elapidae). J. Herpetol. 12: 1–11.

Groombridge, B. 1986. Comments on the *M. pterygoideus glandulae* of crotaline snakes (Reptilia: Viperidae). Herpetologica 42: 449–457.

Groves, J. D. 1977. Aquatic behavior in the northern copperhead, *Agkistrodon contortrix mokasen*. Bull. Maryland Herpetol. Soc. 13: 114–115.

Grow, D. T., and J. Branham. 1996. Reproductive husbandry of the Gila monster (*Heloderma suspectum*), pp. 57–64. *In* P. D. Strimple, Advances in herpetoculture. Intl. Herpetol. Symp. Spec. Pub. 1.

Guan, A. L., A. D. Retzios, G. N. Henderson, and F. S. Markland, Jr. 1991. Purification and characterization of a fibrinolytic enzyme from venom of the southern copperhead snake (*Agkistrodon contortrix contortrix*). Arch. Biochem. Biophys. 289: 197–207.

Guidry, E. V. 1953. Herpetological notes from southeastern Texas. Herpetologica 9: 49–56.

Guilday, J. E. 1962. The Pleistocene local fauna of the Natural Chimneys, Augusta County, Virginia. Ann. Carnegie Mus. 36: 87–122.

Guilday, J. E., P. S. Martin, and A. D. McCrady. 1964. New Paris 4: A Pleistocene cave deposit in Bedford County, Pennsylvania. Natl. Speleol. Soc. Bull. 26: 121–194.

Gulland, J. M., and E. M. Jackson. 1938a. 77. Phosphomonoesterases of bone and snake venoms. Biochem. J. 32: 590–596.

———. 1938b. 78. 5′-Nucleotidase. Biochem J. 32: 597–601.

Gundy, G. C., and G. Z. Wurst. 1976. The occurrence of parietal eyes in recent Lacertilia (Reptilia). J. Herpetol. 10: 113–121.

Günther, A. C. L. G. 1863. Third account of new species of snakes in the collection of the British Museum. Ann. Mag. Nat. Hist. (London) 3: 348–365.

Günther, A. C. L. G. 1885–1902. Reptilia and Batrachia, pp. xx–366. *In* F. D. Goodman and O. Slavin (eds.), Biologia Centrali-Americana: Zoology. Dulau, London.

Gut, H. J., and C. E. Ray. 1963. The Pleistocene vertebrate fauna of Reddick, Florida. Quart. J. Florida Acad. Sci. 26: 315–328.

Gutberlet, R. L., Jr., and M. B. Harvey. 2002. Phylogenetic relationships of New World pitvipers as inferred from anatomical evidence, pp. 51–68. *In* G. W. Schuett, M. Höggren, M. E. Douglas, and H. W. Greene (eds.), Biology of the vipers. Eagle Mountain Publ., Eagle Mountain, Utah.

Guthrie, J. E. 1927. Rattlesnake eggs in Iowa. Copeia 1927(162): 12–14.

Gutiérrez, J. M., and R. Bolaños. 1980. Karyotype of the yellow-bellied sea snake, *Pelamis platurus*. J. Herpetol. 14: 161–165.

Gutzke, W. H. N. 2001. Field observations confirm laboratory reports of defense responses by prey snakes to the odors of predatory snakes, pp. 285–288. *In* A. Marchlewska-Koj, J. J. Lepri, and D. Müller-Schwarze (eds.), Chemical signals in vertebrates 9. Kluwer Acad., New York.

Gutzke, W. H. N., C. Tucker, and R. T. Mason. 1993. Chemical recognition of kingsnakes by crotalines: Effects of size on the ophiophage defensive response. Brain Behav. Evol. 41: 234–238.

Gwade, S. P. 2007. Snakoid instead of venomoid. J .Venom Anim. Toxins Ind. Trop. Dis. 13: 430.

Haast, W. E., and R. Anderson. 1981. Complete guide to snakes of Florida. Phoenix Publ. Co., Miami, Florida.

Hachimori, Y., M. A. Wells, and D. J. Hanahan. 1971. Observations on the phospholipase A_2 of *Crotalus atrox*: Molecular weight and other properties. Biochemistry 10: 4084–4089.

Hadidian, Z. 1956. Proteolytic activity and physiologic and pharmacologic actions of *Agkistrodon piscivorus* venom, pp. 205–215. *In* E. E. Buckley and N. Porges (eds.), Venoms, papers presented at the First International Conference on Venoms, American Association for the Advancement of Science (AAAS), Washington, D.C.

Halama, K. J., A. J. Malisch, M. Aspell, J. T. Rotenberry, and M. F. Allen. 2008. Modeling the landscape niche characteristics of red diamond rattlesnakes (*Crotalus ruber*): Implications for biology and conservation, pp. 463–472. *In* W. K. Hayes, K. R. Beaman, M. D. Cardwell, and S. P. Bush (eds.), The biology of rattlesnakes. Loma Linda Univ. Press, Loma Linda, California.

Hall, D. W. 2008. *Agkistrodon piscivorus leucostoma* (western cottonmouth): Scavenging. Herpetol. Rev. 39: 465.

Hall, E. L. 2001. Role of surgical intervention in the management of crotaline snake envenomation. Ann. Emerg. Med. 37: 175–180.

Hall, R. J. 1980. Effects of environmental contaminants on reptiles: A review. U.S. Fish Wildl. Serv. Spec. Sci. Rep. Wildl. 228: 1–12.

Halstead, B. W. 1970. Poisonous and venomous marine animals of the world, vol. 3: Vertebrates continued. U.S. Gov't. Print. Off., Washington, D.C.

Halter, C. R. 1923. The venomous coral snake. Copeia 1923(123): 105–107.

Hamel, P. B. 1996. *Agkistrodon piscivorus leucostoma* (western cottonmouth). Carrion feeding. Herpetol. Rev. 27: 143.

Hamilton, B. 2005. Use of solar models and GIS to evaluate potential hibernacula in the Great Basin rattlesnake (*Crotalus lutosus*), pp. 30–31. *In* Biology of the Rattlesnakes Symposium (program abstracts). Loma Linda Univ., Loma Linda, California.

Hamilton, W. J., Jr., and J. A. Pollack. 1955. The food of some crotalid snakes from Fort Benning, Georgia. Chicago Acad. Sci. Nat. Hist. Misc. 140: 1–4.

Hammerson, G. A. 1999. Amphibians and reptiles in Colorado. 2nd ed. Univ. Press Colorado, Niwott.

Hampton, P. M. 2005. *Agkistrodon piscivorus leucostoma* (western cottonmouth): Morphology. Herpetol. Rev. 36: 454–455.

Hampton, P. M., and N. B. Ford. 2005. *Agkistrodon piscivorus leucostoma* (western cottonmouth): Reproduction. Herpetol. Rev. 36: 455.

Hanel, R., D. J. Heard, G. A. Ellis, and A. Nguyen. 1999. Isolation of *Clostridium* spp. from the blood of captive lizards: Real or pseudobacteremia? Bull. Assoc. Rept. Amphib. Vet. 9(2): 4–8.

Hantgan, R. R., M. C. Stahle, J. H. Connor, D. S. Lyles, D. A. Horita, M. Rocco, C. Nagaswami, J. W. Weisel, and M. A. McLane. 2004. The disintegrin echistatin stabilizes integrin $\alpha_{IIb}\beta_3$'s open conformation and promotes its oligomerization. J. Mol. Biol. 342: 1625–1636.

Harding, K. A. 1984. A brief review of colubrid snakebite. J. Assoc. Stud. Rept. Amphib. 2: 65–72.

Harding, K. A., and K. R. G. Welch. 1980. Venomous snakes of the world: A checklist. Pergamon Press, New York.

Hardy, D. L. 1990. Venoms and envenomation: A selected bibliography of the recent literature on *Agkistrodon* and its allies, pp. 553–571. *In* H. K. Gloyd and R. Conant, Snakes of the *Agkistrodon* complex: A monographic review. Soc. Stud. Amphib. Rept. Contrib. Herpetol. 6.

Hardy, D. L. 1992. A review of first aid measures for pitviper bite in North America with an appraisal of Extractor™ suction and stun gun electroshock, pp. 405–414. *In* J. A. Campbell and E. D. Brodie, Jr. (eds.), Biology of the pitvipers. Selva, Tyler, Texas.

Hardy, L. M., and R. W. McDiarmid. 1969. The amphibians and reptiles of Sinaloa, México. Univ. Kansas Publ. Mus. Nat. Hist. 18: 39–252.

Harlan, R. 1827. Genera of North American Reptilia and a synopsis of the species. J. Acad. Nat. Sci. Philadelphia 5: 317–372.

Harper, G. R., Jr., and D. W. Pfennig. 2007. Mimicry on the edge: Why do mimics vary in resemblance to their model in different parts of their geographical range? Proc. Royal Soc. B 274: 1955–1961.

———. 2008. Selection overrides gene flow to break down maladaptive mimicry. Nature (London) 451: 1103–1106.

Harris, D. M. 1985. Infralingual plicae: Support for Boulenger's Teiidae (Sauria). Copeia 1985: 560–565.

Harris, H. S., Jr. 1965. Case reports of two dusky pigmy rattlesnake bites (*Sistrurus miliarius barbouri*). Bull. Maryland Herpetol. Soc. 2: 8–10.

———. 2005. Scanning electron microscopy: Scale topography in *Crotalus* and *Sistrurus*. Bull. Maryland Herpetol. Soc. 41: 101–115.

———. 2006. Some notes on color phases in serpents. Bull. Maryland Herpetol. Soc. 42: 181–183.

Harris, H. S., Jr., and R. S. Simmons. 1977. Additional notes concerning cannibalism in pit vipers. Bull. Maryland Herpetol. Soc. 13: 121–122.

Harris, L. E., Jr. 1999. The unpredictable copperhead: A study in comparative behavior. Herpetology 29(2): 7–8.

Harrison, H. H. 1971. The world of the snake. Lippincott, New York.

Hartdegen, R. W., and D. Chiszar. 2001. Discrimination of prey-derived chemical cues by the Gila monster (*Heloderma suspectum*) and lack of effect of a putative masking odor. Amphibia-Reptilia 22: 249–253.

Hartline, P. H. 1974. Thermoreception in snakes, pp. 297–312. *In* A. Fessard (ed.), Handbook of sensory physiology, vol. 3. Springer-Verlag, New York.

Hartline, P. H., L. Kass, and M. S. Loop. 1978. Merging of modalities in the optic tectum: Infrared and visual integration in rattlesnakes. Science (New York) 199: 1225–1229.

Hartnett, W. G. 1931. Poisonous snake bite: Etiology and treatment with case report. Ohio St. Med. J. 27: 636–639.

Harvey, A. L., K. N. Bradley, S. A. Cochran, E. G. Rowan, J. A. Pratt, J. A. Quillfeldt, and D. A. Jerusalinsky. 1998. What can toxins tell us for drug discovery? Toxicon 36: 1635–1640.

Harvey, A. L., E. Kornisiuk, K. N. Bradley, C. Cervenansky, R. Duran, M. Adrover, G. Sanchez, and D. Jerusalinsky. 2002. Effects of muscarinic toxins MT1 and MT2 from green mamba on different muscarinic cholinoceptors. Neurochem. Res. 11: 1543–1554.

Harvey, D. S. 2005. Detectability of a large-bodied snake (*Sistrurus c. catenatus*) by time-constrained searching. Herpetol. Rev. 36: 413–415.

Harvey, D. S., and P. J. Weatherhead. 2006a. Hibernation site selection by eastern massasauga rattlesnakes (*Sistrurus catenatus catenatus*) near their northern range limit. J. Herpetol. 40: 66–73.

———. 2006b. A test of the hierarchical model of habitat selection using eastern massasauga rattlesnakes (*Sistrurus c. catenatus*). Biol. Conserv. 130: 206–216.

Harwood, P. D. 1932. The helminths parasitic in the Amphibia and Reptilia of Houston, Texas and vicinity. Proc. U.S. Natl. Mus. 81: 1–71.

Hatkin, J. 1984. Concurrent filarial and mycotic infections in a Gila monster (*Heloderma suspectum*). Vet. Med. 79: 795–798.

Hawgood, B. J., and H. A. Reid. 1998. OBE MD: Investigation and treatment of snake bite. Toxicon 36: 431–446.

Hawkins, L. 2005. Pacific gull attempting to eat a snake. Bird Observer 836: 34.

Hayes, W. K., S. S. Herbert, G. C. Rehling, and J. F. Gennaro. 2002. Factors that influence venom expenditure in viperids and other snake species during predatory and defensive contexts, pp. 207–233. *In* G. W. Schuett, M. Höggren, M. E. Douglas, and H. W. Greene (eds.), Biology of the vipers. Eagle Mountain Publ., Eagle Mountain, Utah.

Hayes, W. K., K. R. Beaman, M. D. Cardwell, and S. P. Bush (eds.). 2008. The biology of rattlesnakes. Loma Linda Univ. Press, Loma Linda, California.

Haynie, M. R., and A. Knight. 1998. Color pattern variants in the Trans-Pecos copperhead. Southwest. Nat. 43: 499–500.

Heath, W. G. 1961. A trailing device for small animals designed for field study of the Gila monster (*Heloderma suspectum*). Copeia 1961: 491–492.

Heatwole, H. 1999. Sea snakes. 2nd ed. Krieger, Malabar, Florida.

Heatwole, H., and E. Davison. 1976. A review of caudal luring in snakes with notes on its occurrence in the Saharan sand viper, *Cerastes vipera*. Herpetologica 32: 332–336.

Heatwole, H., and E. P. Finnie. 1980. Seal predation on a sea snake. Herpetofauna 11: 24.

Heatwole, H., N. Poran, and P. King. 1999. Ontogenetic changes in the resistance of bullfrogs (*Rana catesbeiana*) to the venom of copperheads (*Agkistrodon contortrix contortrix*) and cottonmouths (*Agkistrodon piscivorus piscivorus*). Copeia 1999: 808–814.

Hecht, M. K., and D. Marien. 1956. The coral snake mimic problem: A reinterpretation. J. Morphol. 98: 335–366.

Hecht, M. K., C. Kropach, and B. M. Hecht. 1974. Distribution of the yellow-bellied sea snake, *Pelamis platurus,* and its significance in relation to the fossil record. Herpetologica 30: 387–396.

Heinrich, G. 1996. *Micrurus fulvius fulvius* (eastern coral snake). Diet. Herpetol. Rev. 27: 25.

Heinrich, G., and K. R. Studenroth, Jr. 1996. *Agkistrodon piscivorus conanti* (Florida cottonmouth): Diet. Herpetol. Rev. 27: 22.

Heitschel, S. 1986. Near death from a Gila monster bite. J. Emerg. Nurs. 12: 259–262.

Henderson, B. M., and E. B. Dujan. 1973. Snake bites in children. J. Ped. Surg. 8: 729–733.

Henderson, J. T., and A. L. Bieber. 1985. Antigenic relatedness of crotalid venoms: The basic subunit of Mojave toxin. Toxicon 23: 23.

———. 1986. Antigenic relationships between Mojave toxin subunits, Mojave toxin and some crotalid venoms. Toxicon 24: 473–479.

Henderson, R. W. 1970. Caudal luring in a juvenile Russell's viper. Herpetologica 26: 276–277.

Henderson, R. W. 1978. Notes on *Agkistrodon bilineatus* (Reptilia, Serpentes, Viperidae) in Belize. J. Herpetol. 12: 412–413.

Hendon, R. A., and A. L. Bieber. 1982. Presynaptic toxins from rattlesnake venoms, pp. 211–246. *In* A. T. Tu (ed.), Rattlesnake venoms: Their actions and treatment. Marcel Dekker, New York.

Hendon, R. A., and A. T. Tu. 1981. Biochemical characterization of the lizard toxin gilatoxin. Biochemistry 20: 3517–3522.

Hensley, M. M. 1949. Mammal diet of *Heloderma*. Herpetologica 5: 152.

———. 1950. Notes on the natural history of *Heloderma suspectum*. Trans. Kansas Acad. Sci. 53: 268–269.

———. 1959. Albinism in North American amphibians and reptiles. Publ. Mus. Michigan St. Univ. Biol. Ser. 1: 133–159.

Henson, P. H., and C. G. Cochrane. 1975. The effect of complement depletion on experimental tissue injury. Ann. New York Acad. Sci. 256: 426–440.

Herbert, S. S. 1998. Factors influencing venom expenditure during defensive bites by cottonmouths (*Agkistrodon piscivorus*) and rattlesnakes (*Crotalus viridis, Crotalus atrox*). Master's thesis, Loma Linda Univ., Loma Linda, California.

Hernandez-Camacho, J. I., R. Alvarez-Leon, and J. M. Renifo-Rey. 2005 (2006). Pelagic sea snake *Pelamis platurus* (Linnaeus, 1766) (Reptilia: Serpentes: Hydrophiidae) is found on the Caribbean coast of Colombia. Memoria de la Fundacion la Salle de Ciencias Naturales 65(164): 143–152.

Hernandez-Plata, G., L. Rey, H. Hernandez, N. Valero, G. Cendali, M. F. Anez, and M. Alvardo. 1993. Contribucion al studio sobre los efectos antineoplasicos del comple crotoxina A-B y cobramina (CFA) en sarcomas de rata, p. 16. *In* Congresso Latino Americano de Herpetologia, 3 Sao Paulo, Univ. Estadual de Campinas.

Herreid, C. F., II. 1961. Snakes as predators of bats. Herpetologica 17: 271–272.

Herrel, A., and F. De Vree. 1999. The cervical musculature in helodermatid lizards. Belg. J. Zool. 129: 175–186.

Herrel, A., I. Wauters, P. Aerts, and F. de Vree. 1997. The mechanics of ovophagy in the beaded lizard (*Heloderma horridum*). J. Herpetol. 31: 383–393.

Herzig, R. H., O. D. Ratnoff, and J. R. Shainoff. 1970. Studies on a procoagulant fraction of southern copperhead snake venom: The preferential release of fibrinopeptide B. J. Lab. Clin. Med. 76: 451–465.

Hill, J. G., III, and S. J. Beaupre. 2008. Body size, growth, and reproduction in a population of western cottonmouths (*Agkistrodon piscivorus leucostoma*) in the Ozark Mountains of northwest Arkansas. Copeia 2008: 105–114.

Hill, J. G., III, I. Hanning, S. J. Beaupre, S. C. Ricke, and M. M. Slavik. 2008. Denaturing gradient gel electrophoresis for the determination of bacterial species diversity in the gastrointestinal tracts of two crotaline snakes. Herpetol. Rev. 39: 433–438.

Hill, R. E., and S. P. Mackessy. 1997. Venom yields from several species of colubrid snakes and differential effects of ketamine. Toxicon 35: 671–678.

Hill, R. E., G. M. Bogdan, and R. C. Dart. 2000. A comparison of reconstitution times for CroTAb and antivenin (Crotalidae) polyvalent [Wyeth]. Toxicologist 54: 171.

Hill, W. H. 1971. Pleistocene snakes from a cave in Kendall County, Texas. Texas J. Sci. 22: 209–216.

Himes, J. G. 2003. Diet composition of *Nerodia sipedon* (Serpentes: Colubridae) and its dietary overlap with, and chemical recognition of *Agkistrodon piscivorus* (Serpentes: Viperidae). Amphibia-Reptilia 24: 181–188.

———. 2004. The non-fish, vertebrate diet of sympatric populations of the cottonmouth (*Agkistrodon piscivorus*) and northern watersnake (*Nerodia sipedon*). Herpetol. Rev. 35: 123–128.

Hirschfeld, S. E. 1968. Vertebrate fauna of Nichol's Hammock, a natural trap. Quart. J. Florida Acad. Sci. 31(3): 177–189.

Hobert, J. P. 1997. The massasauga rattlesnake (*Sistrurus catenatus*) in Colorado. Master's thesis, Univ. Northern Colorado, Greeley.

Hobert, J. P., C. E. Montgomery, and S. P. Mackessy. 2004. Natural history of the massasauga, *Sistrurus catenatus edwardsii*, in southeastern Colorado. Southwest. Nat. 49: 321–326.

Hoff, G. L., and F. H. White. 1977. *Salmonella* in reptiles: Isolation from free-ranging lizards (Reptilia, Lacertilia) in Florida. J. Herpetol. 11: 123–129.

Hoffman, A., and O. Sánchez. 1980. Género y especie nuevos de un ácaro parásito de lagartijas (Acarida: Pterygosomidae). An. Esc. Nac. Cienc. Biol. México 23(1–4): 97–107.

Hoffstetter, R. 1939. Contribution à l'étude des élapidae actuels et fossils et de l'ostéologie des ophidiens. Arch. Mus. Hist. Nat. Lyon 15: 1–78.

———. 1957. Un Saurien hélodermatidé (*Eurheloderma gallicum* nov. gen. et sp.) dans la faune fossile des Phosphorites du Quercy. Bull. Géol. Soc. France 7(6): 775–786.

Hoge, A. R., and S. A. R. W. D. L. Romano. 1971. Neotropical pit vipers, sea snakes, and coral snakes, pp. 211–293. *In* W. Bücherl and E. E. Buckley (eds.), Venomous animals and their venoms, vol. II: Venomous vertebrates. Academic Press, New York.

Hogue-Angeletti, R. A., and R. A. Bradshaw. 1979. Nerve growth factors in snake venoms, pp. 276–294. *In* C. Y. Lee (ed.), Handbook of experimental pharmacology, vol. 52: Snake venoms. Springer-Verlag, Berlin.

Holbrook, J. E. 1836–1842. North American herpetology; or, a description of the reptiles inhabiting the United States. Vols. 1–33. J. Dobson, Philadelphia.

Holl, F. J., and L. N. Allison. 1935. A new trematode *Dasymetria nicolli* from a snake. Trans. Am. Micro. Soc. 54: 226–228.

Holland, D. R., L. L. Clancy, S. W. Muchmore, T. J. Ryde, H. M. Einspahr, B. C. Finzel, R. L. Heinrikson, and K. D. Watenpaugh. 1990. The crystal structure of a lysine 49 phospholipase A_2 from the venom of the cottonmouth snake at 2.0-Å resolution. J. Biol. Chem. 265: 17649–17656.

Holleman, W. H., and L. J. Weiss. 1976. The thrombin-like enzyme from *Bothrops atrox* snake venom: Properties of the enzyme purified by affinity chromatography on p-aminobenzamidine-substituted agarose. J. Biol. Chem. 251: 1663–1669.

Holman, J. A. 1958. The Pleistocene herpetofauna of Saber-tooth Cave, Citrus County, Florida. Copeia 1958: 276–280.

———. 1959a. Amphibians and reptiles from the Pleistocene (Illinoian) of Williston, Florida. Copeia 1959: 96–102.

———. 1959b. A Pleistocene herpetofauna near Orange Lake, Florida. Herpetologica 15: 121–124.

———. 1963. Late Pleistocene amphibians and reptiles of the Clear Creek and Ben Franklin local faunas of Texas. J. Grad. Res. Cent. Southern Methodist Univ. 31: 152–167.

———. 1964. Pleistocene amphibians and reptiles from Texas. Herpetologica 20: 73–83.

———. 1967. A Pleistocene herpetofauna from Ladd's Georgia. Bull. Georgia Acad. Sci. 25: 154–166.

———. 1968. A Pleistocene herpetofauna from Kendall County, Texas. Quart. J. Florida Acad. Sci. 31: 165–172.

———. 1969a. The Pleistocene amphibians and reptiles of Texas. Publ. Mus. Michigan St. Univ. Biol. Ser. 4: 161–192.

———. 1969b. Herpetofauna of the Slaton local fauna of Texas. Southwest. Nat. 14: 203–212.

———. 1972. Herpetofauna of the Kanapolis local fauna (Pleistocene: Yarmouth) of Kansas. Michigan Acad. 5: 87–98.

———. 1977. Upper Miocene snakes (Reptilia, Serpentes) from southeastern Nebraska. J. Herpetol. 11: 323–335.

———. 1978. The late Pleistocene herpetofauna of Devil's Den Sinkhole, Levy County, Florida. Herpetologica 34: 228–237.

———. 1979a. A review of North American tertiary snakes. Publ. Mus. Michigan St. Univ. Paleontol. Ser. 1: 201–260.

———. 1979b. Herpetofauna of the Nash Local Fauna (Pleistocene: Aftonian) of Kansas. Copeia 1979: 747–749.

———. 1980. Paleoclimatic implications of Pleistocene herpetofauanas of eastern and central North America. Trans. Nebraska Acad. Sci. 8: 131–140.

———. 1981. A review of North American Pleistocene snakes. Publ. Mus. Michigan St. Univ. Paleontol. Ser. 1: 261–306.

———. 1982. The Pleistocene (Kansas) herpetofauna of Trout Cave, West Virginia. Ann. Carnegie Mus. 51: 391–404.

———. 1995. Pleistocene amphibians and reptiles in North America. Oxford Univ. Press, New York.

———. 1996. The large Pleistocene (Sangamonian) herpetofauna of the Williston IIIA Site, North-central Florida. Herpetol. Nat. Hist. 4: 35–47.

———. 2000. Fossil snakes of North America: Origin, evolution, distribution, paleoecology. Indiana Univ. Press, Bloomington.

Holman, J. A., and F. Grady. 1987. Herpetofauna of New Trout Cave. Natl. Geogr. Res. 3: 305–317.

Holman, J. A., and M. E. Schloeder. 1991. Fossil herpetofauna of the Lisco C Quarries (Pliocene: early Blancan) of Nebraska. Trans. Nebraska Acad. Sci. 18: 19–29.

Holman, J. A., and A. J. Winkler. 1987. A mid-Pleistocene (Irvingtonian) herpetofauna from a cave in southcentral Texas. Texas Mem. Mus. Univ. Texas Pearce-Sellards Ser. 44: 1–17.

Holman, J. A., G. Bell, and J. Lamb. 1990. A Late Pleistocene herpetofauna from Bell Cave, Alabama. Herpetol. J. 1: 521–529.

Holycross, A. T. 2002. *Sistrurus catenatus edwardsii* (Desert massasauga): Maximum length. Herpetol. Rev. 33: 59.

———. 2003. Desert massasauga (*Sistrurus catenatus edwardsii*). Sonoran Herpetol. 16: 30–32.

Holycross, A. T., and S. P. Mackessy. 2002. Variation in the diet of *Sistrurus catenatus* (massasauga), with emphasis on *Sistrurus catenatus edwardsii* (desert massasauga). J. Herpetol. 36: 454–464.

Holycross, A. T., T. G. Anton, M. E. Douglas, and D. R. Frost. 2008. The type localities of *Sistrurus catenatus* and *Crotalus viridis* (Serpentes: Viperidae), with the unraveling of a most unfortunate tangle of names. Copeia 2008: 421–424.

Holz, G. G., and J. F. Habener. 1998. Black widow spider α-latrotoxin: A presynaptic neurotoxin that shares structural homology with the glucagon-like peptide-1 family of insulin secretagogic hormones. Comp. Biochem. Physiol. 121B: 177–184.

Hook, L. 1936. Notes on the food habits of the copperhead, *Agkistrodon mokasen mokasen* Beauvois and the timber rattlesnake, *Crotalus horridus* Linnaeus. Master's thesis, Cornell Univ., Ithaca, New York.

Hooker, K. R., and E. M. Caravati. 1994. Gila monster envenomation. Ann. Emerg. Med. 24: 731–735.

Howarth, B. 1974. Sperm storage: As a function of the female reproductive system, pp. 237–270. *In* A. D. Johnson and C. W. Foley (eds.), The oviduct and its functions. Academic Press, New York.

Hoy, P. R. 1883. Catalogue of the cold-blooded vertebrates of Wisconsin. Geol. Wisconsin 1: 422–426.

Hsu, C. C., W. B. Wu, Y. H. Chang, H. L. Kuo, and T. F. Huang. 2007. Antithrombotic effect of a protein-type I class snake venom metalloproteinase, kistomin, is mediated by affecting glycoprotein Ib-von Willebrand factor interaction. Mol. Pharmacol. 72: 984–992.

Hsu, C. C., W. B. Wu, and T. F. Huang. 2008. A snake venom metalloproteinase, kistomin, cleaves platelet glycoprotein VI and impairs platelet functions. J. Thromb. Haemostas. 6: 1578–1585.

Huang, S. Y., and J. C. Perez. 1980. Comparative study on hemorrhagic and proteolytic activities of snake venoms. Toxicon 18: 421–426.

Huang, T. T., S. J. Blackwell, and S. R. Lewis. 1981. Tissue necrosis in snakebite. Texas Med. 77(9): 53–58.

Huang, W.-S. 1996. Sexual size dimorphism of sea snakes in Taiwan. Bull. Ntl. Mus. Nat. Sci. 7: 113–120.

Hudnall, J. A. 1979. Surface activity and horizontal movements in a marked population of *Sistrurus miliarius barbouri*. Bull. Maryland Herpetol. Soc. 15: 134–138.

Hughes, R. C., J. H. Baker, and G. B. Dawson. 1941. The tapeworms of reptiles: Part II, Host catalogue. Wasmann Collector 4: 97–104.

Huheey, J. E., and A. Stupka. 1967. Amphibians and reptiles of Great Smoky Mountains National Park. Univ. Tennessee Press, Knoxville.

Hulbert, R. C., Jr., and G. S. Morgan. 1989. Stratigraphy, paleoecology, and vertebrate fauna of the Leisley Shell Pit local fauna, early Pleistocene (Irvingtonian) of southwestern Florida. Pap. Florida Paleontol. 2: 1–19.

Hulme, J. H. 1952. Observation of a snake bite by a cottonmouth moccasin. Herpetologica 8: 51.

Hunsaker, D., II, and C. Johnson. 1959. Internal pigmentation and ultraviolet transmission of the integument in amphibians and reptiles. Copeia 1959: 311–315.

Hurter, J. 1911. Herpetology of Missouri. Trans. St. Louis Acad. Sci. 20: 59–274.

Hutchison, R. H. 1929. On the incidence of snake-bite poisoning in the United States and the results of the new methods of treatment. Bull. Antiv. Inst. Am. 3: 43–57.

———. 1930 . Further notes on the incidence of snakebite poisoning in the United States. Bull Antiv. Inst. Am. 4: 40–43.

Imai, K., T. Nikai, H .Sugihara, and C. L. Ownby. 1989. Hemorrhagic toxin from the venom of *Agkistrodon bilineatus* (common cantil). Intl. J. Biochem. 21: 667–673.

Ineich, I., X. Bonnet, R. Shine, T. Shine, F. Brischoux, M. Lebreton, and L. Chirio. 2006. What, if anything, is a "typical" viper? Biological attributes of basal viperid snakes (genus *Causus* Wagler, 1830). Biol. J. Linnean Soc. 89: 575–588.

Ippen, R. 1972. Ein beitrag zu den spontantumoren bei reptilian. *In* Verhandlungsbericht des 14th Internationalen Symposium Uber die Erkrankungen der Zootiere, Wroclaw, 1972. Akademic-Verlag, Berlin.

Ireland, L. C., and C. Gans. 1977. Optokinetic behavior of the tuatara, *Sphenodon punctatus*. Herpetologica 33: 339–344.

Irvine, F. R. 1954. Snakes as food for man. British J. Herpetol. 1: 183–189.

Ivanyi, C., and W. Altimari. 2004. Venomous reptile bites in academic research. Herpetol. Rev. 35: 49–50.

Iverson, J. B. 1978. Reproductive notes on Florida snakes. Florida Scient. 41: 201–207.

Iwanaga, S., and T. Suzuki. 1979. Enzymes in snake venom, pp. 61–158. In C. Y. Lee (ed.), Handbook of experimental pharmacology, vol. 52: Snake venoms. Springer-Verlag, Berlin.

Jackson, D. D. 2005. Proteases and other enzymes in dusky pigmy rattlesnake (*Sistrurus miliarius barbouri*) venom, p. 32. *In* Biology of the rattlesnakes symposium (program abstracts). Loma Linda Univ., Loma Linda, California.

Jackson, D. L., and R. Franz. 1981. Ecology of the eastern coral snake *(Micrurus fulvius)* in northern peninsular Florida. Herpetologica 37: 213–228.

Jackson, J. F., and D. L. Martin. 1980. Caudal luring in the dusky pygmy rattlesnake, *Sistrurus miliarius barbouri*. Copeia 1980: 926–927.

Jackson, J. J. 1983. Snakes of the southeastern United States. Georgia Ext. Serv., Athens.

Jackson, K. 2003. The evolution of venom-delivery systems in snakes. Zoo. J. Linnean Soc. 137: 337–354.

Jackson, S., and A. D. O'Connor. 2007. Anaphylaxis following re-exposure to CroFab®. *In* Venom week 2007 (program abstracts). Univ. Arizona Colleges of Medicine, Arizona Health Sci. Center, Off. Continuing Medical Educ., Tucson, Arizona.

Jacob, J. S. 1981. Population density and ecological requirements of the western pygmy rattlesnake in Tennessee. U.S. Fish Wildl. Serv., Denver.

Jacob, J. S., and S. L. Carroll. 1982. Effect of temperature on the heart rate-ventilatory response in the copperhead, *Agkistrodon contortrix* (Reptilia: Viperidae). J. Therm. Biol. 7: 117–120.

Jacobson, E. R. 1980. Reptile neoplasms, pp. 255–265. *In* J. B. Murphy and J. T. Collins (eds.), Reproductive biology and diseases of captive reptiles. Soc. Stud. Amphib. Rept. Contrib. Herpetol. 1.

———. 1981. Neoplastic diseases, vol. 2, pp. 429–468. *In* J. E. Cooper and O. F. Jackson (eds.), Diseases in Reptilia. 2 vols. Academic Press, New York.

———. 1984. *Pseudomonas,* pp. 37–47. *In* G. L. Hoff, F. L. Fyre, and E. R. Jacobson (eds.), Diseases of amphibians and reptiles. Plenum, New York.

Jaffe, F. A. 1957. A fatal case of snakebite. Can. Med. Assoc. J. 76: 641–643.

Jaksztien, K. P., and H. G. Petzold. 1959. Durch *Salmonella* infection bedingte Schwierigkeiten bei der Aufzucht von Schlangen und ihre Behandlung. BL. Aquar. Terrar. 6: 79–80.

Jan, G. 1858. Plan d'une iconographie descriptive des ophidiens et description sommaire de nouvelles espèces de serpents. Rev. Mag. Zool., ser. 2, 10: 438–449, 514–527.

Jang, Y. J., D. S. Kim, O. H. Jeon, and D. S. Kim. 2007. Saxatilin suppresses tumor-induced angiogenesis by regulating VEGF expression in NCI-H460 human lung cancer cells. J. Biochem. Mol. Biol. 40: 439–443.

Jansen, D. W. 1987. The myonecrotic effect of Duvernoy's gland secretion of the snake *Thamnophis elegans vagrans*. J. Herpetol. 21: 81–83.

Jaques, R. 1955. Vergleichende Fermentuntersuchungen an tierischen Giften (Cholinesterase, «Lecithinase» Hyaluronidase). Helv. Physiol. Pharmac. Acta 13: 113–120.

Jellen, B. C., and M. J. Kowalski. 2007. Movement and growth of neonate eastern massasaugas (*Sistrurus catenatus*). Copeia 2007: 994–1000.

Jellen, B. C., C. A. Phillips, M. J. Dreslik, and D. B. Shepard. 2001. Reproductive ecology of the eastern massasauga rattlesnake, *Sistrurus catenatus catenatus*. Trans. Illinois St. Acad. Sci. 94 (suppl.): 75.

Jellen, B. C., C. A. Phillips, and M. J. Dreslik. 2007a. *Sistrurus catenatus* (eastern massasauga): Reproduction. Herpetol. Rev. 38: 343.

Jellen, B. C., D. B. Shepard, M. J. Dreslik, and C. A. Phillips. 2007b. Male movement and body size affect mate acquisition in the eastern massasauga (*Sistrurus catenatus*). J. Herpetol. 41: 451–457.

Jemison, S. C., L. A. Bishop, P. G. May, and T. M. Farrell. 1995. The impact of PIT-tags on growth and movement of the rattlesnake, *Sistrurus miliarius*. J. Herpetol. 29: 129–132.

Jenkins, C. L., and C. R. Peterson. 2008. A trophic-based approach to the conservation biology of rattlesnakes: Linking landscape disturbance to rattlesnake populations, pp. 265–274. *In* W. K. Hayes, K. R. Beaman, M. D. Cardwell, and S. P. Bush (eds.), The biology of rattlesnakes. Loma Linda Univ. Press, Loma Linda, California.

Jennings, M. R. 1984. Longevity records for lizards of the family Helodermatidae. Bull. Maryland Herpetol. Soc. 20: 22–23.

Jensen, J. B. 1999. *Terrapene carolina carolina* (eastern box turtle): Diet. Herpetol. Rev. 30: 95.

Jensen, J. B., C. D. Camp, W. Gibbons, and M. J. Elliott (eds.). 2008. Amphibians and reptiles of Georgia. Univ. Georgia Press, Athens.

Jia, Y., B. A. Cantu, E. E. Sánchez, and J. C. Pérez. 2008. Complementary DNA sequencing and identification of mRNAs from the venomous gland of *Agkistrodon piscivorus leucostoma*. Toxicon 51: 1457–1466.

Jiménez-Porras, J. M. 1961. Biochemical studies on venom of the rattlesnake, *Crotalus atrox atrox*. J. Exp. Zool. 148: 251–258.

———. 1971. Biochemistry of snake venoms, pp. 43–85. *In* S. A. Minton (ed.), Snake venoms and envenomation. Marcel Dekker, New York.

Jochimsen, D., and C. R. Peterson. 2004. Factors influencing the road mortality of snakes on the eastern Snake River Plain, p. 18. *In* Proceedings of the Snake Ecology Group 2004 Conference, Jackson County, Illinois.

John-Alder, H. B., C. H. Lowe, and A. F. Bennett. 1983. Thermal dependence of locomotory energetics and aerobic capacity of the Gila monster (*Heloderma suspectum*). J. Comp. Physiol. 151: 119–126.

Johnson, B. D. 1968. Selected Crotalidae venom properties as a source of taxonomic criteria. Toxicon 6: 5–10.

Johnson, B. D., J. C. Tullar, and H. L. Stahnke. 1966. A quantitative protozoan bio-assay method for determining venom potencies. Toxicon 3: 297–300.

Johnson, E. 1974. Zoogeography of the Lubbock Lake site. Mus. Bull. West Texas Mus. Assoc. 15: 107–122.

Johnson, E. K., and C. L. Ownby. 1993a. Isolation of a myotoxin from the venom of *Agkistrodon contortrix laticinctus* (broad-banded copperhead) and pathogenesis of myonecrosis induced by it in mice. Toxicon 31: 243–255.

———. 1993b. Isolation of a hemorrhagic toxin from the venom of *Agkistrodon contortrix laticinctus* (broad-banded copperhead) and pathogenesis of the hemorrhage induced by the toxin in mice. Intl. J. Biochem. 25: 267–278.

Johnson, E. K., K. V. Kardong, and C. L. Ownby. 1987. Observations on white and yellow venoms from an individual southern Pacific rattlesnake (*Crotalus viridis helleri*). Toxicon 25: 1169–1180.

Johnson, E. M. 1975. Faunal and floral material from a Kansas City Hopewell site: Analysis and interpretation. Occ. Pap. Mus. Texas Tech. Univ. 36: 1–37.

Johnson, G. 1992. Swamp rattler. Conservationist (New York) 47(2): 26–33.

———. 2000. Spatial ecology of the eastern massasauga *(Sistrurus c. catenatus)* in a New York peatland. J. Herpetol. 34: 186–192.

Johnson, G., and D. J. Leopold. 1998. Habitat management for the eastern massasauga in a central New York peatland. J. Wildl. Manage. 62: 84–97.

Johnson, J. E., Jr. 1948. Copperhead in a tree. Herpetologica 4: 214.

Johnson, J. P., and C. Ivanyi. 2004. Beaded lizard (*Heloderma horridum*). North American Stud Book, Arizona-Sonora Desert Museum, Tucson, Arizona.

Johnson, L. F., J. S. Jacob, and P. Torrence. 1982. Annual testicular and androgenic cycles of the cottonmouth (*Agkistrodon piscivorus*) in Alabama. Herpetologica 38: 16–25.

Johnson, R. G. 1955. The adaptive and phylogenetic significance of vertebral form in snakes. Evolution 9: 367–388.

———. 1956. The origin and evolution of the venomous snakes. Evolution 10: 56–65.

Johnson, T. B., and G. S. Mills. 1982. A preliminary report on the status of *Crotalus lepidus, C. pricei and C. willardi* in southeastern Arizona. U.S. Fish Wildl. Serv. Contr. 14-16-0002-81-224.

Johnson, T. R. 1987. The amphibians and reptiles of Missouri. Missouri Department of Conservation, Jefferson City.

Johnston, T. H., and P. M. Mawson. 1948. Some new records of nematodes from Australian snakes. Rec. South Australian Mus. 8: 547–553.

Jones, H. I. 2003. Parasitic worms in reptiles from Tasmania and the islands of Bass Strait. Pap. Proc. Royal Soc. Tasmania 137: 7–12.

Jones, J. M. 1976. Variations of venom proteins in *Agkistrodon* snakes from North America. Copeia 1976: 558–562.

Jones, J. M., and P. M. Burchfield. 1971. Relationship of specimen size to venom extracted from the copperhead, *Agkistrodon contortrix*. Copeia 1971: 162–163.

Jones, K. B. 1983. Movement patterns and foraging ecology of Gila monsters (*Heloderma suspectum* Cope) in northwestern Arizona. Herpetologica 39: 247–253.

Joseph, J. S., M. C. M. Chung, P. J. Mirtschin, and R. M. Kini. 2002. Effect of snake venom procoagulants on snake plasma: Implications for the coagulation cascade of snakes. Toxicon 40: 175–183.

Juárez, P., I. Comas, F. González-Candelas, and J. J. Calvete. 2008. Evolution of snake venom disintegrins by positive Darwinian selection. Mol. Biol. Evol. 25: 2391–2407.

Juckett, G., and J. G. Hancox. 2002. Venomous snakebites in the United States: Management review and update. Am. Fam. Physician 65(7): 1367–1374, 1377.

Juliá-Zertuche, J. 1981. Reptiles mexicanos de importancia para la salud pública y su distribución geográfica. Salva Pública México 23: 329–343.

Jungnickel, J. 2002. Zur Haltung und Vermehrung der Mexikanischen Mokassinotter *Agkistrodon taylori* Burger & Robertson, 1951. Sauria (Berlin) 25: 19–22.

Jurkovich, G. J., A. Luterman, K. McCullar, M. L. Ramenofsky, and P. W. Curreri. 1988. Complications of Crotalidae antivenin therapy. J. Trauma 28: 1032–1037.

Kaiser, E., and H. Michl. 1958. Die Biochemie der tierischen Gifte. Franz Deuticke, Wien.

Kallech-Ziri, O., J. Luis, M. El Ayeb, and N. Marrakchi. 2007. Snake venom disintegrins: Classification and therapeutic potential. Arch. Inst. Pasteur Tunis 84(1–4): 29–37.

Kardong, K. V. 1974. Kinesis of the jaw apparatus during the strike in the cottonmouth snake, *Agkistrodon piscivorus*. Forma et Functio 7: 327–354.

———. 1975. Prey capture in the cottonmouth snake (*Agkistrodon piscivorus*). J. Herpetol. 9: 169–175.

———. 1977. Kinesis of the jaw apparatus during swallowing in the cottonmouth snake, *Agkistrodon piscivorus*. Copeia 1977: 338–348.

———. 1979. "Protovipers" and the evolution of snake fangs. Evolution 33: 433–443.

———. 1982a. The evolution of the venom apparatus in snakes from colubrids to viperids & elapids. Mem. Inst. Butantan 46: 105–118.

———. 1982b. Comparative study of changes in prey capture behavior of the cottonmouth (*Agkistrodon piscivorus*) and Egyptian cobra (*Naja haje*). Copeia 1982: 337–343.

———. 1990. General skull, bone, and muscle variation in *Agkistrodon* and related genera, pp. 573–581. *In* H. K. Gloyd and R. Conant, Snakes of the *Agkistrodon* complex: A monographic review. Soc. Stud. Amphib. Rept. Contrib. Herpetol. 6.

Kardong, K. V., and V. L. Bels. 1998. Rattlesnake strike behavior: Kinematics. J. Exp. Biol. 201: 837–850.

Karstad, L. 1961. Reptiles as possible reservoir hosts for eastern encephalitis virus. Proc. N. Am. Wildl. Nat. Res. Conf. 26: 186–202.

Kasturiratne, A., A. R. Wickremasinghe, N. de Silva, N. K. Gunawardena, A. Pathmeswaran, R. Premaratna, L. Savioli, D. G. Lalloo, and H. J. de Silva. 2008. The global burden of snakebite: A literature analysis and modeling based on regional estimates of envenoming and deaths. PLoS Med. 5(11): 1591–1604.

Kauffeld, C. F. 1939. If you like danger—there are snakes. Outdoor Life 83(3): 32–33, 67–68.

———. 1943. Field notes on some Arizona reptiles and amphibians. Am. Midl. Nat. 29: 342–359.

———. 1955. Off with their skins. Animaland 12(5): 1–4.

———. 1957. Snakes and snake hunting. Hanover House, Garden City, New York.

———. 1969. Snakes: The keeper and the kept. Doubleday, Garden City, New York.

Kauffman, F. E. 1928. The pygmy or ground rattler. Bull. Antiv. Inst. Am. 1: 118.

Keegan, H. L. 1944. Indigo snakes feeding upon poisonous snakes. Copeia 1944: 59.

Keegan, H. L., and T. F. Andrews. 1942. Effects of crotalid venom of North American snakes. Copeia 1942: 251–254.

Keenlyne, K. D. 1978. Reproductive cycles in two rattlesnakes. Am. Midl. Nat. 100: 368–375.

Keenlyne, K. D., and J. R. Beer. 1973. Food habits of *Sistrurus catenatus catenatus*. J. Herpetol. 7: 382–384.

Keiser, E. D., Jr. 1971. The poisonous snakes of Louisiana and the emergency treatment of their bites. Louisiana Wildl. Fish. Comm., Baton Rouge.

Keiser, E. D., Jr. 1993. *Agkistrodon piscivorus leucostoma* (western cottonmouth). Behavior. Herpetol. Rev. 24: 34.

Keiser, E. D., Jr., and L. D. Wilson. 1979. Checklist and key to the herpetofauna of Louisiana. 2nd ed. Lafayette Natur. Hist. Mus. Tech. Bull. 1: 1–49.

Keith, J., R. Lee, and D. Chiszar. 1985. Spatial orientation by cottonmouths (*Agkistrodon piscivorus*) after detecting prey. Bull. Maryland Herpetol. Soc. 21: 145–149.

Kennedy, J. P. 1964. Natural history notes on some snakes of eastern Texas. Texas J. Sci. 16: 210–215.

Kennicott, R. 1860. Descriptions of new species of North American serpents in the Museum of the Smithsonian Institution, Washington. Proc. Acad. Nat. Sci. Philadelphia 12: 328–338.

———. 1861. On three new forms of rattlesnakes. Proc. Acad. Nat. Sci. Philadelphia 13: 206–208.

Keogh, J. S. 1998. Molecular phylogeny of elapid snakes and a consideration of their biogeographic history. Biol. J. Linnean Soc. 63: 177–203.

Keqin, G., and R. C. Fox. 1996. Taxonomy and evolution of late Cretaceous lizards (Reptilia: Squamata) from western Canada. Bull. Carnegie Mus. Nat. Hist. 33: 1–108.

Kerchove, C. M., M. S. A. Luna, M. B. Zablith, M. F. M. Lazari, S. S. Smaili, and N. Yamanouye. 2008. α_1-adrenoceptors trigger the snake venom production cycle in secretory cells by activating phosphatidylinositol 4,5-bisphosphate hydrolysis and ERK signaling pathway. Comp. Biochem. Physiol. 150A: 431–437.

Kerfoot, W. C. 1969. Selection of an appropriate index for the study of variability of lizard and snake body scale counts. Syst. Zool. 18: 53–62.

Kerman, K. 1942. Rattlesnake religion, pp. 93–102. *In* L. N. Jones (ed.), Eve's stepchildren: A collection of folk Americana. Caldwell, Idaho.

Keyler, D. E. 1986a. Venomous snakebites in Minnesota—1985. Minnesota Herpetol. Soc. Newsl. 6: 7–8.

———. 1986b. Case of legitimate snakebite. Minnesota Herpetol. Soc. Newsl. 6: 2.

———. 1986c. Dead snake bite. Minnesota Herpetol. Soc. Newsl. 6: 4.

———. 1986d. Copperhead envenomation. Minnesota Herpetol. Soc. Newsl. 6: 3–4.

———. 1988. Pigmy rattlesnake envenomation. Minnesota Herpetol. Soc. Newsl. 8: 7.

———. 2005. Venomous snakebites: Minnesota & Upper Mississippi River Valley 1982–2002. Minnesota Herpetol. Soc. Occ. Pap. 7: 1–28.

———. 2008. Timber rattlesnake (*Crotalus horridus*) envenomations in the upper Mississippi River Valley, pp. 569–580. *In* W. K. Hayes, K. R. Beaman, M. D. Cardwell, and S. P. Bush (eds.), The biology of rattlesnakes. Loma Linda Univ. Press, Loma Linda, California.

Kharin, V. E. 2007. On the second record of yellow-bellied sea snake *Pelamis platurus* (Linnaeus, 1766) from Russia. Russian J. Herpetol. 14: 45–49.

Khole, V. 1991. Toxicities of snake venoms and their components, pp. 405–470. *In* A. T. Tu (ed.), Reptile venoms and toxins: Handbook of natural toxins, vol. 5. Marcel Dekker, New York.

Kilmon, J., and H. Shelton (eds.). 1981. Rattlesnakes in America. Shelton Press, Sweetwater, Texas.

King, R., C. Berg, and B. Hay. 2004. A repatriation study of the eastern massasauga (*Sistrurus catenatus catenatus*) in Wisconsin. Herpetologica 60: 429–437.

King, R. S. 1999. Habitat use and movement patterns of the eastern massasauga in Wisconsin, p. 80. *In* B. Johnson, and M. Wright (eds.), Second international symposium and workshop on conservation of the eastern massasauga rattlesnake, *Sistrurus catenatus catenatus*: Population and habitat management issues in urban, bog, prairie, and forested ecosystems. Toronto Zoo, Scarborough, Ontario, Canada.

Kinghorn, J. R. 1956. The snakes of Australia. Angus and Robertson, London.

Kingsbury, B. A. 1996. Status of the eastern massasauga, *Sistrurus c. catenatus*, in Indiana with management recommendations for recovery. Proc. Indiana Acad. Sci. 105: 195–205.

Kini, R. M. 1998. Proline brackets and identification of potential functional sites in proteins: Toxins to therapeutics. Toxicon 36: 1659–1670.

———. 2002. Molecular moulds with multiple missions: Functional sites in three-finger toxins. Clin. Exp. Pharmacol. Physiol. 29: 815–822.

———. 2005. Structure-function relationships and mechanism of anticoagulant phospholipase A_2 enzymes from snake venoms. Toxicon 45: 1147–1161.

Kirkpatrick, C. H. 1991. Allergic histories and reactions of patients treated with digoxin immune Fab (ovine) antibody. Am. J. Emerg. Med. 9(suppl. 1): 7–10.

Kirkwood, J. K., and C. Gili. 1994. Food consumption in relation to bodyweight in captive snakes. Res. Vet. Sci. 57: 35–38.

Kirtland, J. P. 1838. Report on the zoology of Ohio. Am. Rep. Geol. Surv. Ohio 2: 157–200.

Kitchens, C. S. 1992. Hemostatic aspects of envenomation by North American snakes. Hematology/Oncology Clinics of North America 6: 1188–1195.

Kitchens, C. S., and L. H. S. Van Mierop. 1987. Envenomation by the eastern coral snake (*Micrurus fulvius fulvius*): A study of 39 victims. J. Am. Med. Assoc. 258: 1615–1618.

Kitchens, C. S., S. Hunter, and L. H. S. Van Mierop. 1987. Severe myonecrosis in a fatal case of envenomation by the canebrake rattlesnake (*Crotalus horridus atricaudatus*). Toxicon 25: 455–458.

Klauber, L. M. 1935. The feeding habits of a sea snake. Copeia 1935: 182.

———. 1936. A statistical study of the rattlesnakes: I. Introduction. II. Sex ratio. III. Birth rate. Occ. Pap. San Diego Soc. Nat. Hist. 1: 1–24.

———. 1939. A statistical study of the rattlesnakes. VI. Fangs. Occ. Pap. San Diego Soc. Nat. Hist. 5: 1–61.

———. 1946. The glossy snake, *Arizona*, with descriptions of new subspecies. Trans. San Diego Soc. Nat. Hist. 10: 311–398.

———. 1956. *Agkistrodon* or *Ancistrodon*? Copeia 1956: 258–259.

———. 1972. Rattlesnakes: Their habits, life histories, and influence on mankind. 2nd ed. Univ. California Press, Berkeley.

Klawe, W. L. 1963. Observations on the spawning of four species of tuna (*Neothunnus macropterus*, *Katsuwonus pelamis*, *Auxis thazard* and *Euthynnus lineatus*) in the eastern Pacific Ocean, based on the distribution of their larvae and juveniles. Inter-Am. Trop. Tuna Comm. Bull. 6: 447–540.

———. 1964. Food of the black-and-yellow sea snake, *Pelamis platurus*, from Ecuadorian coastal waters. Copeia 1964: 712–713.

Klemens, M. W. 1993. Amphibians and reptiles of Connecticut and adjacent regions. Bull. Connecticut St. Geol. Nat. Hist. Surv. 112: 1–318.

Klimstra, W. D. 1959. Food habits of the cottonmouth in southern Illinois. Chicago Acad. Sci. Nat. Hist. Misc. 168: 1–8.

Klingenberg, R. J. 1993. Understanding reptile parasites: A basic manual for herpetoculturists & veterinarians. Spl. Ed. Adv. Vivar. Syst. Herpetol. Libr., Lakeside, California.

Kloog, Y., J. Ambar, M. Sokolovsky, E. Kochva, Z. Wollberg, and A. Bdolah. 1988. Sarafotoxin, a novel vasoconstrictor peptide: Phosphoinositide hydrolysis in rat heart and brain. Science (New York) 242(4876): 268–270.

Knight, A., L. D. Densmore, III, and E. D. Rael. 1992. Molecular systematics of the *Agkistrodon* complex, pp. 49–70. *In* J. A. Campbell and E. D. Brodie, Jr. (eds.), Biology of the pitvipers. Selva, Tyler, Texas.

Knopf, G. N., and D. W. Tinkle. 1961. The distribution and habits of *Sistrurus catenatus* in northwest Texas. Herpetologica 17: 126–131.

Kocholaty, W. F., E. B. Ledford, J. G. Daly, and T. A. Billings. 1971. Toxicity and some enzymatic properties and activities of venoms of Crotalidae, Elapidae and Viperidae. Toxicon 9: 131–138.

Kochva, E., C. C. Viljoen, and D. P. Botes. 1982. A new type of toxin in the venom of snakes of the genus *Atractaspis* (Atractaspidinae). Toxicon 20: 581–592.

Kofron, C. P. 1978. Foods and habitats of aquatic snakes (Reptilia, Serpentes) in a Louisiana swamp. J. Herpetol. 12: 543–554.

———. 1979. Reproduction in aquatic snakes in south-central Louisiana. Herpetologica 35: 44–50.

Koh, D. C. I., A. Armugam, and K. Jeyaseelan. 2006. Snake venom components and their applications in biomedicine. Cell. Mol. Life Sci. 63: 3030–3041.

Komori, Y., and T. Nikai. 1998. Chemistry and biochemistry of kallikrein-like enzyme from snake venoms. J. Toxicol. Toxin Rev. 17: 261–277.

Komori, Y., T. Nikai, A. Ohara, S. Yagihashi, and H. Sugihara. 1993. Effect of bilineobin, a thrombin-like proteinase from the venom of common cantil (*Agkistrodon bilineatus*). Toxicon 31: 257–270.

Kornalik, F. 1990. Toxins affecting blood coagulation and fibrinolysis, pp. 683–759. *In* W. T. Shier and D. Mebs (eds.), Handbook of toxinology. Marcel Dekker, New York.

Kornalik, F., and J. Hladovec. 1975. The effect of ecarin-defibrinating enzyme isolated from *Echis carinatus* on experimental thrombosis. Thromb. Res. 7: 611–621.

Kraus, F., D. G. Mink, and W. M. Brown. 1996. Crotaline intergeneric relationships based on mitochondrial DNA sequence data. Copeia 1996: 763–773.

Krebs, J., T. G. Curro, and L. G. Simmons. 2006. The use of a venomous reptile restraining box at Omaha's Henry Doorly Zoo, p. 40. *In* S. A. Seifert (ed.), Snakebites in the new millennium: A state-of-the-art symposium, University of Nebraska Medical Center, 21–23 October 2005, Omaha, Nebraska. J. Med. Toxicol. 2: 29–45.

Krem, M. M., and E. Di Cera. 2001. Molecular markers of serine protease evolution. EMBO J. 20: 3036–3045.

Krempels, D. M. 1984. Near infrared reflectance by coral snakes: Aposematic coloration? Progr. 6th Ann. Mtng. Am. Soc. Ichthyol. Herpetol. 142.

Krochmal, A. R., and G. S. Bakken. 2003. Thermoregulation is the pits: Use of thermal radiation for retreat site selection by rattlesnakes. J. Exp. Biol. 206: 2539–2545.

———. 2005. Einstein and Klauber road cruise in heaven: How understanding physics can unlock the secrets of rattlesnake biology, p. 35. *In* Biology of the Rattlesnakes Symposium (program abstracts). Loma Linda Univ., Loma Linda, California.

Krochmal, A. R., G. S. Bakken, and T. J. LaDuc. 2004. Heat in evolution's kitchen: Evolutionary perspectives on the functions and origin of the facial pit of pitvipers (Viperidae: Crotalinae). J. Exp. Biol. 207: 4231–4238.

Kropach, C. 1971a. Sea snake *(Pelamis platurus)* aggregations on slicks in Panama. Herpetologica 27: 131–135.

———. 1971b. Another color variety of the sea snake *Pelamis platurus* from Panama Bay. Herpetologica 27: 326–327.

———. 1972. *Pelamis platurus* as a potential colonizer of the Caribbean Sea. Bull. Biol. Soc. Washington 2: 267–269.

———. 1975. The yellow-bellied sea snake, *Pelamis*, in the eastern Pacific, pp. 185–213. *In* W. A. Dunson (ed.), The biology of sea snakes. Univ. Park Press, Baltimore, Maryland.

Kropach, C., and J. D. Soule. 1973. An unusual association between an ectoproct and a sea snake. Herpetologica 29: 17–19.

Krysko, K. L., and K. R. Abdelfattah. 2002. *Micrurus fulvius* (eastern coral snake): Prey. Herpetol. Rev. 33: 57–58.

Kumar, V., T. A. Rejent, and W. B. Elliott. 1973. Anticholinesterase activity of elapid venoms. Toxicon 11: 131–138.

Kunkel, D. B., S. C. Curry, M. V. Vance, and P. J. Ryan. 1984. Reptile envenomations. J. Toxicol. Clin. Toxicol. 21: 503–526.

Kunzé, R. E. 1883. The copperhead. Am. Nat. 17: 1229–1238.

Kuppusamy, U. R., and N. P. Das. 1993. Protective effects of tannic acid and related natural compounds on *Crotalus adamenteus* subcutaneous poisoning in mice. Pharmacol. Toxicol. 72: 290–295.

Kuzmin, Y., V. V. Tkach, and S. D. Snyder. 2003. The nematode genus *Rhabdias* (Nematoda: Rhabdiasidae) from amphibians and reptiles of the Nearctic. Comp. Parasitol. 70: 101–114.

Kwok, H. F., and C. Ivanyi. 2008. A minimally invasive method for obtaining venom from helodermatid lizards. Herpetol. Rev. 39: 179–181.

La Barre, W. 1969. They shall take up serpents: Psychology of the southern snake-handling cult. Schocken Books, New York.

Lacépède, B. G. E. 1788–1789. Historie naturelle des quadrupeds ovipares et des serpens. 2 vols. Académie Royal des Sciences, Paris.

Lagesse, L. A., and N. B. Ford. 1996. Ontogenetic variation in the diet of the southern copperhead, *Agkistrodon contortrix*, in northeastern Texas. Texas J. Sci. 48: 48–54.

Landberg, K. H., B. D. Johnson, and D. H. Sifford. 1980. Isolation of phospholipase A_2 from *Agkistrodon bilineatus* venom. Proc. Arkansas Acad. Sci. 34: 119–122.

Langford, G. J., and J. A. Borden. 2007. *Amphiuma tridactylum* (three-toed amphiuma): Cottonmouth envenomation. Herpetol. Rev. 38: 173.

Langley, R. L., and W. E. Morrow. 1997. Deaths resulting from animal attacks in the United States. Wild. Environ. Med. 8: 8–16.

Lanza, B., and S. Boscherini. 2000. The gender of the genera *Podarcis* Wagler 1830 (Lacertidae), *Pelamis* Daudin 1803 (Hydrophiidae) and *Uropeltis* Cuvier 1829 (Uropeltidae). Trop. Zool. 13: 327–329.

La Pointe, J. 1953. Case report of a bite from the massasauga, *Sistrurus catenatus catenatus*. Copeia 1953: 128–129.

Lardie, G. E., and R. L. Lardie. 1976. Striped pattern in a western massasauga. Bull. Oklahoma Herpetol. Soc. 1(3): 40.

Lardie, R. L. 1976. Large centipede eaten by a western massasauga. Bull. Oklahoma Herpetol. Soc. 1(3): 40.

Larson, P., and L. Larson. 1990. The deserts of the Southwest: A Sierra Club naturalist's guide. 2nd ed. Sierra Club Books, San Francisco.

Laszlo, J. 1975. Probing as a practical method of sex recognition in snakes. Intl. Zoo Yearbk. 15: 178–179.

Lathan, L. O., and S. L. Staggers. 1996. Ancrod: The use of snake venom in the treatment of patients with heparin-induced thrombocytopenia and thrombosis undergoing coronary artery bypass grafting; Nursing management. Heart Lung 25: 451–460.

Laughlin, H. E. 1959. Stomach contents of some aquatic snakes from Lake McAlester, Pittsbourgh County, Oklahoma. Texas J. Sci. 11: 83–85.

Laughlin, H. E., and B. J. Wilks. 1962. The use of sodium pentobarbital in population studies of poisonous snakes. Texas J. Sci. 14: 188–191.

Laure, C. J. 1975. Die Primärstruktur des Crotamins. Hoppe-Seyler's Z. Physiol. Chem. 356: 213–215.

Lavín-Murcio, P. A., and J. R. Dixon. 2004. A new species of coral snake (Serpentes, Elapidae) from the Sierra de Tamaulipas, Mexico. Phyllomedusa 3: 3–7.

Lavonas, E. J., C. J. Gerardo, G. O'Malley, T. C. Arnold, S. P. Bush, W. Banner, Jr., M. Steffens, and W. P. Kerns, II. 2004. Initial experience with Crotalidae polyvalent immune Fab (ovine) antivenom in the treatment of copperhead snakebite. Ann. Emerg. Med. 43: 200–206.

Lawrence, J. M. 2007. First record of the yellow-bellied sea snake (*Pelamis platurus*) from Cousine Island, Seychelles. Phelsuma 15: 90–91.

Lawson, R., J. B. Slowinski, B. I. Crother, and F. T. Burbrink. 2005. Phylogeny of the Colubroidea (Serpentes): New evidence from mitochrondrial and nuclear genes. Mol. Phylogenet. Evol. 37: 581–601.

Lazcano, D., M. A. Salinas-Camarena, and J. A. Contreras-Lozano. 2009. Notes on Mexican herpetofauna 12: Are roads in Nuevo León, Mexico, taking their toll on snake populations? Bull. Chicago Herp. Soc. 44: 69–75.

Leavitt, B. B. 1957. Water moccasin preys on pied-billed grebe. Wilson Bull. 69: 112–113.

Lee, D. S. 1968. Herpetofauna associated with Central Florida mammals. Herpetologica 24: 83–84.

———. 1969. Captive snakes preyed upon by fruit flies. Bull. Maryland Herpetol. Soc. 5: 83.

Lee, C.-Y. 1971. Elapid neurotoxins and their mode of action, pp. 111–126. *In* S. A. Minton (ed.), Snake venoms and envenomation. Marcel Dekker, New York.

———. 1979. Snake venoms. Springer-Verlag, New York.

Lee, J. R. 1996. *Agkistrodon piscivorus piscivorus* (eastern cottonmouth): Coloration. Herpetol. Rev. 27: 22.

Lee, J. R., R. A. Seigel, and F. E. Durbian. 2004. *Sistrurus catenatus* (massasauga): Aggressive behavior. Herpetol. Rev. 35: 72–73.

Lee, J. R., A. L. Holbrook, and J. W. Frey. 2008. *Agkistrodon contortrix* (copperhead): Diet. Herpetol. Rev. 39: 97.

Lee, Y. M., L. G. Sargent, and B. A. Kingsbury. 2005. Developing a conservation strategy for the eastern massasauga in Michigan, pp. 36–37. *In* Biology of the Rattlesnakes Symposium (program abstracts). Loma Linda Univ., Loma Linda, California.

Le Galliard, J. F., and T. Tully. 1998. Collaboration internationale à des programmes de conservation dans le nordde la Sierra Madre occidentale (Mexique, Etats Unisd'Amerique du Nord): Les crotales Mexicains. Rapp. Magist. Biol.-Biochem. Ecole Norm. Supér. (Paris).

Leiper, R. T., and E. L. Atkinson. 1914. Helminthes of the British Antarctic Exhibition, 1910–1913. Proc. Zool. Soc. London 1914: 222–226.

Lemos-Espinal, J. A., and H. M. Smith. 2008a. Amphibians & reptiles of the state of Colima, Mexico. Bibliomania, Salt Lake City, Utah.

———. 2008b. Amphibians & reptiles of the state of Chihuahua, Mexico. Bibliomania, Salt Lake City, Utah.

Lemos-Espinal, J. A., D. Chiszar, and H. M. Smith. 2003. Presence of the Río Fuerte beaded lizard (*Heloderma horridum exasperatum*) in western Chihuahua. Bull. Maryland Herpetol. Soc. 39: 47–51.

Lemos-Espinal, J. A., H. M. Smith, and D. Chiszar. 2006. *Agkistrodon b. bilineatus* in western Chihuahua. Bull. Maryland Herpetol. Soc. 42: 173–175.

Leopold, R. S., G. S. Huber, and R. H. Kathan. 1957. An evaluation of the mechanical treatment of snake bite. Military Med. 120: 414–416.

Le Ray, W. J. 1930. The rattlesnake *Sistrurus catenatus* in Ontario. Can. Field-Nat. 44: 201–203.

Leviton, A. E. 1972. Reptiles and amphibians of North America. Doubleday, Garden City, New York.

Lewis, W., and M. Elvin-Lewis. 1977. Medicinal botany: Plants affecting man's health. Wiley-Interscience, New York.

Li, M., B. G. Fry, and R. M. Kini. 2005. Putting the brakes on snake venom evolution: The unique molecular evolutionary patterns of *Aipysurus eydouxii* (marbled sea snake) phospholipase A_2 toxins. Mol. Biol. Evol. 22: 934–941.

Li, Q., and C. L. Ownby. 1994. Development of an enzyme-linked immunosorbent assay (ELISA) for identification of venoms from snakes in the *Agkistrodon* genus. Toxicon 32: 1315–1325.

Li, Q., T. R. Colberg, and C. L. Ownby. 1993. Cross-reactivities of monoclonal antibodies to a myotoxin from the venom of the broad-banded copperhead (*Agkistrodon contortrix laticinctus*). Toxicon 31: 1187–1196.

Lillywhite, H. B. 1993a. Subcutaneous compliance and gravitational adaptation in snakes. J. Exp. Zool. 267: 557–562.

———. 1993b. Orthostatic intolerance of viperid snakes. Physiol. Zool. 66: 1000–1014.

———. 2008. Dictionary of herpetology. Krieger, Melbourne, Florida.

Lillywhite, H. B., and A. Smits. 1992. The cardiovascular adaptations of viperid snakes, pp. 143–154. *In* J. A. Campbell and E. D. Brodie, Jr. (eds.), Biology of the pitvipers. Selva, Tyler, Texas.

Lillywhite, H. B., and F. Zaidan, III. 2004. Physiological ecology of pit vipers on Gulf Coast islands, p. 19. *In* Proceedings of the Snake Ecology Group 2004 Conference, Jackson County, Illinois.

Lillywhite, H. B., C. M. Sheehy, III, and M. D. McCue. 2002. Scavenging behaviors of cottonmouth snakes at island bird rookeries. Herpetol. Rev. 33: 259–261.

Limpus, C. J. 2001. A breeding population of the yellow-bellied sea snake, *Pelamis platurus*, in the Gulf of Carpentaria. Mem. Queensland Mus. 46: 629–630.

Lindner, D. 1962. Feeding observations on *Micruroides*. Bull. Philadelphia Herpetol. Soc. 10(2–3): 31.

Lindsey, P. 1979. Combat behavior in the dusky pygmy rattlesnake, *Sistrurus miliarius barbouri*, in captivity. Herpetol. Rev. 10: 93.

Liner, E. A. 1994. Scientific and common names for the amphibians and reptiles of Mexico in English and Spanish. Soc. Stud. Amphib. Rept. Herpetol. Circ. 23.

Linnaeus, C. 1766. Systema naturae sive regna tria naturae, secundum classes, ordines, genera, species, cum characteribus, differentiis, synonymis, locis. 12th ed. Stockholm, Sweden.

Lintner, C. P., D. E. Keyler, and E. F. Bilden. 2006. Prairie rattlesnake (*Crotalus viridis viridis*) envenomation: Recurrent coagulopathy in a child treated with Immune Fab, pp. 33–34. *In* S. A. Seifert (ed.), Snakebites in the new millennium: A state-of-the-art symposium, University of Nebraska Medical Center, 21–23 October 2005, Omaha, Nebraska. J. Med. Toxicol. 2: 29–45.

Linzey, D. W. 1979. Snakes of Alabama. Strode, Huntsville, Alabama.

Litovitz, T. L., W. Klein-Schwartz, G. C. Rodgers Jr., D. J. Cobaugh, J. Youniss, J. C. Omslaer, M. E. May, A. D. Woolf, and B. E. Benson. 2002. 2001 Annual report of the American Association of Poison Control Centers toxic exposure surveillance system. Am. J. Emerg. Med. 20: 391–451.

Little, M. D. 1966. Seven new species of *Strongyloides* (Nematoda) from Louisiana. J. Parasitol. 52: 85–97.

Liu, C.-S., C.-L. Wang, and R. Q. Blackwell. 1975. Isolation and partial characterization of pelamitoxin A from *Pelamis platurus* venom. Toxicon 13: 31–36.

Liu, C.-Z., H.-C. Peng, and T.-F. Huang. 1995. Crotavirin, a potent platelet aggregation inhibitor purified from the venom of the snake *Crotalus viridis*. Toxicon 33: 1289–1298.

Livezey, R. L. 1949. An aberrant pattern of *Agkistrodon mokeson austrinus*. Herpetologica 5: 93.

Loeb, L. (ed.). 1913. The venom of *Heloderma*. Carnegie Inst. Washington Publ. 177: 1–244.

Lohoefener, R., and R. Altig. 1983. Mississippi herpetology. Mississippi St. Univ. Res. Cent. Nat. Space Tech. Lab. Bull. 1: 1–66.

Lomonte, B., E. Moreno, and J. M. Gutierrez. 1987. Detection of proteins antigenically related to *Bothrops asper* myotoxin in crotaline snake venoms. Toxicon 25: 947–955.

Lomonte, B., Y. Angulo, S. Rufini, W. Cho, J. R. Giglio, M. Ohno, J. J. Daniele, P. Geoghegan, and J. M. Gutiérrez. 1999. Comparative study of the cytolytic activity of myotoxic phospholipases A_2 on mouse endothelial (tEnd) and skeletal muscle (C2C12) cells *in vitro*. Toxicon 37: 145–158.

Lomonte, B., Y. Angulo, and C. Santamaría. 2003a. Comparative study of synthetic peptides corresponding to region 115–129 in Lys^{49} myotoxic phospholipases A_2 from snake venoms. Toxicon 42: 307–312.

Lomonte, B., Y. Angulo, and L. Calderón. 2003b. An overview of lysine-49 phospholipase A_2 myotoxins from crotalid snake venoms and their structural determinants of myotoxic action. Toxicon 42: 885–901.

Lönnberg, A. J. E. 1894. Notes on reptiles and batrachians collected in Florida in 1892 and 1893. Proc. U.S. Natl. Mus. 17(1003): 317–339. (Separates issued 15 November 1894, and completed volume in 1895.)

Loomis, R. B. 1948. Notes on the herpetology of Adams County, Iowa. Herpetologica 4: 121–122.

———. 1956. The chigger mites of Kansas (Acarina, Trombiculidae). Univ. Kansas Sci. Bull. 37, pt. 2(19): 1195–1443.

Lopoo, J. B., J. F. Bealer, P. C. Mantor, and D. W. Tuggle. 1998. Treating the snakebitten child in North America: A study of pit viper bites. J. Ped. Surg. 33: 1593–1595.

Loprinzi, C. L., J. Hennessee, L. Tamsky, and T. E. Johnson. 1983. Snake antivenin administration in a patient allergic to horse serum. South. Med. J. 76: 501–502.

Lougheed, S. C., H. L. Gibbs, K. A. Prior, and P. J. Weatherhead. 2000. A comparison of RAPD versus microsatellite DNA markers in population studies of the massasauga rattlesnake. J. Heredity 91(6): 458–463.

Lovecchio, F., and D. M. Debus. 2001. Snakebite envenomation in children: A 10-year retrospective review. Wild. Environ. Med. 12: 184–189.

Loveridge, A. 1928. Note on *Agkistrodon bilineatus* Günther. Bull. Antiv. Inst. Am. 2: 52.

———. 1938. Food of *Micrurus fulvius fulvius*. Copeia 1938: 201–202.

———. 1944. Cannibalism in the common coral snake. Copeia 1944: 254.

Lovich, J. E., and K. R. Beaman. 2007. A history of Gila monster (*Heloderma suspectum cinctum*): Records from California with comments on factors affecting their distribution. Bull. So. California Acad. Sci. 106: 39–58.

Lowe, C. H. 1948. Effect of venom of *Micruroides* upon *Xantusia vigilis*. Herpetologica 4: 136.

———. 1964. The vertebrates of Arizona. Univ. Arizona Press, Tucson.

Lowe, C. H., D. S. Hinds, P. J. Lardner, and K. E. Justice. 1967. Natural free-running period in vertebrate animal populations. Science (New York) 156(3774): 531–534.

Lowe, C. H., C. R. Schwalbe, and T. B. Johnson. 1986. The venomous reptiles of Arizona. Arizona Game Fish Dept., Phoenix.

Lutterschmidt, W. I., R. L. Nydam, and H. W. Greene. 1996a. County record for the woodland vole, *Microtus pinetorum* (Rodentia: Muridae), LeFlore County, OK, with natural history notes on a predatory snake. Proc. Oklahoma Acad. Sci. 76: 93–94.

Lutterschmidt, W. I., J. J. Lutterschmidt, and H. K. Reinert. 1996b. An improved timing device for monitoring pulse frequency of temperature-sensing transmitters in free-ranging animals. Am. Midl. Nat. 136: 172–180.

Lutterschmidt, W. I., E. D. Roth, K. G. Wunch, E. Levin, and L. H. James. 2007. Bacterial microflora of the anterior digestive tract of two *Agkistrodon* species: Additional evidence for food partitioning? Herpetol. Rev. 38: 33–35.

Lynch, V. J. 2007. Inventing an arsenal: Adaptive evolution and neofunctionalization of snake venom phospholipase A_2 genes. BMC Evol. Biol. 7(2): 1–14.

Lynn, W. G. 1929. A case of delayed birth in *Agkistrodon mokasen*. Bull. Antiv. Inst. Am. 2: 97.

———. 1931. The structure and function of the facial pit of the pitvipers. Am. J. Anat. 49: 97–139.

Lyon, M. W., and C. Bishop. 1936. Bite of the prairie rattlesnake *Sistrurus catenatus*. Raf. Proc. Indiana Acad. Sci. 45: 253–256.

MacCallum, G. G. 1921. Studies in helminthology: Part 1, trematodes; Part 2, cestodes; Part 3, nematodes. Zoopathologica 1: 135–294.

Macchiavelli, G., and V. P. Mascagni. 1990. Reproduction of *Agkistrodon bilineatus bilineatus*. Litt. Serpent. Engl. Ed. 10: 221–224.

Mace, G. M., and R. Lande. 1991. Assessing extinction threats: Towards a reevaluation of IUCN threatened species categories. Conserv. Biol. 5: 148–147.

MacGregor, G. A., and H. K. Reinert. 2001. The use of passive integrated transponders (PIT tags) in snake foraging studies. Herpetol. Rev. 32: 170–172.

Machotka, S. W., II. 1984. Neoplasia in reptiles, pp. 1–18. *In* G. L. Hoff, F. L. Frye, and E. R. Jacobson (eds.), Diseases of amphibians and reptiles. Plenum, New York.

Macht, D. I. 1940. New development in the pharmacology and therapeutics of cobra venom. Trans. Am. Ther. Soc. 40: 62.

———. 1947. Straub phenomenon produced by venom of *Micrurus fulvius*. Copeia 1947: 269–270.

Mackessy, S. P. 1991. Morphology and ultrastructure of the venom glands of the northern Pacific rattlesnake *Crotalus viridis oreganus*. J. Morphol. 208: 109–128.

———. 1993. Fibrinogenolytic proteases from the venoms of juvenile and adult northern Pacific rattlesnakes (*Crotalus viridis oreganus*). Comp. Biochem. Physiol. 106B: 181–189.

———. 1997. Colubrid snake venoms: Composition, biological roles and evolution. *In* American Society of Ichthyologists and Herpetologists Conference Proceedings (abstract), Univ. Washington, Seattle.

———. 1998a. A survey of the herpetofauna of southeastern Colorado with a focus on the current status of two candidates for Protected Species status: The massasauga rattlesnake and the Texas horned lizard. Colorado Div. Wildl. Rep.

———. 1998b. Phosphodiesterases, ribonucleases and deoxyribonucleases, pp. 361–404. *In* G. S. Bailey (ed.), Enzymes from snake venom. Alaken, Fort Collins, Colorado.

———. 2002. Biochemistry and pharmacology of colubrid snake venoms. J. Toxicol. Toxin Rev. 21(1–2): 43–83.

———. 2005. Desert massasauga rattlesnake (*Sistrurus catenatus edwardsii*): A technical conservation assessment. U.S.D.A. For. Serv. Rocky Mt. Reg. Online.

———. 2008. Venom composition in rattlesnakes: Trends and biological significance, pp. 495–510. *In* W. K. Hayes, K. R. Beaman, M. D. Cardwell, and S. P. Bush (eds.), The biology of rattlesnakes. Loma Linda Univ. Press, Loma Linda, California.

Mackessy, S. P., and L. M. Baxter. 2006. Bioweapons synthesis and storage: The venom gland of front-fanged snakes. Zool. Anz. 245: 147–159.

Mackessy, S. P., and A. T. Tu. 1993. Biology of the sea snakes and biochemistry of their venoms, pp. 305–351. *In* A. T. Tu (ed.), Toxin-related diseases: Poisons originating from plants, animals and spoilage. Oxford and IBH Publishers, New Delhi.

Maeda, N., N. Tamiya, T. R. Pattabhiraman, and F. E. Russell. 1978. Some chemical properties of the venom of the rattlesnake, *Crotalus viridis helleri*. Toxicon 16: 431–441.

Maher, B. W., and G. Sievert. 2006. *Agkistrodon contortrix phaeogaster* (Osage copperhead). Herpetol. Rev. 37: 496.

Mahrt, J. L. 1979. Hematozoa of lizards from southeastern Arizona and Isla San Pedro Nolasco, Gulf of California, Mexico. J. Parasitol. 65: 972–975.

Malnate, E. V. 1944. Notes on South Carolinian reptiles. Am. Midl. Nat. 32: 728–731.

———. 1990. A review and comparison of hemipenial structure in the genus *Agkistrodon* (*sensu lato*), pp. 583–588. *In* H. K. Gloyd and R. Conant (eds.), Snakes of the *Agkistrodon* complex: A monographic review. Soc. Stud. Amphib. Rept. Contrib. Herpetol. 6.

Malz, S. 1967. Snake-bite in pregnancy. J. Obstet. Gynaec. Brit. Cwlth. 74: 935–937.

Manning, M. C. 1995. Sequence analysis of fibrolase, a fibrinolytic metalloproteinase from *Agkistrodon contortrix contortrix*. Toxicon 33: 1189–1200.

Mao, S.-H., and B.-Y. Chen. 1980. Sea snakes of Taiwan: A natural history of sea snakes. Natl. Sci. Counc. Republic China, Taipei, Taiwan.

Mao, S.-H., B.-Y. Chen, F. Y. Yin, and Y. W. Guo. 1983. Immunotaxonomic relationships of sea snakes to terrestrial elapids. Comp. Biochem. Physiol. 74A: 869–872.

Maple, W. T., and L. P. Orr. 1968. Overwintering adaptations of *Sistrurus catenatus* in northeastern Ohio. J. Herpetol. 2: 179–180.

Mara, W. P. 1993. Venomous snakes of the world. T. F. H. Publ., Neptune City, New Jersey.

Maraganore, J. M., and R. L. Heinrikson. 1986. The lysine-49 phospholipase A_2 from the venom of *Agkistrodon piscivorus piscivorus*. J. Biol. Chem. 261: 4797–4804.

Maraganore, J. M., G. Merutka, W. Cho, W. Welches, F. J. Kézdy, and R. L. Heinrikson. 1984. A new class of phospholipases A_2 with lysine in place of aspartate 49: Functional consequences for calcium and substrate binding. J. Biol. Chem. 259: 13839–13843.

Marchisin, A. 1978. Observation on an audiovisual "warning" signal in the pigmy rattlesnake, *Sistrurus miliarius* (Reptilia, Serpentes, Crotalidae). Herpetol. Rev. 9: 92–93.

———. 1980. Predator-prey interactions between snake-eating snakes. Ph.D. diss., Rutgers Univ., Newark, New Jersey.

Markland, F. S., Jr. 1986. Antitumor action of crotalase, a defibrinogenating snake venom enzyme. Semin. Thromb. Hemostas. 12: 284–290.

———. 1988. Fibrin(ogen)olytic enzymes from snake venoms, pp. 149–172. *In* H. Pirkle and F. S. Markland, Jr. (eds.), Hemostatis and animal venoms (Hematology vol. 7). Dekker, New York.

———. 1990. Effect of snake venom proteins on tumor growth, pp. 175–200. *In* K. F. Stocker (ed.), Medical use of snake venom proteins. CRC Press, Boca Raton, Florida.

———. 1991. Inventory of α- and β-fibrinogenases from snake venoms: For the Subcommittee on Nomenclature of Exogenous Hemostatic Factors of the Scientific and Standardization Committee of the International Society on Thrombosis and Haemostasis. Thromb. Haemostas. 65: 438–443.

———. 1997. Snake venoms. Drugs 54(3): 1–10.

———. 1998a. Snake venom fibrinogenolytic and fibrinolytic enzymes: An updated inventory. Thromb. Haemostas. 79: 668–674.

———. 1998b. Snake venoms and the hemostatic system. Toxicon 36: 1749–1800.

Markland, F. S., and P. S. Damus. 1971. Purification and properties of a thrombin-like enzyme from the venom of *Crotalus adamanteus* (eastern diamondback rattlesnake). J. Biol. Chem. 246: 6460–6473.

Markland, F. S., Jr., and H. Pirkle. 1977. Biological activities and biochemical properties of thrombin-like enzymes from snake venoms, pp. 71–89. *In* R. L. Lundbold, J. W. Fenton, and K. G. Mann (eds.), Chemistry and biology of thrombin. Ann Arbor Science, Ann Arbor, Michigan.

Markland, F. S., Jr., K. N. N. Reddy, and L.-F. Guan. 1988. Purification and characterization of a direct-acting fibrinolytic enzyme from southern copperhead venom, pp. 173–189. *In* H. Pirkle and F. S. Markland, Jr. (eds.), Hemostatis and animal venoms. Marcel Dekker, New York.

Markland, F. S., S. Swenson, F. Costa, R. Minea, and R. P. Sherwin. 2005. A snake venom disintegrin with potent antitumor and antiangiogenic activity. Toxin Rev. 24: 113–142.

Markowitz, J., H. E. Essex, and F. C. Mann. 1931. Studies on immunity to rattlesnake venom (crotalin). Bull. Antiv. Inst. Am. 5: 28–29.

Marsh, N., and V. Williams. 2005. Practical applications of snake venom toxins in haemostasis. Toxicon 45: 1171–1181.

Marshall, J. C., Jr., J. V. Manning, and B. A. Kingsbury. 2006. Movement and macrohabitat selection of the eastern massasauga in a fen habitat. Herpetologica 62: 141–150.

Martin, B. 1974. Distribution and habitat adaptations in rattlesnakes of Arizona. HERP: Bull. New York Herpetol. Soc. 10(3–4): 3–12.

Martin, B. E. 1976. A reproductive record for the New Mexican ridge-nosed rattlesnake (*Crotalus willardi obscurus*). Bull. Maryland Herpetol. Soc. 12: 126–128.

Martin, D. L. 1982. Home range, movements, and activity patterns of the cottonmouth, *Agkistrodon piscivorus*, in southern Louisiana. Master's thesis, Univ. Southwestern Louisiana, Lafayette.

———. 1984. An instance of sexual defense in the cottonmouth, *Agkistrodon piscivorus*. Copeia 1984: 772–774.

Martin, P. S. 1958. A biogeography of reptiles and amphibians in the Gomez Farias region, Tamaulipas, Mexico. Univ. Michigan Mus. Zool. Misc. Publ. 101: 1–102.

Martin, R. A. 1974. Fossil vertebrates from the Haile XIVA fauna, Alachua County, Florida, pp. 100–113. *In* S. D. Webb (ed.), Pleistocene mammals of Florida. Univ. Press Florida, Gainesville.

Martin, W. H. 1990. The timber rattlesnake, *Crotalus horridus*, in the Appalachian Mountains of eastern North America. Catesbeiana 10: 49.

Martin, W. H., W. S. Brown, E. Possardt, and J. B. Sealy. 2008. Biological variation, management units, and a conservation action plan for the timber rattlesnake (*Crotalus horridus*), pp. 447–462. *In* W. K. Hayes, K. R. Beaman, M. D. Cardwell, and S. P. Bush (eds.), The biology of rattlesnakes. Loma Linda Univ. Press, Loma Linda, California.

Martof, B. S. 1956. Amphibians and reptiles of Georgia, a guide. Univ. Georgia Press, Athens.

Martof, B. S., W. M. Palmer, J. R. Bailey, J. R. Harrison, III, and J. Dermid. 1980. Amphibians and reptiles of the Carolinas and Virginia. Univ. North Carolina Press, Chapel Hill.

Marx, H., J. S. Ashe, and L. E. Watrous. 1988. Phylogeny of the viperine snakes (Viperinae): Part I. Character analysis. Fieldiana: Zool. 51: i–iv, 1–16.

Marx, N., and G. B. Rabb. 1972. Phyletic analysis of fifty characters of advanced snakes. Fieldiana: Zool. 63: 1–321.

Mashiko, H., and H. Takahashi. 1998. Haemorrhagic factors from snake venoms: II. Structures of haemorrhagic factors and types and mechanisms of haemorrhage. J. Toxicol. Toxin Rev. 17: 493–512.

Matthey, R. 1931a. Chromosomes de sauriens: Helodermatidae, Varanidae, Xantusiidae, Anniellidae, Anguidae. Bull. Soc. Vaudoise Sci. Nat. 57: 269.

———. 1931b. Chromosomes de reptiles: Sauriens, ophidiens, cheloniens. L'evolution la formule chromosomiale chez les sauriens. Suisse Zool. 38(9): 117–186.

Mattison, C. 1986. Snakes of the world. Facts on File Publ., New York.

———. 1988. Keeping and breeding snakes. Blandford Press, London.

Mauger, D., and T. P. Wilson. 1999. Population characteristics and seasonal activity of *Sistrurus catenatus catenatus* in Will County, Illinois: Implications for management and monitoring, pp. 110–124. *In* B. Johnson and M. Wright (eds.), Second international symposium and workshop on the conservation of the eastern massasauga rattlesnake, *Sistrurus catenatus catenatus*: Population and habitat management issues in urban, bog, prairie, and forested ecosystems. Toronto Zoo, Scarborough, Ontario, Canada.

———. 2005. *Sistrurus catenatus* (eastern massasauga): Mating activity. Herpetol. Rev. 36: 327–328.

May, P. G., and T. M. Farrell. 1998. Florida's flatwoods cottonmouths: Observations of feeding behavior. Rept. Amphib. Mag. 56: 18–24.

May, P. G., T. M. Farrell, S. T. Heulett, M. A. Pilgrim, L. A. Bishop, D. J. Spence, A. M. Rabatsky, M. G. Campbell, A. D. Aycrigg, and W. E. Richardson, II. 1996. Seasonal abundance and activity of a rattlesnake *(Sistrurus miliarius barbouri)* in central Florida. Copeia 1996: 389–401.

May, P. G., S. T. Heulett, T. M. Farrell, and M. A. Pilgrim. 1997. Live fast, love hard, & die young: The ecology of pigmy rattlesnakes. Rept. Amphib. Mag. January–February 1997: 36–49.

McAllister, C. T., S. E. Trauth, and L. D. Gage. 1995a. Vertebrate fauna of abandoned mines at Gold Mine Springs, Independence County, Arkansas. Proc. Arkansas Acad. Sci. 49: 184–187.

McCauley, R. H., Jr. 1945. The reptiles of Maryland and the District of Columbia. Privately published, Hagerstown, Maryland.

McCollough, N. C., and J. F. Gennaro. 1963a. Evaluation of the venomous snakebite in the southern United States from parallel clinical and laboratory investigations. J. Florida Med. Assn. 49: 959–967.

———. 1963b. Coral snake bites in the United States. J. Florida Med. Assn. 49: 968–972.

———. 1968. Diagnosis, symptoms, treatment and sequelae of envenomation by *Crotalus adamanteus* and genus *Ancistrodon*. J. Florida Med. Assoc. 55: 327–329.

———. 1970. Treatment of venomous snakebite in the United States. Clin. Toxicol. 3: 483–500.

McCoy, C. J. 1961. Birth season and young of *Crotalus scutulatus* and *Agkistrodon contortrix laticinctus*. Herpetologica 17: 140.

McCoy, C. J., and D. E. Hahn. 1979. The yellow-bellied sea snake, *Pelamis platurus* (Reptilia: Hydrophiidae), in the Philippines. Ann. Carnegie Mus. 48: 231–234.

McCoy, C. J., and W. L. Minckley. 1969. *Sistrurus catenatus* (Reptilia: Crotalidae) from the Cuatro Ciénegas Basin, Coahuila, Mexico. Herpetologica 25: 152–153.

McCranie, J. R. 1988. Description of the hemipenis of *Sistrurus ravus* (Serpentes: Viperidae). Herpetologica 44: 123–126.

McCrystal, H. K., and R. J. Green. 1986. *Agkistrodon contortrix pictigaster* (Trans-Pecos copperhead): Feeding. Herpetol. Rev. 17: 61.

McCrystal, H. K., and C. S. Ivanyi. 2008. Translocation of venomous reptiles in the Southwest: A solution—or part of the problem? pp. 395–402. *In* W. K. Hayes, K. R. Beaman, M. D. Cardwell, and S. P. Bush (eds.), The biology of rattlesnakes. Loma Linda Univ. Press, Loma Linda, California.

McCue, M. D. 2005. Enzyme activities and biological functions of snake venoms. Appl. Herpetol. 2: 109–123.

———. 2006. Cost of producing venom in three North American pitviper species. Copeia 2006: 818–825.

McCue, M. D., and H. B. Lillywhite. 2002. Oxygen consumption and the energetics of island-dwelling Florida cottonmouth snakes. Physiol. Biochem. Zool. 75(2): 165–178.

McDiarmid, R. W. 1963. A collection of reptiles and amphibians from the highland faunal assemblage of western Mexico. Contrib. Sci. Nat. Hist. Mus. Los Angeles Co. 68: 1–15.

McDiarmid, R. W., J. A. Campbell, and T'S. A. Touré. 1999. Snake species of the world: A taxonomic and geographic reference, vol. 1. Herpetologists' League, Washington, D.C.

McDowell, S. B. 1972. The genera of sea-snakes of the *Hydrophis* group (Serpentes: Elapidae). Trans. Zool. Soc. London 32: 189–247.

———. 1986. The architecture of the corner of the mouth of colubroid snakes. J. Herpetol. 20: 353–407.

McDuffie, G. T. 1961. Notes on the ecology of the copperhead in Ohio. J. Ohio Herpetol. Soc. 3: 26–27.

———. 1963. Studies on the size, pattern and coloration of the northern copperhead (*Agkistrodon contortrix mokeson* Daudin) in Ohio. J. Ohio Herpetol. Soc. 4: 15–22.

McGurty, B. 2002. *Heloderma suspectum* (Gila monster). Egg predation by juveniles. Herpetol. Rev. 33: 205.

McIntosh, A. 1939. Description of a plagiorchroid trematode, *Leptophyllum tamiamiensis*, n. sp. from a poisonous snake. Proc. Helminthol. Soc. Washington 6: 92–94.

McKeller, M. R., and J. C. Pérez. 2002. The effects of western diamondback rattlesnake (*Crotalus atrox*) venom on the production of antihemorrhagins and/or antibodies in the Virginia opossum (*Didelphis virginiana*). Toxicon 40: 427–439.

McKeown, S. 1996. A field guide to reptiles and amphibians in the Hawaiian Islands. Diamond Head Publ. Co., Los Osos, California.

McKinney, C. O., and R. E. Ballinger. 1966. Snake predators of lizards in western Texas. Southwest. Nat. 11: 410–412.

McKinney, P. E. 2001. Out-of-hospital and interhospital management of crotaline snakebite. Ann. Emerg. Med. 37: 168–174.

McKinstry, D. M. 1978. Evidence of toxic saliva in some colubrid snakes of the United States. Toxicon 16: 523–534.

———. 1983. Morphologic evidence of toxic saliva in colubrid snakes: A checklist of world genera. Herpetol. Rev. 14: 12–15.

McMullen, B. A., K. Fujikawa, and W. Kisiel. 1989. Primary structure of a protein C activator from *Agkistrodon contortrix contortrix* venom. Biochemistry 28: 674–679.

McNally, J., P. Chase, L. Mazullo, F. Sharazi, F. Walter, and M. Vahedian. 2007. Lesser known venomous creatures: Interesting cases. *In* Venom week 2007 (program abstracts). Univ. Arizona Colleges of Medicine, Arizona Health Sci. Center, Off. Continuing Medical Educ., Tucson, Arizona.

Meachem, A., and C. W. Myers. 1961. An exceptional pattern variant of the coral snake, *Micrurus fulvius* (Linnaeus). Quart. J. Florida Acad. Sci. 24: 56–58.

Mead, J. I., and A. M. Phillips, III. 1981. The late Pleistocene and Holocene fauna and flora of Vulture Cave, Grand Canyon, Arizona. Southwest. Nat. 26: 257–288.

Mead, J. I., E. L. Roth, T. R. Van Devender, and D. W. Steadman. 1984. The late Wisconsinan vertebrate fauna from Deadman Cave, southern Arizona. Trans. San Diego Soc. Nat. Hist. 20: 247–276.

Meade, G. P. 1940. Observations on Louisiana captive snakes. Copeia 1940: 165–168.

Meaume, J., Y. Izard, and P. Boquet. 1966. Mise en evidence d'une activitée coagulante dans le venin de *Naja nigricollis*. C. r. hebd. Séanc. Acad. Sci., Paris 262: 1650–1653.

Mebs, D. 1968. Some studies on the biochemistry of the venom gland of *Heloderma horridum*. Toxicon 5: 225–226.

———. 1969. Isolierung und Eigenschaften eines Kallikreins aus dem Gift der Krustenechse *Heloderma suspectum*. Hoppe-Seyler's Zeitschrift für Physiologische Chemie 350: 821–826.

———. 1970a. Biochemistry of kinin releasing enzymes in the venom of the viper *Bitis gabonica* and of the lizard *Heloderma suspectum,* pp. 107–117. *In* F. Sicuteri, M. Rocha e Silva, and N. Back (eds.), Bradykinin and related kinins; Proceedings Symposium on the Cardiovascular and Neural Actions of Bradykinin and Related Kinins. Plenum Press, New York.

———. 1970b. A comparative study of enzyme activities in snake venoms. Intl. J. Biochem. 1: 335–342.

———. 1972. Biochemistry of *Heloderma* venom, pp. 499–513. *In* A. de Vries and E. Kochva (eds.), Toxins of animal and plant origin, vol. 2. Gordon and Breach, New York.

———. 1978. Pharmacology of reptilian venoms, pp. 437–560. *In* C. Gans and K. A. Gans (eds.), Biology of the Reptilia, vol. 8: Physiology B. Academic Press, London.

———. 1990. Use of toxins in neurobiology and muscle research, pp. 57–78. *In* K. F. Stocker (ed.), Medical use of snake venom proteins. CRC Press, Boca Raton, Florida.

———. 1995. Clinical toxicology of Helodermatidae lizard bites, pp. 361–366. *In* J. Meier and J. White (eds.), Handbook of clinical toxicology of animal venoms and poisons. CRC Press, Boca Raton, Florida.

———. 1998. Enzymes in snake venoms: An overview, pp. 1–10. *In* G. S. Bailey (ed.), Enzymes from snake venom. Alakan, Fort Collins, Colorado.

———. 2002. Venomous and poisonous animals: A handbook for biologists, toxicologists and toxinologists, physicians and pharmacists. CRC Press, Boca Raton, Florida.

Mebs, D., and H. W. Raudonat. 1966. Biochemical investigations on *Heloderma* venom. Mem. Inst. Butantan 33: 907–912.

Mebs, D., and Y. Samejima. 1986. Isolation and characterization of myotoxic phospholipases A_2 from crotalid venoms. Toxicon 24: 161–168.

Megonigal, J. P. 1985. *Agkistrodon contortrix mokeson* (northern copperhead) and *Lampropeltis getulus getulus* (eastern kingsnake). Catesbeiana 5(1): 16.

Mehrtens, J. M. 1987. Living snakes of the world in color. Sterling Publ. Co., New York.

Meier, J., and K. Stocker. 1984. Beeinflussung der Toxizität von *Bothrops atrox*-Gift durch Eingriffe in das Gerinnungs-und Kallikreinsystem von Beutetiern. Folia Haematol. Int. Mag. Klin. Morphol. Blutforsch. 111:877–882.

Meier, J., and K. F. Stocker. 1991. Snake venom protein C activators, pp. 265–280. *In* A. T. Tu (ed.), Handbook of natural toxins, vol. 5: Reptile venoms and toxins. Marcel Dekker, New York.

———. 1995. Biology and distribution of venomous snakes of medical importance and the composition of snake venoms, pp. 367–412. *In* J. Meier and J. White (eds.), Handbook of clinical toxicology of animal venoms and poisons. CRC Press, New York.

Meik, J. M., E. N. Smith, and A. A. M. Hernández. 2007. Rediscovery of the rare coralsnake *Micruroides euryxanthus neglectus* (Serpentes: Elapidae). Herpetol. Rev. 38: 293–294.

Menne, H. A. L. 1959. Lets over het voorkomen van ratelslangen in Canada. Lacerta 18: 4–6.

Mertens, R. 1956. Das Problem der Mimikry bei Korallenschlangen. Zool. Jahrb., System. 84: 541–576.

———. 1957. Gibt es ein Mimikry bei Korallenschlangen? Natur u. Volk 87: 56–66.

Meshaka, W. E., S. E. Trauth, B. P. Butterfield, and A. B. Bevill. 1989. Litter size and aberrant pattern in the southern copperhead, *Agkistrodon contortrix contortrix*, from northeastern Arkansas. Bull. Chicago Herpetol. Soc. 24: 91–92.

Messenger, K. 2008. *Sistrurus miliarius miliarius* (Carolina pigmy rattlesnake): Morphology. Herpetol. Rev. 39: 473.

Meszler, R. M., and D. B. Webster. 1968. Histochemistry of the rattlesnake facial pit. Copeia 1968: 722–728.

Meszler, R. M., C. R. Auker, and D. O. Carpenter. 1981. Fine structure and organization of the infrared receptor relay, the lateral descending nucleus of the trigeminal nerve in pit vipers. J. Comp. Neurol. 196: 571–584.

Metcalf, C. K., F. Pezold, and B. G. Crump. 2007. *Agkistrodon piscivorus leucostoma* (western cottonmouth): Activity. Herpetol. Rev. 38: 465–466.

Meylan, P. A. 1982. The squamate reptiles of the Ingles IA Fauna (Irvingtonian: Citrus County, Florida). Bull. Florida St. Mus. Biol. Sci. 27: 1–85.

———. 1995. Pleistocene amphibians and reptiles from the Leisey Shell Pit, Hillsborough County, Florida. Bull. Florida St. Mus. Nat. Hist. 37: 273–297.

Miller, L. R., and W. H. N. Gutzke. 1999. The role of the vomeronasal organ of crotalines (Reptilia: Serpentes: Viperidae) in predator detection. Anim. Behav. 58: 53–57.

Miller, M. F. 1995. Gila monster envenomation. Ann. Emerg. Med. 25: 720.

Minckley, C. O., and W. E. Rinne. 1972. Another massasauga from Mexico. Texas J. Sci. 23: 432–433.

Minea, R., S. Swenson, F. Costa, T. C. Chen, and F. S. Markland. 2005. Development of a novel recombinant disintegrin, contortrostatin, as an effective anti-tumor and anti-angiogenic agent. Pathophysiol. Haemostas. Thromb. 34: 177–183.

Minton, J. E. 1949. Coral snake preyed upon by a bullfrog. Copeia 1949: 288.

Minton, S. A., Jr. 1944. Introduction to the study of the reptiles of Indiana. Am. Midl. Nat. 32: 438–477.

———. 1953. Variation in venom samples from copperheads (*Agkistrodon contortrix mokeson*) and timber rattlesnakes (*Crotalus horridus horridus*). Copeia 1953: 212–215.

———. 1956. Some properties of North American pit viper venoms and their correlation with phylogeny, pp. 145–151. *In* E. Buckley and N. Porges (eds.), Venoms. American Association for the Advancement of Science, Washington, D.C.

———. 1966. A contribution to the herpetology of West Pakistan. Bull. Am. Mus. Nat. Hist. 134: 27–184.

———. 1967. Observations on toxicity and antigenic makeup of venoms from juvenile snakes. Toxicon 4: 294.

———. 1971. Snake venoms and envenomation. Marcel Dekker, New York.

———. 1972. Amphibians and reptiles of Indiana. Indiana Acad. Sci., Monogr. 3: 1–346.

———. 1974. Venom diseases. Charles C. Thomas, Springfield, Illinois.

———. 1976. A list of colubrid envenomations. Kentucky Herp. 7: 4.

———. 1979. Beware: Nonpoisonous snakes. Clin. Toxicol. 15: 259–265.

———. 1980a. Snakebites in the U.S.A. The Snake 12: 141.

———. 1980b. Paraspecific neutralization by antivenoms. The Snake 12: 165.

———. 1982. Snake bite, pp. 283–301. *In* G. V. Hillyer and C. E. Hopla (eds.), Parasitic zoonoses. CRC Press, Boca Raton, Florida.

———. 1983. *Sistrurus catenatus*. Cat. Am. Amphib. Rept. 332: 1–2.

———. 1986. Non-poisonous snake bite. Herpetology 16: 14.

———. 1990a. Venomous bites by nonvenomous snakes: An annotated bibliography of colubrid envenomation. J. Wildl. Med. 1: 119–127.

———. 1990b. Immunologic relationships in *Agkistrodon* and related genera, pp. 589–600. *In* H. K. Gloyd and R. Conant (eds.), Snakes of the *Agkistrodon* complex: A monographic review. Soc. Stud. Amphib. Rept. Contrib. Herpetol. 6.

———. 1992. Serologic relationships among pitvipers: Evidence from plasma albumins and immunodiffusion, pp. 155–161. *In* J. A. Campbell and E. D. Brodie, Jr. (eds.), Biology of the pitvipers. Selva, Tyler, Texas.

———. 1996. Are there any nonvenomous snakes? An update on colubrid envenoming. Adv. Herpetoculture, Spec. Publ. Intl. Herpetol. Symp. 1: 127–134.

Minton, S. A., Jr., and M. R. Minton. 1969. Venomous reptiles. Charles Scribner's, New York.

Minton, S. A., Jr., and S. A. Weinstein. 1986. Geographic and ontogenetic variation in venom of the western diamondback rattlesnake (*Crotalus atrox*). Toxicon 24: 71–80.

———. 1987. Colubrid snake venoms: Immunologic relationships, electrophoretic patterns. Copeia 1987: 993–1000.

Minton, S. A., Jr., H. G. Dowling, and F. E. Russell. 1968. Poisonous snakes of the world: A manual for use by U.S. amphibious forces. U.S. Gov't. Print. Off., Washington, D.C.

Mitchell, J. C. 1975. Notes on a cottonmouth from Petersburg, Virginia. Virginia Herpetol. Soc. Bull. 75: 5.

———. 1977. An instance of cannibalism in *Agkistrodon contortrix* (Serpentes: Viperidae). Bull. Maryland Herpetol. Soc. 13: 119.

———. 1980. Viper's brood. A guide to identifying some of Virginia's juvenile snakes. Virginia Wildl. 41(9): 8–10.

———. 1981. Notes on male combat in two Virginia snakes, *Agkistrodon contortrix* and *Elaphe obsoleta*. Catesbeiana 1(1): 7–9.

———. 1986. Cannibalism in reptiles: A worldwide review. Soc. Stud. Amphib. Rept. Herpetol. Circ. 15: i–iii, 1–37.

———. 1994. The reptiles of Virginia. Smithsonian Inst. Press, Washington, D.C.

Mitchell, J. C., and M. Fieg. 1996. *Agkistrodon contortrix mokasen* (northern copperhead): Bicephaly. Herpetol. Rev. 27: 202–203.

Mitchell, J. C., and W. H. Martin, III. 1981. Where the snakes are. Virginia Wildl. 42(6): 8–9.

Mitchell, J. C., and D. Schwab. 1991. Canebrake rattlesnake, *Crotalus horridus atricaudatus* Latreille, pp. 462–464. *In* K. Terwilliger (ed.), Virginia's endangered species: Proceedings of a symposium. McDonald Woodward Publ. Co., Blacksburg, Virginia.

Mitchell, J. D. 1903. The poisonous snakes of Texas, with notes on their habits. Texas Med. News 12: 411–437.

Mitchell, S. W. 1860. Researches upon the venom of the rattlesnake: With an investigation of the anatomy and physiology of the organs concerned. Smithsonian Contrib. Knowl. 12(6).

Mitchell, S. W., and E. T. Reichert. 1886. Researches upon the venoms of poisonous serpents. Publ. no. 647, Smithsonian Contrib. Knowl. 26(1).

Mittleman, M. B., and R. C. Goris. 1978. Death caused by the bite of the Japanese colubrid snake *Rhabdophis tigrinus* (Boie) (Reptilia, Serpentes, Colubridae). J. Herpetol. 12: 109–111.

Mochca-Morales, J., B. M. Martin, and L. D. Possani. 1990. Isolation and characterization of Helothermine, a novel toxin from *Heloderma horridum horridum* (Mexican beaded lizard) venom. Toxicon 28: 299–309.

Moen, D. S., C. T. Winne, and R. N. Reed. 2005. Habitat-mediated shifts and plasticity in the evaporative water loss rates of two congeneric pit vipers (Squamata, Viperidae, *Agkistrodon*). Evol. Ecol. Res. 7: 759–766.

Moerman, D. E. 1986. Medicinal plants of Native America. 2 vols. Museum of Anthropology, Univ. Michigan, Ann Arbor.

Moksi, H. 1954. A large litter of copperheads (*Agkistrodon contortrix mokeson*). Copeia 1954: 67.

Molenaar, G. J. 1974. An additional trigeminal system in certain snakes possessing infrared receptors. Brain Res. 78: 340–344.

———. 1992. Anatomy and physiology of infrared sensitivity of snakes, pp. 367–435. *In* C. Gans and P. S. Ulinski (eds.), Biology of the Reptilia, vol. 17. Univ. Chicago Press, Chicago, Illinois.

Monroy-Vilchis, O., O. Hernández-Gallegos, and F. Rodríguez-Romero. 2005. *Heloderma horridum horridum* (Mexican beaded lizard): Unusual habitat. Herpetol. Rev. 36: 450.

Montgomery, C. E., N. E. Haertle, and S. J. Beaupre. 2004. *Agkistrodon contortrix* (copperhead): Predation. Herpetol. Rev. 35: 271.

Montgomery, W. B., and G. W. Shuett. 1989. Autumnal mating with subsequent production of offspring in the rattlesnake *Sistrurus miliarius streckeri*. Bull. Chicago Herpetol. Soc. 24: 205–207.

Moon, B. R. 2001. Muscle physiology and the evolution of the rattling system in rattlesnakes. J. Herpetol. 35: 497–500.

Moon, B. R., P. M. Conn, and A. M. Rabatsky. 2004b. *Agkistrodon contortrix* (copperhead): Maximum prey size. Herpetol. Rev. 35: 174.

Moore, J. A., and J. C. Gillingham. 2006. Spatial ecology and multi-scale habitat selection by a threatened rattlesnake: The eastern massasauga (*Sistrurus catenatus catenatus*). Copeia 2006: 742–751.

Moran, J. B., and C. R. Geren. 1979. A comparison of biological and chemical properties of three North American (Crotalidae) snake venoms. Toxicon 17: 237–244.

Morgan, D. L., D. J. Borys, and J. D. Barry. 2007. Coral snake antivenom: Clinical differences between cases receiving it and those that did not. *In* Venom week 2007 (program abstracts). Univ. Arizona Colleges of Medicine, Arizona Health Sci. Center, Off. Continuing Medical Educ., Tucson, Arizona.

Mori, N., H. Ishizaki, and A. T. Tu. 1989. Isolation and characterization of *Pelamis platurus* (yellow-bellied sea snake) postsynaptic neurotoxin. J. Pharm. Pharmacol. 41: 331–334.

Morris, M. A. 1985. Envenomation from the bite of *Heterodon nasicus* (Serpentes: Colubridae). Herpetologica 41: 361–363.

Morris, P. J., and A. C. Alberts. 1996. Determination of sex in white-throated monitors (*Varanus albigularis*), Gila monsters (*Heloderma suspectum*), and beaded lizards (*H. horridum*) using two-dimensional ultrasound imaging. J. Zoo. Wildl. Med. 27(3): 371–377.

Morris, P. J., and C. Henderson. 1998. Gender determination in mature Gila monsters, *Heloderma suspectum*, and Mexican beaded lizards, *Heloderma horridum*, by ultrasound imaging of the ventral tail. Bull. Assoc. Reptilian Amphibian Veterinarians 8(4): 4–5.

Morrissette, J., J. Kratzschmar, B. Haendler, R. Elhayek, J. Mochca-Morales, B. M. Martin, J. R. Patel, R. L. Moss, W. D. Schleuning, R. Coronado, and L. D. Possani. 1995. Primary structure and properties of helothermine, a peptide toxin that blocks ryanodine receptors. Biophysical J. 68 (6): 2280–2288.

Moss, S. T., G. Bogdan, R. C. Dart, S. P. Nordt, S. R. Williams, and R. F. Clark. 1997. Association of rattlesnake bite location with severity of clinical manifestations. Ann. Emerg. Med. 30: 58–61.

Mount, R. H. 1975. Reptiles and amphibians of Alabama. Auburn Univ. Agric. Exp. Stat., Auburn, Alabama.

———. 1981. The red imported fire ant, *Solenopsis invicta* (Hymenoptera: Formicidae) as a possible serious predator on some native southeastern vertebrates: Direct observations and subjective impressions. J. Alabama Acad. Sci. 52: 71–78.

——— (ed.). 1984. Vertebrate wildlife of Alabama. Alabama Agric. Exp. St., Auburn.

Mount, R. H., and J. Cecil. 1982. *Agkistrodon piscivorus* (cottonmouth). Hybridization. Herpetol. Rev. 13: 95–96.

Munekiyo, S. M., and S. P. Mackessy. 2005. Presence of peptide inhibitors in rattlesnake venoms and their effects on endogenous metalloproteases. Toxicon 45: 255–263.

Muniz, J. R. C., A. L. B. Ambrosio, H. S. Selistre-de-Araujo, M. R. Cominetti, A. M. Moura-da-Silva, G. Oliva, R. C. Garratt, and D. H. F. Souza. 2008. The three-dimensional structure of bothropasin, the main hemorrhagic factor from *Bothrops jararaca* venom: Insights for a new classification of snake venom metalloprotease subgroups. Toxicon 52: 807–816.

Munro, D. F. 1947. Effect of a bite by *Sistrurus* on *Crotalus*. Herpetologica 4: 57.

———. 1949a. Effect of DDT powder on small cottonmouths. Herpetologica 5: 71–72.

———. 1949b. Vertical position of the pupil in the Crotalidae. Herpetologica 5: 106–108.

———. 1950. Additional observations on head-bobbing by snakes. Herpetologica 6: 88.

Murakami, M. T., and R. K. Arni. 2005. Thrombomodulin-independent activation of protein C and specificity of hemostatically active snake venom serine proteinases: Crystal structures of native and inhibited *Agkistrodon contortrix contortrix* protein C activator. J. Biol. Chem. 280: 39309–39315.

Murphy, J. B. 1973. A review of diseases and treatment of captive chelonians: Bacterial and viral infections, part two of a series. HISS News-J. 1(3): 77–81.

Murphy, J. B., and B. L. Armstrong. 1978. Maintenance of rattlesnakes in captivity. Univ. Kansas Mus. Nat. Hist. Spec. Publ. 3: 1–40.

Murphy, J. B., and D. E. Jacques. 2006. Death from snakebite: The entwined histories of Grace Olive Wiley and Wesley H. Dickinson. Bull. Chicago Herpetol. Soc. Spl. Suppl.: 1–20.

Murphy, J. C. 1990. A model for regional herpetological society field studies: The snakes of southern California, a proposal, and some thoughts on collecting. Bull. Chicago Herpetol. Soc. 25: 42–45.

Murphy, R. W. 1983. The reptiles: Origin and evolution, pp. 130–158. *In* T. J. Case and M. L. Cody (eds.), Island biogeography in the Sea of Cortéz. Univ. California Press, Berkeley.

———. 1988. The problematic phylogenetic analysis of interlocus heteropolymer isozyme characters: A case study from sea snakes and cobras. Can. J. Zool. 66: 2628–2633.

Murphy, R. W., and C. B. Crabtree. 1988. Genetic identification of a natural hybrid rattlesnake: *Crotalus scutulatus scutulatus* × *C. viridis viridis*. Herpetologica 44: 119–123.

Murphy, R. W., and J. R. Ottley. 1984. Distribution of amphibians and reptiles on islands in the Gulf of California. Ann. Carnegie Mus. 53: 207–230.

Murphy, R. W., J. Fu, A. Lathrop, J. V. Feltham, and V. Kovac. 2002. Phylogeny of the rattlesnakes (*Crotalus* and *Sistrurus*) inferred from sequences of five mitochondrial DNA genes, pp. 69–92. *In* G. W. Schuett, M. Höggren, M. E. Douglas, and H. W. Greene (eds.), Biology of the vipers. Eagle Mountain Publ., Eagle Mountain, Utah.

Murphy, T. D. 1964. Box turtle, *Terrapine carolina*, in stomach of copperhead, *Agkistrodon contortrix*. Copeia 1964: 221.

Myers, C. W. 1965. Biology of the ringsnake, *Diadophis punctatus,* in Florida. Bull. Florida St. Mus. Biol. Sci. 10: 43–90.

Myers, G. S. 1945. Nocturnal observations on sea-snakes in Bahia Honda, Panama. Herpetologica 3: 22–23.

Nair, B. C., C. Nair, and W. B. Elliott. 1979. Isolation and partial characterization of a phospholipase A_2 from the venom of *Crotalus scutulatus salvini*. Toxicon 17: 557–569.

Nauck, E. G. 1929. Untersuchungen über des Gift einer Seeschlange (*Hydrus platurus*) des Pazifischen Ozeans. Deutsche Tropenmed. Zeit. 33: 167–170.

Neill, W. T. 1947. Size and habits of the cottonmouth moccasin. Herpetologica 3: 203–205.

———. 1948. Hibernation of amphibians and reptiles in Richmond County, Georgia. Herpetologica 4: 107–114.

———. 1949. Increased rate of ecdysis in injured snakes. Herpetologica 5: 115–116.

———. 1957. Some misconceptions regarding the eastern coral snake, *Micrurus fulvius*. Herpetologica 13: 111–118.

———. 1960. The caudal lure of various juvenile snakes. Quart. J. Florida Acad. Sci. 23: 173–200.

———. 1968. Snake eats snake. Florida Wildl. 21: 22–25.

Neill, W. T., and E. R. Allen. 1956. Secondarily ingested food items in snakes. Herpetologica 12: 172–174.

Nellis, D. W. 1997. Poisonous plants and animals of Florida and the Caribbean. Pineapple Press, Sarasota, Florida.

Newman, E. A., and P. N. Hartline. 1981. Integration of visual and infrared information in bimodal neurons of the rattlesnake optic tectum. Science (New York) 213: 789–791.

Nichols, A. 1986. Envenomation by a bluestripe garter snake, *Thamnophis sirtalis sirtalis*. Herpetol. Rev. 17: 6.

Nikai, T., E. Katano, Y. Komori, and H. Sugihara. 1988a. β-Fibrinogenase from the venom of *Agkistrodon p. piscivorus*. Comp. Biochem. Physiol. 89B: 509–515.

Nikai, T., E. Katano, Y. Komori, and H. Sugihara. 1988b. Kallikrein-like enzyme from the venom of *Agkistrodon p. piscivorus*. Intl. J. Biochem. 20: 1239–1245.

Nikai, T., K. Imai, H. Sugihara, and A. T. Tu. 1988c. Isolation and characterization of *horridum* toxin with aginine ester hydrolase activity from *Heloderma horridum* (beaded lizard) venom. Arch. Biochem. Biophysiol. 264(1): 270–280.

Nikai, T., K. Imai, Y. Komori, and H. Sugihara. 1992. Isolation and characterization of arginine ester hydrolase from *Heloderma horridum* (beaded lizard) venom. Intl. J. Biochem. 24: 415–420.

Nikai, T., Y. Komori, S. Yagihashi, A. Ohara, Y. Ohizumi, and H. Sugihara. 1993. Isolation and characterization of phospholipase A_2 from *Agkistrodon bilineatus* (common cantil) venom. Intl. J. Biochem. 25: 879–884.

Nikai, T., A. Ohara, Y. Komori, J. W. Fox, and H. Sugihara. 1995. Primary structure of a coagulant enzyme, bilineobin, from *Agkistrodon bilineatus* venom. Arch. Biochem. Biophysics 318(1): 89–96.

Nirthanan, S., and M. C. E. Gwee. 2004. Three-finger α-neurotoxins and the nicotinic acetylcholine receptor, forty years on. J. Pharmacol. Sci. 94: 1–17.

Nirthanan, S., P. Gopalakrishnakone, M. C. E. Gwee, H. E. Khoo, and R. M. Kini. 2003. Nonconventional toxins from elapid venoms. Toxicon 41: 397–407.

Nkinin, S. W., J. P. Chippaux, D. Piétin, Y. Doljansky, O. Trémeau, and A. Ménez. 1997. Genetic origin of the variability of venoms: Impact on the preparation of antivenom sera (SAV). Bull. Soc. Path. Ex. 90: 277–281.

Nobile, M., V. Magnelli, L. Lagostena, J. Mochca-Morales, and L. D. Possani. 1994. The toxin helothermine affects potassium currents in newborn rat cerebellar granule cells. J. Membrane Bio. 139(1): 49–55.

Nobile, M., F. Noceti, G. Prestipino, and L. D. Possani. 1996. Helothermine, a lizard venom toxin, inhibits calcium current in cerebellar granules. Exp. Brain Res. 110(1): 15–20.

Noble, G. K., and A. Schmidt. 1937. The structure and function of the facial and labial pits of snakes. Proc. Am. Philos. Soc. 77: 263–288.

Norell, M. A., and G. Keqin. 1997. Braincase and phylogenetic relationships of *Estesia mongoliensis* from the late Cretaceous of the Gobi Desert and the recognition of a new clade of lizards. Am. Mus. Novit. 3211: 1–25.

Norell, M. A., M. C. McKenna, and M. J. Novacek. 1992. *Estesia mongoliensis*, a new fossil varanoid from the Late Cretaceous Barun Goyot Formation of Mongolia. Am. Mus. Nat. Hist. Novit. 3045: 1–24.

Norris, R. L., Jr. 1995. Bite marks and the diagnosis of venomous snakebite. Wild. Environ. Med. 6: 159–161.

Norris, R. 2004. Venom poisoning by North American reptiles, pp. 683–708. *In* J. A. Campbell and W. W. Lamar (eds.), The venomous reptiles of the western hemisphere. Comstock Publ. Assoc., Cornell Univ. Press, Ithaca, New York.

Norris, R. L. 2008. Snake envenomations: Coral. e Med. Emerg. Med. www.emedicine.com/emerg/topic542.htm.

Norris, R. L., Jr., and S. P. Bush. 2001. North American venomous reptile bites, pp. 896–926. *In* P. S. Auerbach (ed.), Wilderness medicine. 4th ed. Mosby, St. Louis, Missouri.

———. 2007. Bites by venomous reptiles in the Americas, pp. 1051–1085. *In* P. S. Auerbach (ed.), Wilderness medicine. 5th ed., Mosby Elsevier, Philadelphia.

Norris, R. L., and R. C. Dart. 1989. Apparent coral snake envenomation in a patient without visible fang marks. Am. J. Emerg. Med. 7: 402–405.

Norris, R. L., J. A. Wilkerson, and J. Feldman. 2007. Syncope, massive aspiration, and sudden death following rattlesnake bite. Wild. Environ. Med. 18: 206–208.

North, G. W. 1996. Captive propagation of the Rio Fuerte beaded lizard (*Heloderma horridum exasperatum*). Rept. Mag. 4(5): 100–106.

Nowak, E. M. 1998. Implications of nuisance rattlesnake relocation at Montezuma Castle National Monument. Sonoran Herpetol. 11: 2–5.

———. 2005. Nuisance rattlesnake movements and ecology in Arizona (USA) national parks, pp. 40–41. *In* Biology of the Rattlesnakes Symposium (program abstracts). Loma Linda Univ., Loma Linda, California.

Nowak, E. M., T. Hare, and J. McNally. 2002. Management of "nuisance" vipers: Effects of translocation on western diamond-back rattlesnakes (*Crotalus atrox*), pp. 533–560. *In* G. W. Schuett, M. Höggren, M. E. Douglas, and H. W. Greene (eds.), Biology of the vipers. Eagle Mountain Publ., Eagle Mountain, Utah.

Núñez, C. E., Y. Angulo, and B. Lomonte. 2001. Identification of the myotoxic site of the Lys^{49} phospholipase A_2 from *Agkistrodon piscivorus piscivorus* snake venom: Synthetic C-terminal peptides from Lys^{49}, but not from Asp49 myotoxins, exert membrane-damaging activities. Toxicon 39: 1587–1594.

Nydam, R. L. 2000. A new taxon of helodermatid-like lizard from the Albian-Cenomanian of Utah. J. Vert. Paleontol. 20: 285–294.

Obrecht, C. B. 1946. Notes on South Carolina reptiles and amphibians. Copeia 1946: 71–74.

O'Connell, B., D. Chiszar, and H. M. Smith. 1981. Poststrike behavior in cottonmouths (*Agkistrodon piscivorus*) feeding on fish and mice. Bull. Philadelphia Herpetol. Soc. 29: 3–7.

O'Donovan, K. 2006. CroFab™: Taking antivenin manufacture into the 21st century, pp. 44–45. *In* S. A. Seifert (ed.), Snakebites in the new millennium: A state-of-the-art symposium, University of Nebraska Medical Center, 21–23 October 2005, Omaha, Nebraska. J. Med. Toxicol. 2: 29–45.

Offerman, S. R., T. S. Smith, and R. W. Derlet. 2001. Does the aggressive use of polyvalent antivenin for rattlesnake bites result in serious acute side effects? West. J. Med. 175(2): 88–91.

Offerman, S. R., S. P. Bush, J. A. Moynihan, and R. F. Clark. 2002. Crotaline Fab antivenom for the treatment of children with rattlesnake envenomation. Pediatrics 110: 968–971.

Ogawa, H., and Y. Sawai. 1986. Fatal bite of the yamakagashi (*Rhabdophis tigrinus*). The Snake 18: 53–54.

Ogilvie, M. L., and T. K. Gartner. 1984. Identification of lectins in snake venoms. J. Herpetol. 18: 285–290.

Ohta, M., K. Ohta, and K. Hayashi. 1991. Medical application of snake venom components, pp. 349–376. *In* A. T. Tu (ed.), Handbook of natural toxins, vol. 5: Reptile venoms and toxins. Marcel Dekker, New York.

Okonogi, T., Z. Hattori, M. Fukami, T. Inque, M. Sato, Y. Ajiki, and Y. Kogure. 1980. A miscellany report of sea snakes. The Snake 12: 19–24.

Oldham, J., E. Chace, and H. M. Smith. 1983. Immediate ISC: A neglected option of choice in snakebite treatment. Bull. Maryland Herpetol. Soc. 19: 95–99.

Oliver, J. A. 1937. Notes on a collection of amphibians and reptiles from the state of Colima, Mexico. Occ. Pap. Mus. Zool. Univ. Michigan 360: 1–30.

———. 1955. The natural history of North American amphibians and reptiles. Van Nostrand, Princeton, New Jersey.

———. 1958. Snakes in fact and fiction. Macmillan, New York.

Olson, R. E., B. Marx, and R. Rome. 1986. Descriptive dentition morphology of lizards of Middle and North America, I: Scincidae, Teiidae, and Helodermatidae. Bull. Maryland Herpetol. Soc. 22: 97–124.

Omori-Satoh, T., J. Lang, H. Breithaupt, and E. Habermann. 1975. Partial amino acid sequence of the basic *Crotalus* phospholipase A. Toxicon 13: 69–71.

Ortenburger, A. I., and R. D. Ortenburger. 1927 [1926]. Field observations on some amphibians and reptiles of Pima County, Arizona. Proc. Oklahoma Acad. Sci. 6: 101–121.

Ortenburger, R. D. 1924. Notes on the Gila monster. Proc. Oklahoma Acad. Sci. 4: 22.

Orth, J. C. 1939. Moth larvae in a copperhead's stomach. Copeia 1939: 54–55.

Orthner, C. L., P. Bhattacharya, and D. K. Strickland. 1988. Characterization of a protein C activator from the venom of *Agkistrodon contortrix contortrix.* Biochemistry 27: 2558–2564.

Oshima, G., and S. Iwanaga. 1969. Occurrence of glycoproteins in various snake venoms. Toxicon 7: 235–238.

Oshima, G., T. Sato-Ohmori, and T. Suzuki. 1969. Proteinase, arginineester hydrolase and a kinin releasing enzyme in snake venoms. Toxicon 7: 229–233.

Otten, E. J., and D. McKimm. 1983. Venomous snakebite in a patient allergic to horse serum. Ann. Emerg. Med. 12: 624–627.

Owens, T. C. 2006. Ex-situ: Notes on reproduction and captive husbandry of the Guatemalan beaded lizard (*Heloderma horridum charlesbogerti*). Iguana 13(3): 212–215.

Ownby, C. L. 1982. Pathology of rattlesnake envenomation, pp. 163–209. *In* A. T. Tu (ed.), Rattlesnake venoms: Their actions and treatment. Marcel Dekker, New York.

———. 1990. Locally acting agents: Myotoxins, hemorrhagic toxins and dermonecrotic factors, pp. 601–654. *In* W. T. Shier and D. Mebs (eds.), Handbook of toxinology, Marcel Dekker, New York.

———. 1998. Structure, function and biophysical aspects of the myotoxins from snake venoms. J. Toxicol. Toxin Rev. 17: 213–238.

Ownby, C. L., and T. R. Colberg. 1990. Comparison of the immunogenicity and antigenic composition of several venoms of snakes in the family Crotalidae. Toxicon 28: 189–199.

Ownby, C. L., H. S. Selistre de Araujo, S. P. White, and J. E. Fletcher. 1999. Lysine 49 phospholipase A_2 proteins. Toxicon 37: 411–445.

Oyama, E., and H. Takahashi. 2007. Distribution of low molecular weight platelet aggregation inhibitors from snake venoms. Toxicon 49: 293–298.

Painter, C. W., and L. Fitzgerald. 1998. Rattlesnake roundups and commercial trade of the western diamondback rattlesnake *Crotalus atrox*. Sonoran Herpetol. 11: 16–18.

Pal, S. K., A. Gomes, S. C. Dasgupta, and A. Gomes. 2002. Snake venom as therapeutic agents: From toxin to drug development. Indian J. Exp. Biol. 40: 1353–1358.

Palis, J. G. 1993. *Agkistrodon piscivorus conanti* (Florida cottonmouth). Prey. Herpetol. Rev. 24: 59, 62.

Palisot de Beauvois, A. M. F. J. 1799. Memoir on Amphibia. Serpents. Trans. Am. Philos. Soc. 4: 362–381.

Palmer, T. 1992. Landscape with reptile: Rattlesnakes in an urban world. Ticknor & Fields, New York.

Palmer, W. M. 1965. Intergradation among the copperheads (*Agkistrodon contortrix* Linnaeus) in the North Carolina coastal plain. Copeia 1965: 246–247.

———. 1971. Distribution and variation of the Carolina pigmy rattlesnake, *Sistrurus miliarius miliarius* Linnaeus, in North Carolina. J. Herpetol. 5: 39–44.

———. 1974. Poisonous snakes of North Carolina. St. Mus. Nat. Hist. North Carolina, Raleigh.

———. 1978. *Sistrurus miliarius*. Cat. Am. Amphib. Rept. 220: 1–2.

Palmer, W. M., and A. L. Braswell. 1995. Reptiles of North Carolina. Univ. North Carolina Press, Chapel Hill.

Palmer, W. M., and G. M. Williamson. 1971. Observations on the natural history of the Carolina pigmy rattlesnake, *Sistrurus miliarius miliarius* Linnaeus. J. Elisha Mitchell Sci. Soc. 87: 20–25.

Pantanowitz, L., and F. Guidozzi. 1996. Management of snake and spider bite in pregnancy. Obstet. Gynecol. Surv. 51: 615–620.

Parent, C., and P. J. Weatherhead. 2000. Behavioral and life history responses of eastern massasauga rattlesnakes (*Sistrurus catenatus catenatus*) to human disturbance. Oecologia (Berlin) 125: 170–178.

Parker, H. W. 1977. Snakes of the world. Dover Publ., New York.

Parkinson, C. L., S. M. Moody, and J. E. Ahlquist. 1997. Phylogenetic relationships of the "*Agkistrodon* complex" based on mitrochondrial DNA sequence data, pp. 63–78. *In* R. S. Thorpe, W. Wüster, and A. Malhotra (eds.), Venomous snakes: Ecology, evolution and snakebite. Clarendon Press, Oxford.

Parkinson, C. L., K. R. Zamudio, and H. W. Greene. 2000. Phylogeny of the pitviper clade *Agkistrodon*: Historical ecology, species status, and conservation of cantils. Mol. Ecol. 9: 411–420.

Parkinson, C. L., J. A. Campbell, and P. T. Chippindale. 2002. Multigene phylogenetic analysis of pitvipers, with comments on their biogeography, pp. 93–110. *In* G. W. Schuett, M. Höggren, M. E. Douglas, and H. W. Greene (eds.), Biology of the vipers. Eagle Mountain Publ., Eagle Mountain, Utah.

Parmley, D. 1986. Herpetofauna of the Rancholabrean Schulze Cave local fauna of Texas. J. Herpetol. 20: 1–10.

———. 1988. Middle Holocene herpetofauna of Klein Cave, Kerr County, Texas. Southwest. Nat. 33: 378–382.

Parmley, D., and J. A. Holman. 1995. Hemphillian (late Miocene) snakes from Nebraska, with comments on Arikareean through Blancan snakes of midcontinental North America. J. Vert. Paleontol. 15: 79–95.

———. 2007. Earliest fossil record of a pigmy rattlesnake (Viperidae: *Sistrurus* Garman). J. Herpetol. 41: 141–144.

Parrish, H., and R. Hayes. 1970. Hospital management of pit viper venenations. Clin. Toxicol. 3: 501.

Parrish, H. M. 1963. Analysis of 460 fatalities from venomous animals in the United States. Am. J. Med. Sci. 245(2): 12–141.

———. 1964a. Snakebite injuries in Louisiana. J. Louisiana St. Med. Soc. 116: 249–257.

———. 1964b. Texas snakebite statistics. Texas St. J. Med. 60: 592–598.

———. 1966. Incidence of treated snakebites in the United States. Public Health Reprints 81: 269–276.

Parrish, H. M., and L. P. Donovan. 1964. Facts about snakebites in Alabama. J. Med. Assoc. St. Alabama 33: 297–305.

Parrish, H. M., and M. S. Kahn. 1967a. Bites by coral snakes: Report of a case and suggested therapy. J. Am. Med. Assoc. 182: 949.

———. 1967b. Bites by coral snakes: Reports of 11 representative cases. Am. J. Med. Sci. 253: 561–568.

Parrish, H. M., A. W. MacLaurin, and R. L. Tuttle. 1956. North American pit vipers: Bacterial flora of the mouths and venom glands. Virginia Med. Monthly 83: 383–385.

Parrish, H. M., J. C. Goldner, and S. L. Silberg. 1965. Comparison between snakebites in children and adults. Pediatrics 36: 251–156.

Parsons, H., and M. Sarell. 2005. Managing a landscape with rattlesnakes, p. 42. *In* Biology of the Rattlesnakes Symposium (program abstracts). Loma Linda Univ., Loma Linda, California.

Pattabhiraman, T. R., and F. E. Russell. 1973. A lethal protein from the venom of the southern Pacific rattlesnake *Crotalus viridis helleri*. Proc. West. Pharmacol. Soc. 16: 107–110.

Pattabhiraman, T. R., D. C. Buffkin, and F. E. Russell. 1974. Some chemical and pharmacological properties of toxic fractions from the venom of the southern Pacific rattlesnake, *Crotalus viridis helleri*—II. Proc. West. Pharmacol. Soc. 17: 223–227.

Patten, T. J., D. D. Fogell, and J. D. Fawcett. 2005. Conservation of the massasauga (*Sistrurus catenatus*) in southeast Nebraska, p. 42. *In* Biology of the Rattlesnakes Symposium (program abstracts). Loma Linda Univ., Loma Linda, California.

Patterson, R. A. 1967a. Some physiological effects caused by venom from the Gila monster, *Heloderma suspectum*. Toxicon 5: 5–10.

———. 1967b. Smooth muscle stimulating action of venom from the Gila monster, *Heloderma suspectum*. Toxicon 5: 11–15.

Patterson, R. A., and I. S. Lee. 1969. Effects of *Heloderma suspectum* venom on blood coagulation. Toxicon 7: 321–324.

Paulson, D. R. 1967. Searching for sea serpents. Sea Frontiers 13: 244–250.

Payne, P. 1977. Manteo cottonmouth on trail of snake of different color. HERP: Bull. New York Herpetol. Soc. 13(2): 39–40.

Pelton, R. W., and K. W. Carden. 1974. Snake handlers: God-fearers? or, fanatics? Thomas Nelson, Nashville, Tennessee.

Penn, G. H. 1943. Herpetological notes from Cameron Parish, Louisiana. Copeia 1943: 58–59.

Pérez, J. C., and E. E. Sánchez. 1999. Natural protease inhibitors to hemorrhagins in snake venoms and their potential use in medicine. Toxicon 37: 703–728.

Pérez, J. C., R. W. Finberg, and E. E. Sánchez. 2008. Important considerations in developing antivenoms, pp. 581–590. *In* W. K. Hayes, K. R. Beaman, M. D. Cardwell, and S. P. Bush (eds.), The biology of rattlesnakes. Loma Linda Univ. Press, Loma Linda, California.

Perry, J. 1978. An observation of "dance" behavior in the western cottonmouth, *Agkistrodon piscivorus leucostoma* (Reptilia, Serpentes, Viperidae). J. Herpetol. 12: 428–429.

Peters, J. A. 1964. Dictionary of herpetology. Hafner Publ. Co., New York.

Peters, J. A., and B. Orejas-Miranda. 1970. Catalogue of the neotropical Squamata Part I. Snakes. Smithsonian Inst. Press, Washington, D.C.

Peters, U. W. 1979. Second generation breeding of the cantil *Agkistrodon bilineatus* at Taronga Zoo. Intl. Zoo Yearbk. 19: 100–101.

Peterson, A. 1990. Ecology and management of a timber rattlesnake (*Crotalus horridus* L.) population in south-central New York state, pp. 255–261. *In* R. S. Mitchell, C. J. Sheviak, and D. J. Leopold (eds.), Ecosystem management: Rare species and significant habitats. Proc. 15th Ann. Nat. Areas Conf., New York St. Mus. Bull. 471.

Peterson, H. W., and H. M. Smith. 1973. Observations on sea snakes in the vicinity of Acapulco, Guerrero, Mexico. Bull. Chicago Herpetol. Soc. 8: 29.

Peterson, K. H. 1982a. Reproduction in captive *Heloderma suspectum*. Herpetol. Rev. 13: 122–124.

———. 1982b [1981]. Reproduction in captive *Agkistrodon bilineatus taylori* at the Houston Zoo. Zool. Consortium, 5th Ann. Rept. Symp. on Captive Propagation and Husbandry [Oklahoma City Zoo, Oklahoma City, Oklahoma, 11–14 June 1981], pp. 68–69.

———. 1990. Conspecific and self-envenomation in snakes. Bull. Chicago Herpetol. Soc. 25: 26–28.

Petzold, H. G., and W. E. Engelmann. 1975. Kleinere mitteilungen: Über zwei weitere fälle von kannibalismus bei schlangen. Zool. Garten N. F., Jena 45(4/6): 513–520.

Pfennig, D. W., W. R. Harcombe, and K. S. Pfennig. 2001. Frequency-dependent Batesian mimicry: Predators avoid look-alikes of venomous snakes only when the real thing is around. Nature (London) 410: 323.

Pfennig, D. W., G. R. Harper, Jr., A. F. Brumo, W. R. Harcombe, and K. S. Pfennig. 2007. Population differences in predation on Batesian mimics in allopatry with their model: Selection against mimics is strongest when they are common. Behav. Ecol. Sociobiol. 61: 505–511.

Phelps, T. 1981. Poisonous snakes. Blandford Press, Poole, Dorset, United Kingdom.

Phillips, C., M. J. Dreslik, T. G. Anton, D. Mauger, T. P. Wilson, D. B. Shepard, A. Resetar, and B. C. Jellen. 2004. Tracking the continued decline of the eastern massasauga, *Sistrurus c. catenatus,* in Illinois, p. 22. *In* Proceedings of the Snake Ecology Group 2004 Conference, Jackson County, Illinois.

Philpot, V. B., Jr. 1954. Neutralization of snake venom in vitro by serum from the nonvenomous Japanese snake *Elaphe quadrivirgata*. Herpetologica 10: 158–160.

Phisalix, M. 1912. Structure et travail sécrétoire de la glande venimeuse de *l'Heloderma suspectum*. Bull. Mus. Nat. Hist. (Paris) 1912: 184–190.

———. 1917. L'appareil venimeux et le venin de *l'Heloderma suspectum* Cope. J. Physiol. Pathol. 17: 15–43.

Piacentine, J., S. C. Curry, and P. J. Ryan. 1986. Life-threatening anaphylaxis following Gila monster bite. Ann. Emerg. Med. 15: 959–961.

Pianka, E. R. 1967. Lizard species diversity. Ecology 48: 333–351.

Pianka, E. R., and L. J. Vitt. 2003. Lizards: Windows to the evolution of diversity. Univ. California Press, Berkeley.

Picado, T. C. 1931. Epidermal microornaments of the Crotalinae. Bull. Antiv. Inst. Am. 4: 104–105.

Pichon-Prum, N., O. Rolland, and L. Debourcieu. 1990. Apport de l'electrofocalisation à la systematique du genre *Agkistrodon* (Crotalinae). Biochem. Syst. Ecol. 18(4): 281–286.

Pickwell, G. V. 1971. Knotting and coiling behavior in the pelagic sea snake *Pelamis platurus* (L.). Copeia 1971: 348–350.

———. 1972. The venomous sea snakes. Fauna 1(4): 17–32.

———. 1978. Comparative immunology of sea snake venoms, p. 105. *In* P. Rosenberg (ed.), Toxins: Animal, plant and microbial. Pergamon Press, Oxford.

Pickwell, G. V., and W. A. Culotta. 1980. *Pelamis, Pelamis platurus*. Cat. Am. Amphib. Rept. 255: 1–4.

Pickwell, G. V., J. A. Vick, W. H. Shipman, and M. M. Grenan. 1972. Production, toxicity, and preliminary pharmacology of venom from the sea snake, *Pelamis platurus*: With observations on its probable threat to man along middle America, pp. 247–265. *In* L. R. Worthen (ed.), Proc. 3rd Food-drugs from the sea conference. Marine Technology Society, Washington, D.C.

Pickwell, G. V., R. L. Bezy, and J. E. Fitch. 1983. Northern occurrences of the sea snake, *Pelamis platurus*, in the eastern Pacific, with a record of predation on the species. California Fish Game 69: 172–177.

Picolo, G., R. Giorgi, M. M. Bernardi, and Y. Cury. 1998. The antinociceptive effect of *Crotalus durissus terrificus* snake venom is mainly due to a supraspinally integrated response. Toxicon 36: 223–227.

Pilgrim, M. A. 2007. Expression of maternal isotopes in offspring: Implications for interpreting ontogenetic shifts in isotopic composition of consumer tissues. Isotopes in Environmental and Health Studies 43: 155–163.

Pinney, R. 1981. The snake book. Doubleday, Garden City, New York.

———. 1994. Sea snakes. Reptile & Amphibian Mag. January–February 1994: 21–24, 26–28, 30–32, 34–35.

Pinou, T., S. Vicario, M. Marschner, and A. Caccone. 2004. Relict snakes of North America and their relationships within Caenophidia, using likelihood-based Bayesian methods on mitochondrial sequences. Mol. Phylogenet. Evol. 32: 563–574.

Pirkle, H., and K. Stocker. 1991. Thrombin-like enzymes from snake venoms: An inventory. [For the Subcommittee on Nomenclature of Exogenous Hemostatic Factors of the Scientific and Standardization Committee of the International Society on Thrombosis and Haemostasis.] Thromb. Haemostas. 65: 444–450.

Pisani, G. R., and H. S. Fitch. 1993. A survey of Oklahoma's rattlesnake roundups. Kansas Herpetol. Soc. Newsl. 92: 7–15.

Placentine, J., S. C. Curry, and P. J. Ryan. 1986. Life-threatening anaphylaxis following Gila monster bite. Ann. Emerg. Med. 15: 959–961.

Platt, D. R., and T. R. Rainwater. 2000. *Agkistrodon piscivorus* (cottonmouth): Diet. Herpetol. Rev. 31: 244.

Platt, S. G., K. R. Russell, W. E. Snyder, L. W. Fontenot, and S. Miller. 1999. Distribution and conservation status of selected amphibians and reptiles in the piedmont of South Carolina. J. Elisha Mitchell Sci. Soc. 115(1): 8–19.

Plowman, D. M., T. L. Reynolds, and S. M. Joyce. 1995. Poisonous snakebite in Utah. West. J. Med. 163: 547–551.

Pokriefka, R. A., C. A. Weatherby, A. F. Ognjan, F. A. Paul, and R. E. Amenta. 1993. Handbook of antimicrobial therapy for reptiles and amphibians. Herpetol. Ichthyol. Infect. Dis. Assoc., Detroit, Michigan.

Pollard, C. B., A. F. Novak, R. W. Harmon, and W. H. Runzler. 1952. A study of the toxicity and stability of dried moccasin (*Agkistrodon piscivorus*) venom. Quart. J. Florida Acad. Sci. 15: 162–164.

Pook, C. E., and R. McEwing. 2005. Mitochondrial DNA sequences from dried snake venom: A DNA barcoding approach to the identification of venom samples. Toxicon 46: 711–715.

Pope, C. H. 1937. Snakes alive and how they live. Viking Press, New York.

———. 1944. The poisonous snakes of the New World. New York Zool. Soc., New York.

———. 1955. The reptile world. Alfred A. Knopf, New York.

———. 1958. Fatal bite of captive African rear-fanged snake (*Dispholidus*). Copeia 1958: 280–282.

Porter, K. R. 1972. Herpetology. Saunders, Philadelphia.

Possani, L. D., A. C. Alagon, P. L. Fletcher, M. J. Varela, and J. Z. Juliá. 1979. Purification and characterization of a phospholipase A_2 from the venom of the coral snake, *Micrurus fulvius microgalbineus* (Brown and Smith). Biochem. J. 179: 603–606.

Possani, L. D., B. P. Sosa, A. C. Alagón, and P. M. Burchfield. 1980. The venom from the snakes *Agkistrodon bilineatus taylori* and *Crotalus durissus totonacus*: Lethality, biochemical and immunological properties. Toxicon 18: 356–360.

Possardt, E. E., W. H. Martin, W. S. Brown, and J. Sealy. 2005. A range wide action plan for the timber rattlesnake (*Crotalus horridus*): Hope for actual conservation progress or more paper? p. 43. *In* Biology of the Rattlesnakes Symposium (program abstracts). Loma Linda Univ., Loma Linda, California.

Pough, F. H. 1988. Mimicry and related phenomena, pp. 153–234. *In* C. Gans and R. B. Huey (eds.), Biology of the Reptilia, vol. 16: Ecology B. Defense and life history. Alan R. Liss, New York.

Pough, F. H., and H. B. Lillywhite. 1984. Blood volume and blood oxygen capacity of sea snakes. Physiol. Zool. 57: 32–39.

Pough, F. H., R. M. Andrews, J. E. Cadle, M. L. Crump, A. H. Savitzky, and K. D. Wells. 2004. Herpetology. 3rd ed. Pearson/Prentice Hall, Upper Saddle River, New Jersey.

Poulin, S., and C. S. Ivanyi. 2003. A technique for manual restraint of helodermatid lizards. Herpetol. Rev. 34: 43.

Pousa, R. H. 1928. Report of four cases of rattlesnake bite. Bull. Antiv. Inst. Am. 2: 61–63.

Powell, R., J. T. Collins, and E. D. Hooper, Jr. 1998. A key to amphibians & reptiles of the continental United States and Canada. Univ. Press Kansas, Lawrence.

Powell, R. L., and C. S. Lieb. 2008. Perspective on venom evolution in *Crotalus,* pp. 551–556. *In* W. K. Hayes, K. R. Beaman, M. D. Cardwell, and S. P. Bush (eds.), The biology of rattlesnakes. Loma Linda Univ. Press, Loma Linda, California.

Powell, R. L., E. E. Sánchez, and J. C. Pérez. 2006. Farming for venom: Survey of snake venom extraction facilities worldwide. Appl. Herpetol. 3: 1–10.

Powell, R. L., C. S. Lieb, and E. D. Rael. 2008. Geographic distribution of Mojave toxin and Mojave toxin subunits among selected *Crotalus* species, pp. 537–550. *In* W. K. Hayes, K. R. Beaman, M. D. Cardwell, and S. P. Bush (eds.), The biology of rattlesnakes. Loma Linda Univ. Press, Loma Linda, California.

Powers, A. 1973. A review of the purpose of the rattle in crotalids as a defensive diversionary mechanism. Bull. Maryland Herpetol. Soc. 9: 30–32.

Pregill, G. K., J. A. Gauthier, and H. W. Greene. 1986. The evolution of helodermatid squamates, with description of a new taxon and an overview of Varanoidea. Trans. San Diego Soc. Nat. Hist. 21: 167–202.

Preston, C. A. 1989. Hypotension, myocardial infarction, and coagulopathy following Gila monster bite. J. Emerg. Med. 7: 37–40.
Preston, R. E. 1979. Late Pleistocene cold-blooded vertebrate fauna from the mid-continental United States. I. Reptilia; Testudines, Crocodilia. Univ. Michigan Mus. Paleontol. Pap. Paleontol. 19: 1–53.
Price, A. H. 1998. Poisonous snakes of Texas. Texas Parks Wildl. Press, Austin.
Price, J. A., III, and C. G. Sanny. 2007. CroFab™ total anti-venom activity measured by SE-HPLC, and anti-PLA_2 activity assayed *in vitro* at physiological pH. Toxicon 49: 848–854.
Price, M. S., and T. J. LaDuc. 2009. *Agkistrodon piscivorus leucostoma* (western cottonmouth): Diet. Herpetol. Rev. 40: 93.
Price, R. 1989. A unified microdermatoglyphic analysis of the genus *Agkistrodon*. The Snake 21: 90–100.
Price, R., and P. Kelly. 1989. Microdermatoglyphics: Basal patterns and transition zones. J. Herpetol. 23: 244–261.
Priede, M. 1990. The sea snakes are coming. New Scientist 128(1742): 29–33.
Printiss, D. J. 1994. *Coluber constrictor priapus* (southern black racer). Prey. Herpetol. Rev. 25: 70.
Prior, K. A., and P. J. Weatherhead. 1994. Response of free-ranging eastern massasauga rattlesnakes to human disturbance. J. Herpetol. 28: 255–257.
Pu, X. C., P. T. Wong, and P. Gopalakrishnakone. 1995. A novel analgesic toxin (hannalgesin) from the venom of king cobra (*Ophiophagus hannah*). Toxicon 33: 1425–1431.
Pycraft, W. P. 1925. Camouflage in nature. 2nd ed. Hutchinson and Co., London.
Quan, A. N., D. Quan, and S. C. Curry. 2007. Improving CroFab reconstitution times. *In* Venom week 2007 (program abstracts). Univ. Arizona Colleges of Medicine, Arizona Health Sci. Center, Off. Continuing Medical Educ., Tucson, Arizona.
Quinn, H. R. 1979a. Reproduction and growth of the Texas coral snake (*Micrurus fulvius tenere*). Copeia 1979: 453–463.
———. 1979b. Sexual dimorphism in tail pattern of Oklahoma snakes. Texas J. Sci. 31: 157–160.
Quintero-Diaz, G., J. Vazquez-Diaz, and H. M. Smith. 2000. *Micrurus distans zweifeli* (Zweifel's coral snake). Herpetol. Rev. 31: 114.
Rabalias, F. C. 1969. A checklist of trematodes from Louisiana snakes. Proc. Louisiana Acad. Sci. 32: 18–20.
Rabatsky, A. M. 2008. Caudal luring as a precursor in the evolution of the rattle: A test using an ancestral rattlesnake, *Sistrurus miliarius barbouri*, pp. 143–154. *In* W. K. Hayes, K. R. Beaman, M. D. Cardwell, and S. P. Bush (eds.), The biology of rattlesnakes. Loma Linda Univ. Press, Loma Linda, California.
Rabatsky, A. M., and T. M. Farrell. 1996. The effects of age and light level on foraging posture and frequency of caudal luring in the rattlesnake, *Sistrurus miliarius barbouri*. J. Herpetol. 30: 558–561.
Rabatsky, A. M., and J. M. Waterman. 2005. Ontogenetic shifts and sex differences in caudal luring in the dusky pygmy rattlesnake, *Sistrurus miliarius barbouri*. Herpetologica 61: 87–91.
Rafinesque, C. S. 1818. Further account of discoveries in natural history in the western states. Am. Month. Mag. Crit. Rev. 4: 39–42.
Rage, J. C. 1984. Serpentes. Handbuch der Palaeoherpetologie. Part II. Gustav Fischer, Stuttgart.
Railliet, A. 1899. Evolution sans hétérogonie d'un Angiostome de la Couleuvre a collier. Compt. Rend. Acad. Sci. Paris 129: 1271–1273.
Rainwater, T. R., K. D. Reynolds, J. E. Cañas, G. P. Cobb, T. A. Anderson, S. T. McMurry, and P. N. Smith. 2005. Organochlorine pesticides and mercury in cottonmouths (*Agkistrodon piscivorus*) from northeastern Texas, USA. Environ. Toxicol. Chem. 24: 665–673.
Rainwater, T. R., T. A. Anderson, S. G. Platt, and P. N. Smith. 2006. *Agkistrodon piscivorus leucostoma* (western cottonmouth): Diet. Herpetol. Rev. 37: 228.
Rainwater, T. R., J. L. Waldron, S. M. Welch, and S. G. Platt. 2007. *Agkistrodon piscivorus piscivorus* (eastern cottonmouth): Predation. Herpetol. Rev. 38: 202.
Raloff, J. 2005. Plants take bite out of deadly snake venoms. Sci. News 167: 206.

Ramírez, M. S., E. E. Sánchez, C. García-Prieto, J. C. Pérez, G. R. Chapa, M. R. McKeller, R. Ramírez, and Y. D. Anda. 1999. Screening for fibrinolytic activity in eight viperid venoms. Comp. Biochem. Physiol. 124C: 91–98.

Ramírez-Bautista, A. 1994. Manual y claves illustradas de los amphibios y reptiles de la región de Chamela, Jalisco, México. Universidad Nacional Autónoma de México, México, D. F.

Ramírez-Velázques, A., and C. A. Guichard-Romero. 1989. El escorpión negro: Combates ritualizados (Combates ritualizados del escropión negro, *Heloderma horridum alvarezi*, en cautividad). Publicación del Instituto de Historia Natural, Tuxtla Gutiérrez, Chiapas, México.

Ramos, O. H. P., and H. S. Selistre-de-Araujo. 2006. Snake venom metalloproteases—structure and function of catalytic and disintegrin domains. Comp. Biochem. Physiol. 142C: 328–346.

Ramsey, H. W., G. K. Snyder, H. Kitchen, and W. J. Taylor. 1972. Fractionation of coral snake venom: Preliminary studies on the separation and characterization of the protein fractions. Toxicon 10: 67–72.

Ramsey, L. W. 1948. Combat dance and range extension of *Agkistrodon piscivorus leucostoma*. Herpetologica 4: 228.

Rasmussen, A. R. 1997. Systematics of sea snakes: A critical review, pp. 15–30. *In* R. S. Thorpe, W. Wüster, and A. Malhotra (eds.), Venomous snakes: Ecology, evolution and snakebite. Clarendon Press, Oxford.

———. 2002. Phylogenetic analysis of the "true" aquatic elapid snakes Hydrophiinae (sensu Smith et al., 1977) indicates two independent radiations into water. Steenstrupia 27(1): 47–63.

Raufman, J. P. 1996. Bioactive peptides from lizard venoms. Regulatory-peptides 61: 1–18.

Raun, G. G. 1965. A guide to Texas snakes. Texas Mem. Mus., Mus. Notes (9): 1–85.

Raun, G. G., and F. R. Gehlbach. 1972. Amphibians and reptiles in Texas. Dallas Mus. Natur. Hist. Bull. (2): 1–61.

Raymond, L. R. 2004. Reproduction of the western pygmy rattlesnake (*Sistrurus miliarius streckeri* Gloyd) under captive conditions in northwestern Louisiana. Proc. Louisiana Acad. Sci. 66: 29–34.

Reams, R. D., C. J. Franklin, and J. M. Davis. 1999. *Micrurus fulvius tener* (Texas coral snake). Diet. Herpetol. Rev. 30: 228– 229.

Reed, R. N. 2003. Interspecific patterns of species richness, geographic range size, and body size among New World venomous snakes. Ecography 26: 107–117.

Reese, A. M. 1947. The hemipenes of copperhead embryos. Herpetologica 3: 206–208.

Rehling, G. C. 2002. Venom expenditure in multiple bites by rattlesnakes and cottonmouths. Master's thesis, Loma Linda Univ., Loma Linda, California.

Reichling, S. B. 2008. Reptiles and amphibians of the southern pine woods. Univ. Press Florida, Gainesville.

Reid, H. A. 1956a. Sea-snake bites. British Med. J. (4984): 73–78.

———. 1956b. Sea-snake bite research. Trans. Roy. Soc. Trop. Med. Hyg. 50(1): 517–542.

———. 1980a. Defibrination syndromes in snake bite. The Snake 12: 160.

———. 1980b. Antivenom treatment of sea snake bites. The Snake 12: 176.

Reid, H. A., and R. D. G. Theakston. 1983. The management of snake bite. Bull. W.H.O. 61: 885–895.

Reid, M., and A. Nichols. 1970. Predation by reptiles on the periodic cicada. Bull. Maryland Herpetol. Soc. 6: 57.

Reinert, H. K. 1978. The ecology and morphological variation of the massasauga rattlesnake (*Sistrurus catenatus*). Master's thesis, Clarion State College, Clarion, Pennsylvania.

———. 1981. Reproduction by the massasauga (*Sistrurus catenatus catenatus*). Am. Midl. Nat. 105: 393–395.

———. 1984a. Habitat separation between sympatric snake populations. Ecology 65: 478–486.

———. 1984b. Habitat variation within sympatric snake populations. Ecology 65: 1673–1682.

———. 1985. Timber rattlesnake, *Crotalus horridus* Linnaeus, pp. 282–285. *In* H. Genoways and F. J. Brenner (eds.), Species of special concern in Pennsylvania. Spec. Publ. Carnegie Mus. Nat. Hist. 11: 1–430.

———. 1990. A profile and impact assessment of organized rattlesnake hunts in Pennsylvania. J. Pennsylvania Acad. Sci. 64: 136–144.

———. 1992. Radiotelometric field studies of pitvipers: Data acquisition and analysis, pp. 185–198. *In* J. A. Campbell and E. D. Brodie, Jr. (eds.), Biology of the pitvipers. Selva, Tyler, Texas.

Reinert, H. K., and W. R. Kodrich. 1982. Movements and habitat utilization by the massasauga, *Sistrurus catenatus catenatus*. J. Herpetol. 16: 162–171.

Reiserer, R. S. 2002. Stimulus control of caudal luring and other feeding responses: A program for research on visual perception in vipers, pp. 361–383. *In* G. W. Schuett, M. Höggren, M. E. Douglas, and H. W. Greene (eds.), Biology of the vipers. Eagle Mountain Publ., Eagle Mountain, Utah.

Repp, R. 1998a. Wintertime observations on five species of reptiles in the Tucson area: Sheltersite selections/Fidelity to sheltersites/Notes on behavior. Bull. Chicago Herpetol. Soc. 33: 49–56.

———. 1998b. Getting hammered: Testosterone + alcohol = snakebite statistic. Sonoran Herpetol. 11: 30–33.

Reyes-Velasco, J., C. I. Grünwald, J. M. Jones, and G. N. Weatherman. 2008. *Crotalus willardi meridionalis* (southern ridge-nosed rattlesnake). Herpetol. Rev. 39: 485.

Reyes-Velasco, J., I. A. Hermosillo-Lopez, C. I. Grünwald, and O. A. Avila-Lopez. 2009. New state records for amphibians and reptiles from Colima, Mexico. Herpetol. Rev. 40: 117–120.

Reynolds, R. P., and G. V. Pickwell. 1984. Records of the yellow-bellied sea snake, *Pelamis platurus*, from the Galápagos Islands. Copeia 1984: 786–789.

Richards, G. M., G. Du Vair, and M. Laskowski, Sr. 1965. Comparison of the levels of phosphodiesterase, endonuclease, and monophosphatases in several snake venoms. Biochemistry 4: 501–503.

Rijst, H. V. D. 1990. Colourful misunderstandings. Litt. Serpent. Engl. Ed. 10: 206–211.

Riley, J., and J. T. Self. 1979. On the systematics of the pentastomid genus *Porocephalus* (Humboldt, 1811) with descriptions of two new species. System. Parasitol. 1: 25–42.

Roberts, A. R., and J. Quarters. 1947. *Sistrurus* in Michigan. Herpetologica 4: 6.

Roberts, L. S. 1956. *Ophiotaenia grandis* LaRue (Cestoda: Proteocephalidae) in McCurtain County, Oklahoma. J. Parasitol. 42: 20.

Roberts, R. S., T. A. Csencsitz, and C. W. Heard. 1985. Upper extremity compartment syndromes following pit viper envenomation. Clinical Orthopaedics and Related Research 193: 184–188.

Rodas, J. T. 1938. Contribución al studio de las serpientes venenosas de Guatemala. Tipogr. Nac., Guatemala City.

Rodrigo, C., and G. de Souza. 2006. A simple restraining device for venomous snakes. Bull. Chicago Herpetol. Soc. 41: 183–184.

Rodríguez-Acosta, A., S. Magaldi, J. C. Pérez, E. E. Sánchez, M. E. Girón, and I. Aguilar. 2007. A protein homology detection between rattlesnakes (Viperidae: Crotalinae) from South and North America deduced from antigenically related metalloproteases. Anim. Biol. 57: 401–407.

Roelke, C. E., and M. J. Childress. 2007. Defensive and infrared reception responses of true vipers, pitvipers, *Azemiops* and colubrids. J. Zool. (London) 273: 421–425.

Rogers, K. L. 1976. Herpetofauna of the Beck Ranch local fauna (upper Pliocene: Blancan) of Texas. Publ. Mus. Michigan St. Univ. Paleontol. Ser. 1: 167–200.

———. 1982. Herpetofaunas of Courland Canal and Hall Ash local faunas (Pleistocene: early Kansas) of Jewell Co., Kansas. J. Herpetol. 16: 174–177.

———. 1984. Herpetofauna of the Big Springs and Hornet's Nest quarries (northeastern Nebraska, Pleistocene: late Blancan). Trans. Nebraska Acad. Sci. 12: 81–94.

Rogers, L. 1903. On the physiological action of the poison of the Hydrophidae. Proc. Royal Soc. London 71: 481–496.

Roller, J. A. 1977. Gila monster bite: A case report. Clin. Toxicol. 10(4): 423–427.

Romer, A. S. 1956. Osteology of the reptiles. Univ. Chicago Press, Chicago.

Rosado-López, L., and F. A. Laviada-Arrigunaga. 1977. Accidente ofídico. Communicación de 38 casos. Presna Méd. Méx. 42: 409–412.

Rose, W. 1950. The reptiles and amphibians of southern Africa. Maskew Miller, Cape Town, South Africa.

Rosenberg, P. 1979. Pharmacology of phospholipase A_2 from snake venoms, pp. 403–447. *In* C. Y. Lee (ed.), Snake venoms: Handbook of experimental pharmacology, vol. 52. Springer-Verlag, Berlin.

———. 1987. The relationship between enzymatic activity and pharmacological properties of phospholipases in natural poisons, pp. 129–184. *In* J. B. Harris (ed.), Natural toxins: Animal, plant and microbial. Oxford Univ. Press, Oxford.

———. 1988. Phospholipase A_2 toxins, p. 27. *In* J. O. Dolly (ed.), Neurotoxins in neurochemistry, Ellis Horwood, Chichester.

———. 1990. Phospholipases, pp. 667–677. *In* W. T. Shier and D. Mebs. (eds.), Handbook of toxinology. Marcel Dekker, New York.

Rosenfeld, G., L. Nahas, and E. M. A. Kelen. 1968. Coagulation and hemolysis by snake venoms, pp. 229–273. *In* W. Bücherl, E. Buckley, and V. Deulofeu (eds.), Venomous animals and their venoms, vol. 1: Venomous vertebrates. Academic Press, New York.

Ross, D. A. 1989. Amphibians and reptiles in the diets of North American raptors. Wisconsin Dept. Nat. Resour. Endang. Resour. Rep. 59: 1–33.

Rossi, J. V. 1992. Snakes of the United States and Canada: Keeping them healthy in captivity, vol. I: Eastern area. Krieger, Malabar, Florida.

Rossi, J. V., and R. Rossi. 1995. Snakes of the United States and Canada, vol. 2: Western area. Krieger, Malabar, Florida.

Roth, E. D. 2003. "Handedness" in snakes? Lateralization of coiling behavior in a cottonmouth, *Agkistrodon piscivorus leucostoma*, population. Anim. Behav. 66: 337–341.

———. 2005. Spatial ecology of a cottonmouth (*Agkistrodon piscivorus*) population in east Texas. J. Herpetol. 39: 308–312.

Roth, E. D., and J. A. Johnson. 2004. Size-based variation in antipredator behavior within a snake (*Agkistrodon piscivorus*) population. Behav. Ecol. 15(2): 365–370.

Roth, E. D., P. G. May, and T. M. Farrell. 1999. Pigmy rattlesnakes use frog-derived chemical cues to select foraging sites. Copeia 1999: 772–774.

Roth, E. D., W. D. W. Ginn, L. J. Vitt, and W. I. Lutterschmidt. 2003. *Agkistrodon piscivorus leucostoma* (western cottonmouth): Diet. Herpetol. Rev. 34: 60.

Roth, E. D., W. I. Lutterschmidt, and D. A. Wilson. 2006. Relative medial and dorsal cortex volume in relation to sex differences in spatial ecology of a snake population. Brain Behav. Evol. 67: 103–110.

Rowe, M. P., T. M. Farrell, and P. M. May. 2002. Rattle loss in pygmy rattlesnakes (*Sistrurus miliarius*): Causes, consequences, and implications for rattle function and evolution, pp. 385–404. *In* G. W. Schuett, M. Höggren, M. E. Douglas, and H. W. Greene (eds.), Biology of the vipers. Eagle Mountain Publ., Eagle Mountain, Utah.

Roze, J. 1967. A check list of the New World venomous coral snakes (Elapidae), with descriptions of new forms. Am. Mus. Novit. 2287: 1–60.

———. 1982. New World coral snakes (Elapidae): A taxonomic and biological summary. Mem. Inst. Butantan 46: 305–338.

———. 1974. *Micruroides, M. euryxanthus*. Cat. Am. Amphib. Rept. 163: 1–4.

———. 1996. Coral snakes of the Americas: Biology, identification, and venoms. Krieger, Malabar, Florida.

Roze, J. A., and G. M. Tilger. 1983. *Micrurus fulvius*. Cat. Am. Amphib. Rept. 316: 1–4.

Ruben, J. A., and C. Geddes. 1983. Some morphological correlates of striking in snakes. Copeia 1983: 221–225.

Rubinoff, I., and C. Kropach. 1970. Differential reactions of Atlantic and Pacific predators to sea snake. Nature (London) 228: 1288–1290.

Rubinoff, I., J. B. Graham, and J. Motta. 1986. Diving of the sea snake *Pelamis platurus* in the Gulf of Panamá. I. Dive depth and duration. Marine Biol. 91: 181–191.

Rubio, M., G. Greer, and R. T. Bryant. 2003. New herpetofaunal county records for Georgia. Herpetol. Rev. 34: 78–80.

Ruff, J. D., B. D. Johnson, and D. H. Sifford. 1980. A glycoprotein proteinase in *Agkistrodon bilineatus* venom. Proc. Arkansas Acad. Sci. 34: 130–131.

Ruick, J. D., Jr., 1948. Collecting coral snakes, *Micrurus fulvius tenere*, in Texas. Herpetologica 4: 215–216.

Ruiz, J. M. 1952. Sobre a distinção genérica dos Crotalidae (Ophidia: Crotaloidea) baseada em alguns caracteres osteológicos. (Nota preliminar). Mem. Inst. Butantan 23: 109–113.

Rundquist, E. M., and J. Triplett. 1993. Additional specimens of the western cottonmouth (*Agkistrodon piscivorus leucostoma*, Reptilia: Squamata) from Kansas. Trans. Kansas Acad. Sci. 96: 148–151.

Russell, F. E. 1967a. Bites by the Sonoran coral snake, *Micruroides euryxanthus*. Toxicon 5: 39–42.

———. 1967b. Pharmacology of animal venoms. Clin. Pharmacol. Therapeu. 8: 849–873.

———. 1969. Clinical aspects of snake venom poisoning in North America. Toxicon 7: 33–37.

———. 1980a. Snake venom poisoning in the United States. Annual Rev. Med. 31: 247–259.

———. 1980b. Snake venom poisoning. Lippincott, Philadelphia.

———. 1983. Snake venom poisoning. Scholium International, Great Neck, New York.

———. 1984. Snake venoms, pp. 469–480. *In* M. W. J. Ferguson (ed.), The structure, development and evolution of reptiles. A Festschrift in honour of Professor A. d'A. Bellairs on occasion of his retirement. Symp. Zool. Soc. London (52), Academic Press, London.

Russell, F. E., and C. M. Bogert. 1981. Gila monster: Its biology, venom and bite—a review. Toxicon 19: 341–359.

Russell, F. E., and A. F. Brodie. 1974. Venoms of reptiles, pp. 449–478. *In* M. Florkin and B. T. Scheer (eds.), Chemical zoology, vol. 9: Amphibia and Reptilia. Academic Press, New York.

Russell, F. E., and J. A. Emery. 1959. Use of the chick in zootoxicologic studies on venoms. Copeia 1959: 73–74.

———. 1961. Incision and suction following the injection of rattlesnake venom. Am. J. Med. Sci. 241: 160–166.

Russell, F. E., and L. Lauritzen. 1966. Antivenins. Trans. Roy. Soc. Trop. Med. Hyg. 60: 797–810.

Russell, F. E., and A. L. Picchioni. 1983. Snake venom poisoning. Clin. Toxicol. Consultant 5: 73–87.

Russell, F. E., and H. W. Puffer. 1970. Pharmacology of snake venoms. Clin. Toxicol. 3: 433–444.

———. 1971. Pharmacology of snake venoms, pp. 87–98. *In* S. A. Minton (ed.), Snake venoms and envenomation. Marcel Dekker, New York.

Russell, F. E., and R. S. Scharffenberg. 1964. Bibliography of snake venoms and venomous snakes. Bibliographic Associates, West Covina, California.

Russell, F. E., J. A. Emery, and T. E. Long. 1960. Some properties of rattlesnake venom following 26 years storage. Proc. Soc. Exp. Biol. Med. 103: 737–739.

Russell, F. E., J. Strassberg, and F. W. Buess. 1962. Zootoxicological properties of venom phosphodiesterase. Fed. Proc. 21: 242.

Russell, F. E., F. W. Buess, and M. Y. Woo. 1963a. Zootoxicological properties of venom phosphodiesterase. Toxicon 1: 99–108.

Russell, F. E., F. W. Buess, M. Y. Woo, and R. Eventov. 1963b. Zootoxicological properties of venom L-amino acid oxidase. Toxicon 1: 229–230.

Russell, F. E., N. Ružić, and H. Gonzalez. 1973. Effectiveness of antivenin (Crotalidae) polyvalent following injection of *Crotalus* venom. Toxicon 11: 461–464.

Russell, F. E., R. W. Carlson, W. Wainschel, and J. Osborne. 1975. Snake venom poisoning in the United States: Experiences with 550 cases. J. Am. Med. Assoc. 233: 341–344.

Russell, F. E., T. R. Pattabhiraman, D. C. Buffkin, R. W. Carlson, R. C. Schaeffer, H. Whigham, M. D. Fairchild, N. Tamiya, N. Maeda, and R. Tanz. 1976. Some chemical and physiopharmacologic properties of the venoms of the rattlesnakes *Crotalus viridis helleri* and *Crotalus scutulatus scutulatus*. United States–Republic of China Binational Seminar on Protein Chemistry, Snake Venoms and Hormonal Proteins, Taipei, Taiwan, March 1976. Toxicon 14: 417–418.

Russell, F. E., J. B. Sullivan, N. B. Egen, W. S. Jeter, F. S. Markland, W. A. Wingert, and D. Bar-Or. 1985. Preparation of a new antivenin by affinity chromatography. Am. J. Trop. Med. Hyg. 34: 141–150.

Ryerson, D. L. 1949. A preliminary study of reptilian blood. J. Entomol. Zool. 41: 49–55.

Sabath, M., and R. Worthington. 1959. Eggs and young of certain Texas reptiles. Herpetologica 15: 31–32.

Saint Girons, H. 1982. Reproductive cycles of male snakes and their relationships with climate and female reproductive cycles. Herpetologica 38: 5–16.

Samejima, Y., Y. Aoki, and D. Mebs. 1991. Amino acid sequence of a myotoxin from venom of the eastern diamondback rattlesnake (*Crotalus adamanteus*). Toxicon 29: 461–468.

Sánchez, E. E., and J. C. Pérez. 2006. Inhibition of two North American coral snake venoms by the United States and Mexican coral snake antivenoms, p. 41. *In* S. A. Seifert (ed.), Snakebites in the new millennium: A state-of-the-art symposium, University of Nebraska Medical Center, 21–23 October 2005, Omaha, Nebraska. J. Med. Toxicol. 2: 29–45.

Sánchez, E. E., M. S. Ramírez, J. A. Galán, G. López, A. Rodríguez-Acosta, and J. C. Pérez. 2003a. Cross reactivity of three antivenoms against North American snake venoms. Toxicon 41: 315–320.

Sánchez, E. E., J. A. Galán, J. C. Perez, A. Rodríguez-Acosta, P. B. Chase, and J. C. Pérez. 2003b. The efficacy of two antivenoms against the venom of North American snakes. Toxicon 41: 357–365.

Sánchez, E. E., J. A. Galán, and J. C. Pérez. 2006. Inhibition of lung tumor formation in BALB/c mice treated with disintegrins isolated from *Crotalus atrox* (western diamondback rattlesnake) venom, pp. 41–42. *In* S. A. Seifert (ed.), Snakebites in the new millennium: A state-of-the-art symposium, University of Nebraska Medical Center, 21–23 October 2005, Omaha, Nebraska. J. Med. Toxicol. 2: 29–45.

Sánchez, E. E., J. C. Lopez-Johnston, A. Rodríguez-Acosta, and J. C. Pérez. 2008. Neutralization of two North American coral snake venoms with United States and Mexican antivenoms. Toxicon 51: 297–303.

Sanders, J. S., and J. S. Jacob. 1981. Thermal ecology of the copperhead (*Agkistrodon contortrix*). Herpetologica 37: 264–270.

Sanders, L. 2009. Venom hunters: Scientists probe toxins, revealing the healing powers of biochemical weapons. Science News 176(4): 16–20.

Sant´ Ana, C. D., F. K. Ticli, L. L. Oliveira, J. R. Giglio, C. G. V. Rechia, A. L. Fuly, H. S. Selistre de Araúja, J. J. Franco, R. G. Stabeli, A. M. Soares, and S. V. Sampaio. 2008. BjussuSP-I: A new thrombin-like enzyme isolated from *Bothrops jararacussu* snake venom. Comp. Biochem. Physiol. 151A: 443–454.

Sanz, L., H. L. Gibbs, S. P. Mackessy, and J. J. Calvete. 2006. Venom proteomes of closely related *Sistrurus* rattlesnakes with divergent diets. J. Proteome Res. 5: 2098–2122.

Sargant, W. 1949. Some cultural group abreactive techniques and their relation to modern treatments. Proc. Roy. Soc. Med. 42: 367–374.

Sasaki, K., A. J. Place, and K. N. Gaylor. 2008. Attitudes toward rattlesnakes by the peoples of North America and implications for rattlesnake conservation, pp. 473–484. *In* W. K. Hayes, K. R. Beaman, M. D. Cardwell, and S. P. Bush (eds.), The biology of rattlesnakes. Loma Linda Univ. Press, Loma Linda, California.

Savage, B., U. M. Marzec, B. H. Chao, L. A. Harker, J. M. Maraganore, and Z. M. Ruggeri. 1990. Binding of the snake venom-derived proteins applaggin and echistatin to argine-glycine-aspartic acid recognition site(s) on platelet glycoprotein IIb·IIIa complex inhibits receptor function. J. Biol. Chem. 20: 11766–11772.

Savage, J. M. 1966. The origins and history of the central American herpetofauna. Copeia 1966: 719–766.

Savage, J. M., and J. B. Slowinski. 1990. A simple consistent terminology for the basic colour patterns of the venomous coral snakes and their mimics. Herpetol. J. 1: 530–532.

———. 1992. The colouration of the venomous coral snakes (family Elapidae) and their mimics (families Aniliidae and Colubridae). Biol. J. Linn. Soc. 45: 235–254.

Savage, T. 1967. The diet of rattlesnakes and copperheads in the Great Smoky Mountains National Park. Copeia 1967: 226–227.

Savitsky, A. H. 1992. Embryonic development of the maxillary and prefrontal bones of crotaline snakes, pp. 119–142. *In* J. A. Campbell and E. D. Brodie, Jr. (eds.), Biology of the pitvipers. Selva, Tyler, Texas.

Savitsky, B. A. C. 1992. Laboratory studies on piscivory in an opportunistic pitviper, the cottonmouth, *Agkistrodon piscivorus*, pp. 347–368. *In* J. A. Campbell and E. D. Brodie, Jr. (eds.), Biology of the pitvipers. Selva, Tyler, Texas.

Say, T. 1823. [Description of *Crotalus tergeminus*], p. 499. *In* E. James (compiler), Account of an expedition from Pittsburgh to the Rocky Mountains, performed in the years 1819 and 1820. Carey and Lea, Philadelphia.

Scalzo, A. J., J. A. Weber, M. W. Thompson, P. B. Johnson, C. M. Blume-Odom, and R. L. Tominack. 2007. Copperhead envenomation: To CroFab™ or not to CroFab™? That is the question. *In* Venom week 2007 (program abstracts). Univ. Arizona Colleges of Medicine, Arizona Health Sci. Center, Off. Continuing Medical Educ., Tucson, Arizona.

Scarborough, R. M., J. W. Rose, M. A. Hsu, D. R. Phillips, V. A. Fried, A. M. Campbell, L. Nannizzi, and I. F. Charo. 1991. Barbourin: A GPHb-IIIa-specific integrin antagonist from the venom of *Sistrurus m. barbouri*. J. Biol. Chem. 266: 9359–9362.

Schad, G. A. 1962. Studies on the genus *Kalicephalus* (Nematoda: Diaphanocephaloidea): II. A taxonomic revision of the genus *Kalicephalus* Molin, 1861. Can. J. Zool. 40: 1035–1165.

Schaefer, N. 1976. The mechanism of venom transfer from the venom duct to the fang in snakes. Herpetologica 32: 71–76.

Schaefer, W. H. 1934. Diagnosis of sex in snakes. Copeia 1934: 181.

Schaeffer, R. C., Jr., T. R. Pattabhiraman, R. W. Carlson, F. E. Russell, and M. H. Weil. 1979. Cardiovascular failure produced by a peptide from the venom of the southern Pacific rattlesnake, *Crotalus viridis helleri*. Toxicon 17: 447–453.

Scharman, E. J., and V. D. Noffsinger. 2001. Copperhead snakebites: Clinical severity of local effects. Ann. Emerg. Med. 38: 55–61.

Scheppegrell, W. 1928. A coral snake record. Bull. Antiv. Inst. Am. 2: 78–79.

Schieck, A., F. Kornalik, and E. Habermann. 1972a. The prothrombin-activating principle from *Echis carinatus* venom: I. Preparation and biochemical properties. Naunyn-Schmiedebergs Arch. Exp. Path. Pharmacol. 272: 402–416.

Schieck, A., E. Habermann, and F. Kornalik. 1972b. The prothrombin-activating principle from *Echis carinatus* venom: II. Coagulation studies in vitro and in vivo. Naunyn-Schmiedebergs Arch. Exp. Path. Pharmacol. 274: 7–17.

Schmidt, C. 2002. A demographical analysis of the prairie rattlesnakes collected for the 2000 and 2001 Sharon Springs, Kansas, rattlesnake roundups. J. Kansas Herpetol. 1: 12–18.

Schmidt, F. L., and W. E. Hubbard. 1940. A new trematode, *Neorenifer serpentis,* from the water moccasin. Am. Midl. Nat. 23: 729–730.

Schmidt, K. P. 1928. Notes on American coral snakes. Bull. Antiv. Inst. Am. 2: 63–64.

———. 1932. Stomach contents of some American coral snakes, with the description of a new species of *Geophis*. Copeia 1932: 6–9.

———. 1953. A check list of North American amphibians and reptiles. 6th ed. Am. Soc. Ichthyol. Herpetol., Chicago.

Schmidt, K. P., and D. D. Davis. 1941. Field book of snakes of the United States and Canada. Putnam, New York.

Schmidt, K. P., and R. F. Inger. 1957. Living reptiles of the world. Hanover House, Garden City, New York.

Schmidt, K. P., and F. A. Shannon. 1947. Notes on amphibians and reptiles of Michoacan, Mexico. Fieldiana: Zool. 31: 63–85.

Schmidt, M. E., Y. Z. Abdelbaki, and A. T. Tu. 1972. Fine structural changes of myelinated nerve associated with copperhead envenomation. Acta Neuropathol. 21: 68–75.

Schmidt, R. S. 1964. Phylogenetic significance of lizard cochlea. Copeia 1964: 542–549.

Schmidt-Nielsen, K., and R. Fange. 1958. Salt glands in marine reptiles. Nature (London) 182: 783–785.

Schneider, K. M. 1941. Über Fettlager im Schwanz der Krusten- (*Heloderma* Wiegm.) und Stutz-Echse (*Trachydosaurus* Gray). Der Zoologische Garten 13: 236–247.

Schoener, T. W. 1977. Competition and the niche, pp. 35–136. *In* C. Gans and D. W. Tinkle (eds.), Biology of the Reptilia, vol. 7. Academic Press, New York.

Schoettler, W. H. A. 1951. On the stability of desiccated snake venoms. J. Immun. 67: 299–304.

Schofer, J. 2005. Population modeling of Arizona black rattlesnakes (*Crotalus oreganus cerberus*) at a den site near Flagstaff, Arizona, p. 47. *In* Biology of the Rattlesnakes Symposium (program abstracts). Loma Linda Univ., Loma Linda, California.

Schuett, G. W. 1982. A copperhead (*Agkistrodon contortrix*) brood produced from autumn copulations. Copeia 1982: 700–702.

———. 1986. Selected topics on reproduction of the copperhead, *Agkistrodon contortrix* (Reptilia, Serpentes, Viperidae). Master's thesis, Central Michigan Univ., Mt. Pleasant.

———. 1992. Is long-term sperm storage an important component of the reproductive biology of temperate pitvipers? pp. 169–184. *In* J. A. Campbell and E. D. Brodie, Jr. (eds.), Biology of the pit vipers. Selva, Tyler, Texas.

———. 1996. Fighting dynamics of male copperheads, *Agkistrodon contortrix* (Serpentes, Viperidae): Stress-induced inhibition of sexual behavior in losers. Zoo Biol. 15: 209–221.

Schuett, G. W., and D. Duvall. 1996. Head lifting by female copperheads, *Agkistrodon contortrix*, during courtship: Potential mate choices. Anim. Behav. 51: 367–373.

Schuett, G. W., and J. C. Gillingham. 1986. Sperm storage and multiple paternity in the copperhead, *Agkistrodon contortrix*. Copeia 1986: 807–811.

———. 1988. Courtship and mating of the copperhead, *Agkistrodon contortrix*. Copeia 1988: 374–381.

———. 1989. Male-male agonistic behavior of the copperhead, *Agkistrodon contortrix*. Amphibia-Reptilia 10: 243–266.

Schuett, G. W., and M. S. Grober. 2000. Post-fight levels of plasma lactate and corticosterone in male copperheads, *Agkistrodon contortrix* (Serpentes, Viperidae): Differences between winners and losers. Physiol. Behav. 71: 335–341.

Schuett, G. W., and F. Kraus. 1982. *Agkistrodon contortrix pictigaster* (Trans-Pecos copperhead). Neonates. Herpetol. Rev. 13: 17.

Schuett, G. W., D. L. Clark, and F. Kraus. 1984. Feeding mimicry in the rattlesnake *Sistrurus catenatus*, with comments on the evolution of the rattle. Anim. Behav. 32(2): 625–626.

Schuett, G. W., H. J. Harlow, J. D. Rose, E. A. Van Kirk, and W. J. Murdoch. 1996. Levels of plasma corticosterone and testosterone in male copperheads (*Agkistrodon contortrix*) following staged fights. Horm. Behav. 30: 60–68.

Schuett, G. W., H. J. Harlow, J. D. Rose, E. A. Van Kirk, and W. J. Murdoch. 1997. Annual cycle of plasma testosterone in male copperheads, *Agkistrodon contortrix* (Serpentes, Viperidae): Relationship to timing of spermatogenesis, mating, and agonistic behavior. Gen. Comp. Endocrinol. 105: 417–424.

Schuett, G. W., M. Höggren, M. E. Douglas, and H. W. Greene (eds.). 2002. Biology of the vipers. Eagle Mountain Publ., Eagle Mountain, Utah.

Schulz, M. 2003. Frequency of sea snake strandings in north-eastern New South Wales. Herpetofauna 33: 109–110.

Schwammer, H. 1983. Herpetologische Beobachtungen aus Colorado/USA. Aquaria (St. Gallen) 30(6): 90–93.

Schwartzwelder, J. 1950. Snake-bite accidents in Louisiana: With data on 306 cases. Am. J. Trop. Med. 30: 575–587.

Schwick, G., and F. Dickgiesser. 1963. Probleme der Antigen- und Fermentanalyse im Zusammenhang mit der Herstellung polyvalenter Schlangengiftseren, pp. 35–66. *In* Die Giftschlangen der Erde. Behringwerk-Mitteilungen, N. G. Elwert, Marburg/Lahn.

Scott, D. E., R. U. Fischer, J. D. Congdon, and S. A. Busa. 1995. Whole body lipid dynamics and reproduction in the eastern cottonmouth, *Agkistrodon piscivorus*. Herpetologica 51: 472–487.

Scott, D. L., A. Achari, M. Zajac, and P. B. Sigler. 1986. Crystallization and preliminary diffraction studies of the Lys-49 phospholipase A_2 from *Agkistrodon piscivorus piscivorus*. J. Biol. Chem. 261: 12337–12338.

Scott, D. L., A. Achari, J. C. Vidal, and P. B. Sigler. 1992. Crystallographic and biochemical studies of the (inactive) Lys-49 phospholipase A_2 from the venom of *Agkistrodon piscivorus piscivorus*. J. Biol. Chem. 267: 22645–22657.

Scrocchi, G. J. 1992. Análisis preliminar de la osteología craneal del género *Micrurus* Wagler (Ophidia: Elapidae). Acta Zool. Iilloana 41: 311–327.

Scudder, K. M., and D. Chiszar. 1977. Effects of six visual stimulus conditions on defensive and exploratory behavior in two species of rattlesnakes. Psychol. Rec. 3: 519–526.

Seelex, S. F. 1963. Interim statement on first-aid therapy for bites by venomous snakes. Toxicon 1: 81–87.

Seibert, H. C. 1965. A snake hibernaculum uncovered in midwinter. J. Ohio Herpetol. Soc. 5: 29.

Seifert, S. A. (ed.). 2006. Snakebites in the new millennium: A state-of-the-art symposium, University of Nebraska Medical Center, 21–23 October 2005, Omaha, Nebraska. J. Med. Toxicol. 2: 29–45.

———. 2007a. NPDS-based characterization of native U.S. elapid and viperid envenomations. *In* Venom week 2007 (program abstracts). Univ. Arizona Colleges of Medicine, Arizona Health Sci. Center, Off. Continuing Medical Educ., Tucson, Arizona.

———. 2007b. Exotic antivenoms in the United States. 2007. *In* Venom week 2007 (program abstracts). Univ. Arizona Colleges of Medicine, Arizona Health Sci. Center, Off. Continuing Medical Educ., Tucson, Arizona.

Seifert, S. A., and L. V. Boyer. 2001. Recurrence phenomena after immunoglobulin therapy for snake envenomations: Part 1. Pharmacokinetics and pharmacodynamics of immunoglobulin antivenoms and related antibodies. Ann. Emerg. Med. 37: 189–195.

Seifert, S. A., L. V. Boyer, R. C. Dart, R. S. Porter, and L. Sjostrom. 1997. Relationship of venom effects to venom antigen and antivenom serum concentrations in a patient with *Crotalus atrox* envenomation treated with a Fab antivenom. Ann. Emerg. Med. 30: 49–53.

Seigel, D. S., and D. M. Sever. 2006. Utero-muscular twisting and sperm storage in viperids. Herpetol. Cons. Bio. 1(2): 87–92.

———. 2008a. Sperm aggregations in female *Agkistrodon piscivorus* (Reptilia: Squamata): A histological and ultrastructural investigation. J. Morphol. 269: 189–206.

———. 2008b. Seasonal variation in the oviduct of female *Agkistrodon piscivorus* (Reptilia: Squamata): An ultrastructural investigation. J. Morphol. 269: 980–997.

Seigel, R. A. 1986. Ecology and conservation of an endangered rattlesnake, *Sistrurus catenatus*, in Missouri, USA. Biol. Conserv. 35: 333–346.

Seigel, R. A., and J. T. Collins (eds.). 1993. Snakes: Ecology & behavior. McGraw-Hill, New York.

——— (eds.). 2002. Snakes: Ecology & behavior. Blackburn Press, Caldwell, New Jersey.

Seigel, R. A., and H. S. Fitch. 1984. Ecological patterns of relative clutch mass in snakes. Oecologia (Berlin) 61: 293–301.

———. 1985. Annual variation in reproduction in snakes in a fluctuating environment. J. Anim. Ecol. 54: 497–505.

Seigel, R. A., and M. A. Pilgrim. 2002. Long-term changes in movement patterns of massasaugas (*Sistrurus catenatus*), pp. 405–412. *In* G. W. Schuett, M. Höggren, M. E. Douglas, and H. W. Greene (eds.), Biology of the vipers. Eagle Mountain Publ., Eagle Mountain, Utah.

Seigel, R. A., and C. A. Sheil. 1999. Population viability analysis: Applications for the conservation of massasaugas, pp. 17–22. *In* B. Johnson and M. Wright (eds.), Second international symposium and workshop on the conservation of the eastern massasauga rattlesnake, *Sistrurus catenatus catenatus*: Population and habitat management issues in urban, bog, prairie and forested ecosystems. Toronto Zoo, Toronto.

Seigel, R. A., L. E. Hunt, J. L. Knight, L. Malaret, and N. L. Zuschlag. 1984. Vertebrate ecology and systematics: A tribute to Henry S. Fitch. Univ. Kansas Mus. Nat. Hist. Spec. Publ. 10: 1–278.

Seigel, R. A., H. S. Fitch, and N. B. Ford. 1986. Variability in relative clutch mass in snakes among and within species. Herpetologica 42: 179–185.

Seigel, R. A., C. A. Sheil, and J. S. Doody. 1998. Changes in a population of an endangered rattlesnake *Sistrurus catenatus* following a severe flood. Biol. Conserv. 83(2): 127–131.

Seigel, R. A., J. T. Collins, and S. S. Novak. 2001. Snakes: Ecology and evolutionary biology. Blackburn Press, Caldwell, New Jersey.

Selistre de Araujo, H. S., S. P. White, and C. L. Ownby. 1996. cDNA cloning and sequence analysis of a lysine-49 phospholipase A_2 myotoxin from *Agkistrodon contortrix laticinctus* snake venom. Arch. Biochem. Biophys. 326: 21–30.

Senanayake, M. P., C. A. Ariaratnam, S. Abeywickrema, and A. Belligaswatte. 2005. Two Sri Lankan cases of identified sea snake bites, without envenoming. Toxicon 45: 861–863.

Senior, K. 1999. Taking the bite out of snake venoms. Lancet 353: 1946–1947.

Serrano, S. M. T., J. Kim, D. Wang, B. Dragulev, J. D. Shannon, H. H. Mann, G. Veit, R. Wagener, M. Koch, and J. W. Fox. 2006. The cysteine-rich domain of snake venom metalloproteinases is a ligand for von Willebrand factor A domains: Role in substrate targeting. J. Biolog. Chem. 281(52): 39746–39756.

Servent, D., V. Winckler-Dietrich, H. Y. Hu, P. Kessler, P. Drevet, D. Bertrand, and A. Menez. 1997. Only snake curaremimetic toxins with a fifth disulfide bond have high affinity for the neuronal alpha-7 nicotinic receptor. J. Biol. Chem. 272: 24279–24286.

Setser, K. 2007. Use of anesthesia increases precision of snake length measurements. Herpetol. Rev. 38: 409–411.

Sever, D. M., D. S. Seigel, A. Bagwill, M. E. Eckstut, L. Alexander, A. Camus, and C. Morgan. 2008. Renal sexual segment of the cottonmouth snake, *Agkistrodon piscivorous* (Reptilia, Squamata, Viperidae). J. Morphol. 269: 640–653.

Sexton, O. J., P. Jacobson, and J. E. Bramble. 1992. Geographic variation in some activities associated with hibernation in Nearctic pitvipers, pp. 337–345. *In* J. A. Campbell and E. D. Brodie, Jr. (eds.), Biology of the pitvipers. Selva, Tyler, Texas.

Seyle, C. W., Jr., and G. K. Williamson. 1982. *Gopherus polyphemus* (gopher tortoise): Burrow associates. Herpetol. Rev. 13: 48.

Seymour, R. S. 1974. How sea snakes may avoid the bends. Nature (London) 250: 489–490.

Seymour, R. S., R. G. Spragg, and M. T. Hartman. 1981. Distribution of ventilation and perfusion in the sea snake, *Pelamis platurus*. J. Comp. Physiol. 145: 109–115.

Shannon, F. A. 1953. Case reports of two Gila monster bites. Herpetologica 9: 125–127.

Shaw, C. E. 1948a. A note on the food habits of *Heloderma suspectum* Cope. Herpetologica 4: 145.

———. 1948b. The male combat "dance" of some crotalid snakes. Herpetologica 4: 137–145.

———. 1950. The Gila monster in New Mexico. Herpetologica 6: 37–39.

———. 1961. Snakes of the sea. Zoo Nooz 34(7): 3–5.

———. 1962. Sea snakes at the San Diego Zoo. Intl. Zoo Yearb. 4: 49–52.

———. 1964. Beaded lizards—dreaded, but seldom deadly. Zoonooz 37: 10–15.

———. 1968. Reproduction of the Gila monster (*Heloderma suspectum*) at the San Diego Zoo. Zool. Gart. 35: 1–6.

———. 1971. The coral snakes, genera *Micrurus* and *Micruroides*, of the United States and northern Mexico, pp. 157–172. *In* W. Bücherl and E. E. Buckley (eds.), Venomous animals and their venoms, vol. 2: Venomous vertebrates. Academic Press, New York.

Shaw, C. E., and S. Campbell. 1974. Snakes of the American West. Alfred E. Knopf, New York.

Sheldon, D. 1929. Another reputed remedy of the American Indians. Bull. Antiv. Inst. Am. 3: 59.

Shelton, D. N. 1996. Mussels, a snake and a wounded collector. Am. Conchologist March 1996: 19.

Shepard, D. B., M J. Dreslik, C. A. Phillips, and B. C. Jellen. 2003. *Sistrurus catenatus catenatus* (eastern massasauga): Male-male aggression. Herpetol. Rev. 34: 155–156.

Shepard, D. B., C. A. Phillips, M. J. Dreslik, and B. C. Jellen. 2004. Prey preference and diet of neonate eastern massasaugas (*Sistrurus c. catenatus*). Am. Midl. Nat. 152: 360–368.

Shepard, D. B., M. J. Dreslik, B. C. Jellen, and C. A. Phillips. 2008a. Reptile road mortality around an oasis in the Illinois Corn Desert with emphasis on the endangered eastern massasauga. Copeia 2008: 350–359.

Shepard, D. B., A. R. Kuhns, M. J. Dreslik, and C. A. Phillips. 2008b. Roads as barriers to animal movement in fragmented landscapes. Anim. Conserv. 11: 288–296.

Sherbrooke, W. C., and M. F. Westphal. 2006. Responses of greater roadrunners during attacks on sympatric venomous and nonvenomous snakes. Southwest. Nat. 51: 41–47.

Shermann, D. G., R. P. Atkinson, T. Chippendale, K. A. Levin, K. Ng, N. Futrell, C. Y. Hsu, and D. E. Levy. 2000. Intravenous ancrod for treatment of acute ischemic stroke: The STAT

(Stroke Treatment with Ancrod Trial) study, a randomized controlled trial. J. Am. Med. Assoc. 283: 2395–2403.

Shiffer, C. N. 1983. The mild-mannered, misunderstood massasauga. Pennsylvania Angler 52(11): 30.

Shine, R. 1978. Sexual size dimorphism and male combat in snakes. Oecologia (Berlin) 33: 269–277.

———. 1991. Australian snakes. A natural history. Cornell Univ. Press, Ithaca, New York.

———. 1993. Sexual dimorphism in snakes, pp. 49–86. *In* R. A. Seigel, and J. T. Collins (eds.), Snakes: Ecology and behavior. McGraw-Hill, New York.

———. 2003. Reproductive strategies in snakes. Proc. Royal Soc. London B 270: 995–1004.

Shine, R., and R. A. Seigel. 1996. A neglected life-history trait: Clutch-size variance in snakes. J. Zool. (London) 239: 209–223.

Shipman, W. H., and G. V. Pickwell. 1973. Venom of the yellow-bellied sea snake (*Pelamis platurus*): Some physical and chemical properties. Toxicon 11: 375–377.

Shively, S. H., and J. C. Mitchell. 1994. Male combat in copperheads *(Agkistrodon contortrix)* from northern Virginia. Banisteria 3: 29–30.

Shu, Y. Y., H. R. Allen, and C. R. Geren. 1988. Isolation and characterization of the weakly acidic hemorrhagins from timber rattlesnake venom, pp. 445–455. *In* H. Pirkle and F. S. Markland, Jr. (eds.), Hemostasis and animal venoms. Marcel Dekker, New York.

Shufeldt, R. W. 1882. The bite of the Gila monster (*Heloderma suspectum*). Am. Nat. 16: 907–909.

———. 1890. Contributions to the study of *Heloderma suspectum*. Proc. Zool. Soc. London 1890: 148–244.

———. 1891. The poison apparatus of the *Heloderma*. Nature (London) 43: 514–515.

Shum, S., J. E. Jaramillo, R. Franklin, and M. Fernandez. 2006. Prospective measurement of intra-compartmental pressure following crotaline snakebites as a determinant for compartment syndrome development, p. 33. *In* S. A. Seifert (ed.), Snakebites in the new millennium: A state-of-the-art symposium, University of Nebraska Medical Center, 21–23 October 2005, Omaha, Nebraska. J. Med. Toxicol. 2: 29–45.

Shupe, K. S. 1977. Unusual coloration in two rattlesnakes: *Crotalus adamanteus* and *Sistrurus miliarius barbouri*. Bull. Philadelphia Herpetol. Soc. 24: 23–24.

Sibley, H. 1951. Snakes are scared of you! Field and Stream 55(9): 46–48.

Sifford, D. H., and B. D. Johnson. 1978. Fractionation of *Agkistrodon bilineatus* venom by ion exchange chromatography, pp. 231–242. *In* P. Rosenberg (ed.), Toxins: Animal, plant and microbial. Pergamon Press, New York.

Siigur, E., K. Tõnismägi, K. Trummal, M. Samel, H. Vija, J. Subbi, and J. Siigur. 2001. Factor X activator from *Vipera lebetina* snake venom, molecular characterization and substrate specificity. Biochim. Biophys. Acta 1568: 90–98.

Siigur, J., and E. Siigur. 1992. The direct acting α-fibrin(ogen)olytic enzymes from snake venoms. J. Toxicol. Toxin Rev. 11: 91–113.

———. 2006. Factor X activating proteases from snake venoms. J. Toxicol. Toxin Rev. 25: 235–255.

Simpson, J. W., and L. J. Rider. 1971. Collagenolytic activity from venom of the rattlesnake *Crotalus atrox*. Proc. Soc. Exp. Biol. Med. 137: 893–895.

Simpson, J. W., A. C. Taylor, and B. M. Levy. 1973. Elastolytic activity from venom of the rattlesnake *Crotalus atrox*. Proc. Soc. Exp. Biol. Med. 144: 380–383.

Sinsheimer, R. L., and J. F. Koerner. 1952. A purification of venom phosphodiesterase. J. Biol. Chem. 198: 293–396.

Six, D. A., and E. A. Dennis. 2000. The expanding superfamily of phospholipase A_2 enzymes: Classification and characterization. Biochim. Biophys. Acta 1488: 1–19.

Slavens, F. L., and K. Slavens. 2000. Reptiles and amphibians in captivity: Breeding, longevity, and inventory current January 1, 1999. Slaveware, Seattle, Washington.

Slotta, K. H., and H. Fraenkel-Conrat. 1938a. Estudos quimicos sobre dos venenos ofidicos: 4. Purificação e cristalização do veneno da cobra cascavel. Mem. Inst. Butantan 12: 505–512.

———. 1938b. Two active proteins from rattlesnake venom. Nature (London) 142: 213.

Slowinski, J. B. 1995. A phylogenetic analysis of the New World coral snakes (Elapidae: *Lepto-*

micrurus, Micruroides, and *Micrurus*) based on allozymic and morphological characters. J. Herpetol. 29: 325–338.

Slowinski, J. B., and R. Lawson. 2005. Elapid relationships, pp. 174–189. *In* M. A. Donnelly, B. Crother, C. Guyer, M. H. Wake, and M. E. White (eds.), Ecology and evolution in the tropics. Univ. Chicago Press, Illinois.

Slowinski, J. B., A. Knight, and A. P. Rooney. 1997. Inferring species trees from gene trees: A phylogenetic analysis of the Elapidae (Serpentes) based on the amino acid sequences of venom proteins. Mol. Phylog. Evol. 8: 349–362.

Smetsers, P. 1990. Care and breeding of *Sistrurus miliaris barbouri*, the Barbour's pygmy rattlesnake. Litt. Serpent. Engl. Ed. 10: 181–189.

———. 1993. *Agkistrodon bilineatus* Gunther: The tropical moccasin. Litt. Serpent. Engl. Ed. 13: 38–40.

Smith, A. G. 1940. Notes on the reproduction of the northern copperhead, *Agkistrodon mokasen cupreus* (Rafinesque). Ann. Carnegie Mus. 28: 77–82.

Smith, A. J. 1910. A new filarial species (*F. mitchelli* n.s.) found in *Heloderma suspectum*, and its larvae in a tick parasite upon the Gila monster. Univ. Pennsylvania Med. Bull. 23: 487–497.

Smith, C. 1997. *Agkistrodon contortrix contortrix* (southern copperhead): Diet. Herpetol. Rev. 28: 153.

Smith, C. F., K. Schwenk, R. L. Earley, and G. W. Schuett. 2008. Sexual size dimorphism of the tongue in a North American pitviper. J. Zool. (London) 274: 367–374.

Smith, H. M. 1935. Miscellaneous notes on Mexican lizards. Univ. Kansas Sci. Bull. 22: 119–155.

———. 1943. Summary of the collections of snakes and crocodilians made in Mexico under the Walter Rathbone Bacon Traveling Scholarship. Proc. U.S. Natl. Mus. 93: 393–504.

———. 1946. Handbook of lizards of the United States and of Canada. Comstock Publ. Co., Ithaca, New York.

———. 1952. A revised arrangement of maxillary fangs of snakes. Turtox News 30: 214–218.

Smith, H. M., and E. D. Brodie, Jr. (eds.). 1982. Reptiles of North America: A guide to field identification. Golden Press, New York.

Smith, H. M., and D. Chiszar. 2001. A new subspecies of cantil (*Agkistrodon bilineatus*) from central Veracruz, Mexico (Reptilia: Serpentes). Bull. Maryland Herpetol. Soc. 37: 130–136.

Smith, H. M., and P. S. Chrapliwy. 1958. New and noteworthy Mexican herptiles from the Lidicker collection. Herpetologica 13: 267–271.

Smith, H. M., and H. K. Gloyd. 1963. Nomenclatural notes on the snake names *Scytale, Boa scytale*, and *Agkistrodon mokasen*. Herpetologica 19: 280–282.

Smith, H. M., and C. Grant. 1958. Noteworthy herptiles from Jalisco, Mexico. Herpetologica 14: 18–23.

Smith, H. M., and O. Sanders. 1952. Distributional data on Texan amphibians and reptiles. Texas J. Sci. 4: 204–219.

Smith, H. M., J. A. Lemos-Espinal, and P. Heimes. 2005. 2005 Amphibians and reptiles from northwestern Mexico. Bull. Chicago Herpetol. Soc. 40: 206–212.

Smith, M. A. 1926. Monograph of the sea-snakes (Hydrophiidae). British Museum Natural History, London.

Smith, P. W. 1961. The amphibians and reptiles of Illinois. Illinois Nat. Hist. Surv. Bull. 28: 1–298.

Smith, P. W., and J. C. List. 1955. Notes on Mississippi amphibians and reptiles. Am. Midl. Nat. 53: 115–125.

Smith, P. W., and L. M. Page. 1972. Repeated mating of a copperhead and timber rattlesnake. Herpetol. Rev. 4: 196.

Smith, S. M. 1975. Innate recognition of coral snake pattern by a possible avian predator. Science (New York) 187: 759–760.

Smith, S. M., and A. M. Mostrom. 1985. "Coral snake" rings: Are they helpful in foraging? Copeia 1985: 384–387.

Smith, S. V., and K. M. Brinkhous. 1991. Inventory of exogenous platelet-aggregating agents derived from venoms. Thromb. Haemostas. 66: 259–263.

Smith, T. 1818. On the structure of the poisonous fangs of serpents. Philos. Trans. Royal Soc. London 1818: 471–476.

Snellings, E., Jr. 1982. The pygmy rattlesnake: Petite but not passive. Florida Nat. 55(2): 12–13.

Snellings, E., Jr., and J. T. Collins. 1996. *Sistrurus miliarius barbouri* (dusky pigmy rattlesnake). Maximum size. Herpetol. Rev. 27: 84.

Snider, A. T., and J. K. Bowler. 1992. Longevity of reptiles and amphibians in North American collections. Soc. Stud. Amphib. Rept. Herpetol. Circ. 21: 1–40.

Snider, A. T., Y. M. Lee, D. Hyde, and R. Christoffel. 2005. The eastern massasauga, *Sistrurus c. catenatus,* in Michigan: Conservation through education, p. 49. *In* Biology of the Rattlesnakes Symposium (program abstracts). Loma Linda Univ., Loma Linda, California.

Snow, F. H. 1906. Is the Gila monster a poisonous reptile? Trans. Kansas Acad. Sci. 20: 218–221.

Snyder, B. 1949. Diamondbacks and dollar bills. Florida Wildl. 4(5): 3–5, 16.

Snyder, C. C., and R. P. Knowles. 1988. Snakebites: Guidelines for practical management. Postgrad. Med. 83: 52–75.

Snyder, C. C., J. E. Pickins, R. P. Knowles, J. L. Emerson, and W. A. Hines. 1968. A definitive study of snakebite. J. Florida Med. Assoc. 55: 330–337.

Snyder, D. H., D. F. Burchfield, and R. W. Nall. 1967. First records of the pigmy rattlesnake in Kentucky. Herpetologica 23: 240–241.

Snyder, G. K., H. W. Ramsey, W. J. Taylor, and C. Y. Chiou. 1973. Neuromuscular blockade of chick biventer cervicis nerve-muscle preparations by a fraction from coral snake venom. Toxicon 11: 505–508.

Soares, A. M., and J. R. Giglio. 2003. Chemical modifications of phospholipases A_2 from snake venoms: Effects on catalytic and pharmacological properties. Toxicon 42: 855–868.

Soares, A. M., F. K. Ticli, S. Marcussi, M. V. Lourenço, A. H. Januário, S. V. Sampaio, J. R. Giglio, B. Lomonte, and P. S. Pereira. 2005. Medicinal plants with inhibitory properties against snake venoms. Curr. Med. Chem. 12: 2625–2641.

Sogandares-Bernal, F., and H. Grenier. 1971. Life cycles and host-specificity of the plagiorchiid trematodes *Ochetosoma kansensis* (Crow, 1913) and *O. latertotrema* (Byrd and Denton, 1938). J. Parastiol. 57: 297.

Solórzano, A. 1995. A case of human bite by the pelagic sea snake, *Pelamis platurus* (Serpentes: Hydrophiidae). Rev. Biol. Trop. 43: 321–322.

———. 2004. Serpientes de Costa Rica. Univ. Costa Rica, San José.

Solórzano, A., M. Romero, J. M. Gutierrez, and M. Sasa. 1999. Venom composition and diet of the cantil *Agkistrodon bilineatus howardgloydi* (Serpentes: Viperidae). Southwest. Nat. 44: 478–483.

Soto, J. G., R. L. Powell, S. R. Reyes, L. Wolana, L. J. Swanson, E. E. Sanchez, and J. C. Perez. 2006. Genetic variation of a disintegrin gene found in the American copperhead snake (*Agkistrodon contortrix*). Gene 373: 1–7.

Soulé, M. E., and A. J. Sloan. 1966. Biogeography and distribution of reptiles and amphibians on the islands in the Gulf of California, Mexico. Trans. San Diego Soc. Nat. Hist. 14: 137–156.

Southall, P. D. 1991. The relationship between wildlife and highways in the Paynes Prairie Basin. Florida Dept. Transportation, District 2, Lake City, Florida.

Speake, D. W., and J. A. McGlincy. 1981. Response of indigo snakes to gassing their dens. Proc. Annu. Conf. Southeast. Assoc. Fish Wildl. Agen. 35: 135–138.

Speake, D. W., and R. H. Mount. 1973. Some possible ecological effects of "rattlesnake roundups" in the southeastern coastal plain. Proc. 27th Ann. Conf. Southeastern Assoc. Game Fish Commissioners 1973: 267–277.

Spiess, P. 1998. The Gila monster (*Heloderma suspectum*). Rept. Amphib. Mag. 54: 22–27.

Spiller, H. A., and G. M. Bosse. 2003. Prospective study of morbidity associated with snakebite envenomation. J. Toxicol. Clin. Toxicol. 41: 125–130.

Sprent, J. F. A. 1978. Ascaridoid nematodes of amphibians and reptiles: *Paraheterotyphlum*. J. Helminthol. 52: 163–170.

———. 1979. Ascaridoid nematodes of amphibians and reptiles: *Terranova*. J. Helminthol. 53: 265–282.

Stabler, R. M. 1939. Frequency of skin shedding in snakes. Copeia 1939: 227–229.

———. 1951. Some observations on two cottonmouth moccasins made during twelve and fourteen years of captivity. Herpetologica 7: 89–92.

Stabler, R. M., and S. M. Schmittner. 1958. A microfilaria from the Gila monster. J. Parasitol. 44 (suppl.): 33–34.
Stadelman, R. E. 1928. The poisoning power of the new-born copperhead with case report. Bull. Antiv. Inst. Am. 2: 67–69.
———. 1929a. Some venom extraction records. Bull. Antiv. Inst. Am. 3: 29.
———. 1929b. Further notes on the venom of newborn copperheads. Bull. Antiv. Inst. Am. 3: 81.
Stafford, D. P., F. W. Plapp, Jr., and R. R. Fleet. 1976. Snakes as indicators of environmental contamination: Relation of detoxifying enzymes and pesticide residues to species occurrence in three aquatic ecosystems. Arch. Environ. Contam. Toxicol. 5: 15–27.
Stafford, P. 2000. Snakes. Smithsonian Inst. Press, Washington, D.C.
Stahnke, H. L. 1950. The food of the Gila monster. Herpetologica 6: 103–106.
———. 1952. A note on the food of the Gila monster, *Heloderma suspectum* Cope. Herpetologica 8: 64–65.
———. 1966. The treatment of venomous bites and stings. Rev. ed. Arizona St. Univ., Tempe.
Stahnke, H. L., W. A. Heffron, and D. L. Lewis. 1970. Bite of the Gila monster. Rocky Mountain Med. J. 67(9): 25–30.
Stanford, C. F., D. Olsen, G. M. Bogdan, and R. C. Dart. 2005. Complications of crotaline antivenom therapy in the United States. *In* Snakebites in the new millennium: A state-of-the-art symposium (program abstracts). Univ. Nebraska Medical Center, Center for Continuing Educ., Omaha.
Stebbins, R. C. 1954. Amphibians and reptiles of western North America. McGraw-Hill, New York.
———. 1985. A field guide to western reptiles and amphibians. Houghton Mifflin, Boston.
———. 2003. A field guide to western reptiles and amphibians. 3rd ed. Houghton Mifflin, Boston.
Steehouder, T. 1988. *Agkistrodon piscivorus*, the cottonmouth. Litt. Serpent. Engl. Ed. 8: 173–181.
Stegall, T. D., D. H. Sifford, and B. D. Johnson. 1994. Proteinase, BAEEase, and TAMEase activities of selected crotalid venoms. SAAS Bull. Biochem. Biotech. 7: 7–13.
Stejneger, L. S. 1895 [1893]. The poisonous snakes of North America, pp. 337–487. *In* Annual Report of the U.S. National Museum, Smithsonian Institution, Washington, D.C.
———. 1898. The poisonous snakes of North America. Smithsonian Institution Report 1898: 338–487.
———. 1902. The reptiles of the Huachuca Mountains, Arizona. Proc. U.S. Natl. Mus. 25(1282): 149–158.
Stevens, M. S. 1977. Further study of Castolon local fauna (early Miocene) Big Bend National Park, Texas. Pearce-Sellards Ser. Texas Mem. Mus. 28: 1–69.
Stewart, B. G. 1984. *Agkistrodon contortrix laticinctus* (broad-banded copperhead) combat. Herpetol. Rev. 15: 17.
Stewart, G. R., and R. S. Daniel. 1975. Microornamentation of lizard scales: Some variations and taxonomic correlations. Herpetologica 31: 117–130.
Stickel, W. H. 1952. Venomous snakes of the United States and treatment of their bites. U.S. Dept. Int., Wildl. Leafl. (339): 1–29.
Stiles, K., D. Chiszar, and H. M. Smith. 2000. Trail-following behavior in copperheads (*Agkistrodon contortrix*). J. Colorado-Wyoming Acad. Sci. 32(1): 19.
Stiles, K., P. Stark, D. Chiszar, and H. M. Smith. 2002. Strike-induced chemosensory searching (SICS) and trail-following behavior in copperheads (*Agkistrodon contortrix*), pp. 413– 418. *In* G. W. Schuett, M. Höggren, M. E. Douglas, and H. W. Greene (eds.), Biology of the vipers. Eagle Mountain Publ., Eagle Mountain, Utah.
Stille, B. 1987. Dorsal scale microdermatoglyphics and rattlesnake (*Crotalus* and *Sistrurus*) phylogeny (Reptilia: Viperidae: Crotalinae). Herpetologica 43: 98–104.
Stimson, A. C., and H. T. Engelhardt. 1960. The treatment of snakebite. J. Occup. Med. 2: 163–168.
Stitt, E. W., C. R. Schwalbe, and D. E. Swann. 2003. Signs of Gila monster predation on desert tortoise nests. Sonoran Herpetol. 16: 113.

Stocker, K., H. Fischer, J. Meier, M. Brogli, and L. Svendsen. 1987. Characterization of the protein C activator Protac® from the venom of the southern copperhead (*Agkistrodon contortrix*) snake. Toxicon 25: 239–252.

Stocker, K. F. 1986. Use of snake venom proteins in the diagnosis and therapy of haemostatic disorders, pp. 9–40. *In* F. Kornalik and D. Mebs (eds.), Proceedings 7th European symposium on animal, plant and microbial toxins. Int. Soc. Toxinology, Prague.

———. 1990a. Medical use of snake venom proteins. CRC Press, Boca Raton, Florida.

———. 1990b. Snake venom proteins affecting hemostasis and fibrinolysis, pp. 97–160. *In* K. F. Stocker (ed.). Medical use of snake venom proteins. CRC Press, Boca Raton, Florida.

Stoddard, H. L., Sr. 1978. Birds of Grady County, Georgia. Bull. Tall Timbers Res. Sta. 21: 1–175.

Storer, D. H. 1839. Reptiles of Massachusetts. Rep. Comm. Zool. Surv. Massachusetts: 203–253.

Storer, T. I. 1931. *Heloderma* poisoning in man. Bull. Antiv. Inst. Am. 5: 12–15.

Straight, R. C., and J. L. Glenn. 1993. Human fatalities caused by venomous animals in Utah, 1900–90. Great Basin Nat. 53: 390–394.

Straight, R. C., J. L. Glenn, T. B. Wolt, and M. C. Wolfe. 1991. Regional differences in content of small basic peptide toxins in the venoms of *Crotalus adamanteus* and *Crotalus horridus*. Comp. Biochem. Physiol. 100B: 51–58.

Strecker, J. K., Jr. 1908. The reptiles and batrachians of McLennan County, Texas. Proc. Biol. Soc. Washington 21: 69–84.

Strecker, J. K., and J. E. Johnson, Jr. 1935. Notes on the herpetology of Wilson County, Texas. Contrib. Baylor Mus. Bull. 38(3): 17–23.

Strecker, J. K., and W. J. Williams. 1928. Field notes on the herpetology of Bowie County, Texas. Contrib. Baylor Univ. Mus. 17: 1–19.

Streiffer, R. H. 1986. Bite of the venomous lizard, the Gila monster. Postgrad. Med. 79: 297–302.

Strimple, P. 1988. Comments on caudal luring in snakes with observations on this behavior in two subspecies of cantils, *Agkistrodon bilineatus* ssp. Notes from NOAH 15(6): 6–10.

———. 1992. Caudal movements of an adult cantil, *Agkistrodon bilineatus bilineatus* Gunther, pp. 55–56. *In* P. D. Strimple and J. L. Strimple (eds.), Contributions to herpetology. Great. Cincinnati Herpetol. Soc. , Cincinnati, Ohio.

———. 1995a. Captive reproduction of Gila monsters: A review. Reptiles 3(7): 16, 26.

———. 1995b. Comments on caudal luring in snakes with observations on this behavior in two subspecies of cantils, *Agkistrodon bilineatus* ssp. Litt. Serpent. Engl. Ed. 15: 74–77.

Strimple, P. D., A. J. Tomassoni, E. J. Otten, and D. Bahner. 1997. Report on envenomation by a Gila monster (*Heloderma suspectum*) with a discussion of venom apparatus, clinical findings, and treatment. Wild. Environ. Med. 8: 111–116.

Stringer, J. M., R. A. Kainer, and A. T. Tu. 1972. Myonecrosis induced by rattlesnake venom: An electron microscope study. Am. J. Pathol. 67: 127–140.

Stubbs. T. H. 1979. Moccasin. Florida Nat. 52(4): 2–4.

Studenroth, K. R. 1991. *Agkistrodon piscivorus conanti* (Florida cottonmouth): Foraging behavior. Herpetol. Rev. 22: 60.

Stürzbecher, J., U. Neumann, and J. Meier. 1991. Inhibition of the protein C activator Protac^R^, a serine proteinase from the venom of the southern copperhead snake *Agkistrodon contortrix contortrix*. Toxicon 29: 151–155.

Suazo-Ortuño, I., O. Flores-Villela, and D. Garcia-Parra. 2004. *Micrurus distans* (west Mexican coral snake): Tree climbing. Herpetol. Rev. 35: 276.

Suchard, J. R., and F. LoVecchio. 1999. Envenomations by rattlesnakes thought to be dead. New England J. Med. 340: 1930.

Sullivan, B. K., G. W. Schuett, and M. A. Kwiatkowski. 2002. Natural history observations: *Heloderma suspectum* (Gila monster). Mortality/predation? Herpetol. Rev. 33: 135–136.

Sullivan, B. K., M. A. Kwiatkowski, and G. W. Schuett. 2004. Translocation of urban Gila monsters: A problematic conservation tool. Biol. Conserv. 117: 235–242.

Sumichrast, F. 1864. Note on the habits of some Mexican reptiles. Ann. Mag. Nat. Hist. Ser. 3: 497–507.

Surface, H. A. 1906. The serpents of Pennsylvania. Bull. Pennsylvania St. Dept. Agric. Div. Zool. 4: 133–208.

Sutherland, S. K., and A. R. Coulter. 1981. Early management of bites by the eastern diamondback rattlesnake (*Crotalus adamanteus*): Studies in monkeys (*Macaca fascicularis*). Am. J. Trop. Med. Hyg. 30: 497–500.

Sutton, W. B., M. G. Bolus, and Y. Wong. 2009. *Lampropeltis getula nigra* (black kingsnake): Ophiophagy. Herpetol. Rev. 40: 231.

Suzuki, T., and S. Iwanaga. 1958a. Studies on snake venom: II. Some observations on the alkaline phosphatases of Japanese and Formosan snake venoms. J. Pharm. Soc. Japan 78: 354–361.

———. 1958b. Studies on snake venom: IV. Purification of the alkaline phosphatases in cobra venom. J Pharm. Soc. Japan 78: 368–375.

Svoboda, P., C. H. Adler, and J. Meier. 1992. Antimicrobial activity of several Elapidae and Viperidae snake venoms. *In* P. Gopalakrishnakone and C. D. Tan (eds.), Proceedings 10th world congress on animal, plant and microbial toxins (3–8 November 1991), Singapore.

Swan, G. 1995. A photographic guide to snakes & other reptiles of Australia. Ralph Curtis Books, Sanibel Island, Florida.

Swannack, T. M., and M. R. J. Forstner. 2003. *Micrurus fulvius tener* (Texas coral snake): Diet. Herpetol. Rev. 34: 376.

Swanson, P. 1952. The reptiles of Venango County, Pennsylvania. Am. Midl. Nat. 47: 161–182.

Swanson, P. L. 1930. Notes on the massasauga. Bull. Antiv. Inst. Am. 4: 70–71.

———. 1933. The size of *Sistrurus catenatus catenatus* at birth. Copeia 1933: 37.

———. 1946. Effects of snake venoms on snakes. Copeia 1946: 242–249.

Swaroop, W. H., and B. Grab. 1954. Snakebite mortality in the World. Bull. World Health Org. 10: 35–76.

Swenson, S., and F. S. Markland, Jr. 2005. Snake venom fibrin(ogen)olytic enzymes. Toxicon 45: 1021–1039.

———. 2006. Fibrolase. J. Toxicol. Toxin Rev. 25: 351–378.

Swenson, S., F. Costa, W. Ernst, G. Fujii, and F. S. Markland. 2005. Contortrostatin, a snake venom disintegrin with anti-angiogenic and anti-tumor activity. Pathophysiol. Haemost. Thromb. 34: 169–176.

Switak, K. H. 2002. Extraordinary feeding behavior for the yellow-bellied seasnake, *Pelamis platurus*. Bull. Chicago Herpetol. Soc. 37: 194–195.

Szyndlar, Z., and J. C. Rage. 2002. Fossil record of the true vipers, pp. 419–444. *In* G. W. Schuett, M. Höggren, M. E. Douglas, and H. W. Greene (eds.), Biology of the vipers. Eagle Mountain Publ., Eagle Mountain, Utah.

Takagi, J., F. Sekiya, K. Kasahara, Y. Inada, and Y. Saito. 1988. Venom from southern copperhead snake (*Agkistrodon contortrix contortrix*): II. A unique phospholipase A_2 that induces platelet aggregation. Toxicon 26: 199–206.

Takahashi, H., and H. Mashiko. 1998. Haemorrhagic factors from snake venoms: I. Properties of haemorrhagic factors and antihaemorrhagic factors. J. Toxicol. Toxin Rev. 17: 315–335.

Takasaki, C. 1998. The toxicology of sea snake venoms. J. Toxicol. Toxin Rev. 17: 361–372.

Takeya, H., and S. Iwanaga. 1998. Proteases that induce hemorrhage, pp. 11–38. *In* G. S. Bailey (ed.), Enzymes from snake venom. Alakan, Fort Collins, Colorado.

Talan, D. A., D. M. Citron, G. D. Overturf, B. Singer, P. Froman, and E. J. C. Goldstein. 1991. Antibacterial activity of crotalid venoms against oral snake flora and other clinical bacteria. J. Infect. Dis. 164: 195–198.

Tamiya, N. 1980. Active sites of the snake neurotoxin molecules. The Snake 12: 147–148.

Tan, N.-H., and G. Ponnudurai. 1990. A comparative study of the biological activities of venoms from snakes of the genus *Agkistrodon* (moccasins and copperheads). Comp. Biochem. Physiol. 95B: 577–582.

———. 1991. A comparative study of the biological activities of rattlesnake (genera *Crotalus* and *Sistrurus*) venom. Comp. Biochem. Physiol. 98C: 455–461.

———. 1992. Comparative study of the enzymatic, hemorrhagic, procoagulant and anticoagulant activities of some animal venoms. Comp. Biochem. Physiol. 103C: 299–302.

Taub, A. M. 1963. On the longevity and fecundity of *Heloderma horridum horridum*. Herpetologica 19: 149.

Taub, A. M., and W. A. Dunson. 1967. The salt gland in a sea snake *(Laticauda)*. Nature (London) 215: 995–996.

Tavano, J. J., A. L. Pitt, and M. A. Nickerson. 2007. *Agkistrodon piscivorus leucostoma* (western cottonmouth): Behavior. Herpetol. Rev. 38: 202.

Tay, J., A. L. Castillo, Z. J. Julia, C. R. Romero, and C. O. Velasco. 1980a. Accidentes por mordedura de animales ponzoñosos: 1a. Parte. Rev. Fac. Med. U. N. A. M. (México) 23: 4–17.

———. 1980b. Accidentes por mordedura de animales ponzoñosos: 2a. Parte. Rev. Fac. Med. U. N. A. M. (México) 23: 18–25.

Taylor, E. H. 1953. Early records of the seasnake *Pelamis platurus* in Latin America. Copeia 1953: 124.

Taylor, K. P. A. 1929. Apparent cure of purpura hemorrhagica by bothropic antivenin. Bull. Antiv. Inst. Am. 3: 42–43.

Tay Zavala, J., L. E. Castillo Alarcón, R. Romero Cabello, and J. Juliá Zertuche. 1979. Reptiles ponzoñosos de importancia medica, I y II parte. Congresso Nac. Zool, México.

Tay Zavala, J., L. E. Castillo Alarcón, J. Juliá Zertuche, R. Romero Cabello, and O. Velazco Castrejón. 1980. Accidentes por mordedura de animales ponzoñosos, 2a. parte. Rev. Fac. Medicina, México.

Telford, S. R., Jr. 1955. A description of the eggs of the coral snake, *Micrurus f. fulvius*. Copeia 1955: 258.

Teng, C. M., and F. N. Ko. 1988. Comparison of the platelet aggregation induced by three thrombin-like enzymes of snake venoms and thrombin. Thromb. Haemostas. 59: 304–309.

Tennant, A. 1984. The snakes of Texas. Texas Monthly Press, Austin.

———. 1985. A field guide to Texas snakes. Texas Monthly Press, Austin.

———. 1997. A field guide to snakes of Florida. Gulf Publ. Co., Houston, Texas.

———. 1998. A field guide to Texas snakes. 2nd ed. Gulf Publ. Co., Houston, Texas.

———. 2003. Snakes of North America: Eastern and central regions. Lone Star Books, Latham, Maryland.

Tennesen, M. 2009. Snakebit: Southern California sees a rise in extratoxic venom. Scientific American April 2009: 27–29.

Thompson, P. E., and C. G. Huff. 1944a. A saurian malarial parasite, *Plasmodium mexicanum,* n. sp., with both elongatum- and gallinaceum-types of exoerythrocytic stages. J. Infect. Dis. 74: 48–67, 78–79.

———. 1944b. Saurian malarial parasites of the United States and Mexico. J. Infect. Dis. 74: 68–79.

Thorpe, R. S., W. Wüster, and A. Malhotra. 1997. Venomous snakes: Ecology, evolution and snakebite. Clarendon Press, Oxford.

Tiersch, T. R., and C. R. Figiel, Jr. 1991. A triplod snake. Copeia 1991: 838–841.

Tiersch, T. R., C. R. Figiel, Jr., R. M. Lee, III, R. W. Chandler, and A. E. Houston. 1990. Use of flow cytometry to screen for the effects of environmental mutagens: Baseline DNA values in cottonmouth snakes. Bull. Environ. Contam. Toxicol. 45: 833–839.

Tihen, J. A. 1962. A review of New World fossil bufonids. Am. Midl. Nat. 68: 1–50.

Timmerman, W. W., and W. H. Martin. 2003. Conservation guide to the eastern diamondback rattlesnake, *Crotalus adamanteus.* Soc. Stud. Amphib. Rept. Herpetol. Circ. 32.

Tinkham, E. R. 1971a. The biology of the Gila monster, pp. 381– 413. *In* W. Bücherl and E. E. Buckley (eds.), Venomous animals and their venoms, vol. II: Venomous vertebrates. Academic Press, New York.

———. 1971b. The venom of the Gila monster, pp. 415–422. *In* W. Bücherl and E. E. Buckley (eds.), Venomous animals and their venoms, vol. II: Venomous vertebrates. Academic Press, New York.

Tinkle, D. W. 1959. Observations of reptiles and amphibians in a Louisiana swamp. Am. Midl. Nat. 62: 189–205.

———. 1967. The life and demography of the side-blotched lizard, *Uta stansburiana*. Misc. Publ. Mus. Zool. Univ. Michigan 132: 1–182.

Tinkle, D. W., and J. W. Gibbons. 1977. The distribution and evolution of viviparity in reptiles. Misc. Publ. Mus. Zool. Univ. Michigan 154: 1–55.

Tipton, B. L. 2005. Snakes of the Americas: Checklist and lexicon. Krieger, Melbourne, Florida.

Todd, R. E., Jr. 1973. The cottonmouth in Kentucky. Kentucky Herpetol. 4(2–4): 5–9.

Tomes, C. S. 1875. On the structure and development of the teeth of Ophidia. Philos. Trans. Royal Soc. London 165: 297–302.

———. 1877. On the development and succession of the poison-fangs of snakes. Philos. Trans. Royal Soc. London 166: 377–385.

Toom, P. M., P. G. Squire, and A. T. Tu. 1969. Characterization of the enzymatic and biological activities of snake venoms by isoelectric focusing. Biochim. Biophys. Acta 181: 339–341.

Toombs, C. F. 2001. Alfimeprase: Pharmacology of a novel fibrinolytic metalloproteinase for thrombolysis. Haemostasis 31: 141–147.

Toombs, C. F., and S. R. Deitcher. 2006. Nonclinical and clinical characterization of a novel acting thrombolytic: Alfimeprase. Toxin Rev. 25: 379–392.

Tovar-Tovar, H., and F. Mendoza-Quijano. 2001. *Agkistrodon taylori* (Taylor's cantil). Herpetol. Rev. 32: 276–277.

Trauth, S. E., and B. G. Cochran. 1991. *Hemidactylium scutatum* (four-toed salamander): Predation. Herpetol. Rev. 22: 55.

Trauth, S. E., and C. T. McAllister. 1995. Vertebrate prey of selected Arkansas snakes. Proc. Arkansas Acad. Sci. 49: 188–192.

Trauth, S. E., H. W. Robison, and M. V. Plummer. 2004. The amphibians and reptiles of Arkansas. Univ. Arkansas Press, Fayetteville.

Travers, S. L., K. L. Krysko, S. J. Allred, and L. R. Tirado. 2008. *Sistrurus miliarius barbouri* (dusky pigmy rattlesnake): Diet. Herpetol. Rev. 39: 473.

Troost, G. 1836. On a new genus of serpents, and new species of the genus *Heterodon*, inhabiting Tennessee. Ann. Lyc. Nat. Hist. New York 3: 174–190.

Trowbridge, A. H. 1937. Ecological observations on Oklahoma amphibians and reptiles. Am. Midl. Nat. 18: 285–303.

True, F. W. 1883. On the bite of the North American coral snakes (genus *Elaps*). Am. Nat. 17: 26–31.

Trutnau, L. 1998. Schlangen im Terrarium, vol. 2: Giftschlangen. Verlag Eugen Ulmer, Stuttgart.

Tryon, B., and H. K. McCrystal. 1982. *Micrurus fulvius tenere* reproduction. Herpetol. Rev. 13: 47–48.

Tsai, F. T. 1961. Studies on the snake venom enzyme XIII: On the ribonuclease activity of Formosan snake venoms. Hukuoka Acta Med. 52: 47–51.

Tsai, I. H., Y. H. Chen, Y. M. Wang, M. C. Tu, and A. T. Tu. 2001. Purification, sequencing, and phylogenetic analyses of novel Lys-49 phospholipases A_2 from the venoms of rattlesnakes and other pit vipers. Arch. Biochem. Biophys. 394: 236–244.

Tsetlin, V. 1999. Snake venom alpha-neurotoxins and other "three-finger" proteins. Eur. J. Biochem. 264: 281–286.

Tsetlin, V. I., and F. Hucho. 2004. Snake and snail toxins acting on nicotinic acetylcholine receptors: Fundamental aspects and medical applications. FEBS Lett. 557: 9–13.

Tu, A. T. 1974. Sea snake investigation in the Gulf of Thailand. J. Herpetol. 8: 201–210.

———. 1976. Investigation of the sea snake, *Pelamis platurus* (Reptilia, Serpentes, Hydrophiidae), on the Pacific Coast of Costa Rica, Central America. J. Herpetol. 10: 13–18.

———. 1977. Venoms: Chemistry and molecular biology. John Wiley and Sons, New York.

——— (ed.). 1982. Rattlesnake venoms: Their actions and treatment. Marcel Dekker, New York.

———. 1988. Overview of snake venom chemistry. Adv. Exp. Med. Biol. 391: 37–62.

———. 1991a. Primary structure of the sea-snake neurotoxins and their modes of attachment to acetylcholine receptor, pp. 87–95. *In* M. F. Thompson, R. Sarojini, and R. Nagabhushanam (eds.), Bioactive compounds from marine organisms: With emphasis on the Indian ocean. An Indo-United States Symposium. Oxford and IBH Publishers, New Delhi.

———. 1991b. A lizard venom: Gila monster (Genus: *Heloderma*), pp. 755–773. *In* A. T. Tu (ed.), Handbook of natural toxins, vol. 5: Reptile venoms and toxins. Marcel Dekker, New York.

Tu, A. T., and B. L. Adams. 1968. Phylogenetic relationships among venomous snakes of the genus *Agkistrodon* from Asia and the North American continent. Nature (London) 217: 760–762.

Tu, A. T., and D. S. Murdock. 1967. Protein nature and some enzymatic properties of the lizard *Heloderma suspectum suspectum* (Gila monster) venom. Comp. Biochem. Physiol. 22: 389–396.

Tu, A. T., P. M. Toom, and D. S. Murdock. 1967. Chemical differences in the venoms of genetically different snakes, pp. 351–362. *In* F. E. Russell and P. Saunders (eds.), Animal toxins. Pergamon Press, New York.

Tu, A. T., T. S. Lin, and A. L. Bieber. 1975. Purification and chemical characterization of the major neurotoxin from the venom of *Pelamis platurus*. Biochemistry 14: 3408–3413.

Tubbs, K. A., R. W. Nelson, J. R. Krone, and A. L. Bieber. 2000. Mass spectral studies of snake venoms and some of their toxins. J. Toxicol. 19: 1–22.

Uddman, R., P. J. Goadsley, I. Jansen-Olesen, and L. Edvinsson. 1999. Helospectin-like peptides: Immunochemical localization and effects on isolated cerebral arteries and on local cerebral blood flow in the cat. J. Cerebral Blood Flow Metab. 19: 61–67.

Uhler, F. M., C. Cottam, and T. E. Clarke. 1939. Food of snakes of the George Washington National Forest, Virginia. Trans. N. Am. Wildl. Conf. 4: 605–622.

Underwood, G. 1967. A contribution to the classification of snakes. Publ. British Mus. Nat. Hist. 653: i–x, 1–179.

Upton, S. J., C. T. McAllister, and C. M. Garrett. 1993. Description of a new species of *Eimeria* (Apicomplexa: Eimeriidae) from *Heloderma suspectum* (Sauria: Helodermatidae). Texas J. Sci. 45: 155–159.

Utaisincharoen, P., S. P. Mackessy, R. A. Miller, and A. T. Tu. 1993. Complete primary structure and biochemical properties of gilatoxin, a serine protease with kallikrein-like and angiotensin-degrading activities. J. Biol. Chem. 268: 21975–21983.

Vaeth, R. H. 1984. A note on courtship and copulatory behavior in *Micrurus fulvius*. Bull. Chicago Herpetol. Soc. 18: 86–88.

Vallarino, O., and P. J. Weldon. 1996. Reproduction in the yellow-bellied sea snake (*Pelamis platurus*) from Panama: Field and laboratory observations. Zoo Biol. 15: 309–314.

Van Bruggen, A. C. 1961. *Pelamis platurus,* an unusual item of food of *Octopus* spec. Banisteria 25: 73–74.

van de Beek, H. 1991. A successful breeding of *Sistrurus catenatus catenatus*. Litt. Serpent. Engl. Ed. 11: 8–12.

van den Bergh, C. J., A. J. Slotboom, H. M. Verheij, and G. H. de Haas. 1988. The role of aspartic acid-49 in the active site of phospholipase A_2: A site-specific mutagenesis study of porcine pancreatic phospholipase A_2 and the rationale of the enzymatic activity of [lysine49] phospholipase A_2 from *Agkistrodon piscivorus piscivorus*' venom. Eur. J. Biochem. 176(2): 353–357.

Van Denburgh, J. 1922. The reptiles of western North America, vol. 2: Snakes and turtles. Occ. Pap. California Acad. Sci. 10: 617–1028.

Van Denburgh, J., and O. B. Wright. 1900. On the physiological action of the poisonous secretion of the Gila monster (*Heloderma suspectum*). Am. J. Physiol. 4: 209–238.

Vanderduys, E., and R. Hobson. 2004. Two accounts of birth, post beach-washing, in yellow-bellied sea snakes, *Pelamis platurus*. Herpetofauna 34: 86–87.

Vandermeers, A., M.-C. Vandermeers-Piret, L. Vigneron, J. Rathe, M. Stievenart, and J. Christophe. 1991. Differences in primary structure among five phospholipases A_2 from *Heloderma suspectum*. Eur. J. Biochem. 196: 537–544.

Van Devender, T. A. 1990. Late quaternary vegetation and climate of the Sonoran desert, United States and Mexico, pp. 134–165. *In* J. L. Betancourt, T. R. Van Devender, and P. S. Martin (eds.). Packrat middens: The last 40,000 years of biotic change. Univ. Arizona Press, Tucson.

Van Devender, T. R., and R. Conant. 1990. Pleistocene forests and copperheads in the eastern United States, and the historical biogeography of new world *Agkistrodon*, pp. 601–614. *In* H. K. Gloyd and R. Conant (eds.), Snakes of the *Agkistrodon* complex: A monographic review. Contributions to Herpetology 6. Soc. Stud. Amphib. Rept., Ithaca, New York.

Van Devender, T. R., and E. F. Enderson. 2007. *Micrurus distans distans* (west Mexican coral snake). Herpetol. Rev. 38: 488.

Vandewalle, T. J. 2005. *Sistrurus catenatus catenatus* (eastern massasauga): Male-male combat. Herpetol. Rev. 36: 196–197.

Vandewalle, T. J., and W. L. Vandewalle. 2008. *Sistrurus catenatus catenatus* (eastern massasauga): Diet. Herpetol. Rev. 39: 358.

Van Mierop, L. H. S. 1976a. Poisonous snakebite: A review. 1. Snakes and their venom. J. Florida Med. Assoc. 63: 191–200.

———. 1976b. Poisonous snakebite: A review. 2. Symptomatology and treatment. J. Florida Med. Assoc. 63: 201–210.

v. d. Velde, H. 1990. *Agkistrodon contortox laticinctus*. Litt. Serpent. Engl. Ed. 10: 149.

Veer, V., D. Chiszar, and H. M. Smith. 1997. *Sonora semiannulata* (ground snake). Antipredation. Herpetol. Rev. 28: 91.

Verkerk, J. W. 1986. Verzorging en kweek van de dwergratelslang, *Sistrurus miliarius barbouri*. Lacerta 45: 15–20.

———. 1987. Enkele aanvullende opmerkingen over het gedrag van de dwergratelslang (*Sistrurus miliarius barbouri*). Lacerta 45: 142–143.

Vermersch, T. G., and R. E. Kuntz. 1986. Snakes of south-central Texas. Eakin Press, Austin, Texas.

Vest, D. K. 1981. Envenomation following the bite of a wandering garter snake (*Thamnophis elegans vagrans*). Clin. Toxicol. 18: 573–579.

Vial, J. L., T. L. Berger, and W. T. McWilliams, Jr. 1977. Quantitative demography of copperheads, *Agkistrodon contortrix* (Serpentes: Viperidae). Res. Popul. Ecol. (Kyoto) 18: 223–234.

Vick, J. A. 1971. Symptomatology of experimental and clinical crotalid envenomation, pp. 71–86. *In* L. L. Simpson (ed.), Neuropoisons: Their pathophysiological actions, vol. 1. Plenum Press, New York.

Vick, J. A., H. P. Ciuchta, and J. H. Manthei. 1967. Pathophysiological studies of ten snake venoms, pp. 269–282. *In* F. E. Russell and P. R. Saunders (eds.), Animal toxins. Pergamon Press, New York.

Vick, J. A., J. von Bredlow, M. M. Grenan, and G. V. Pickwell. 1975. Sea snake antivenin and experimental envenomation therapy, pp. 463–485. *In* W. A. Dunson (ed.), The biology of sea snakes. University Park Press, Baltimore, Maryland.

Vidal, J. C., and A. O. M. Stoppani. 1971. Isolation and properties of an inhibitor of phospholipase A from *Bothrops neuwiedi* venom. Arch. Biochem. Biophys. 147: 66–76.

Vidal Breard, J. J. 1950. Iron in serpent venoms. Arch. Farm. Bioquím. (Tucuman) 4: 321.

Vigle, G. O., and H. Heatwole. 1978. A bibliography of the Hydrophiidae. Smithsonian Herpetol. Inform. Serv. (41): 1–20.

Villa [Rivas], J. 1962. Las serpientes venenosas de Nicaragua. Managua, Nicaragua, Editorial Novedades: 1–90 [91–94].

Villeneuve, M., and D. Rivard. 1986. Massasauga rattlesnake management at Georgian Bay Islands National Park, Ontario, Canada. Newslett. Bull. Can. Soc. Environ. Biol. 42(4): 27–29.

Vincent, S. E., A. Herrel, and D. J. Irschick. 2004a. Sexual dimorphism in head shape and diet in the cottonmouth snake (*Agkistrodon piscivorus*). J. Zool. (London) 264: 53–59.

———. 2004b. Ontogeny of intersexual head shape and prey selection in the pitviper *Agkistrodon piscivorus*. Biol. J. Linnaean Soc. 81: 151–159.

———. 2005. Comparisons of aquatic versus terrestrial predatory strikes in the pitviper, *Agkistrodon piscivorus*. J. Exp. Zool. 303A: 476–488.

Visser, G. 2002. Hatchling Gila monster (*Heloderma suspectum*) in Rotterdam Zoo. Podarcis 3(3): 114–117.

Visser, J. 1967. Color varieties, brood size, and food of South African *Pelamis platurus* (Ophidia: Hydrophiidae). Copeia 1967: 219.

Vital Brazil, O. 1987. Coral snake venoms: Mode of action and pathophysiology of experimental envenomation. Rev. Inst. Med. Trop. Sao Paulo 29: 119–126.

Vitt, L. J., and A. C. Hulse. 1973. Observations on feeding habits and tail display of the Sonoran coral snake, *Micruroides euryxanthus*. Herpetologica 29: 302–304.

Vitt, L. J., and H. J. Price. 1982. Ecological and evolutionary determinants of relative clutch mass in lizards. Herpetologica 38: 237–255.

Vogel, C. W. 1991. Cobra venom factor: The complement-activating protein of cobra venom, pp. 147–188. *In* A. T. Tu (ed.), Handbook of natural toxins, vol. 5: Reptile venoms and toxins. Marcel Dekker, New York.

Vogt, R. C. 1981. Natural history of amphibians and reptiles of Wisconsin. Milwaukee Public Museum, Milwaukee, Wisconsin.

Vogt, W. 1990. Snake venom constituents affecting the complement system, pp. 79–96. *In* K. F. Stocker (ed.), Medical use of snake venom proteins. CRC Press, Boca Raton, Florida.

Vorhies, C. T. 1929. Feeding of *Micruroides euryxanthus*, the Sonoran coral snake. Bull. Antiv. Inst. Am. 2: 98.

Voris, H. K. 1951. Miscellaneous notes on the eggs and young of Texan and Mexican reptiles. Zoologica 36: 37–48.

———. 1971. New approaches to character analysis applied to the sea snakes (Hydrophiidae). Syst. Biol. 20: 442–458.

———. 1972. The role of sea snakes (Hydrophiidae) in the tropical structure of coastal ocean communities. J. Mar. Biol. Assoc. India 14: 429–442.

———. 1977. A phylogeny of the sea snakes (Hydrophiidae). Fieldiana: Zool. 70: 79–166.

———. 1983. *Pelamis platurus* (culebra del mar, pelagic sea snake), pp. 411–412. *In* D. H. Janzen (ed.), Costa Rican natural history. Univ. Chicago Press, Chicago, Illinois.

Voris, H. K., and B. C. Jayne. 1976. The costocutaneous muscles in some sea snakes (Reptilia, Serpentes). J. Herpetol. 10: 175–180.

Voris, H. K., and H. H. Voris. 1983. Feeding strategies in marine snakes: An analysis of evolutionary, morphological, behavioral and ecological relationships. Am. Zool. 23: 411–425.

Voris, H. K., H. Voris, and W. B. Jeffries. 1983. Sea snakes: Mark-release-recapture. Field Mus. Nat. Hist. Bull. 54(9): 5–10.

Wadsworth, J. R. 1960. Tumors and tumor-like lesions of snakes. J. Am. Vet. Med. Assoc. 137: 419–420.

Wagler, J. 1824. Serpentum brasiliensium species novae ou historie naturelle des espèces nouvelles de serpens, recueillies et observées pendent le voyage dans l'intérieuer du Brésil dans les annees 1817, 1818, 1819, 1820. . . . Publiée par Jean de Spix. Monachii: Typis Francisci Seraphici Hübschmanni.

———. 1830. Natürliches System der Amphibien, mit voranfehender Classification der Säugthiere und Vogel. München, Stuttgart, and Tubingen.

Wagner, C. W., and E. S. Golladay. 1989. Crotalid envenomation in children: Selective conservative management. J. Ped. Surg. 24: 128–131.

Wagner, E., R. Smith, and F. Slavens. 1976. Breeding the Gila monster *Heloderma suspectum* in captivity. Intl. Zoo Yearbk. 16: 74–78.

Wagner, F. W., and J. M. Prescott. 1966. A comparative study of proteolytic activities in the venoms of some North American snakes. Comp. Biochem. Physiol. 17: 191–201.

Wagner, F. W., A. M. Speekherman, and J. M. Prescott. 1968. *Leucostoma* peptidase A: Isolation and physical properties. J. Biol. Chem. 243: 4486–4493.

Wagner, R. T. 1962. Notes on the combat dance in *Crotalus adamanteus*. Bull. Philadelphia Herpetol. Soc. 10(1): 7–8.

Walker, K. J. 2003. An illustrated guide to trunk vertebrae of cottonmouth (*Agkistrodon piscivorus*) and diamondback rattlesnake (*Crotalus adamanteus*) in Florida. Bull. Florida Mus. Nat. Hist. 44: 91–100.

Wall, F. 1921. Snakes of Ceylon. H. R. Cottle, Gov't. Print., Colombo, Ceylon.

Wallach, V. 1990. A record brood for the southern copperhead, *Agkistrodon c. contortrix* Linnaeus. Bull. Maryland Herpetol. Soc. 26: 17–20.

———. 1998. The lungs of snakes, pp. 93–295. *In* C. Gans and A. S. Gaunt (eds.), Biology of the Reptilia, vol. 19: Morphology G. Soc. Stud. Amphib. Rept., Ithaca, New York.

Walls, G. L. 1932. Pupil shapes in reptilian eyes. Bull. Antiv. Inst. Am. 5: 68–70.

Walsh, E., S. Gabrey, and M. L. McCallum. 2005. *Agkistrodon piscivorus* (western cottonmouth): Foraging behavior. Herpetol. Rev. 36: 186.

Walter, F. G., P. B. Chase, M. D. Fernandez, and J. McNally. 2007. Venomous snakes: North American Crotalinae envenomation, pp. 399–422. *In* M. Shannon, S. Borron, and M. Burns (eds.), Haddad and Winchester's clinical management of poisoning and drug overdose. 4th ed. Sanders Elsevier, Philadelphia.

Walters, A. C., and W. Card. 1996. *Agkistrodon piscivorus conanti* (Florida cottonmouth). Herpetol. Rev. 27: 203.

Walters, A. C., D. T. Roberts, and C. V. Covell, Jr. 1996. *Agkistrodon contortrix contortrix* (southern copperhead). Prey. Herpetol. Rev. 27: 202.

Warkentin, K. M. 1995. Adaptive plasticity in hatching age: A response to predation risk trade-offs. Proc. Natl. Acad. Sci. USA 92: 3507–3510.

Warrell, D. A. 1997. Geographical and intraspecies variation in the clinical manifestations of envenoming by snakes, pp. 189–203. *In* R. S. Thorpe, W. Wüster, and A. Malhotra (eds.), Venomous snakes: Ecology, evolution and snakebite. Clarendon Press, Oxford.

———. 2007. World antivenom shortage. *In* Venom week 2007 (program abstracts). Univ. Arizona Colleges of Medicine, Arizona Health Sci. Center, Off. Continuing Medical Educ., Tucson, Arizona.

Watson, S. H. 2006. *Agkistrodon piscivorus piscivorus* (eastern cottonmouth). Catesbeiana 26(2): 75.

Watt, C. H., Jr. 1985. Treatment of poisonous snakebite with emphasis on digit dermotomy. South. Med. J. 78: 694–699.

Weatherhead, P. J., and K. A. Prior. 1992. Preliminary observations of habitat use and movements of the eastern massasauga rattlesnake (*Sistrurus c. catenatus*). J. Herpetol. 26: 447–452.

Webb, R. G. 1970. Reptiles of Oklahoma. Univ. Oklahoma Press, Norman.

Webb, S. D., and E. Simons. 2006. Vertebrate paleontology, pp. 215–246. *In* S. D. Webb (ed.), First Floridians and last mastodons: The Page-Ladson Site in the Aucilla River. Springer, Dordrecht, The Netherlands.

Weber, R. A., and R. R. White IV. 1993. Crotalidae envenomation in children. Ann. Plast. Surg. 31: 141–145.

Wechter, W. J., A. J. Mikulski, and M. Laskowski, Sr. 1968. Gradation of specificity with regard to sugar among nucleases. Biochem. Biophys. Res. Commun. 30: 318–322.

Weigel, R. D. 1962. Fossil vertebrates of Vero, Florida. Florida Geol. Surv. Spec. Publ. 10: 1–59.

Weigmann, A. F. A. 1829a. Über dei Gesetzlichkeit in der geographischen Verbreitung der Saurier. Isis von Oken 22: 418–428.

———. 1829b. Üeber das Acaltetepan oder Temaculcahua des Hernandez, eine neue Gattung der Saurer, *Heloderma*. Isis von Oken 22: 624–629.

Weinstein, S. A., S. A. Minton, and C. E. Wilde. 1985. The distribution among ophidian venoms of a toxin isolated from the venom of the Mojave rattlesnake (*Crotalus scutulatus scutulatus*). Toxicon 23: 825–844.

Weis, R., and R. J. McIsaac. 1971. Cardiovascular and muscular effects of venom from coral snake, *Micrurus fulvius*. Toxicon 9: 219–228.

Weiser, E., Z. Wollberg, E. Kochva, and S. Y. Lee. 1984. Cardiotoxic effects of the venom of the burrowing asp *Atractaspis engaddensis* (Atractaspididae, Ophidia). Toxicon 22: 767–774.

Weiss, R. F. 1988. Herbal medicine (translated from 6th German ed. of Lehrbuch der Phytotherepie by A. R. Meuss). Beaconsfield Publ., Beaconsfield, England.

Welches, W., I. Reardon, and R. L. Heinrikson. 1993. An examination of structural interactions presumed to be of importance in the stabilization of phospholipase A_2 dimers based upon comparative protein sequence analysis of a monomeric and dimeric enzyme from the venom of *Agkistrodon p. piscivorus*. J. Protein Chem. 12(2): 187–193.

Weldon, P. J. 1988. Feeding responses of Pacific snappers (genus *Lutjanus*) to the yellow-bellied sea snake *(Pelamis platurus)*. Zool. Sci. (Tokyo) 5: 443–448.

Weldon, P. J., and G. M. Burghardt. 1979. The ophiophage defensive response in crotaline snakes: Extension to new taxa. J. Chem. Ecol. 5: 141–151.

Weldon, P. J., and O. Vallarino. 1988. Wounds on the yellow-bellied sea snake *(Pelamis platurus)* from Panama: Evidence of would-be predators? Biotropica 20: 174–176.

Weldon, P. J., H. W. Sampson, L. Wong, and H. A. Lloyd. 1991. Histology and biochemistry of the scent glands of the yellow-bellied sea snake (*Pelamis platurus*: Hydrophiidae). J. Herpetol. 25: 367–370.

Weldon, P. J., R. Ortiz, and T. R. Sharp. 1992. The chemical ecology of crotaline snakes, pp. 309–319. *In* J. A. Campbell and E. D. Brodie, Jr. (eds.), Biology of the pitvipers. Selva, Tyler, Texas.

Wellborn, S., K. M. Scudder, H. M. Smith, K. Stimac, and D. Chiszar. 1982. Investigatory behavior in snakes III: Effects of familiar odors on investigation of clean cages. Psychol. Rec. 32: 169–177.

Wells, M. A., and D. J. Hanahan. 1969. Studies on phospholipase A: I. Isolation and characterization of two enzymes from *Crotalus adamanteus* venom. Biochemistry, N.Y. 8: 414–424.

Welsh, J. H. 1966. Serotonin and related tryptamine derivatives in snake venoms. Mem. Inst. Butantan 33: 509–518.

Werler, J. E. 1950. The poisonous snakes of Texas and the first aid treatment of their bites. Texas Game Fish Bull. 31: 1–40.

Werler, J. E., and D. M. Darling. 1950. A case of poisoning from the bite of a coral snake, *Micrurus f. tenere* Baird and Girard. Herpetologica 6: 197–199.

Werler, J. E., and J. R. Dixon. 2000. Texas snakes: Identification, distribution, and natural history. Univ. Texas Press, Austin.

Werman, S. D. 2008. Phylogeny and the evolution of β-neurotoxic phospholipases A_2 (PLA_2) in the venoms of rattlesnakes, *Crotalus* and *Sistrurus* (Serpentes: Viperidae), pp. 511–536. *In* W. K. Hayes, K. R. Beaman, M. D. Cardwell, and S. P. Bush (eds.), The biology of rattlesnakes. Loma Linda Univ. Press, Loma Linda, California.

Wermelinger, L. S., D. L. S. Dutra, A. L. Oliveira-Carvalho, M. R. Soares, C. Bloch, Jr., and R. B. Zingali. 2005. Fast analysis of low molecular mass compounds present in snake venom: Identification of ten new pyroglutamate-containing peptides. Rapid Commun. Mass Spectom. 19: 1703–1708.

Werner, R. M., and R. E. Faith. 1978. Decrease in lethal effect of snake venom by serum of the opossum, *Didelphis marsupialis*. Lab. Anim. Sci. 28: 710–713.

West, G. S. 1895. On the buccal glands and teeth of certain poisonous snakes. Proc. Zool. Soc. London 1895: 812–826.

West, J. 1990. The biology of captive sea snakes. Herpetofauna 20: 28–31.

West, L. W. 1981. Notes on captive reproduction and behavior in the Mexican cantil (*Agkistrodon bilineatus*). Herpetol. Rev. 12: 86–87.

Wetmore, A. 1965. The birds of the Republic of Panama. Part 1. Tinamidae (Tinamous) to Rynchopidae (Skimmers). Smithsonian Misc. Coll. 150: 1–483.

Wharton, C. H. 1960. Birth and behavior of a brood of cottonmouths, *Agkistrodon piscivorus piscivorus*, with notes on tail-luring. Herpetologica 16: 124–129.

———. 1966. Reproduction and growth in the cottonmouths, *Agkistrodon piscivorus* Lacépède, of Cedar Keys, Florida. Copeia 1966: 149–161.

———. 1969. The cottonmouth moccasin on Sea Horse Key, Florida. Bull. Florida St. Mus. Biol. Sci. 14: 227–272.

White, A. M. 1979. An unusually large brood of northern copperheads *(Agkistrodon contortrix mokeson)* from Ohio. Ohio J. Sci. 79: 78.

White, B. D., G. C. Rodgers, Jr., N. J. Matyunas, and F. Allen. 1988. Copperhead snakebites reported to the Kentucky Regional Poison Center 1986: Epidemiology and treatment suggestions. J. Kentucky Med. Assoc. 86: 61–66.

White, J. 1995. Clinical toxicology of sea snake bites, pp. 159–170. *In* J. Meier and J. White (eds.), Handbook of clinical toxicology of animal venoms and poisons. CRC Press, New York.

———. 2005. Snake venoms and coagulopathy. Toxicon 45: 951–967.

White, R. R., and R. A. Weber. 1991. Poisonous snakebite in central Texas. Ann. Surg. 213: 466–472.

Whitehead, D. 1940. Handling rattlesnakes to demonstrate one's faith. Syndicated by Associated Press in September.

Whitley, R. E. 1996. Conservative treatment of copperhead snakebites without antivenin. J. Trauma Inj. Infect. Crit. Care 41: 219–221.

Whitlow, K. S., A. A. DiMaio, and S. R. Rose. 2007. Systemic effects after copperhead envenomation. *In* Venom week 2007 (program abstracts). Univ. Arizona Colleges of Medicine, Arizona Health Sci. Center, Off. Continuing Medical Educ., Tucson, Arizona.

Whorley, J. R. 2000. Keys to partial mammals: A method for identifying prey items from snakes. Herpetol. Rev. 31: 227–229.

Wickler, W. 1968. Mimicry in plants and animals. McGraw Hill, New York.

Wiegmann, A. F. A. 1829a. Über die Gesetzlichkeit in der geographischen Verbreitung der Saurier. Isis von Oken 22: 418–428.

———. 1829b. Ueber das Acaltetepan oder Temaculcahua des Hernandez, eine neue Gattung der Saurer, *Heloderma*. Isis von Oken 22: 624–629.

Wijeyewickrema, L. C., E. E. Gardiner, M. Moroi, M. C. Berndt, and R. K. Andrews. 2007. Snake venom metalloproteinases, crotarhagin and alborhagin, induce ectodomain shedding of the platelet collagen receptor, glycoprotein VI. Thromb. Haemostas. 98: 1285–1290.

Wiley, E. O. 1972. The Pleistocene herpetofauna of Dark Canyon Cave, New Mexico. Herpetol. Rev. 4: 128.

Williams, E. J., S. C. Sung, and M. Laskowski, Sr. 1961. Action of venom phosphodiesterase on deoxyribonucleic acid. J. Biol. Chem. 236: 1130–1134.

Williams, M. I., R. D. Birkhead, P. R. Moosman, Jr., and S. M. Boback. 2004. *Agkistrodon piscivorus* (cottonmouth): Diet. Herpetol. Rev. 35: 271–272.

Willis, T. W., and A. T. Tu. 1988. Purification and biochemical characterization of atroxase, a nonhemorrhagic fibrinolytic protease from western diamondback rattlesnake venom. Biochemistry 27: 4769–4777.

Willis, T. W., A. T. Tu, and C. W. Miller. 1989. Thrombolysis with a snake venom protease in a rat model of venous thrombosis. Thromb. Res. 53: 19–29.

Willson, J. D., C. T. Winne, M. E. Dorcas, and J. W. Gibbons. 2006. Post-drought responses of semi-aquatic snakes inhabiting an isolated wetland: Insights on different strategies for persistence in a dynamic habitat. Wetlands 26: 1071–1078.

Willson, P. 1908. Snake poisoning in the United States: A study based on an analysis of 740 cases. Arch. Intern. Med. 1: 516–570.

Wilson, A. B., and S. A. Minton. 1983. *Agkistrodon piscivorus leucostoma* (western cottonmouth). U.S.A.: Indiana. Herpetol. Rev. 14: 84.

Wilson, L. D., and L. Porras. 1983. The ecological impact of man on the south Florida herpetofauna. Univ. Kansas Mus. Nat. Hist. Spec. Publ. 9: i–vi, 1–89.

Wilson, T. P., and D. Mauger. 1999. Home range and habitat use of *Sistrurus catenatus catenatus* in eastern Will County, Illinois, pp. 125–134. *In* B. Johnson and M. Wright (eds.), Second international symposium and workshop on the conservation of the eastern massasauga rattlesnake, *Sistrurus catenatus catenatus*: Population and habitat management issues in urban, bog, prairie, and forested ecosystems. Toronto Zoo, Scarborough, Ontario, Canada.

Wingert, W. A., and L. Chan. 1988. Rattlesnake bites in southern California and rationale for recommended treatment. West. J. Med. 148: 37–44.

Wingert, W. A., and J. Wainschel. 1975. Diagnosis and management of envenomation by poisonous snakes. South. Med. J. 68: 1015–1026.

Wingert, W. A., T. R. Pattabhiraman, R. Cleland, P. Meyer, R. Pattabhiraman, and F. A. Russell. 1980. Distribution and pathology of copperhead (*Agkistrodon contortrix*) venom. Toxicon 18: 591–601.

Winkel, K. 2007. U.S. snakebite mortality, 1979–2005. *In* Venom week 2007 (program abstracts). Univ. Arizona Colleges of Medicine, Arizona Health Sci. Center, Off. Continuing Medical Educ., Tucson, Arizona.

Winsor, H. 1948. Hosts of eustrongyloid worms from Fairmount Park Aquarium and Philadelphia Zoo. Proc. Pennsylvania Acad. Sci. 22: 68–72.

Wittner, D. 1978. A discussion of venomous snakes of North America: Distribution and occurrence of North American snakes of the genera *Crotalus, Sistrurus*, and *Agkistrodon*. HERP: Bull. New York Herpetol. Soc. 14: 12–17.

Witz, B. W., D. S. Wilson, and M. D. Palmer. 1991. Distribution of *Gopherus polyphemus* and its vertebrate symbionts in three burrow categories. Am. Midl. Nat. 126: 152–158.

Wolff, N. O., and T. S. Githens. 1939a. Record venom extraction from the water moccasin. Copeia 1939: 52.

———. 1939b. Yield and toxicity of venom from snakes extracted over a period of two years. Copeia 1939: 234.

Wollberg, Z., R. Shabo-Shina, N. Intrator, A. Bdolah, E. Kochva, G. Shavit, Y. Oron, B. A. Vidne, and S. A. Gitter. 1988. A novel cardiotoxic polypeptide from the venom of *Atractaspis engaddensis* (burrowing asp): Cardiac effects in mice and isolated rat and human heart preparations. Toxicon 26: 525–534.

Wood, J. T. 1954. The distribution of poisonous snakes in Virginia. Virginia J. Sci. 5: 152–167.

Woodburne, M. O. 1956. Notes on the snake, *Sistrurus catenatus tergeminus*, in southwestern Kansas and northwestern Oklahoma. Copeia 1956: 125–126.

Woodin, W. H., III. 1953. Notes on some reptiles from the Huachuca area of southeastern Arizona. Bull. Chicago Acad. Sci. 9: 285–296.

Woodson, W. D. 1947. Toxicity of *Heloderma* venom. Herpetologica 4: 31–33.

———. 1949a. Diseases of *Heloderma*. Herpetologica 5: 91.

———. 1949b. Summary of *Heloderma*'s food habits. Herpetologica 5: 91–92.

World Health Organization. 2007. Rabies and envenomations: A neglected public health issue. Report of a Consultative Meeting, World Health Organization, Geneva, 10 January 2007. WHO Press, Geneva, Switzerland.

Wozniak, E. J., J. Wisser, and M. Schwartz. 2006. Venomous adversaries: A reference to snake identification, field safety, and bite-victim first aid for disaster-response personnel deploying into the hurricane-prone regions of North America. Wild. Environ. Med. 17: 246–266.

Wright, A. H., and S. C. Bishop. 1915. A biological reconnaissance of the Okefinokee Swamp in Georgia. II. Snakes. Proc. Acad. Nat. Sci. Philadelphia 67: 139–192.

Wright, A. H., and A. A. Wright. 1957. Handbook of snakes of the United States and Canada, vols. 1 and 2. Comstock Publ. Associates, Cornell Univ. Press, Ithaca, New York.

———. 1962. Handbook of snakes of the United States and Canada, vol. 3: Bibliography. Edwards Brothers, Ann Arbor, Michigan.

Wright, B. A. 1941. Habit and habitat studies of the massasauga rattlesnake (*Sistrurus catenatus catenatus* Raf.) in northeastern Illinois. Am. Midl. Nat. 25: 659–672.

Wright, R. A. S. 1987. Natural history observations on venomous snakes near the Peaks of Otter, Bedford County, Virginia. Catesbeiana 7(2): 2–9.

Wright, R. A. S., and W. P. Gray. 2000. Copperheads on the York-James Peninsula, Virginia. Catesbeiana 20(1): 22–31.

Wyeth-Ayerst Laboratories. 1998. Antivenin (Crotalidae) polyvalent, pp. 3009–3010. *In* Physicians' desk reference, 52nd ed. Medical Economics, Montvale, New Jersey.

Xiong, Y., W. Wand, X. Pu, and J. Song. 1992. Preliminary study on the mechanism of using snake venoms to substitute for morphine. Toxicon 30: 567.

Yamaguti, S. 1961. Systema Helminthum: The nematodes of vertebrates, part I. Interscience Publ., New York.

Yamazaki, Y., and T. Morita. 2004. Structure and function of snake venom cysteine-rich secretory proteins. Toxicon 44: 227–231.

Yamazaki, Y., Y. Matsunaga, Y. Nakano, and T. Morita. 2005. Identification of vascular endothelial growth factor receptor-binding protein in the venom of eastern cottonmouth: A new role of snake venom myotoxic Lys^{49}-phospholipase A_2. J. Biol. Chem. 280: 29989–29992.

Yatkola, D. A. 1976. Fossil *Heloderma* (Reptilia, Helodermatidae). Occ. Pap. Mus. Nat. Hist. Univ. Kansas 51: 1–14.

Yeh, C.-H., H.-C. Peng, R.-S. Yang, and T.-F. Huang. 2001. Rhodostomin, a snake venom disintegrin, inhibits angiogenesis elicited by basic fibroblast growth factor and suppresses tumor growth by a selective $\alpha_V\beta_3$ blockade of endothelial cells. Mol. Pharmacol. 59: 1333–1342.

Yerger, R. W. 1953. Yellow bullhead preyed upon by cottonmouth moccasin. Copeia 1953: 115.

Young, B. A. 2008. Perspectives on the regulation of venom expulsion in snakes, pp. 181–190. *In* W. K. Hayes, K. R. Beaman, M. D. Cardwell, and S. P. Bush (eds.), The biology of rattlesnakes. Loma Linda Univ. Press, Loma Linda, California.

Young, B. A., and I. P. Brown. 1993. On the acoustic profile of the rattlesnake rattle. Amphibia-Reptilia 14: 373–380.

Young, B. A., and K. Jackson. 2008. Functional specialization of the extrinsic venom gland musculature within the Crotaline snakes (Reptilia: Serpentes) and the role of the M. Pterygoideus Glandulae, pp. 47–54. *In* W. K. Hayes, K. R. Beaman, M. D. Cardwell, and S. P. Bush (eds.), The biology of rattlesnakes. Loma Linda Univ. Press, Loma Linda, California.

Young, B. A., K. Meltzer, C. Marsit, and G. Abishahin. 1999. Cloacal popping in snakes. J. Herpetol. 33: 557–566.

Young, B. A., A. Aguiar, and H. Lillywhite. 2008. Foraging cues used by insular Florida cottonmouths, *Agkistrodon piscivorus conanti*. South American J. Herpetol. 3: 135–144.

Young, N. 1940. Snakebite: Treatment and nursing care. Am. J. Nurs. 40: 657–660.

Yousef, G. M., M. B. Elliott, A. D. Kopolovic, E. Serry, and E. P. Diamandis. 1995. Families and clans of serine peptidases. Arch. Biochem. Biophys. 318: 247–250.

———. 2004. Sequence and evolutionary analysis of the human trypsin subfamily of serine peptidases. Biochim. Biophys. Acta 1698: 77–86.

Zaidan, F., III. 2001. Western cottonmouth (*Agkistrodon piscivorus leucostoma*) sexual dimorphism and dichromatism in northwestern Arkansas. Herpetol. Nat. Hist. 8: 79–82.

———. 2003. Variation in cottonmouth (*Agkistrodon piscivorus leucosoma*) resting metabolic rates. Comp. Biochem. Physiol. 134A: 511–523.

Zaidan, F., III, D. L. Kreider, and S. J. Beaupre. 2003. Testosterone cycles and reproductive energetics: Implications for northern range limits of the cottonmouth (*Agkistrodon piscivorus leucostoma*). Copeia 2003: 231–240.

Zamudio, K. R., D. L. Hardy, Sr., M. Martins, and H. W. Greene. 2000. Fang tip spread, puncture distance, and suction for snake bite. Toxicon 38: 723–728.

Zann, L. P., R. J. Cuffey, and C. Kropach. 1975. Fouling organisms and parasites associated with the skin of sea snakes, pp. 251–265. *In* W. A. Dunson (ed.), The biology of sea snakes. Univ. Park Press, Baltimore, Maryland.

Zarafonetis, C. J. D., and J. P. Kalas. 1960a. Some hematologic and biochemical findings in *Heloderma horridum*, the Mexican beaded lizard. Copeia 1960: 240–241.

———. 1960b. Serotonin, catechol amines, and amine oxidase activity in the venoms of certain reptiles. Am. J. Med. Sci. 240: 764–768.

Zegel, J. C. 1975. Notes on collecting and breeding the eastern coral snake, *Micrurus fulvius fulvius*. Bull. Southwest. Herpetol. Soc. 1: 9–10.

Zehnder, M. P., and J. Mariaux. 1999. Molecular systematic analysis of the order Proteocephalidea (Eucestoda) based on mitochondrial and nuclear rDNA sequences. Int. J. Parasitol. 29: 1841–1852.

Zeiller, W. 1969. Maintenance of the yellow-bellied seasnake, *Pelamis platurus,* in captivity. Copeia 1969: 407–408.

Zeller, E. A. 1948. Enzymes of snake venoms and their biological significance, pp. 459–495. *In* F. F. Nora (ed.), Advances in enzymology, vol. 8. Interscience, New York.

———. 1950a. Über Phosphatasen II. Über eine neue Adenosintriphosphatase. Helv. Chim. Acta 33: 821–833.

———. 1950b. The formation of pyrophosphate from adenosine triphosphate in the presence of snake venom. Arch. Biochem. 28: 138–139.

Zhang, L., and W. T. Wu. 2003. Progress in studies on anticancer components from snake venom. Chinese J. Zool. 38: 120–124.

Zhong, X. Y., G. F. Liu, and Q. C. Wang. 1993. Purification and anticancer activity of cytotoxin-14 from venom of *Naja naja atra*. Zhongguo Yao Li Xue Bao 14: 279–282.

Zimmerman, E. G., and C. W. Kilpatrick. 1973. Karyology of North American crotaline snakes (family Viperidae) of the genera *Agkistrodon, Sistrurus,* and *Crotalus*. Can. J. Genet. Cytol. 15(3): 389–395.

Zimmermann, A. A., and C. H. Pope. 1948. Development and growth of the rattle of rattlesnakes. Fieldiana: Zool. 32: 355–413.

Zingali, R. B. 2007. Interaction of snake-venom proteins with blood coagulation factors: Mechanisms of anticoagulant activity. Toxin Rev. 26: 25–46.

Zippel, K. C., H. B. Lillywhite, and C. R. J. Mladinich. 2001. New vascular system in reptiles: Anatomy and postural hemodynamics of the vertebral venous plexus in snakes. J. Morphol. 250: 173–184.

Zozaya, J., and R. E. Stadelman. 1930. Hypersensitiveness to snake venom proteins: A case report. Bull. Antiv. Inst. Am. 3: 93–95.

Zug, G. R., and C. H. Ernst. 2004. Smithsonian answer book: Snakes. Smithsonian Books, Washington, D.C.

Zug, G. R., L. J. Vitt, and J. P. Caldwell. 2001. Herpetology: An introductory biology of amphibians and reptiles. 2nd ed. Academic Press, New York.

Zugaib, M., A. C. S. D. de Barros, R. E. Bittar, E. de A. Burdmann, and B. Neme. 1985. Abruptio placentae following snake bite. Am. J. Obstet. Gynecol. 151: 754–755.

Zweifel, R. G. 1960. Results of the Puritan-American Museum of Natural History expedition to western Mexico: 9. Herpetology of the Tres Marías Islands. Bull. Am. Mus. Nat. Hist. 119: 77–128.

Zweifel, R. G., and K. S. Norris. 1955. Contribution to the herpetology of Sonora, Mexico: Descriptions of new subspecies of snakes (*Micruroides euryxanthus* and *Lampropeltis getulus*) and miscellaneous collecting notes. Am. Midl. Nat. 54: 230–249.

Zylstra, E., J. Capps, and B. Weise. 2005. Field observations of interactions between the desert tortoise and the Gila monster. Sonoran Herpetol. 18: 30–31.

Index to Common and Scientific Names